D0138282

Jin Ho Kwak
Sungpyo Hong

Linear Algebra

Birkhäuser
Boston • Basel • Berlin

Jin Ho Kwak
Sungpyo Hong
Department of Mathematics
Pohang University of Science and Technology
Pohang, The Republic of Korea

Library of Congress Cataloging-in-Publication Data

Kwak, Jin Ho, 1948-
Linear Algebra / Jin Ho Kwak, Sungpyo Hong.
p. cm.
Includes index.
ISBN 0-8176-3999-3 (alk. paper). -- ISBN 3-7643-3999-3 (alk. paper)
1. Algebras, Linear. I. Hong, Sungpyo, 1948- . II. Title.
QA184.K94 1997
512'.5--dc21 97-9062
CIP

Printed on acid-free paper

Birkhäuser

ISBN 0-8176-3999-3
ISBN 3-7643-3999-3
Typesetting by the authors in LaTeX
Printed and bound by Hamilton Printing, Rensselaer, NY
Printed in the U.S.A.

9 8 7 6 5 4 3 2 1

Preface

Linear algebra is one of the most important subjects in the study of science and engineering because of its widespread applications in social or natural science, computer science, physics, or economics. As one of the most useful courses in undergraduate mathematics, it has provided essential tools for industrial scientists. The basic concepts of linear algebra are vector spaces, linear transformations, matrices and determinants, and they serve as an abstract language for stating ideas and solving problems.

This book is based on the lectures delivered several years in a sophomore-level linear algebra course designed for science and engineering students. The primary purpose of this book is to give a careful presentation of the basic concepts of linear algebra as a coherent part of mathematics, and to illustrate its power and usefulness through applications to other disciplines. We have tried to emphasize the computational skills along with the mathematical abstractions, which have also an integrity and beauty of their own. The book includes a variety of interesting applications with many examples not only to help students understand new concepts but also to practice wide applications of the subject to such areas as differential equations, statistics, geometry, and physics. Some of those applications may not be central to the mathematical development and may be omitted or selected in a syllabus at the discretion of the instructor. Most basic concepts and introductory motivations begin with examples in Euclidean space or solving a system of linear equations, and are gradually examined from different points of views to derive general principles.

For those students who have completed a year of calculus, linear algebra may be the first course in which the subject is developed in an abstract way, and we often find that many students struggle with the abstraction and miss the applications. Our experience is that, to understand the material, students should practice with many problems, which are sometimes omitted because of a lack of time. To encourage the students to do repeated practice, we placed in the middle of the text not only many examples but also some carefully selected problems, with answers or helpful hints. We have tried to make this book as easily accessible and clear as possible, but certainly there may be some awkward expressions in several ways. Any criticism or comment from the readers will be appreciated.

We are very grateful to many colleagues in Korea, especially to the faculty members in the mathematics department at Pohang University of Science and Technology (POSTECH), who helped us over the years with various aspects of this book. For their valuable suggestions and comments, we would like to thank the students at POSTECH, who have used photocopied versions of the text over the past several years. We would also like to acknowledge the invaluable assistance we have received from the teaching assistants who have checked and added some answers or hints for the problems and exercises in this book. Our thanks also go to Mrs. Kathleen Roush who made this book much more legible with her grammatical corrections in the final manuscript. Our thanks finally go to the editing staff of Birkhäuser for gladly accepting our book for publication.

Jin Ho Kwak
Sungpyo Hong
E-mail: jinkwak@postech.ac.kr
sungpyo@postech.ac.kr

April 1997, in Pohang, Korea

"Linear algebra is the mathematics of our modern technological world of complex multivariable systems and computers"

– Alan Tucker –

"We (Halmos and Kaplansky) share a love of linear algebra. I think it is our conviction that we'll never understand infinite-dimensional operators properly until we have a decent mastery of finite matrices. And we share a philosophy about linear algebra: we think basis-free, we write basis-free, but when the chips are down we close the office door and compute with matrices like fury"

– Irving Kaplansky –

Contents

Linear Algebra

Chapter 1

Linear Equations and Matrices

1.1 Introduction

One of the central motivations for linear algebra is solving systems of linear equations. We thus begin with the problem of finding the solutions of a system of m linear equations in n unknowns of the following form:

$$\begin{cases} a_{11}x_1 & + & a_{12}x_2 & + & \cdots & + & a_{1n}x_n & = & b_1 \\ a_{21}x_1 & + & a_{22}x_2 & + & \cdots & + & a_{2n}x_n & = & b_2 \\ & & & & & & & \vdots & \\ a_{m1}x_1 & + & a_{m2}x_2 & + & \cdots & + & a_{mn}x_n & = & b_m, \end{cases}$$

where $x_1, x_2, \ldots, x_n$ are the unknowns and a_{ij}'s and b_i's denote constant (real or complex) numbers.

A sequence of numbers $(s_1, s_2, \ldots, s_n)$ is called a **solution** of the system if $x_1 = s_1, x_2 = s_2, \ldots, x_n = s_n$ satisfy each equation in the system simultaneously. When $b_1 = b_2 = \cdots = b_m = 0$, we say that the system is **homogeneous**.

The central topic of this chapter is to examine whether or not a given system has a solution, and to find a solution if it has one. For instance, any homogeneous system always has at least one solution $x_1 = x_2 = \cdots = x_n = 0$, called the **trivial solution**. A natural question is whether such a homogeneous system has a nontrivial solution. If so, we would like to have a systematic method of finding all the solutions. A system of linear equations is said to be **consistent** if it has at least one solution, and **inconsistent** if it has no solution. The following example gives us an idea how to answer the above questions.

Example 1.1 When $m = n = 2$, the system reduces to two equations in two unknowns x and y:

$$\begin{cases} a_1x + b_1y = c_1 \\ a_2x + b_2y = c_2. \end{cases}$$

Geometrically, each equation in the system represents a straight line when we interpret x and y as coordinates in the xy-plane. Therefore, a point $P = (x, y)$ is a solution if and only if the point P lies on both lines. Hence there are three possible types of solution set:

(1) the empty set if the lines are parallel,

(2) only one point if they intersect,

(3) a straight line: *i.e.*, infinitely many solutions, if they coincide.

The following examples and diagrams illustrate the three types:

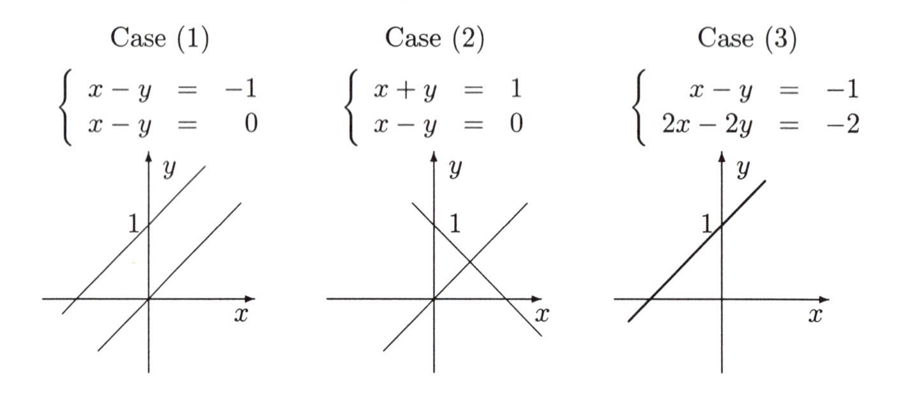

To decide whether the given system has a solution and to find a general method of solving the system when it has a solution, we repeat here a well-known elementary method of *elimination* and *substitution*.

Suppose first that the system consists of only one equation $ax + by = c$. Then the system has either infinitely many solutions (*i.e.*, points on the straight line $x = -\frac{b}{a}y + \frac{c}{a}$ or $y = -\frac{a}{b}x + \frac{c}{b}$ depending on whether $a \neq 0$ or $b \neq 0$) or no solutions when $a = b = 0$ and $c \neq 0$.

We now assume that the system has two equations representing two lines in the plane. Then clearly the two lines are parallel with the same slopes if and only if $a_2 = \lambda a_1$ and $b_2 = \lambda b_1$ for some $\lambda \neq 0$, or $a_1b_2 - a_2b_1 = 0$. Furthermore, the two lines either coincide (infinitely many solutions) or are distinct and parallel (no solutions) according to whether $c_2 = \lambda c_1$ holds or not.

Suppose now that the lines are not parallel, or $a_1b_2 - a_2b_1 \neq 0$. In this case, the two lines cross at a point, and hence there is exactly one solution: For instance, if the system is homogeneous, then the lines cross at the origin, so $(0, 0)$ is the only solution. For a nonhomogeneous system, we may find the solution as follows: Express x in terms of y from the first equation, and then substitute it into the second equation (*i.e.*, eliminate the variable x from the second equation) to get

$$(b_2 - \frac{a_2}{a_1}b_1)y = c_2 - \frac{a_2}{a_1}c_1.$$

Since $a_1b_2 - a_2b_1 \neq 0$, this can be solved as

$$y = \frac{a_1c_2 - a_2c_1}{a_1b_2 - a_2b_1},$$

which is in turn substituted into one of the equations to find x and give a complete solution of the system. In detail, the process can be summarized as follows:

(1) Without loss of generality, we may assume $a_1 \neq 0$ since otherwise we can interchange the two equations. Then the variable x can be eliminated from the second equation by adding $-\frac{a_2}{a_1}$ times the first equation to the second, to get

$$\begin{cases} a_1x + b_1y = c_1 \\ (b_2 - \frac{a_2}{a_1}b_1)y = c_2 - \frac{a_2}{a_1}c_1. \end{cases}$$

(2) Since $a_1b_2 - a_2b_1 \neq 0$, y can be found by multiplying the second equation by a nonzero number $\frac{a_1}{a_1b_2 - a_2b_1}$ to get

$$\begin{cases} a_1x + b_1y = c_1 \\ y = \frac{a_1c_2 - a_2c_1}{a_1b_2 - a_2b_1}. \end{cases}$$

(3) Now, x is solved by substituting the value of y into the first equation, and we obtain the solution to the problem:

$$\begin{cases} x &= \dfrac{b_2c_1 - b_1c_2}{a_1b_2 - a_2b_1} \\[2ex] y &= \dfrac{a_1c_2 - a_2c_1}{a_1b_2 - a_2b_1}. \end{cases}$$

Note that the condition $a_1b_2 - a_2b_1 \neq 0$ is necessary for the system to have only one solution. □

In this example, we have changed the original system of equations into a simpler one using certain operations, from which we can get the solution of the given system. That is, if (x, y) satisfies the original system of equations, then x and y must satisfy the above simpler system in (3), and vice versa.

It is suggested that the readers examine a system of three equations in three unknowns, each equation representing a plane in the 3-dimensional space $\mathbb{R}^3$, and consider the various possible cases in a similar way.

Problem 1.1 For a system of three equations in three unknowns

$$\begin{cases} a_{11}x + a_{12}y + a_{13}z = b_1 \\ a_{21}x + a_{22}y + a_{23}z = b_2 \\ a_{31}x + a_{32}y + a_{33}z = b_3, \end{cases}$$

describe all the possible types of the solution set in $\mathbb{R}^3$.

1.2 Gaussian elimination

As we have seen in Example 1.1, a basic idea for solving a system of linear equations is to change the given system into a simpler system, keeping the solutions unchanged; the example showed how to change a general system to a simpler one. In fact, the main operations used in Example 1.1 are the following three operations, called **elementary operations**:

(1) multiply a nonzero constant throughout an equation,

(2) interchange two equations,

(3) change an equation by adding a constant multiple of another equation.

After applying a finite sequence of these elementary operations to the given system, one can obtain a simpler system from which the solution can be derived directly.

Note also that each of the three elementary operations has its *inverse* operation which is also an elementary operation:

(1)′ divide the equation with the same nonzero constant,

(2)′ interchange two equations again,

(3)′ change the equation by subtracting the same constant multiple of the same equation.

By applying these inverse operations in reverse order to the simpler system, one can recover the original system. This means that a solution of the original system must also be a solution of the simpler one, and *vice versa.*

These arguments can be formalized in mathematical language. Observe that in performing any of these basic operations, only the coefficients of the variables are involved in the calculations and the variables $x_1, \ldots, x_n$ and the equal sign "=" are simply repeated. Thus, keeping the order of the variables and "=" in mind, we just extract the coefficients only from the equations in the given system and make a rectangular array of numbers:

$$\begin{bmatrix} a_{11} & a_{12} & \cdots & a_{1n} & b_1 \\ a_{21} & a_{22} & \cdots & a_{2n} & b_2 \\ \vdots & \vdots & & \vdots & \vdots \\ a_{m1} & a_{m2} & \cdots & a_{mn} & b_m \end{bmatrix}.$$

This matrix is called the **augmented matrix** for the system. The term *matrix* means just any rectangular array of numbers, and the numbers in this array are called the *entries* of the matrix. To explain the above operations in terms of matrices, we first introduce some terminology even though in the following sections we shall study matrices in more detail.

Within a matrix, the horizontal and vertical subarrays

$$[a_{i1}\ a_{i2}\ \cdots\ a_{in}\ b_i] \quad \text{and} \quad \begin{bmatrix} a_{1j} \\ a_{2j} \\ \vdots \\ a_{mj} \end{bmatrix}$$

are called the i-th *row* (matrix) and the j-th *column* (matrix) of the augmented matrix, respectively. Note that the entries in the j-th column are just the coefficients of j-th variable x_j, so there is a correspondence between columns of the matrix and variables of the system.

Since each row of the augmented matrix contains all the information of the corresponding equation of the system, we may deal with this augmented matrix instead of handling the whole system of linear equations.

The elementary operations to a system of linear equations are rephrased as the **elementary row operations** for the augmented matrix, as follows:

(1) multiply a nonzero constant throughout a row,

(2) interchange two rows,

(3) change a row by adding a constant multiple of another row.

The *inverse* operations are

(1)′ divide the row by the same constant,

(2)′ interchange two rows again,

(3)′ change the row by subtracting the same constant multiple of the other row.

Definition 1.1 Two augmented matrices (or systems of linear equations) are said to be **row-equivalent** if one can be transformed to the other by a finite sequence of elementary row operations.

If a matrix B can be obtained from a matrix A in this way, then we can obviously recover A from B by applying the inverse elementary row operations in reverse order. Note again that an elementary row operation does not alter the solution of the system, and we can formalize the above argument in the following theorem:

Theorem 1.1 *If two systems of linear equations are row-equivalent, then they have the same set of solutions.*

The general procedure for finding the solutions will be illustrated in the following example:

Example 1.2 Solve the system of linear equations:

$$\left\{\begin{array}{rcrcrcr} & & 2y & + & 4z & = & 2 \\ x & + & 2y & + & 2z & = & 3 \\ 3x & + & 4y & + & 6z & = & -1\,. \end{array}\right.$$

Solution: We could work with the augmented matrix alone. However, to compare the operations on systems of linear equations with those on the augmented matrix, we work on the system and the augmented matrix in parallel. Note that the associated augmented matrix of the system is

$$\begin{bmatrix} 0 & 2 & 4 & 2 \\ 1 & 2 & 2 & 3 \\ 3 & 4 & 6 & -1 \end{bmatrix}.$$

(1) Since the coefficient of x in the first equation is zero while that in the second equation is not zero, we interchange these two equations:

$$\begin{cases} x + 2y + 2z = 3 \\ \quad\; 2y + 4z = 2 \\ 3x + 4y + 6z = -1 \end{cases} \qquad \begin{bmatrix} 1 & 2 & 2 & 3 \\ 0 & 2 & 4 & 2 \\ 3 & 4 & 6 & -1 \end{bmatrix}.$$

(2) Add -3 times the first equation to the third equation:

$$\begin{cases} x + 2y + 2z = 3 \\ \quad\; 2y + 4z = 2 \\ \quad - 2y \qquad = -10 \end{cases} \qquad \begin{bmatrix} 1 & 2 & 2 & 3 \\ 0 & 2 & 4 & 2 \\ 0 & -2 & 0 & -10 \end{bmatrix}.$$

The coefficient 1 of the first unknown x in the first equation (row) is called the **pivot** in this first elimination step.

Now the second and the third equations involve only the two unknowns y and z. Leave the first equation (row) alone, and the same elimination procedure can be applied to the second and the third equations (rows): The pivot for this step is the coefficient 2 of y in the second equation (row). To eliminate y from the last equation,

(3) Add 1 times the second equation (row) to the third equation (row):

$$\begin{cases} x + 2y + 2z = 3 \\ \quad\; 2y + 4z = 2 \\ \qquad\quad 4z = -8 \end{cases} \qquad \begin{bmatrix} 1 & 2 & 2 & 3 \\ 0 & 2 & 4 & 2 \\ 0 & 0 & 4 & -8 \end{bmatrix}.$$

The elimination process done so far to obtain this result is called a **forward elimination**: *i.e.*, elimination of x from the last two equations (rows) and then elimination of y from the last equation (row).

Now the pivots of the second and third rows are 2 and 4, respectively. To make these entries 1,

(4) Divide each row by the pivot of the row:

$$\begin{cases} x + 2y + 2z = 3 \\ \quad\ \ y + 2z = 1 \\ \qquad\quad\ \ z = -2 \end{cases} \qquad \begin{bmatrix} 1 & 2 & 2 & 3 \\ 0 & 1 & 2 & 1 \\ 0 & 0 & 1 & -2 \end{bmatrix}.$$

The resulting matrix on the right side is called a **row-echelon form** of the matrix, and the 1's at the leftmost entries in each row are called the **leading** 1's. The process so far is called a **Gaussian elimination**.

We now want to eliminate numbers above the leading 1's;

(5) Add -2 times the third row to the second and the first rows,

$$\begin{cases} x + 2y \qquad = 7 \\ \quad\ \ y \qquad = 5 \\ \qquad\quad z = -2 \end{cases} \qquad \begin{bmatrix} 1 & 2 & 0 & 7 \\ 0 & 1 & 0 & 5 \\ 0 & 0 & 1 & -2 \end{bmatrix}.$$

(6) Add -2 times the second row to the first row:

$$\begin{cases} x \qquad\qquad = -3 \\ \quad\ \ y \qquad = 5 \\ \qquad\quad z = -2 \end{cases} \qquad \begin{bmatrix} 1 & 0 & 0 & -3 \\ 0 & 1 & 0 & 5 \\ 0 & 0 & 1 & -2 \end{bmatrix}.$$

This matrix is called the **reduced row-echelon form**. The procedure to get this reduced row-echelon form from a row-echelon form is called the **back substitution**. The whole process to obtain the reduced row-echelon form is called a **Gauss-Jordan elimination**.

Notice that the corresponding system to this reduced row-echelon form is row-equivalent to the original one and is essentially a solved form: *i.e.*, the solution is $x = -3,\ y = 5,\ z = -2$. □

In general, a matrix of **row-echelon form** satisfies the following properties.

(1) The first nonzero entry of each row is 1, called a leading 1.

(2) A row containing only 0's should come after all rows with some nonzero entries.

(3) The leading 1's appear from left to the right in successive rows. That is, the leading 1 in the lower row occurs farther to the right than the leading 1 in the higher row.

Moreover, the matrix of the **reduced row-echelon form** satisfies

(4) Each column that contains a leading 1 has zeros everywhere else, in addition to the above three properties.

Note that an augmented matrix has only one reduced row-echelon form while it may have many row-echelon forms. In any case, the number of nonzero rows containing leading 1's is equal to the number of columns containing leading 1's. The variables in the system corresponding to columns with the leading 1's in a row-echelon form are called the **basic variables**. In general, the reduced row-echelon form U may have columns that do not contain leading 1's. The variables in the system corresponding to the columns without leading 1's are called **free variables**. Thus the sum of the number of basic variables and that of free variables is precisely the total number of variables.

For example, the first two matrices below are in reduced row-echelon form, and the last two just in row-echelon form.

$$\begin{bmatrix} 1 & 0 & 0 \\ 0 & 1 & 0 \\ 0 & 0 & 0 \end{bmatrix}, \quad \begin{bmatrix} 0 & 1 & 5 & 0 & 6 \\ 0 & 0 & 0 & 1 & 1 \\ 0 & 0 & 0 & 0 & 0 \end{bmatrix}, \quad \begin{bmatrix} 1 & 2 & 3 & 2 \\ 0 & 1 & 4 & 5 \\ 0 & 0 & 1 & 7 \end{bmatrix}, \quad \begin{bmatrix} 1 & 1 & 2 & 6 \\ 0 & 1 & 1 & 0 \\ 0 & 0 & 1 & 3 \end{bmatrix}.$$

Notice that in an augmented matrix $[A\ \mathbf{b}]$, the last column $\mathbf{b}$ does not correspond to any variable. Hence, if we consider the four matrices above as augmented matrices for some systems, then the systems corresponding to the first and the last two augmented matrices have only basic variables but no free variables. In the system corresponding to the second augmented matrix, the second and the forth variables, x_2 and x_4, are basic, and the first and the third variables, x_1 and x_3, are free variables. These ideas will be used in later chapters.

In summary, by applying a finite sequence of elementary row operations, the augmented matrix for a system of linear equations can be changed to its reduced row-echelon form which is row-equivalent to the original one. From the reduced row-echelon form, we can decide whether the system has a solution, and find the solution of the given system if it has one.

Example 1.3 Solve the following system of linear equations by Gauss-Jordan elimination.

$$\left\{ \begin{array}{rcrcrcrcr} x_1 & + & 3x_2 & - & 2x_3 & & & = & 3 \\ 2x_1 & + & 6x_2 & - & 2x_3 & + & 4x_4 & = & 18 \\ & & x_2 & + & x_3 & + & 3x_4 & = & 10. \end{array} \right.$$

Solution: The augmented matrix for the system is

$$\begin{bmatrix} 1 & 3 & -2 & 0 & 3 \\ 2 & 6 & -2 & 4 & 18 \\ 0 & 1 & 1 & 3 & 10 \end{bmatrix}.$$

The Gaussian elimination begins with:

(1) Adding -2 times the first row to the second produces

$$\begin{bmatrix} 1 & 3 & -2 & 0 & 3 \\ 0 & 0 & 2 & 4 & 12 \\ 0 & 1 & 1 & 3 & 10 \end{bmatrix}.$$

(2) Note that the coefficient of x_2 in the second equation is zero and that in the third equation is not. Thus, interchanging the second and the third rows produces

$$\begin{bmatrix} 1 & 3 & -2 & 0 & 3 \\ 0 & 1 & 1 & 3 & 10 \\ 0 & 0 & 2 & 4 & 12 \end{bmatrix}.$$

(3) The pivot in the third row is 2. Thus, dividing the third row by 2 produces a row-echelon form

$$\begin{bmatrix} 1 & 3 & -2 & 0 & 3 \\ 0 & 1 & 1 & 3 & 10 \\ 0 & 0 & 1 & 2 & 6 \end{bmatrix}.$$

This is a row-echelon form, and we now continue the back-substitution:

(4) Adding -1 times the third row to the second, and 2 times the third row to the first produces

$$\begin{bmatrix} 1 & 3 & 0 & 4 & 15 \\ 0 & 1 & 0 & 1 & 4 \\ 0 & 0 & 1 & 2 & 6 \end{bmatrix}.$$

(5) Finally, adding -3 times the second row to the first produces the reduced row-echelon form:

$$\begin{bmatrix} 1 & 0 & 0 & 1 & 3 \\ 0 & 1 & 0 & 1 & 4 \\ 0 & 0 & 1 & 2 & 6 \end{bmatrix}.$$

The corresponding system of equations is

$$\begin{cases} x_1 & & & + & x_4 & = & 3 \\ & x_2 & & + & x_4 & = & 4 \\ & & x_3 & + & 2x_4 & = & 6. \end{cases}$$

Since x_1, x_2, and x_3 correspond to the columns containing leading 1's, they are the basic variables, and x_4 is the free variable. Thus by solving this system for the basic variables in terms of the free variable x_4, we have the system of equations in a solved form:

$$\begin{cases} x_1 & = & 3 & - & x_4 \\ x_2 & = & 4 & - & x_4 \\ x_3 & = & 6 & - & 2x_4. \end{cases}$$

By assigning an arbitrary value t to the free variable x_4, the solutions can be written as

$$(x_1,\ x_2,\ x_3,\ x_4) = (3 - t,\ 4 - t,\ 6 - 2t,\ t),$$

for any $t \in \mathbb{R}$, where $\mathbb{R}$ denotes the set of real numbers. □

Remark: Consider a homogeneous system

$$\begin{cases} a_{11}x_1 & + & a_{12}x_2 & + & \cdots & + & a_{1n}x_n & = & 0 \\ a_{21}x_1 & + & a_{22}x_2 & + & \cdots & + & a_{2n}x_n & = & 0 \\ & & & & & & & \vdots & \\ a_{m1}x_1 & + & a_{m2}x_2 & + & \cdots & + & a_{mn}x_n & = & 0, \end{cases}$$

with the number of unknowns greater than the number of equations: that is, $m < n$. Since the number of basic variables cannot exceed the number of rows, a free variable always exists as in Example 1.3, so by assigning an arbitrary value to each free variable we can always find infinitely many nontrivial solutions.

Problem 1.2 Suppose that the augmented matrix for a system of linear equations has been reduced to the reduced row-echelon form below by elementary row operations. Solve the systems:

$$(1) \begin{bmatrix} 1 & 0 & 0 & 5 \\ 0 & 1 & 0 & -2 \\ 0 & 0 & 0 & 4 \end{bmatrix}, \qquad (2) \begin{bmatrix} 1 & 0 & 0 & 4 & -1 \\ 0 & 1 & 0 & 2 & 6 \\ 0 & 0 & 1 & 3 & 2 \end{bmatrix}.$$

We note that if a row-echelon form of an augmented matrix has a row of the type $[\,0\ 0\ \cdots\ 0\ b\,]$ with $b \neq 0$, then it represents an equation of the form $0x_1 + 0x_2 + \cdots + 0x_n = b$ with $b \neq 0$. In this case, the system has no solution. If $b = 0$, then it has a row containing only 0's that can be neglected. Hence, when we deal with a row-echelon form, we may assume that the zero rows are deleted. Note also that, as in Example 1.3, if there exists at least one free variable in the row-echelon form, then the system has infinitely many solutions. On the other hand, if the system has no free variable, the system has a unique solution.

To study systems of linear equations in terms of matrices systematically, we will develop some general theories of matrices in the following sections.

Problem 1.3 Solve the following systems of equations by Gaussian elimination. What are the pivots?

$$(1)\ \left\{\begin{array}{rcrcrcl} -x & + & y & + & 2z & = & 0 \\ 3x & + & 4y & + & z & = & 0 \\ 2x & + & 5y & + & 3z & = & 0. \end{array}\right. \qquad (2)\ \left\{\begin{array}{rcrcrcl} & & 2y & - & z & = & 1 \\ 4x & - & 10y & + & 3z & = & 5 \\ 3x & - & 3y & & & = & 6. \end{array}\right.$$

$$(3)\ \left\{\begin{array}{rcrcrcrcl} w & + & x & + & y & & & = & 3 \\ -3w & - & 17x & + & y & + & 2z & = & 1 \\ 4w & - & 17x & + & 8y & - & 5z & = & 1 \\ & - & 5x & - & 2y & + & z & = & 1. \end{array}\right.$$

Problem 1.4 Determine the condition on b_i so that the following system has a solution.

$$(1)\ \left\{\begin{array}{rcrcrcl} x & + & 2y & + & 6z & = & b_1 \\ 2x & - & 3y & - & 2z & = & b_2 \\ 3x & - & y & + & 4z & = & b_3. \end{array}\right. \qquad (2)\ \left\{\begin{array}{rcrcrcl} x & + & 3y & - & 2z & = & b_1 \\ 2x & - & y & + & 3z & = & b_2 \\ 4x & + & 2y & + & z & = & b_3. \end{array}\right.$$

1.3 Matrices

Rectangular arrays of real numbers arise in many real-world problems. Historically, it was the English mathematician A. Cayley who first introduced the word "matrix" in the year 1858. The meaning of the word is "that within which something originates," and he used matrices simply as a source for rows and columns to form squares.

In this section we are interested only in very basic properties of such matrices.

Definition 1.2 An m by n (written $m\times n$) **matrix** is a rectangular array of numbers arranged into m (horizontal) rows and n (vertical) columns. The

size of a matrix is specified by the number m of rows and the number n of columns.

In general, a matrix is written in the following form:

$$A = \begin{bmatrix} a_{11} & a_{12} & \cdots & a_{1n} \\ a_{21} & a_{22} & \cdots & a_{2n} \\ \vdots & \vdots & & \vdots \\ a_{m1} & a_{m2} & \cdots & a_{mn} \end{bmatrix} = [a_{ij}]_{m \times n},$$

or just $A = [a_{ij}]$ if the size of the matrix is clear from the context. The number a_{ij} is called the (i, j)-**entry** of the matrix A, and can be also written as $a_{ij} = [A]_{ij}$.

An $m \times 1$ matrix is called a **column (matrix)** or sometimes a **column vector**, and a $1 \times n$ matrix is called a **row (matrix)**, or a **row vector**. These special cases are important, as we will see throughout the book. We will generally use capital letters like A, B, C for matrices and small boldface letters like $\mathbf{x}$, $\mathbf{y}$, $\mathbf{z}$ for columns or row vectors.

Definition 1.3 Let $A = [a_{ij}]$ be an $m \times n$ matrix. The **transpose** of A is the $n \times m$ matrix, denoted by A^T, whose j-th column is taken from the j-th row of A: That is, $A^T = [b_{ij}]$ with $b_{ij} = a_{ji}$.

For example, if $A = \begin{bmatrix} 1 & 3 & 5 \\ 2 & 4 & 6 \end{bmatrix}$, then $A^T = \begin{bmatrix} 1 & 2 \\ 3 & 4 \\ 5 & 6 \end{bmatrix}$.

In particular, the transpose of a column vector is a row vector and vice versa. For example, for an $n \times 1$ column vector

$$\mathbf{x} = \begin{bmatrix} x_1 \\ x_2 \\ \vdots \\ x_n \end{bmatrix},$$

its transpose $\mathbf{x}^T = [x_1 \; x_2 \; \cdots \; x_n]$ is a row vector.

Definition 1.4 Let $A = [a_{ij}]$ be an $m \times n$ matrix.

(1) A is called a **square matrix of order** n if $m = n$.

In the following, we assume that A is a square matrix of order n.

(2) The entries a_{11}, a_{22}, $\ldots$, a_{nn} are called the **diagonal entries** of A.

(3) A is called a **diagonal** matrix if all the entries except for the diagonal entries are zero.

(4) A is called an **upper (lower) triangular** matrix if all the entries below (above, respectively) the diagonal are zero.

The following matrices U and L are the general forms of the upper triangular and lower triangular matrices, respectively:

$$U = \begin{bmatrix} a_{11} & a_{12} & \cdots & a_{1n} \\ 0 & a_{22} & \cdots & a_{2n} \\ \vdots & & \ddots & \vdots \\ 0 & 0 & \cdots & a_{nn} \end{bmatrix}, \quad L = \begin{bmatrix} a_{11} & 0 & \cdots & 0 \\ a_{21} & a_{22} & \cdots & 0 \\ \vdots & & \ddots & \vdots \\ a_{n1} & a_{n2} & \cdots & a_{nn} \end{bmatrix}.$$

Note that a matrix which is both upper and lower triangular must be a diagonal matrix, and the transpose of an upper (lower) triangular matrix is lower (upper, respectively) triangular matrix.

Definition 1.5 Two matrices $A = [a_{ij}]$ and $B = [b_{ij}]$ are said to be **equal**, written $A = B$, if they have the same size and corresponding entries are equal: *i.e.*, $a_{ij} = b_{ij}$ for all i and j.

This definition allows us to write matrix equations. A simple example is $(A^T)^T = A$ by definition.

Let $M_{m\times n}(\mathbb{R})$ denote the set of all $m \times n$ matrices with entries of real numbers. Among the elements of $M_{m\times n}(\mathbb{R})$, we can define two operations, called scalar multiplication and the sum of matrices, as follows:

Scalar multiplication: Given an $m \times n$ matrix $A = [a_{ij}]$ and a *scalar* k (which is simply a real number), the scalar multiplication kA of k and A is defined to be the matrix $kA = [ka_{ij}]$: *i.e.*, in an expanded form:

$$k \begin{bmatrix} a_{11} & \cdots & a_{1n} \\ \vdots & \ddots & \vdots \\ a_{m1} & \cdots & a_{mn} \end{bmatrix} = \begin{bmatrix} ka_{11} & \cdots & ka_{1n} \\ \vdots & \ddots & \vdots \\ ka_{m1} & \cdots & ka_{mn} \end{bmatrix}.$$

Sum of matrices: If $A = [a_{ij}]$ and $B = [b_{ij}]$ are two matrices of the same size, then the **sum** $A + B$ is defined to be the matrix $A + B = [a_{ij} + b_{ij}]$:

i.e., in an expanded form:

$$\begin{bmatrix} a_{11} & \cdots & a_{1n} \\ \vdots & \ddots & \vdots \\ a_{m1} & \cdots & a_{mn} \end{bmatrix} + \begin{bmatrix} b_{11} & \cdots & b_{1n} \\ \vdots & \ddots & \vdots \\ b_{m1} & \cdots & b_{mn} \end{bmatrix} = \begin{bmatrix} a_{11}+b_{11} & \cdots & a_{1n}+b_{1n} \\ \vdots & \ddots & \vdots \\ a_{m1}+b_{m1} & \cdots & a_{mn}+b_{mn} \end{bmatrix}.$$

Note that matrices of different sizes cannot be added. It is quite clear that $A + A = 2A$, and $A + (A + A) = (A + A) + A = 3A$. Thus, inductively we define $nA = (n-1)A + A$ for any positive integer n. If B is any matrix, then $-B$ is by definition the multiplication $(-1)B$. Moreover, if A and B are two matrices of the same size, then the difference $A - B$ is by definition the sum $A+(-1)B = A+(-B)$. A matrix whose entries are all zero is called a **zero matrix**, denoted by the symbol $\mathbf{0}$ (or $\mathbf{0}_{m\times n}$ when we emphasize the number of rows and columns).

Clearly, matrix addition has the same properties as the addition of real numbers. The real numbers in the context here are traditionally called **scalars** even though "numbers" is a perfectly good name and "scalar" sounds more technical. The following theorem lists some basic rules of these operations.

Theorem 1.2 *Suppose that the sizes of A, B and C are the same. Then the following rules of matrix arithmetic are valid:*

(1) $(A + B) + C = A + (B + C)$, (*written as* $A + B + C$) (Associativity),

(2) $A + \mathbf{0} = \mathbf{0} + A = A$,

(3) $A + (-A) = (-A) + A = \mathbf{0}$,

(4) $A + B = B + A$, (Commutativity),

(5) $k(A + B) = kA + kB$,

(6) $(k + \ell)A = kA + \ell A$,

(7) $(k\ell)A = k(\ell A)$.

Proof: We prove only (5) and the remaining are left for exercises. For any (i, j),

$$[k(A + B)]_{ij} = k[A + B]_{ij} = k([A]_{ij} + [B]_{ij}) = [kA]_{ij} + [kB]_{ij}.$$

Consequently, $k(A + B) = kA + kB$. □

Definition 1.6 A square matrix A is said to be **symmetric** if $A^T = A$, or **skew-symmetric** if $A^T = -A$.

For example, the matrices A and B below

$$A = \begin{bmatrix} 1 & a & b \\ a & 3 & c \\ b & c & 5 \end{bmatrix}, \quad B = \begin{bmatrix} 0 & 1 & 2 \\ -1 & 0 & 3 \\ -2 & -3 & 0 \end{bmatrix}$$

are symmetric and skew-symmetric, respectively. Notice here that all the diagonal entries of a skew-symmetric matrix must be zero, since $a_{ii} = -a_{ii}$.

By a direct computation, one can easily verify the following rules of the transpose of matrices:

Theorem 1.3 *Let A and B be $m \times n$ matrices. Then*

$$(kA)^T = kA^T, \quad \textit{and} \quad (A+B)^T = A^T + B^T.$$

Problem 1.5 Prove the remaining parts of Theorem 1.2.

Problem 1.6 Find a matrix B such that $A + B^T = (A - B)^T$, where

$$A = \begin{bmatrix} 2 & -3 & 0 \\ 4 & -1 & 3 \\ -1 & 0 & 1 \end{bmatrix}.$$

Problem 1.7 Find a, b, c and d such that

$$\begin{bmatrix} a & b \\ c & d \end{bmatrix} = 2\begin{bmatrix} a & 3 \\ 2 & a+c \end{bmatrix} + \begin{bmatrix} 2+b & a+9 \\ c+d & b \end{bmatrix}.$$

1.4 Products of matrices

We introduced the operations sum and scalar multiplication of matrices in Section 1.3. In this section, we introduce the product of matrices. Unlike the sum of two matrices, the product of matrices is a little bit more complicated, in the sense that it is defined for two matrices of different sizes or for square matrices of the same order. We define the product of matrices in three steps:

(1) For a $1 \times n$ row matrix $\mathbf{a} = [a_1 \ \cdots \ a_n]$ and an $n \times 1$ column matrix $\mathbf{x} = [x_1 \ \cdots \ x_n]^T$, the **product** $\mathbf{ax}$ is a 1×1 matrix (*i.e.*, just a number) defined by the rule

$$\mathbf{ax} = [a_1 \ a_2 \ \cdots \ a_n] \begin{bmatrix} x_1 \\ x_2 \\ \vdots \\ x_n \end{bmatrix} = [a_1x_1 + a_2x_2 + \cdots + a_nx_n] = \left[\sum_{i=1}^{n} a_i x_i\right].$$

Note that the number of columns of the first matrix must be equal to the number of rows of the second matrix to have entrywise multiplications of the entries.

(2) For an $m \times n$ matrix

$$A = \begin{bmatrix} a_{11} & a_{12} & \cdots & a_{1n} \\ a_{21} & a_{22} & \cdots & a_{2n} \\ \vdots & \vdots & & \vdots \\ a_{m1} & a_{m2} & \cdots & a_{mn} \end{bmatrix} = \begin{bmatrix} \mathbf{a}_1 \\ \mathbf{a}_2 \\ \vdots \\ \mathbf{a}_m \end{bmatrix},$$

where $\mathbf{a}_i$'s denote the row vectors, and for an $n \times 1$ column matrix $\mathbf{x} = [x_1 \ \cdots \ x_n]^T$, the **product** $A\mathbf{x}$ is by definition an $m \times 1$ matrix defined by the rule:

$$A\mathbf{x} = \begin{bmatrix} \mathbf{a}_1\mathbf{x} \\ \mathbf{a}_2\mathbf{x} \\ \vdots \\ \mathbf{a}_m\mathbf{x} \end{bmatrix} = \begin{bmatrix} \sum_{i=1}^{n} a_{1i}x_i \\ \sum_{i=1}^{n} a_{2i}x_i \\ \vdots \\ \sum_{i=1}^{n} a_{mi}x_i \end{bmatrix},$$

or in an expanded form

$$\begin{bmatrix} a_{11} & a_{12} & \cdots & a_{1n} \\ a_{21} & a_{22} & \cdots & a_{2n} \\ \vdots & \vdots & & \vdots \\ a_{m1} & a_{m2} & \cdots & a_{mn} \end{bmatrix} \begin{bmatrix} x_1 \\ x_2 \\ \vdots \\ x_n \end{bmatrix} = \begin{bmatrix} a_{11}x_1 + a_{12}x_2 + \cdots + a_{1n}x_n \\ a_{21}x_1 + a_{22}x_2 + \cdots + a_{2n}x_n \\ \vdots \\ a_{m1}x_1 + a_{m2}x_2 + \cdots + a_{mn}x_n \end{bmatrix},$$

which is just an $m \times 1$ column matrix of the form $[b_1 \, b_2 \, \cdots \, b_m]^T$.

Therefore, for a system of m linear equations in n unknowns, by writing the n unknowns as an $n \times 1$ column matrix $\mathbf{x}$ and the coefficients as an $m \times n$ matrix A the system may be expressed as a matrix equation $A\mathbf{x} = \mathbf{b}$. Notice that this looks just like the usual linear equation in one variable: $ax = b$.

(3) Product of matrices: Let A be an $m \times n$ matrix and B an $n \times r$ matrix. The **product** AB is defined to be an $m \times r$ matrix whose columns are the products of A and the columns of B in corresponding order.

Thus if A is $m \times n$ and B is $n \times r$, then B has r columns and each column of B is an $n \times 1$ matrix. If we denote them by $\mathbf{b}^1, \ldots, \mathbf{b}^r$, or $B = [\mathbf{b}^1 \cdots \mathbf{b}^r]$, then

$$AB = \begin{bmatrix} A\mathbf{b}^1 & A\mathbf{b}^2 & \cdots & A\mathbf{b}^r \end{bmatrix} = \begin{bmatrix} \mathbf{a}_1\mathbf{b}^1 & \mathbf{a}_1\mathbf{b}^2 & \cdots & \mathbf{a}_1\mathbf{b}^r \\ \mathbf{a}_2\mathbf{b}^1 & \mathbf{a}_2\mathbf{b}^2 & \cdots & \mathbf{a}_2\mathbf{b}^r \\ \vdots & & \ddots & \vdots \\ \mathbf{a}_m\mathbf{b}^1 & \mathbf{a}_m\mathbf{b}^2 & \cdots & \mathbf{a}_m\mathbf{b}^r \end{bmatrix},$$

which is an $m \times r$ matrix. Therefore, the (i,j)-entry $[AB]_{ij}$ of AB is the i-th entry of the j-th column matrix

$$A\mathbf{b}^j = \begin{bmatrix} \mathbf{a}_1\mathbf{b}^j \\ \mathbf{a}_2\mathbf{b}^j \\ \vdots \\ \mathbf{a}_m\mathbf{b}^j \end{bmatrix},$$

i.e., for $i = 1, \ldots, m$ and $j = 1, \ldots, r$, it is the product of i-th row and j-th column of A:

$$[AB]_{ij} = \mathbf{a}_i\mathbf{b}^j = \sum_{k=1}^{n} a_{ik}b_{kj}.$$

Example 1.4 Consider the matrices

$$A = \begin{bmatrix} 2 & 3 \\ 4 & 0 \end{bmatrix}, \qquad B = \begin{bmatrix} 1 & 2 & 0 \\ 5 & -1 & 0 \end{bmatrix}.$$

The columns of AB are the product of A and each column of B:

$$\begin{bmatrix} 2 & 3 \\ 4 & 0 \end{bmatrix}\begin{bmatrix} 1 \\ 5 \end{bmatrix} = \begin{bmatrix} 2\cdot 1 + 3\cdot 5 \\ 4\cdot 1 + 0\cdot 5 \end{bmatrix} = \begin{bmatrix} 17 \\ 4 \end{bmatrix},$$

$$\begin{bmatrix} 2 & 3 \\ 4 & 0 \end{bmatrix}\begin{bmatrix} 2 \\ -1 \end{bmatrix} = \begin{bmatrix} 2\cdot 2 + 3\cdot(-1) \\ 4\cdot 2 + 0\cdot(-1) \end{bmatrix} = \begin{bmatrix} 1 \\ 8 \end{bmatrix},$$

$$\begin{bmatrix} 2 & 3 \\ 4 & 0 \end{bmatrix}\begin{bmatrix} 0 \\ 0 \end{bmatrix} = \begin{bmatrix} 2\cdot 0 + 3\cdot 0 \\ 4\cdot 0 + 0\cdot 0 \end{bmatrix} = \begin{bmatrix} 0 \\ 0 \end{bmatrix}.$$

Therefore, AB is

$$\begin{bmatrix} 2 & 3 \\ 4 & 0 \end{bmatrix} \begin{bmatrix} 1 & 2 & 0 \\ 5 & -1 & 0 \end{bmatrix} = \begin{bmatrix} 17 & 1 & 0 \\ 4 & 8 & 0 \end{bmatrix}.$$

Since A is a 2×2 matrix and B is a 2×3 matrix, the product AB is a 2×3 matrix. If we concentrate, for example, on the $(2,1)$-entry of AB, we single out the second row from A and the first column from B, and then we multiply corresponding entries together and add them up, *i.e.*, $4\cdot1+0\cdot5=4$. □

Note that the product AB of A and B is not defined if the number of columns of A and the number of rows of B are not equal.

Remark: In step (2), we could have defined for a $1\times n$ row matrix A and an $n\times r$ matrix B using the same rule defined in step (1). And then in step (3) an appropriate modification produces the same definition of the product of matrices. We suggest the readers verify this (see Example 1.6).

The **identity matrix** of order n, denoted by I_n (or I if the order is clear from the context), is a diagonal matrix whose diagonal entries are all 1, *i.e.*,

$$I_n = \begin{bmatrix} 1 & 0 & \cdots & 0 \\ 0 & 1 & & \vdots \\ \vdots & & \ddots & 0 \\ 0 & \cdots & 0 & 1 \end{bmatrix}.$$

By a direct computation, one can easily see that $AI_n = A = I_nA$ for any $n\times n$ matrix A.

Many, but not all, of the rules of arithmetic for real or complex numbers also hold for matrices with the operations of scalar multiplication, the sum and the product of matrices. The matrix $\mathbf{0}_{m\times n}$ plays the role of the number 0, and I_n that of the number 1 in the set of real numbers.

The rule that does not hold for matrices in general is the commutativity $AB = BA$ of the product, while the commutativity of the matrix sum $A+B=B+A$ does hold in general. The following example illustrates the noncommutativity of the product of matrices.

Example 1.5 Let $A = \begin{bmatrix} 1 & 0 \\ 0 & -1 \end{bmatrix}$ and $B = \begin{bmatrix} 0 & 1 \\ 1 & 0 \end{bmatrix}$. Then,

$$AB = \begin{bmatrix} 0 & 1 \\ -1 & 0 \end{bmatrix}, \quad BA = \begin{bmatrix} 0 & -1 \\ 1 & 0 \end{bmatrix}.$$

Thus the matrices A and B in this example satisfy $AB \neq BA$. □

The following theorem lists some rules of ordinary arithmetic that do hold for matrix operations.

Theorem 1.4 *Let A, B, C be arbitrary matrices for which the matrix operations below are defined, and let k be an arbitrary scalar. Then*

(1) $A(BC) = (AB)C$, (*written as* ABC) (Associativity),

(2) $A(B + C) = AB + AC$, *and* $(A + B)C = AC + BC$, (Distributivity),

(3) $IA = A = AI$,

(4) $k(BC) = (kB)C = B(kC)$,

(5) $(AB)^T = B^T A^T$.

Proof: Each equality can be shown by direct calculations of each entry of both sides of the equalities. We illustrate this by proving (1) only, and leave the others to the readers.

Assume that $A = [a_{ij}]$ is an $m\times n$ matrix, $B = [b_{k\ell}]$ is an $n\times p$ matrix, and $C = [c_{st}]$ is a $p\times r$ matrix. We now compute the (i, j)-entry of each side of the equation. Note that BC is an $n\times r$ matrix whose (i, j)-entry is $[BC]_{ij} = \sum_{\lambda=1}^{p} b_{i\lambda}c_{\lambda j}$. Thus

$$[A(BC)]_{ij} = \sum_{\mu=1}^{n} a_{i\mu}[BC]_{\mu j} = \sum_{\mu=1}^{n} a_{i\mu} \sum_{\lambda=1}^{p} b_{\mu\lambda}c_{\lambda j} = \sum_{\mu=1}^{n} \sum_{\lambda=1}^{p} a_{i\mu}b_{\mu\lambda}c_{\lambda j}.$$

Similarly, AB is an $m\times p$ matrix with the (i, j)-entry $[AB]_{ij} = \sum_{\mu=1}^{n} a_{i\mu}b_{\mu j}$, and

$$[(AB)C]_{ij} = \sum_{\lambda=1}^{p} [AB]_{i\lambda}c_{\lambda j} = \sum_{\lambda=1}^{p} \sum_{\mu=1}^{n} a_{i\mu}b_{\mu\lambda}c_{\lambda j} = \sum_{\mu=1}^{n} \sum_{\lambda=1}^{p} a_{i\mu}b_{\mu\lambda}c_{\lambda j}.$$

This clearly shows that $[A(BC)]_{ij} = [(AB)C]_{ij}$ for all i, j, and consequently $A(BC) = (AB)C$ as desired. □

Problem 1.8 Prove or disprove: If A is not a zero matrix and $AB = AC$, then $B = C$.

Problem 1.9 Show that any triangular matrix A satisfying $AA^T = A^T A$ is a diagonal matrix.

Problem 1.10 For a square matrix A, show that

(1) AA^T and $A + A^T$ are symmetric,

(2) $A - A^T$ is skew-symmetric, and

(3) A can be expressed as the sum of symmetric part $B = \frac{1}{2}(A + A^T)$ and skew-symmetric part $C = \frac{1}{2}(A - A^T)$, so that $A = B + C$.

As an application of our results on matrix operations, we shall prove the following important theorem:

Theorem 1.5 *Any system of linear equations has either no solution, exactly one solution, or infinitely many solutions.*

Proof: We have already seen that a system of linear equations may be written as $A\mathbf{x} = \mathbf{b}$, which may have no solution or exactly one solution. Now assume that the system $A\mathbf{x} = \mathbf{b}$ of linear equations has more than one solution and let $\mathbf{x}_1$ and $\mathbf{x}_2$ be two different solutions so that $A\mathbf{x}_1 = \mathbf{b}$ and $A\mathbf{x}_2 = \mathbf{b}$. Let $\mathbf{x}_0 = \mathbf{x}_1 - \mathbf{x}_2 \neq \mathbf{0}$. Since $A\mathbf{x}$ is just a particular case of a matrix product, Theorem 1.4 gives us

$$A(\mathbf{x}_1 + k\mathbf{x}_0) = A\mathbf{x}_1 + kA\mathbf{x}_0 = \mathbf{b} + k(A\mathbf{x}_1 - A\mathbf{x}_2) = \mathbf{b},$$

for any real number k. This says that $\mathbf{x}_1 + k\mathbf{x}_0$ is also a solution of $A\mathbf{x} = \mathbf{b}$ for any k. Since there are infinitely many choices for k, $A\mathbf{x} = \mathbf{b}$ has infinitely many solutions. $\square$

Problem 1.11 For which values of a does each of the following systems have no solution, exactly one solution, or infinitely many solutions?

$$(1)\ \left\{\begin{array}{rcrcrcl} x & + & 2y & - & 3z & = & 4 \\ 3x & - & y & + & 5z & = & 2 \\ 4x & + & y & + & (a^2-14)z & = & a+2. \end{array}\right.$$

$$(2)\ \left\{\begin{array}{rcrcrcl} x & - & y & + & z & = & 1 \\ x & + & 3y & + & az & = & 2 \\ 2x & + & ay & + & 3z & = & 3. \end{array}\right.$$

1.5 Block matrices

In this section we introduce some techniques that will often be very helpful in manipulating matrices. A **submatrix** of a matrix A is a matrix obtained from A by deleting certain rows and/or columns of A. Using a system of horizontal and vertical lines, we can partition a matrix A into submatrices, called **blocks**, of A as follows: Consider a matrix

$$A = \left[\begin{array}{ccc|c} a_{11} & a_{12} & a_{13} & a_{14} \\ a_{21} & a_{22} & a_{23} & a_{24} \\ \hline a_{31} & a_{32} & a_{33} & a_{34} \end{array}\right],$$

divided up into four blocks by the dotted lines shown. Now, if we write

$$A_{11} = \begin{bmatrix} a_{11} & a_{12} & a_{13} \\ a_{21} & a_{22} & a_{23} \end{bmatrix}, \quad A_{12} = \begin{bmatrix} a_{14} \\ a_{24} \end{bmatrix},$$
$$A_{21} = \begin{bmatrix} a_{31} & a_{32} & a_{33} \end{bmatrix}, \quad A_{22} = \begin{bmatrix} a_{34} \end{bmatrix},$$

then A can be written as

$$A = \begin{bmatrix} A_{11} & A_{12} \\ A_{21} & A_{22} \end{bmatrix},$$

called a **block matrix**.

The product of matrices partitioned into blocks also follows the matrix product formula, as if the A_{ij} were numbers:

$$\begin{aligned} A &= \begin{bmatrix} A_{11} & A_{12} \\ A_{21} & A_{22} \end{bmatrix}, \qquad B = \begin{bmatrix} B_{11} & B_{12} \\ B_{21} & B_{22} \end{bmatrix}; \\ AB &= \begin{bmatrix} A_{11}B_{11} + A_{12}B_{21} & A_{11}B_{12} + A_{12}B_{22} \\ A_{21}B_{11} + A_{22}B_{21} & A_{21}B_{12} + A_{22}B_{22} \end{bmatrix}, \end{aligned}$$

provided that the number of columns in A_{ik} is equal to the number of rows in B_{kj}. This will be true only if the columns of A are partitioned in the same way as the rows of B.

It is not hard to see that the matrix product by blocks is correct. Suppose, for example, that we have a 3×3 matrix A and partition it as

$$A = \left[\begin{array}{cc|c} a_{11} & a_{12} & a_{13} \\ a_{21} & a_{22} & a_{23} \\ \hline a_{31} & a_{32} & a_{33} \end{array}\right] = \begin{bmatrix} A_{11} & A_{12} \\ A_{21} & A_{22} \end{bmatrix},$$

and a 3×2 matrix B which we partition as

$$B = \left[\begin{array}{cc} b_{11} & b_{12} \\ b_{21} & b_{22} \\ \hline b_{31} & b_{32} \end{array}\right] = \left[\begin{array}{c} B_{11} \\ B_{21} \end{array}\right].$$

Then the entries of $C = [c_{ij}] = AB$ are

$$c_{ij} = (a_{i1}b_{1j} + a_{i2}b_{2j}) + a_{i3}b_{3j}\,.$$

The quantity $a_{i1}b_{1j} + a_{i2}b_{2j}$ is simply the $(i\,j)$-entry of $A_{11}B_{11}$ if $i \le 2$, and the $(i\,j)$-entry of $A_{21}B_{11}$ if $i = 3$. Similarly, $a_{i3}b_{3j}$ is the $(i\,j)$-entry of $A_{12}B_{21}$ if $i \le 2$, and of $A_{22}B_{21}$ if $i = 3$. Thus AB can be written as

$$AB = \left[\begin{array}{c} C_{11} \\ C_{12} \end{array}\right] = \left[\begin{array}{c} A_{11}B_{11} + A_{12}B_{21} \\ A_{21}B_{11} + A_{22}B_{21} \end{array}\right].$$

In particular, if an $m \times n$ matrix A is partitioned into blocks of column vectors: *i.e.*, $A = [\mathbf{a}^1\ \mathbf{a}^2\ \cdots\ \mathbf{a}^n]$, where each block $\mathbf{a}^j$ is the j-th column, then the product $A\mathbf{x}$ with $\mathbf{x} = [x_1\ \cdots\ x_n]^T$ is the sum of the block matrices (or column vectors) with coefficients x_j's:

$$A\mathbf{x} = [\mathbf{a}^1\ \mathbf{a}^2\ \cdots\ \mathbf{a}^n]\left[\begin{array}{c} x_1 \\ x_2 \\ \vdots \\ x_n \end{array}\right] = x_1\mathbf{a}^1 + x_2\mathbf{a}^2 + \cdots + x_n\mathbf{a}^n,$$

where $x_j\mathbf{a}^j = x_j[a_{1j}\ a_{2j}\ \cdots\ a_{nj}]^T$.

Example 1.6 Let A be an $m \times n$ matrix partitioned into the row vectors $\mathbf{a}_1,\ \mathbf{a}_2,\ \ldots,\ \mathbf{a}_n$ as its blocks, and let B be an $n \times r$ matrix so that their product AB is well-defined. By considering the matrix B as a block, the product AB can be written

$$AB = \left[\begin{array}{c} \mathbf{a}_1 \\ \mathbf{a}_2 \\ \vdots \\ \mathbf{a}_m \end{array}\right] B = \left[\begin{array}{c} \mathbf{a}_1 B \\ \mathbf{a}_2 B \\ \vdots \\ \mathbf{a}_m B \end{array}\right] = \left[\begin{array}{cccc} \mathbf{a}_1\mathbf{b}^1 & \mathbf{a}_1\mathbf{b}^2 & \cdots & \mathbf{a}_1\mathbf{b}^r \\ \mathbf{a}_2\mathbf{b}^1 & \mathbf{a}_2\mathbf{b}^2 & \cdots & \mathbf{a}_2\mathbf{b}^r \\ \vdots & & \ddots & \vdots \\ \mathbf{a}_m\mathbf{b}^1 & \mathbf{a}_m\mathbf{b}^2 & \cdots & \mathbf{a}_m\mathbf{b}^r \end{array}\right],$$

where $\mathbf{b}^1,\ \mathbf{b}^2,\ \cdots,\ \mathbf{b}^r$ denote the columns of B. Hence, the row vectors of AB are the products of the row vectors of A and B.

Problem 1.12 Compute AB using block multiplication, where

$$A = \left[\begin{array}{rr|rr} 1 & 2 & 1 & 0 \\ -3 & 4 & 0 & 1 \\ \hline 0 & 0 & 2 & -1 \end{array}\right], \quad B = \left[\begin{array}{rr|r} 1 & 0 & 2 \\ 0 & 1 & 3 \\ \hline 2 & 3 & 4 \\ 3 & -2 & 1 \end{array}\right].$$

1.6 Inverse matrices

As we saw in Section 1.4, a system of linear equations can be written as $A\mathbf{x} = \mathbf{b}$ in matrix form. This form resembles one of the simplest linear equation in one variable $ax = b$ whose solution is simply $x = a^{-1}b$ when $a \neq 0$. Thus it is tempting to write the solution of the system as $\mathbf{x} = A^{-1}\mathbf{b}$. However, in the case of matrices we first have to have a precise meaning of A^{-1}. To discuss this we begin with the following definition.

Definition 1.7 For an $m \times n$ matrix A, an $n \times m$ matrix B is called a **left inverse** of A if $BA = I_n$, and an $n \times m$ matrix C is called a **right inverse** of A if $AC = I_m$.

Example 1.7 From a direct calculation for two matrices

$$A = \begin{bmatrix} 1 & 2 & -1 \\ 2 & 0 & 1 \end{bmatrix} \quad \text{and} \quad B = \begin{bmatrix} 1 & -3 \\ -1 & 5 \\ -2 & 7 \end{bmatrix},$$

we have $AB = I_2$, and $BA = \begin{bmatrix} -5 & 2 & -4 \\ 9 & -2 & 6 \\ 12 & -4 & 9 \end{bmatrix} \neq I_3$.

Thus, the matrix B is a right inverse but not a left inverse of A, while A is a left inverse but not a right inverse of B. Since $(AB)^T = B^T A^T$ and $I^T = I$, a matrix A has a right inverse if and only if A^T has a left inverse. □

However, if A is a square matrix and has a left inverse, then we prove later (Theorem 1.8) that it has also a right inverse, and vice versa. Moreover, the following lemma shows that the left inverses and the right inverses of a square matrix are all equal. (This is not true for nonsquare matrices, of course).

Lemma 1.6 *If an $n \times n$ square matrix A has a left inverse B and a right inverse C, then B and C are equal, i.e., $B = C$.*

Proof: A direct calculation shows that

$$B = BI = B(AC) = (BA)C = IC = C.$$

Now any two left inverses must be both equal to a right inverse C, and hence to each other, and any two right inverses must be both equal to a left inverse B, and hence to each other. So there exist only one left and only one right inverse for a square matrix A ifit is known that A has both left and right inverses. Furthermore, the left and right inverses are equal. □

This theorem says that if a matrix A has both a right inverse and a left inverse, then they must be the same. However, we shall see in Chapter 3 that any $m \times n$ matrix A with $m \neq n$ cannot have both a right inverse and a left inverse: that is, a nonsquare matrix may have only a left inverse or only a right inverse. In this case, the matrix may have many left inverses or many right inverses.

Example 1.8 A nonsquare matrix $A = \begin{bmatrix} 1 & 0 \\ 0 & 1 \\ 0 & 0 \end{bmatrix}$ can have more than one left inverse. In fact, for any $x, y \in \mathbb{R}$, one can easily check that the matrix $B = \begin{bmatrix} 1 & 0 & x \\ 0 & 1 & y \end{bmatrix}$ is a left inverse of A. □

Definition 1.8 An $n \times n$ square matrix A is said to be **invertible** (or **nonsingular**) if there exists a square matrix B of the same size such that

$$AB = I = BA.$$

Such a matrix B is called the **inverse** of A, and is denoted by A^{-1}. A matrix A is said to be **singular** if it is not invertible.

Note that Lemma 1.6 shows that if a square matrix A has both left and right inverses, then it must be unique. That is why we call B "the" inverse of A. For instance, consider a 2×2 matrix $A = \begin{bmatrix} a & b \\ c & d \end{bmatrix}$. If $ad - bc \neq 0$, then it is easy to verify that

$$A^{-1} = \frac{1}{ad-bc}\begin{bmatrix} d & -b \\ -c & a \end{bmatrix} = \begin{bmatrix} \dfrac{d}{ad-bc} & \dfrac{-b}{ad-bc} \\ \dfrac{-c}{ad-bc} & \dfrac{a}{ad-bc} \end{bmatrix},$$

since $AA^{-1} = I_2 = A^{-1}A$. (Check this product of matrices for practice!) Note that any zero matrix is singular.

Problem 1.13 Let A be an invertible matrix and k any nonzero scalar. Show that
(1) A^{-1} is invertible and $(A^{-1})^{-1} = A$;
(2) the matrix kA is invertible and $(kA)^{-1} = \frac{1}{k}A^{-1}$;
(3) A^T is invertible and $(A^T)^{-1} = (A^{-1})^T$.

Theorem 1.7 *The product of invertible matrices is also invertible, whose inverse is the product of the individual inverses in reverse order:*

$$(AB)^{-1} = B^{-1}A^{-1}.$$

Proof: Suppose that A and B are invertible matrices of the same size. Then $(AB)(B^{-1}A^{-1}) = A(BB^{-1})A^{-1} = AIA^{-1} = AA^{-1} = I$, and similarly $(B^{-1}A^{-1})(AB) = I$. Thus AB has the inverse $B^{-1}A^{-1}$. □

We have written the inverse of A as "A to the power -1", so we can give the meaning of A^k for any integer k: Let A be a square matrix. Define $A^0 = I$. Then, for any positive integer k, we define the power A^k of A inductively as

$$A^k = A(A^{k-1}).$$

Moreover, if A is invertible, then the negative integer power is defined as

$$A^{-k} = (A^{-1})^k \quad \text{for } k > 0.$$

It is easy to check that with these rules we have $A^{k+\ell} = A^k A^\ell$ whenever the right hand side is defined. (If A is not invertible, $A^{3+(-1)}$ is defined but A^{-1} is not.)

Problem 1.14 Prove:
(1) If A has a zero row, so does AB.
(2) If B has a zero column, so does AB.
(3) Any matrix with a zero row or a zero column cannot be invertible.

Problem 1.15 Let A be an invertible matrix. Is it true that $(A^k)^T = (A^T)^k$ for any integer k? Justify your answer.

1.7 Elementary matrices

We now return to the system of linear equations $A\mathbf{x} = \mathbf{b}$. If A has a right inverse B such that $AB = I_m$, then $\mathbf{x} = B\mathbf{b}$ is a solution of the system since

$$A\mathbf{x} = A(B\mathbf{b}) = (AB)\mathbf{b} = \mathbf{b}.$$

In particular, if A is an invertible square matrix, then it has only one inverse A^{-1} by Lemma 1.6, and $\mathbf{x} = A^{-1}\mathbf{b}$ is the only solution of the system. In this section, we discuss how to compute A^{-1} when A is invertible.

Recall that Gaussian elimination is a process in which the augmented matrix is transformed into its row-echelon form by a finite number of elementary row operations. In the following, we will show that each elementary row operation can be expressed as a nonsingular matrix, called an *elementary matrix*, and hence the process of Gaussian elimination is simply multiplying a finite sequence of corresponding elementary matrices to the augmented matrix.

Definition 1.9 A matrix E obtained from the identity matrix I_n by executing only one elementary row operation is called an **elementary matrix**.

For example, the following matrices are three elementary matrices corresponding to each type of the three elementary row operations.

(1) $\begin{bmatrix} 1 & 0 \\ 0 & -5 \end{bmatrix}$: multiply the second row of I_2 by -5;

(2) $\begin{bmatrix} 1 & 0 & 0 & 0 \\ 0 & 0 & 0 & 1 \\ 0 & 0 & 1 & 0 \\ 0 & 1 & 0 & 0 \end{bmatrix}$: interchange the second and the fourth rows of I_4;

(3) $\begin{bmatrix} 1 & 0 & 3 \\ 0 & 1 & 0 \\ 0 & 0 & 1 \end{bmatrix}$: add 3 times the third row to the first row of I_3.

It is an interesting fact that, if E is an elementary matrix obtained by executing a certain elementary row operation on the identity matrix I_m, then for any $m \times n$ matrix A, *the product EA is exactly the matrix that is obtained when the same elementary row operation in E is executed on A.* The following example illustrates this argument. (Note that AE is not what we want. For this, see Problem 1.17).

Example 1.9 For simplicity, we work on a 3×1 column matrix $\mathbf{b}$. Suppose that we want to do the operation "adding $(-2) \times$ the first row to the second row" on matrix $\mathbf{b}$. Then, we execute this operation on the identity matrix I first to get an elementary matrix E:

$$E = \begin{bmatrix} 1 & 0 & 0 \\ -2 & 1 & 0 \\ 0 & 0 & 1 \end{bmatrix}.$$

Multiplying the elementary matrix E to $\mathbf{b}$ on the left produces the desired result:

$$E\mathbf{b} = \begin{bmatrix} 1 & 0 & 0 \\ -2 & 1 & 0 \\ 0 & 0 & 1 \end{bmatrix} \begin{bmatrix} b_1 \\ b_2 \\ b_3 \end{bmatrix} = \begin{bmatrix} b_1 \\ b_2 - 2b_1 \\ b_3 \end{bmatrix},$$

Similarly, the operation "interchanging the first and third rows" on the matrix $\mathbf{b}$ can be achieved by multiplying a *permutation matrix* P, which is an elementary matrix obtained from I_3 by interchanging two rows, to $\mathbf{b}$ on the left:

$$P\mathbf{b} = \begin{bmatrix} 0 & 0 & 1 \\ 0 & 1 & 0 \\ 1 & 0 & 0 \end{bmatrix} \begin{bmatrix} b_1 \\ b_2 \\ b_3 \end{bmatrix} = \begin{bmatrix} b_3 \\ b_2 \\ b_1 \end{bmatrix}.$$

□

Recall that each elementary row operation has an inverse operation, which is also an elementary operation, that brings the matrix back to the original one. Thus, suppose that E denotes an elementary matrix corresponding to an elementary row operation, and let E' be the elementary matrix corresponding to its "inverse" elementary row operation in E. Then,

(1) if E multiplies a row by $c \neq 0$, then E' multiplies the same row by $\frac{1}{c}$;

(2) if E interchanges two rows, then E' interchanges them again;

(3) if E adds a multiple of one row to another, then E' subtracts it back from the same row.

Thus, for any $m \times n$ matrix A, $E'EA = A$, and $E'E = I = EE'$. That is, *every elementary matrix is invertible so that* $E^{-1} = E'$, *which is also an elementary matrix.*

For instance, if

$$E_1 = \begin{bmatrix} 1 & 0 & 0 \\ 0 & c & 0 \\ 0 & 0 & 1 \end{bmatrix}, \ E_2 = \begin{bmatrix} 1 & 0 & 0 \\ 0 & 1 & 0 \\ 3 & 0 & 1 \end{bmatrix}, \ E_3 = \begin{bmatrix} 0 & 1 & 0 \\ 1 & 0 & 0 \\ 0 & 0 & 1 \end{bmatrix}, \text{ then}$$

$$E_1^{-1} = \begin{bmatrix} 1 & 0 & 0 \\ 0 & 1/c & 0 \\ 0 & 0 & 1 \end{bmatrix}, \ E_2^{-1} = \begin{bmatrix} 1 & 0 & 0 \\ 0 & 1 & 0 \\ -3 & 0 & 1 \end{bmatrix}, \ E_3^{-1} = \begin{bmatrix} 0 & 1 & 0 \\ 1 & 0 & 0 \\ 0 & 0 & 1 \end{bmatrix}.$$

Definition 1.10 A **permutation matrix** is a square matrix obtained from the identity matrix by permuting the rows.

Problem 1.16 Prove:

(1) A permutation matrix is the product of a finite number of elementary matrices each of which is corresponding to the "row-interchanging" elementary row operation.

(2) Any permutation matrix P is invertible and $P^{-1} = P^T$.

(3) The product of any two permutation matrices is a permutation matrix.

(4) The transpose of a permutation matrix is also a permutation matrix.

Problem 1.17 Define the **elementary column operations** for a matrix by just replacing "row" by "column" in the definition of the elementary row operations. Show that if A is an $m \times n$ matrix and if E is an elementary matrix obtained by executing an elementary column operation on I_n, then AE is exactly the matrix that is obtained from A when the same column operation is executed on A.

The next theorem establishes some fundamental relationships between $n \times n$ square matrices and systems of n linear equations in n unknowns.

Theorem 1.8 *Let A be an $n \times n$ matrix. The following are equivalent:*

(1) *A has a left inverse;*

(2) *$A\mathbf{x} = \mathbf{0}$ has only the trivial solution $\mathbf{x} = \mathbf{0}$;*

(3) *A is row-equivalent to I_n;*

(4) *A is a product of elementary matrices;*

(5) *A is invertible;*

(6) *A has a right inverse.*

Proof: **(1) $\Rightarrow$ (2)** : Let $\mathbf{x}$ be a solution of the homogeneous system $A\mathbf{x} = \mathbf{0}$, and let B be a left inverse of A. Then

$$\mathbf{x} = I_n\mathbf{x} = (BA)\mathbf{x} = BA\mathbf{x} = B\mathbf{0} = \mathbf{0}.$$

(2) ⇒ (3) : Suppose that the homogeneous system $A\mathbf{x} = \mathbf{0}$ has only the trivial solution $\mathbf{x} = \mathbf{0}$:

$$\begin{cases} x_1 & & & = & 0 \\ & x_2 & & = & 0 \\ & & \ddots & & \\ & & x_n & = & 0. \end{cases}$$

This means that the augmented matrix $[A\ \mathbf{0}]$ of the system $A\mathbf{x} = \mathbf{0}$ is reduced to the system $[I_n\ \mathbf{0}]$ by Gauss-Jordan elimination. Hence, A is row-equivalent to I_n.

(3) ⇒ (4) : Assume A is row-equivalent to I_n, so that A can be reduced to I_n by a finite sequence of elementary row operations. Thus, we can find elementary matrices $E_1, E_2, \ldots, E_k$ such that

$$E_k \cdots E_2 E_1 A = I_n.$$

Since $E_1, E_2, \ldots, E_k$ are invertible, by multiplying both sides of this equation on the left successively by $E_k^{-1}, \ldots, E_2^{-1}, E_1^{-1}$, we obtain

$$A = E_1^{-1} E_2^{-1} \cdots E_k^{-1} I_n = E_1^{-1} E_2^{-1} \cdots E_k^{-1},$$

which expresses A as the product of elementary matrices.

(4) ⇒ (5) is trivial, because any elementary matrix is invertible. In fact, $A^{-1} = E_k \cdots E_2 E_1$.

(5) ⇒ (1) and **(5) ⇒ (6)** are trivial.

(6) ⇒ (5) : If B is a right inverse of A, then A is a left inverse of B and we can apply (1) ⇒ (2) ⇒ (3) ⇒ (4) ⇒ (5) to B and conclude that B is invertible, with A as its unique inverse. That is, B is the inverse of A and so A is invertible. □

This theorem shows that a square matrix is invertible if it has a one-side inverse. In particular, if a square matrix A is invertible, then $\mathbf{x} = A^{-1}\mathbf{b}$ is a unique solution to the system $A\mathbf{x} = \mathbf{b}$.

Problem 1.18 Find the inverse of the product

$$\begin{bmatrix} 1 & 0 & 0 \\ 0 & 1 & 0 \\ 0 & -c & 1 \end{bmatrix} \begin{bmatrix} 1 & 0 & 0 \\ 0 & 1 & 0 \\ -b & 0 & 1 \end{bmatrix} \begin{bmatrix} 1 & 0 & 0 \\ -a & 1 & 0 \\ 0 & 0 & 1 \end{bmatrix}.$$

As an application of the preceding theorem, we give a practical method for finding the inverse A^{-1} of an invertible $n \times n$ matrix A. If A is invertible, there are elementary matrices $E_1, E_2, \ldots, E_k$ such that

$$E_k \cdots E_2 E_1 A = I_n.$$

Hence,

$$A^{-1} = E_k \cdots E_2 E_1 = E_k \cdots E_2 E_1 I_n.$$

It follows that *the sequence of row operations that reduces an invertible matrix A to I_n will resolve I_n to A^{-1}*. In other words, let $[A \mid I]$ be the augmented matrix with the columns of A on the left half, the columns of I on the right half. A Gaussian elimination, applied to both sides, by some elementary row operations reduces the augmented matrix $[A \mid I]$ to $[U \mid K]$, where U is a row-echelon form of A. Next, the back substitution process by another series of elementary row operations reduces $[U \mid K]$ to $[I \mid A^{-1}]$:

$$\begin{aligned} [A \mid I] &\rightarrow [E_\ell \cdots E_1 A \mid E_\ell \cdots E_1 I] = [U \mid K] \\ &\rightarrow [F_k \cdots F_1 U \mid F_k \cdots F_1 K] = [I \mid A^{-1}], \end{aligned}$$

where $E_\ell \cdots E_1$ represents a Gaussian elimination and $F_k \cdots F_1$ represents the back substitution. The following example illustrates the computation of an inverse matrix.

Example 1.10 Find the inverse of

$$A = \begin{bmatrix} 1 & 2 & 3 \\ 2 & 3 & 5 \\ 1 & 0 & 2 \end{bmatrix}.$$

We apply Gauss-Jordan elimination to

$$\begin{aligned} [A \mid I] &= \left[\begin{array}{ccc|ccc} 1 & 2 & 3 & 1 & 0 & 0 \\ 2 & 3 & 5 & 0 & 1 & 0 \\ 1 & 0 & 2 & 0 & 0 & 1 \end{array}\right] \begin{array}{l} \\ (-2)\text{row } 1 + \text{row } 2 \\ (-1)\text{row } 1 + \text{row } 3 \end{array} \\ &\rightarrow \left[\begin{array}{ccc|ccc} 1 & 2 & 3 & 1 & 0 & 0 \\ 0 & -1 & -1 & -2 & 1 & 0 \\ 0 & -2 & -1 & -1 & 0 & 1 \end{array}\right] (-1)\text{row } 2 \\ &\rightarrow \left[\begin{array}{ccc|ccc} 1 & 2 & 3 & 1 & 0 & 0 \\ 0 & 1 & 1 & 2 & -1 & 0 \\ 0 & -2 & -1 & -1 & 0 & 1 \end{array}\right] \begin{array}{l} \\ \\ (2)\text{row } 2 + \text{row } 3 \end{array} \end{aligned}$$

$$\rightarrow \left[\begin{array}{ccc|ccc} 1 & 2 & 3 & 1 & 0 & 0 \\ 0 & 1 & 1 & 2 & -1 & 0 \\ 0 & 0 & 1 & 3 & -2 & 1 \end{array}\right].$$

This is $[U \mid K]$ obtained by Gaussian elimination. Now continue the back substitution to reduce $[U \mid K]$ to $[I \mid A^{-1}]$

$$\begin{aligned} [U \mid K] &= \left[\begin{array}{ccc|ccc} 1 & 2 & 3 & 1 & 0 & 0 \\ 0 & 1 & 1 & 2 & -1 & 0 \\ 0 & 0 & 1 & 3 & -2 & 1 \end{array}\right] \begin{array}{l} (-1)\text{row } 3 + \text{row } 2 \\ (-3)\text{row } 3 + \text{row } 1 \end{array} \\ &\rightarrow \left[\begin{array}{ccc|ccc} 1 & 2 & 0 & -8 & 6 & -3 \\ 0 & 1 & 0 & -1 & 1 & -1 \\ 0 & 0 & 1 & 3 & -2 & 1 \end{array}\right] \quad (-2)\text{row } 2 + \text{row } 1 \\ &\rightarrow \left[\begin{array}{ccc|ccc} 1 & 0 & 0 & -6 & 4 & -1 \\ 0 & 1 & 0 & -1 & 1 & -1 \\ 0 & 0 & 1 & 3 & -2 & 1 \end{array}\right] = \left[I \mid A^{-1}\right]. \end{aligned}$$

Thus, we get

$$A^{-1} = \begin{bmatrix} -6 & 4 & -1 \\ -1 & 1 & -1 \\ 3 & -2 & 1 \end{bmatrix}.$$

(The reader should verify that $AA^{-1} = I = A^{-1}A$.) □

Note that if A is not invertible, then, at some step in Gaussian elimination, a zero row will show up on the left side in $[U \mid K]$. For example, the matrix $A = \begin{bmatrix} 1 & 6 & 4 \\ 2 & 4 & -1 \\ -1 & 2 & 5 \end{bmatrix}$ is row-equivalent to $\begin{bmatrix} 1 & 6 & 4 \\ 0 & -8 & -9 \\ 0 & 0 & 0 \end{bmatrix}$ which is a noninvertible matrix.

Problem 1.19 Write A^{-1} as a product of elementary matrices for A in Example 1.10.

of A by using Gaussian elimination.

Problem 1.20 Find the inverse of each of the following matrices:

$$A = \begin{bmatrix} 1 & -1 & 2 \\ -1 & 0 & 2 \\ -6 & 4 & 11 \end{bmatrix}, B = \begin{bmatrix} 1 & 0 & 0 & 0 \\ 1 & 2 & 0 & 0 \\ 1 & 2 & 4 & 0 \\ 1 & 2 & 4 & 8 \end{bmatrix}, C = \begin{bmatrix} k & 0 & 0 & 0 \\ 1 & k & 0 & 0 \\ 0 & 1 & k & 0 \\ 0 & 0 & 1 & k \end{bmatrix} \quad (k \neq 0).$$

Problem 1.21 When is a diagonal matrix $D = \begin{bmatrix} d_1 & & 0 \\ & \ddots & \\ 0 & & d_n \end{bmatrix}$ nonsingular, and what is D^{-1}?

From Theorem 1.8, a square matrix A is nonsingular if and only if $A\mathbf{x} = \mathbf{0}$ has only the trivial solution. That is, a square matrix A is singular if and only if $A\mathbf{x} = \mathbf{0}$ has a nontrivial solution, say $\mathbf{x}_0$. Now, for any column vector $\mathbf{b} = [b_1 \ \cdots \ b_n]^T$, if $\mathbf{x}_1$ is a solution of $A\mathbf{x} = \mathbf{b}$ for a singular matrix A, then so is $k\mathbf{x}_0 + \mathbf{x}_1$ for any k:

$$A(k\mathbf{x}_0 + \mathbf{x}_1) = k(A\mathbf{x}_0) + A\mathbf{x}_1 = k\mathbf{0} + \mathbf{b} = \mathbf{b}.$$

This argument strengthens Theorem 1.5 as follows when A is a square matrix:

Theorem 1.9 *If A is an invertible $n \times n$ matrix, then for any column vector $\mathbf{b} = [b_1 \ \cdots \ b_n]^T$, the system $A\mathbf{x} = \mathbf{b}$ has exactly one solution $\mathbf{x} = A^{-1}\mathbf{b}$. If A is not invertible, then the system has either no solution or infinitely many solutions according to whether or not the system is inconsistent.* □

Problem 1.22 Write the system of linear equations

$$\begin{cases} x + 2y + 2z = 10 \\ 2x - 2y + 3z = 1 \\ 4x - 3y + 5z = 4 \end{cases}$$

in matrix form $A\mathbf{x} = \mathbf{b}$ and solve it by finding $A^{-1}\mathbf{b}$.

1.8 *LDU* factorization

Recall that a basic method of solving a linear system $A\mathbf{x} = \mathbf{b}$ is by Gauss-Jordan elimination. For a fixed matrix A, if we want to solve more than one system $A\mathbf{x} = \mathbf{b}$ for various values of $\mathbf{b}$, then the same Gaussian elimination on A has to be repeated over and over again. However, this repetition may be avoided by expressing Gaussian elimination as an invertible matrix which is a product of elementary matrices.

We first assume that no permutations of rows are necessary throughout the whole process of Gaussian elimination on $[A \ \mathbf{b}]$. Then the forward elimination is just to multiply finitely many elementary matrices $E_k, \ \ldots, \ E_1$ to the augmented matrix $[A \ \mathbf{b}]$: that is,

$$[E_k \cdots E_1 A \quad E_k \cdots E_1 \mathbf{b}] = [U \ \mathbf{c}],$$

where each E_i is a lower triangular elementary matrix whose diagonal entries are all 1's and $[U\ \mathbf{c}]$ is the augmented matrix of the system obtained after forward elimination on $A\mathbf{x} = \mathbf{b}$ (Note that U need not be an upper triangular matrix if A is not a square matrix). Therefore, if we set $L = (E_k \cdots E_1)^{-1} = E_1^{-1} \cdots E_k^{-1}$, then $A = LU$ and

$$\mathbf{c} = U\mathbf{x} = E_k \cdots E_1 A\mathbf{x} = E_k \cdots E_1 \mathbf{b} = L^{-1}\mathbf{b}.$$

Note that L is a lower triangular matrix whose diagonal entries are all 1's (see Problem 1.24). Now, for any column matrix $\mathbf{b}$, the system $A\mathbf{x} = LU\mathbf{x} = \mathbf{b}$ can be solved in two steps: first compute $\mathbf{c} = L^{-1}\mathbf{b}$ which is a forward elimination, and then solve $U\mathbf{x} = \mathbf{c}$ by the back substitution.

This means that, to solve the ℓ-systems $A\mathbf{x} = \mathbf{b}_i$ for $i = 1, \ldots, \ell$, we first find the matrices L and U such that $A = LU$ by performing forward elimination on A, and then compute $\mathbf{c}_i = L^{-1}\mathbf{b}_i$ for $i = 1, \ldots, \ell$. The solutions of $A\mathbf{x} = \mathbf{b}_i$ are now those of $U\mathbf{x} = \mathbf{c}_i$.

Example 1.11 Consider the system of linear equations

$$A\mathbf{x} = \begin{bmatrix} 2 & 1 & 1 & 0 \\ 4 & 1 & 0 & 1 \\ -2 & 2 & 1 & 1 \end{bmatrix} \begin{bmatrix} x_1 \\ x_2 \\ x_3 \end{bmatrix} = \begin{bmatrix} 1 \\ -2 \\ 7 \end{bmatrix} = \mathbf{b}.$$

The elementary matrices for Gaussian elimination of A are easily found to be

$$E_1 = \begin{bmatrix} 1 & 0 & 0 \\ -2 & 1 & 0 \\ 0 & 0 & 1 \end{bmatrix}, \quad E_2 = \begin{bmatrix} 1 & 0 & 0 \\ 0 & 1 & 0 \\ 1 & 0 & 1 \end{bmatrix}, \quad \text{and} \quad E_3 = \begin{bmatrix} 1 & 0 & 0 \\ 0 & 1 & 0 \\ 0 & 3 & 1 \end{bmatrix},$$

so that

$$E_3E_2E_1A = \begin{bmatrix} 2 & 1 & 1 & 0 \\ 0 & -1 & -2 & 1 \\ 0 & 0 & -4 & 4 \end{bmatrix} = U.$$

Note that U is the matrix obtained from A after forward elimination, and $A = LU$ with

$$L = E_1^{-1}\, E_2^{-1}\, E_3^{-1} = \begin{bmatrix} 1 & 0 & 0 \\ 2 & 1 & 0 \\ -1 & -3 & 1 \end{bmatrix},$$

which is a lower triangular matrix with 1's on the diagonal. Now, the system

$$L\mathbf{c} = \mathbf{b} : \quad \left\{ \begin{array}{rcrcrcr} c_1 & & & & & = & 1 \\ 2c_1 & + & c_2 & & & = & -2 \\ -c_1 & - & 3c_2 & + & c_3 & = & 7 \end{array} \right.$$

resolves to $\mathbf{c} = (1, -4, -4)$ and the system

$$U\mathbf{x} = \mathbf{c}: \quad \left\{ \begin{array}{rcrcrcrcr} 2x_1 & + & x_2 & + & x_3 & & & = & 1 \\ & - & x_2 & - & 2x_3 & + & x_4 & = & -4 \\ & & & - & 4x_3 & + & 4x_4 & = & -4 \end{array} \right.$$

resolves to

$$\mathbf{x} = \begin{bmatrix} -1 + t \\ 2 + 3t \\ 1 - t \\ t \end{bmatrix} = \begin{bmatrix} -1 \\ 2 \\ 1 \\ 0 \end{bmatrix} + t \begin{bmatrix} 1 \\ 3 \\ -1 \\ 1 \end{bmatrix},$$

for $t \in \mathbb{R}$. It is suggested that the readers find the solutions for various values of $\mathbf{b}$. □

Problem 1.23 Determine an LU decomposition of the matrix

$$A = \begin{bmatrix} 1 & -1 & 0 \\ -1 & 2 & -1 \\ 0 & -1 & 2 \end{bmatrix},$$

and then find solutions of $A\mathbf{x} = \mathbf{b}$ for (1) $\mathbf{b} = [1\ 1\ 1]^T$ and (2) $\mathbf{b} = [2\ 0\ -1]^T$.

Problem 1.24 Let A, B be two lower triangular matrices. Prove that
(1) their product is also a lower triangular matrix;
(2) if A is invertible, then its inverse is also a lower triangular matrix;
(3) if the diagonal entries are all 1's, then the same holds for their product and their inverses.

Note that the same holds for upper triangular matrices, and for the product of more than two matrices.

Now suppose that A is a *nonsingular square* matrix with $A = LU$ in which no row interchanges were necessary. Then the pivots on the diagonal of U are all nonzero, and the diagonal of L are all 1's. Thus, by dividing each i-th row of U by the nonzero pivot d_i, the matrix U is factorized into a diagonal matrix D whose diagonals are just the pivots $d_1, d_2, \ldots, d_n$ and a new upper triangular matrix, denoted again by U, whose diagonals are all 1's so that $A = LDU$. For example,

$$\begin{bmatrix} d_1 & r & \cdots & s \\ 0 & d_2 & & t \\ \vdots & & \ddots & u \\ 0 & & \cdots & d_n \end{bmatrix} = \begin{bmatrix} d_1 & 0 & \cdots & 0 \\ 0 & d_2 & & 0 \\ \vdots & & \ddots & 0 \\ 0 & & \cdots & d_n \end{bmatrix} \begin{bmatrix} 1 & r/d_1 & & s/d_1 \\ 0 & 1 & & t/d_2 \\ \vdots & & \ddots & u/d_{n-1} \\ 0 & & \cdots & 1 \end{bmatrix}.$$

This decomposition of A is called the LDU **factorization** of A. Note that, in this factorization, U is just a row-echelon form of A (with leading 1's on the diagonal) after Gaussian elimination and before back substitution.

In Example 1.11, we found a factorization of A as

$$A = \begin{bmatrix} 1 & 0 & 0 \\ 2 & 1 & 0 \\ -1 & -3 & 1 \end{bmatrix} \begin{bmatrix} 2 & 1 & 1 \\ 0 & -1 & -2 \\ 0 & 0 & -4 \end{bmatrix}.$$

This can be further factored as $A = LDU$ by taking

$$\begin{bmatrix} 2 & 1 & 1 \\ 0 & -1 & -2 \\ 0 & 0 & -4 \end{bmatrix} = \begin{bmatrix} 2 & 0 & 0 \\ 0 & -1 & 0 \\ 0 & 0 & -4 \end{bmatrix} \begin{bmatrix} 1 & 1/2 & 1/2 \\ 0 & 1 & 2 \\ 0 & 0 & 1 \end{bmatrix} = DU.$$

Suppose now that during forward elimination row interchanges are necessary. In this case, we can first do all the row interchanges before doing any other type of elementary row operations, since the interchange of rows can be done at any time, before or after the other operations, with the same effect on the solution. Those "row-interchanging" elementary matrices altogether form a permutation matrix P so that no more row interchanges are needed during Gaussian elimination of PA. So PA has an LDU factorization.

Example 1.12 Consider a square matrix $A = \begin{bmatrix} 0 & 1 & 2 \\ 0 & 1 & 0 \\ 1 & 0 & 0 \end{bmatrix}$. For Gaussian elimination, it is clearly necessary to interchange the first row with the third row, that is, we need to multiply the permutation matrix $P = \begin{bmatrix} 0 & 0 & 1 \\ 0 & 1 & 0 \\ 1 & 0 & 0 \end{bmatrix}$ to A so that

$$PA = \begin{bmatrix} 1 & 0 & 0 \\ 0 & 1 & 0 \\ 0 & 1 & 2 \end{bmatrix} = \begin{bmatrix} 1 & 0 & 0 \\ 0 & 1 & 0 \\ 0 & -1 & 1 \end{bmatrix} \begin{bmatrix} 1 & 0 & 0 \\ 0 & 1 & 0 \\ 0 & 0 & 2 \end{bmatrix} = LU.$$

□

Of course, if we choose a different permutation P', then the LDU factorization of $P'A$ may be different from that of PA, even if there is another permutation matrix P'' that changes $P'A$ to PA. However, if we fix a permutation matrix P when it is necessary, the uniqueness of the LDU factorization of A can be proved.

Theorem 1.10 *For an invertible matrix A, the LDU factorization of A is unique up to a permutation: that is, for a fixed P the expression $PA = LDU$ is unique.*

Proof: Suppose that $A = L_1D_1U_1 = L_2D_2U_2$, where the L's are lower triangular, the U's are upper triangular, all with 1's on the diagonal, and the D's are diagonal matrices with no zeros on the diagonal. We need to show $L_1 = L_2$, $D_1 = D_2$, and $U_1 = U_2$.

Note that the inverse of a lower (upper) triangular matrix is also a lower (upper) triangular matrix. And the inverse of a diagonal matrix is also diagonal. Therefore, by multiplying $(L_1D_1)^{-1} = D_1^{-1}L_1^{-1}$ on the left and U_2^{-1} on the right, our equation $L_1D_1U_1 = L_2D_2U_2$ becomes

$$U_1U_2^{-1} = D_1^{-1}L_1^{-1}L_2D_2 .$$

The left side is an upper triangular matrix, while the right side is a lower triangular matrix. Hence, both sides must be diagonal. However, since the diagonal entries of the upper triangular matrix $U_1U_2^{-1}$ are all 1's, it must be the identity matrix I (see Problem 1.24). Thus $U_1U_2^{-1} = I$, *i.e.*, $U_1 = U_2$. Similarly, $L_1^{-1}L_2 = D_1D_2^{-1}$ implies that $L_1 = L_2$ and $D_1 = D_2$. □

In particular, if A is symmetric (*i.e.*, $A = A^T$), and if it can be factored into $A = LDU$ without row interchanges, then we have

$$LDU = A = A^T = (LDU)^T = U^TD^TL^T = U^TDL^T,$$

and thus, by the uniqueness of factorizations, we have $U = L^T$ and $A = LDL^T$.

Problem 1.25 Find the factors L, D, and U for $A = \begin{bmatrix} 2 & -1 & 0 \\ -1 & 2 & -1 \\ 0 & -1 & 2 \end{bmatrix}$.

What is the solution to $A\mathbf{x} = \mathbf{b}$ for $\mathbf{b} = [1\,0\,-1]^T$?

Problem 1.26 For all possible permutation matrices P, find the LDU factorization of PA for $A = \begin{bmatrix} 1 & 2 & 3 \\ 2 & 4 & 2 \\ 1 & 1 & 1 \end{bmatrix}$.

1.9 Application: Linear models

(1) In an electrical network, a simple current flow may be illustrated by a diagram like the one below. Such a network involves only voltage sources, like batteries, and resistors, like bulbs, motors, or refrigerators. The voltage is measured in *volts*, the resistance in *ohms*, and the current flow in amperes (*amps*, in short). For such an electrical network, current flow is governed by the following three laws:

- **Ohm's Law:** The voltage drop V across a resistor is the product of the current I and the resistance R: $V = IR$.
- **Kirchhoff's Current Law (KCL):** The current flow into a node equals the current flow out of the node.
- **Kirchhoff's Voltage Law (KVL):** The algebraic sum of the voltage drops around a closed loop equals the total voltage sources in the loop.

Example 1.13 Determine the currents in the network given in the above figure.

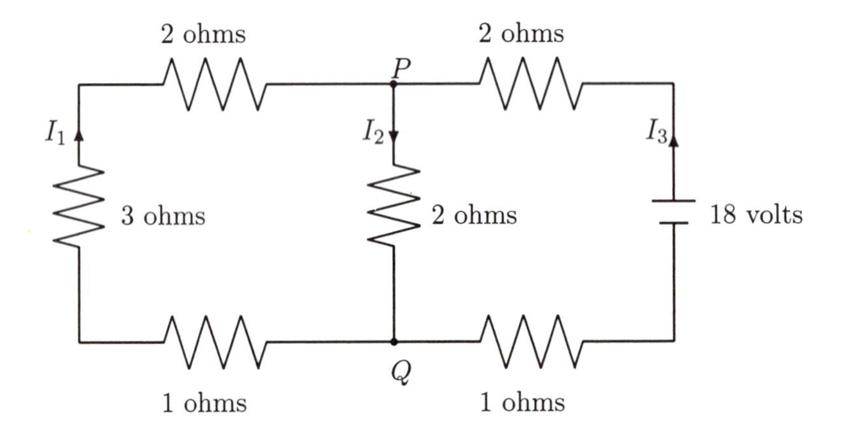

Solution: By applying KCL to nodes P and Q, we get equations

$$\begin{aligned} I_1 + I_3 &= I_2 \text{ at } P, \\ I_2 &= I_1 + I_3 \text{ at } Q. \end{aligned}$$

Observe that both equations are the same, and one of them is redundant. By applying KVL to each of the loops in the network clockwise direction, we get

$$\begin{aligned} 6I_1 + 2I_2 &= 0 \text{ from the left loop,} \\ 2I_2 + 3I_3 &= 18 \text{ from the right loop.} \end{aligned}$$

Collecting all the equations, we get a system of linear equations:

$$\begin{cases} I_1 - I_2 + I_3 = 0 \\ 6I_1 + 2I_2 = 0 \\ 2I_2 + 3I_3 = 18. \end{cases}$$

By solving it, the currents are $I_1 = -1$ amp, $I_2 = 3$ amps and $I_3 = 4$ amps. The negative sign for I_1 means that the current I_1 flows in the direction opposite to that shown in the figure. □

Problem 1.27 Determine the currents in the following networks.

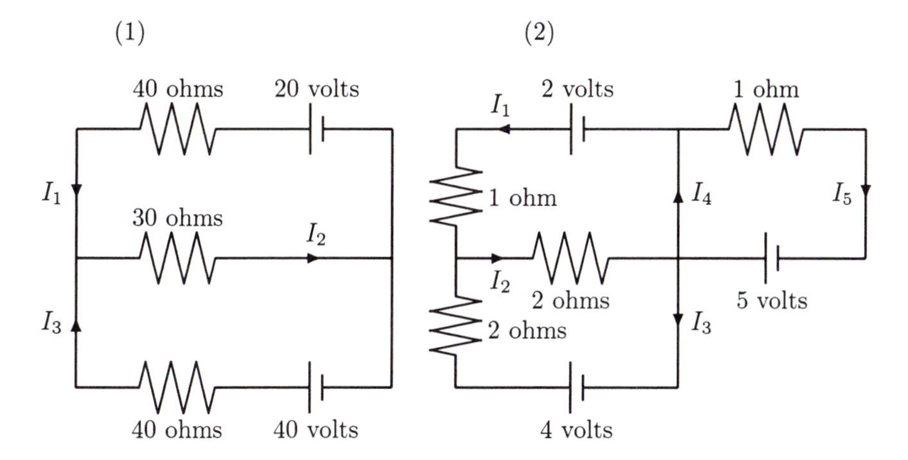

(2) Cryptography is the study of sending messages in disguised form (secret codes) so that only the intended recipients can remove the disguise and read the message; modern cryptography uses advanced mathematics. As another application of invertible matrices, we introduce a simple coding. Suppose we associate a prescribed number with every letter in the alphabet; for example,

$$\begin{array}{cccccccccc} A & B & C & D & \cdots & X & Y & Z & \text{Blank} & ? & ! \\ \updownarrow & \updownarrow & \updownarrow & \updownarrow & & \updownarrow & \updownarrow & \updownarrow & \updownarrow & \updownarrow & \updownarrow \\ 0 & 1 & 2 & 3 & \cdots & 23 & 24 & 25 & 26 & 27 & 28. \end{array}$$

Suppose that we want to send the message "GOOD LUCK". Replace this message by

$$6,\ 14,\ 14,\ 3,\ 26,\ 11,\ 20,\ 2,\ 10$$

according to the preceding substitution scheme. A code of this type could be cracked without difficulty by a number of techniques of statistical methods, like the analysis of frequency of letters. To make it difficult to crack the code, we first break the message into six vectors in $\mathbb{R}^3$, each with 3 components (optional), by adding extra blanks if necessary:

$$\begin{bmatrix} 6 \\ 14 \\ 14 \end{bmatrix}, \quad \begin{bmatrix} 3 \\ 26 \\ 11 \end{bmatrix}, \quad \begin{bmatrix} 20 \\ 2 \\ 10 \end{bmatrix}.$$

Next, choose a nonsingular 3×3 matrix A, say

$$A = \begin{bmatrix} 1 & 0 & 0 \\ 2 & 1 & 0 \\ 1 & 1 & 1 \end{bmatrix},$$

which is supposed to be known to *both* sender and receiver. Then as a linear transformation A translates our message into

$$A\begin{bmatrix} 6 \\ 14 \\ 14 \end{bmatrix} = \begin{bmatrix} 6 \\ 26 \\ 34 \end{bmatrix}, \quad A\begin{bmatrix} 3 \\ 26 \\ 11 \end{bmatrix} = \begin{bmatrix} 3 \\ 32 \\ 40 \end{bmatrix}, \quad A\begin{bmatrix} 20 \\ 2 \\ 10 \end{bmatrix} = \begin{bmatrix} 20 \\ 42 \\ 32 \end{bmatrix}.$$

By putting the components of the resulting vectors consecutively, we transmit

$$6,\ 26,\ 34,\ 3,\ 32,\ 40,\ 20,\ 42,\ 32.$$

To decode this message, the receiver may follow the following process. Suppose that we received the following reply from our correspondent:

$$19,\ 45,\ 26,\ 13,\ 36,\ 41.$$

To decode it, first break the message into two vectors in $\mathbb{R}^3$ as before:

$$\begin{bmatrix} 19 \\ 45 \\ 26 \end{bmatrix}, \quad \begin{bmatrix} 13 \\ 36 \\ 41 \end{bmatrix}.$$

We want to find two vectors $\mathbf{x}_1$, $\mathbf{x}_2$ such that $A\mathbf{x}_i$ is the i-th vector of the above two vectors: *i.e.*,

$$A\mathbf{x}_1 = \begin{bmatrix} 19 \\ 45 \\ 26 \end{bmatrix}, \quad A\mathbf{x}_2 = \begin{bmatrix} 13 \\ 36 \\ 41 \end{bmatrix}.$$

Since A is invertible, the vectors $\mathbf{x}_1$, $\mathbf{x}_2$ can be found by multiplying the inverse of A to the two vectors given in the message. By an easy computation, one can find

$$A^{-1} = \begin{bmatrix} 1 & 0 & 0 \\ -2 & 1 & 0 \\ 1 & -1 & 1 \end{bmatrix}.$$

Therefore,

$$\mathbf{x}_1 = \begin{bmatrix} 1 & 0 & 0 \\ -2 & 1 & 0 \\ 1 & -1 & 1 \end{bmatrix} \begin{bmatrix} 19 \\ 45 \\ 26 \end{bmatrix} = \begin{bmatrix} 19 \\ 7 \\ 0 \end{bmatrix}, \quad \mathbf{x}_2 = \begin{bmatrix} 13 \\ 10 \\ 18 \end{bmatrix}.$$

The numbers one obtains are

$$19,\ 7,\ 0,\ 13,\ 10,\ 18.$$

Using our correspondence between letters and numbers, the message we have received is "THANKS".

Problem 1.28 Encode "TAKE UFO " using the same matrix A used in the above example.

(3) Another significant application of linear algebra is to a mathematical model in economics. In most nations, an economic society may be divided into many sectors that produce goods or services, such as the automobile industry, oil industry, steel industry, communication industry, and so on. Then a fundamental problem in economics is to find the *equilibrium* of the supply and the demand in the economy.

There are two kinds of demands for goods: the *intermediate demand* from the industries themselves (or the sectors) that are needed as inputs for their own production, and the *extra demand* from the consumer, the governmental use, surplus production, or exports. Practically, the interrelation between the sectors is very complicated, and the connection between the extra demand and the production is unclear. A natural question is *whether there is a production level such that the total amounts produced (or supply) will exactly balance the total demand for the production*, so that the equality

$$\begin{aligned}\{Total\ output\} &= \{Total\ demand\} \\ &= \{Intermediate\ demand\} + \{Extra\ demand\}\end{aligned}$$

holds. This problem can be described by a system of linear equations, which is called the *Leontief Input-Output Model.* To illustrate this, we show a simple example.

Suppose that a nation's economy consists of three sectors: I_1 = automobile industry, I_2 = steel industry, and I_3 = oil industry.

Let $\mathbf{x} = [x_1\, x_2\, x_3]^T$ denote the production vector (or production level) in $\mathbb{R}^3$, where each entry x_i denotes the total amount (in a common unit such as "dollars" rather than quantities such as "tons" or "gallons") of the output that the industry I_i produces per year.

The intermediate demand may be explained as follows. Suppose that, for the total output x_2 units of the steel industry I_2, 20% is contributed by the output of I_1, 40% by that of I_2 and 20% by that of I_3. Then we can write this as a column vector, called a *unit consumption vector* of I_2:

$$\mathbf{c}_2 = \begin{bmatrix} 0.2 \\ 0.4 \\ 0.2 \end{bmatrix}.$$

For example, if I_2 decides to produce 100 units per year, then it will order (or demand) 20 units from I_1, 40 units from I_2, and 20 units from I_3: *i.e.*, the consumption vector of I_2 for the production $x_2 = 100$ units can be written as a column vector: $100\mathbf{c}_2 = [20\ 40\ 20]^T$. From the concept of the consumption vector, it is clear that the sum of decimal fractions in the column $\mathbf{c}_2$ must be ≤ 1.

In our example, suppose that the demands (inputs) of the outputs are given by the following matrix, called an *input-output matrix*:

$$A = \text{input}\ \begin{matrix} \\ I_1 \\ I_2 \\ I_3 \\ \\ \end{matrix} \overset{\displaystyle\text{output}}{\begin{matrix} I_1 & I_2 & I_3 \\ \left[\begin{matrix} 0.3 \\ 0.1 \\ 0.3 \end{matrix}\right. & \begin{matrix} 0.2 \\ 0.4 \\ 0.2 \end{matrix} & \left.\begin{matrix} 0.3 \\ 0.1 \\ 0.3 \end{matrix}\right] \\ \uparrow & \uparrow & \uparrow \\ \mathbf{c}_1 & \mathbf{c}_2 & \mathbf{c}_3 \end{matrix}}.$$

In this matrix, an industry looks down a column to see how much it needs from where to produce its total output, and it looks across a row to see how much of its output goes to where. For example, the second row says that, out of the total output x_2 units of the steel industry I_2, as the intermediate demand, the automobile industry I_1 demands 10% of the output x_1, the steel industry I_2 demands 40% of the output x_2 and the oil industry I_3 demands 10% of the output x_3. Therefore, it is now easy to see that the intermediate demand of the economy can be written as

$$A\mathbf{x} = \begin{bmatrix} 0.3 & 0.2 & 0.3 \\ 0.1 & 0.4 & 0.1 \\ 0.3 & 0.2 & 0.3 \end{bmatrix} \begin{bmatrix} x_1 \\ x_2 \\ x_3 \end{bmatrix} = \begin{bmatrix} 0.3x_1 + 0.2x_2 + 0.3x_3 \\ 0.1x_1 + 0.4x_2 + 0.1x_3 \\ 0.3x_1 + 0.2x_2 + 0.3x_3 \end{bmatrix}.$$

Suppose that the extra demand in our example is given by $\mathbf{d} = [d_1, d_2, d_3]^T = [30, 20, 10]^T$. Then the problem for this economy is to find the production vector $\mathbf{x}$ satisfying the following equation:

$$\mathbf{x} = A\mathbf{x} + \mathbf{d}.$$

Another form of the equation is $(I - A)\mathbf{x} = \mathbf{d}$, where the matrix $I - A$ is called the *Leontief matrix*. If $I - A$ is not invertible, then the equation may have no solution or infinitely many solutions depending on what $\mathbf{d}$ is. If $I - A$ is invertible, then the equation has the unique solution $\mathbf{x} = (I - A)^{-1}\mathbf{d}$. Now, our example can be written as

$$\begin{bmatrix} x_1 \\ x_2 \\ x_3 \end{bmatrix} = \begin{bmatrix} 0.3 & 0.2 & 0.3 \\ 0.1 & 0.4 & 0.1 \\ 0.3 & 0.2 & 0.3 \end{bmatrix} \begin{bmatrix} x_1 \\ x_2 \\ x_3 \end{bmatrix} + \begin{bmatrix} 30 \\ 20 \\ 10 \end{bmatrix}.$$

In this example, it turns out that the matrix $I - A$ is invertible and

$$(I - A)^{-1} = \begin{bmatrix} 2.0 & 1.0 & 1.0 \\ 0.5 & 2.0 & 0.5 \\ 1.0 & 1.0 & 2.0 \end{bmatrix}.$$

Therefore,

$$\mathbf{x} = (I - A)^{-1}\mathbf{d} = \begin{bmatrix} 90 \\ 60 \\ 70 \end{bmatrix},$$

which gives the total amount of product x_i of the industry I_i for one year to meet the required demand.

Remark: **(1)** Under the usual circumstances, the sum of the entries in a column of the consumption matrix A is less than one because a sector should require less than one unit's worth of inputs to produce one unit of output. This actually implies that $I - A$ is invertible and the production vector $\mathbf{x}$ is feasible in the sense that the entries in $\mathbf{x}$ are all nonnegative as the following argument shows.

(2) In general, by using induction one can easily verify that for any $k = 1, 2, \ldots,$

$$(I - A)(I + A + \cdots + A^{k-1}) = I - A^k.$$

If the sums of column entries of A are all strictly less than one, then $\lim_{k\to\infty} A^k = \mathbf{0}$ (see Section 6.6 for the limit of a sequence of matrices). Thus, we get $(I - A)(I + A + \cdots + A^k + \cdots) = I$, that is,

$$(I - A)^{-1} = I + A + \cdots + A^k + \cdots.$$

This also shows a practical way of computing $(I - A)^{-1}$ since by taking k sufficiently large the right side may be made very close to $(I - A)^{-1}$. In Chapter 6, an easier method of computing A^k will be shown.

In summary, if A and $\mathbf{d}$ have nonnegative entries and if the sum of the entries of each column of A is less than one, then $I - A$ is invertible and the inverse is given as the above formula. Moreover, as the formula shows the entries of the inverse are all nonnegative, and so are those of the production vector $\mathbf{x} = (I - A)^{-1}\mathbf{d}$.

Problem 1.29 Determine the total demand for industries I_1, I_2 and I_3 for the input-output matrix A and the extra demand vector $\mathbf{d}$ given below:

$$A = \begin{bmatrix} 0.1 & 0.7 & 0.2 \\ 0.5 & 0.1 & 0.6 \\ 0.4 & 0.2 & 0.2 \end{bmatrix} \quad \text{with } \mathbf{d} = \mathbf{0}.$$

Problem 1.30 Suppose that an economy is divided into three sectors: I_1 = services, I_2 = manufacturing industries, and I_3 = agriculture. For each unit of output, I_1 demands no services from I_1, 0.4 units from I_2, and 0.5 units from I_3. For each unit of output, I_2 requires 0.1 units from sector I_1 of services, 0.7 units from other parts in sector I_2, and no product from sector I_3. For each unit of output, I_3 demands 0.8 units of services I_1, 0.1 units of manufacturing products from I_2, and 0.1 units of its own output from I_3. Determine the production level to balance the economy when 90 units of services, 10 units of manufacturing, and 30 units of agriculture are required as the extra demand.

1.10 Exercises

1.1. Which of the following matrices are in row-echelon form or in reduced row-echelon form?

$$A = \begin{bmatrix} 1 & 0 & 0 & 0 & -3 \\ 0 & 0 & 1 & 0 & 4 \\ 0 & 0 & 0 & 1 & 2 \end{bmatrix}, \quad B = \begin{bmatrix} 1 & 0 & 0 & 0 & 2 \\ 0 & 0 & 1 & 0 & 0 \\ 0 & 0 & 0 & 1 & 3 \\ 0 & 0 & 0 & 0 & 0 \end{bmatrix},$$

$$C = \begin{bmatrix} 0 & 0 & 0 & 0 & 0 \\ 0 & 0 & 1 & 2 & -3 \\ 0 & 0 & 0 & 1 & 0 \\ 0 & 0 & 0 & 0 & 0 \end{bmatrix}, \quad D = \begin{bmatrix} 0 & 1 & 0 & 0 & 5 \\ 0 & 0 & 1 & 1 & -4 \\ 0 & 0 & 0 & 1 & 3 \end{bmatrix},$$

$$E = \begin{bmatrix} 0 & 1 & 0 & 0 & 5 \\ 0 & 0 & 1 & 0 & 4 \\ 0 & 1 & 0 & -2 & 3 \end{bmatrix}, \quad F = \begin{bmatrix} 1 & 0 & 0 & 0 & 1 \\ 0 & 1 & 0 & 0 & 2 \\ 0 & 0 & 0 & 1 & -1 \\ 0 & 0 & 0 & 0 & 0 \end{bmatrix}.$$

1.2. Find a row-echelon form of each matrix.

$$(1) \begin{bmatrix} 1 & -3 & 2 & 1 & 2 \\ 3 & -9 & 10 & 2 & 9 \\ 2 & -6 & 4 & 2 & 4 \\ 2 & -6 & 8 & 1 & 7 \end{bmatrix} \quad (2) \begin{bmatrix} 1 & 2 & 3 & 4 & 5 \\ 2 & 3 & 4 & 5 & 1 \\ 3 & 4 & 5 & 1 & 2 \\ 4 & 5 & 1 & 2 & 3 \\ 5 & 1 & 2 & 3 & 4 \end{bmatrix}$$

1.3. Find the reduced row-echelon form of the matrices in Exercise **1.2**.

1.4. Solve the systems of equations by Gauss-Jordan elimination.

$$(1) \begin{cases} x_1 + x_2 + x_3 - x_4 = -2 \\ 2x_1 - x_2 + x_3 + x_4 = 0 \\ 3x_1 + 2x_2 - x_3 - x_4 = 1 \\ x_1 + x_2 + 3x_3 - 3x_4 = -8\,. \end{cases}$$

$$(2) \begin{cases} 2x - 3y = 8 \\ 4x - 5y + z = 15 \\ 2x + 4z = 1\,. \end{cases}$$

What are the pivots in each elimination step?

1.5. Which of the following systems has a nontrivial solution?

$$(1) \begin{cases} x + 2y + 3z = 0 \\ 2y + 2z = 0 \\ x + 2y + 3z = 0. \end{cases} \quad (2) \begin{cases} 2x + y - z = 0 \\ x - 2y - 3z = 0 \\ 3x + y - 2z = 0. \end{cases}$$

1.6. Determine all values of the b_i that make the following system consistent:

$$\begin{cases} x + y - z = b_1 \\ 2y + z = b_2 \\ y - z = b_3\,. \end{cases}$$

1.7. Determine the condition on b_i so that the following system has no solution:

$$\begin{cases} 2x + y + 7z = b_1 \\ 6x - 2y + 11z = b_2 \\ 2x - y + 3z = b_3 . \end{cases}$$

1.8. Let A and B be matrices of the same size.

(1) Show that, if $A\mathbf{x} = 0$ for all $\mathbf{x}$, then A is the zero matrix.

(2) Show that, if $A\mathbf{x} = B\mathbf{x}$ for all $\mathbf{x}$, then $A = B$.

1.9. Compute ABC and CAB, for

$$A = \begin{bmatrix} 2 & -1 & 1 \\ 1 & 2 & 1 \end{bmatrix}, \quad B = \begin{bmatrix} 3 \\ 1 \\ -1 \end{bmatrix}, \quad C = \begin{bmatrix} 1 & -1 \end{bmatrix}.$$

1.10. Prove that if A is a 3×3 matrix such that $AB = BA$ for every 3×3 matrix B, then $A = kI_3$ for some constant k.

1.11. Let $A = \begin{bmatrix} 1 & 2 & 0 \\ 0 & 1 & 3 \\ 0 & 0 & 1 \end{bmatrix}$. Find A^k for all integers k.

1.12. Compute $(2A - B)C$ and CC^T for

$$A = \begin{bmatrix} 1 & 0 & 0 \\ 0 & 1 & 0 \\ 1 & 0 & 1 \end{bmatrix}, \quad B = \begin{bmatrix} 1 & 0 & 0 \\ -2 & 1 & 0 \\ 0 & 0 & 1 \end{bmatrix}, \quad C = \begin{bmatrix} 2 & 1 & 1 \\ 4 & 1 & 0 \\ -2 & 2 & 1 \end{bmatrix}.$$

1.13. Let $f(x) = a_n x^n + a_{n-1}x^{n-1} + \cdots + a_1 x + a_0$ be a polynomial. For any square matrix A, a *matrix polynomial* $f(A)$ is defined as $f(A) = a_n A^n + a_{n-1}A^{n-1} + \cdots + a_1 A + a_0 I$. For $f(x) = 3x^3 + x^2 - 2x + 3$, find $f(A)$ for

$$(1)\ A = \begin{bmatrix} 1 & 2 & 0 \\ -3 & 4 & 0 \\ 0 & 0 & 5 \end{bmatrix}, \qquad (2)\ A = \begin{bmatrix} 1 & -1 & 2 \\ 0 & 2 & -1 \\ 0 & 0 & 3 \end{bmatrix}.$$

1.14. Find the symmetric part and the skew-symmetric part of each of the following matrices.

$$(1)\ A = \begin{bmatrix} 1 & 3 & 3 \\ 2 & 5 & 9 \\ -1 & 3 & 2 \end{bmatrix} \qquad (2)\ A = \begin{bmatrix} 1 & 3 & 4 \\ 0 & 2 & -1 \\ 0 & 0 & 3 \end{bmatrix}$$

1.15. Find AA^T and A^TA for the matrix $A = \begin{bmatrix} 1 & -1 & 0 & 2 \\ 2 & 1 & 3 & 1 \\ 2 & 8 & 4 & 0 \end{bmatrix}$.

1.16. Let $A^{-1} = \begin{bmatrix} 1 & 1 & 2 \\ 0 & 1 & 3 \\ 4 & 2 & 1 \end{bmatrix}$.

(1) Find a matrix B such that $AB = \begin{bmatrix} 1 & 2 \\ 0 & 1 \\ 4 & 1 \end{bmatrix}$.

(2) Find a matrix C such that $AC = A^2 + A$.

1.17. Find all possible choices of a, b and c so that $A = \begin{bmatrix} a & b \\ c & 0 \end{bmatrix}$ has an inverse matrix such that $A^{-1} = A$.

1.18. Decide whether or not each of the following matrices is invertible. Find the inverses for invertible ones.

$$A = \begin{bmatrix} 1 & 2 & 3 & 4 \\ 0 & 2 & 3 & 4 \\ 0 & 0 & 3 & 4 \\ 0 & 0 & 0 & 4 \end{bmatrix} \quad B = \begin{bmatrix} 1 & 1 & 1 \\ 0 & 2 & 3 \\ 5 & 5 & 1 \end{bmatrix} \quad C = \begin{bmatrix} 1 & 2 & -1 \\ 3 & 2 & 3 \\ 2 & 2 & 1 \end{bmatrix}$$

1.19. Suppose A is a 2×1 matrix and B is a 1×2 matrix. Prove that the product AB is not invertible.

1.20. Find three matrices which are row equivalent to $A = \begin{bmatrix} 2 & -1 & 3 & 4 \\ 0 & 1 & 2 & -1 \\ 5 & 2 & -3 & 4 \end{bmatrix}$.

1.21. Write the following systems of equations as matrix equations $A\mathbf{x} = \mathbf{b}$ and solve them by computing $A^{-1}\mathbf{b}$:

$$(1) \begin{cases} 2x_1 - x_2 + 3x_3 = 2 \\ \phantom{2x_1 -{}} x_2 - 4x_3 = 5 \\ 2x_1 + x_2 - 2x_3 = 7, \end{cases} \quad (2) \begin{cases} x_1 - x_2 + x_3 = 5 \\ x_1 + x_2 - x_3 = -1 \\ 4x_1 - 3x_2 + 2x_3 = -3\,. \end{cases}$$

1.22. Find the LDU factorization for each of the following matrices:

(1) $A = \begin{bmatrix} 2 & 1 \\ 8 & 7 \end{bmatrix}$, (2) $A = \begin{bmatrix} 1 & 0 \\ 8 & 1 \end{bmatrix}$.

1.23. Find the LDL^T factorization of the following symmetric matrices:

(1) $A = \begin{bmatrix} 1 & 2 & 3 \\ 2 & 6 & 8 \\ 3 & 8 & 10 \end{bmatrix}$, (2) $A = \begin{bmatrix} a & b \\ b & d \end{bmatrix}$.

1.24. Solve $A\mathbf{x} = \mathbf{b}$ with $A = LU$, where L and U are given as

$$L = \begin{bmatrix} 1 & 0 & 0 \\ -1 & 1 & 0 \\ 0 & -1 & 1 \end{bmatrix}, \quad U = \begin{bmatrix} 1 & -1 & 0 \\ 0 & 1 & -1 \\ 0 & 0 & 1 \end{bmatrix}, \quad \mathbf{b} = \begin{bmatrix} 2 \\ -3 \\ 4 \end{bmatrix}.$$

Forward elimination is the same as $L\mathbf{c} = \mathbf{b}$, and back-substitution is $U\mathbf{x} = \mathbf{c}$.

1.25. Let $A = \begin{bmatrix} 1 & 1 & 1 \\ 1 & 4 & 5 \\ 1 & 4 & 7 \end{bmatrix}$ and $\mathbf{b} = \begin{bmatrix} 1 \\ 3 \\ 5 \end{bmatrix}$.

(1) Solve $A\mathbf{x} = \mathbf{b}$ by Gauss-Jordan elimination.
(2) Find the LDU factorization of A.
(3) Write A as a product of elementary matrices.
(4) Find the inverse of A.

1.26. A square matrix A is said to be *nilpotent* if $A^k = \mathbf{0}$ for a positive integer k.

(1) Show that an invertible matrix is not nilpotent.
(2) Show that any triangular matrix with zero diagonal is nilpotent.
(3) Show that if A is a nilpotent with $A^k = \mathbf{0}$, then $I - A$ is invertible with its inverse $I + A + \cdots + A^{k-1}$.

1.27. A square matrix A is said to be *idempotent* if $A^2 = A$.

(1) Find an example of an idempotent matrix other than $\mathbf{0}$ or I.
(2) Show that, if a matrix A is both idempotent and invertible, then $A = I$.

1.28. Determine whether the following statements are true or false, in general, and justify your answers.

(1) Let A and B be row-equivalent square matrices. Then A is invertible if and only if B is invertible.
(2) Let A be a square matrix such that $AA = A$. Then A is the identity.
(3) If A and B are invertible matrices such that $A^2 = I$ and $B^2 = I$, then $(AB)^{-1} = BA$.
(4) If A and B are invertible matrices, $A + B$ is also invertible.
(5) If A, B and AB are symmetric, then $AB = BA$.
(6) If A and B are symmetric and the same size, then AB is also symmetric.
(7) Let $AB^T = I$. Then A is invertible if and only if B is invertible.
(8) If a square matrix A is not invertible, then neither is AB for any B.
(9) If E_1 and E_2 are elementary matrices, then $E_1E_2 = E_2E_1$.
(10) The inverse of an invertible upper triangular matrix is upper triangular.
(11) Any invertible matrix A can be written as $A = LU$, where L is lower triangular and U is upper triangular.
(12) If A is invertible and symmetric, then A^{-1} is also symmetric.

Chapter 2

Determinants

2.1 Basic properties of determinant

Our primary interest in Chapter 1 was in the solvability or solutions of a system $A\mathbf{x} = \mathbf{b}$ of linear equations. For an invertible matrix A, Theorem 1.8 shows that the system has a unique solution $\mathbf{x} = A^{-1}\mathbf{b}$ for any $\mathbf{b}$.

Now the question is how to decide whether or not a square matrix A is invertible. In this section, we introduce the notion of *determinant* as a real-valued function of square matrices that satisfies certain axiomatic rules, and then show that a square matrix A is invertible if and only if the determinant of A is not zero. In fact, we saw in Chapter 1 that a 2×2 matrix $A = \begin{bmatrix} a & b \\ c & d \end{bmatrix}$ is invertible if and only if $ad - bc \neq 0$. This number is called the determinant of A, and is defined formally as follows:

Definition 2.1 For a 2×2 matrix $A = \begin{bmatrix} a & b \\ c & d \end{bmatrix} \in M_{2\times 2}(\mathbb{R})$, the **determinant** of A is defined as $\det A = ad - bc$.

In fact, it turns out that geometrically the determinant of a 2×2 matrix A represents, up to sign, the area of a parallelogram in the xy-plane whose edges are constructed by the row vectors of A (see Theorem 2.9), so it will be very nice if we can have the same idea of determinant for higher order matrices. However, the formula itself in Definition 2.1 does not provide any clue of how to extend this idea of determinant to higher order matrices. Hence, we first examine some fundamental properties of the determinant function defined in Definition 2.1.

By a direct computation, one can easily verify that the function det in Definition 2.1 satisfies the following three fundamental properties:

(1) $\det\begin{bmatrix} 1 & 0 \\ 0 & 1 \end{bmatrix} = 1.$

(2) $\det\begin{bmatrix} c & d \\ a & b \end{bmatrix} = bc - ad = -(ad - bc) = -\det\begin{bmatrix} a & b \\ c & d \end{bmatrix}.$

(3) $$\begin{aligned}\det\begin{bmatrix} ka + \ell a' & kb + \ell b' \\ c & d \end{bmatrix} &= (ka + \ell a')d - (kb + \ell b')c \\ &= k(ad - bc) + \ell(a'd - b'c) \\ &= k\det\begin{bmatrix} a & b \\ c & d \end{bmatrix} + \ell\det\begin{bmatrix} a' & b' \\ c & d \end{bmatrix}.\end{aligned}$$

Actually all the important properties of the determinant function can be derived from these three properties. We will show in Lemma 2.3 that if a function $f : M_{2\times 2}(\mathbb{R}) \to \mathbb{R}$ satisfies the properties (1), (2) and (3) above, then it must be of the form $f(A) = ad - bc$. An advantage of looking at these properties of the determinant rather than looking at the explicit formula given in Definition 2.1 is that these three properties enable us to define the determinant function for any $n \times n$ square matrices.

Definition 2.2 A real-valued function $f : M_{n\times n}(\mathbb{R}) \to \mathbb{R}$ of all $n \times n$ square matrices is called a **determinant** if it satisfies the following three rules:

($\mathbf{R}_1$) the value of f of the identity matrix is 1, *i.e.*, $f(I_n) = 1$;

($\mathbf{R}_2$) the value of f changes sign if any two rows are interchanged;

($\mathbf{R}_3$) f is linear in the first row: that is, by definition,

$$f\left(\begin{bmatrix} k\mathbf{r}_1 + \ell\mathbf{r}'_1 \\ \mathbf{r}_2 \\ \vdots \\ \mathbf{r}_n \end{bmatrix}\right) = kf\left(\begin{bmatrix} \mathbf{r}_1 \\ \mathbf{r}_2 \\ \vdots \\ \mathbf{r}_n \end{bmatrix}\right) + \ell f\left(\begin{bmatrix} \mathbf{r}'_1 \\ \mathbf{r}_2 \\ \vdots \\ \mathbf{r}_n \end{bmatrix}\right),$$

where $\mathbf{r}_i$'s denote the row vectors of a matrix.

It is already shown that the det on 2×2 matrices satisfies these rules. We will show later that for each positive integer n there always exists such a function $f : M_{n\times n}(\mathbb{R}) \to \mathbb{R}$ satisfying the three rules in the definition, and, moreover, it is unique. Therefore, we say "the" determinant and designate it as "det" in any order.

Let us first derive some direct consequences of the three rules in the definition (the readers are suggested to verify that det of 2×2 matrices also satisfies the following properties):

Theorem 2.1 *The determinant satisfies the following properties.*

(1) *The determinant is linear in each row, i.e., for each row the rule* $(\mathbf{R}_3)$ *also holds.*

(2) *If A has either a zero row or two identical rows, then* $\det A = 0$.

(3) *The elementary row operation that adds a constant multiple of one row to another row leaves the determinant unchanged.*

Proof: **(1)** Any row can be placed in the first row with a change of sign in the determinant by rule $(\mathbf{R}_2)$, and then use rules $(\mathbf{R}_3)$ and $(\mathbf{R}_2)$.

(2) If A has a zero row, then the row is zero times the zero row. If A has two identical rows, then interchanging these identical rows changes only the sign of the determinant, but not A itself. Thus we get $\det A = -\det A$.

(3) By a direct computation using (1), we get

$$f\left(\begin{bmatrix} \vdots \\ \mathbf{r}_i + k\mathbf{r}_j \\ \vdots \\ \mathbf{r}_j \\ \vdots \end{bmatrix}\right) = f\left(\begin{bmatrix} \vdots \\ \mathbf{r}_i \\ \vdots \\ \mathbf{r}_j \\ \vdots \end{bmatrix}\right) + kf\left(\begin{bmatrix} \vdots \\ \mathbf{r}_j \\ \vdots \\ \mathbf{r}_j \\ \vdots \end{bmatrix}\right),$$

in which the second term on the right side is zero by (2). □

It is now easy to see the effect of elementary row operations on evaluations of the determinant. The first elementary row operation that "multiplies a constant k to a row" changes the determinant to k times the determinant by (1) of Theorem 2.1. The rule $(\mathbf{R}_2)$ in the definition explains the effect of the elementary row operation that "interchanges two rows". The last elementary row operation that "adds a constant multiple of a row to another" is explained in (3) of Theorem 2.1.

Example 2.1 Consider a matrix

$$A = \begin{bmatrix} 1 & 1 & 1 \\ a & b & c \\ b+c & c+a & b+a \end{bmatrix}.$$

If we add the second row to the third, then the third row becomes

$$[a+b+c \quad a+b+c \quad a+b+c],$$

which is a scalar multiple of the first row. Thus, $\det A = 0$. □

Problem 2.1 Show that, for an $n \times n$ matrix A and $k \in \mathbb{R}$, $\det(kA) = k^n \det A$.

Problem 2.2 Explain why $\det A = 0$ for

$$(1)\ A = \begin{bmatrix} a+1 & a+4 & a+7 \\ a+2 & a+5 & a+8 \\ a+3 & a+6 & a+9 \end{bmatrix}, \qquad (2)\ A = \begin{bmatrix} a & a^4 & a^7 \\ a^2 & a^5 & a^8 \\ a^3 & a^6 & a^9 \end{bmatrix}.$$

Recall that any square matrix can be transformed to an upper triangular matrix by forward eliminations. Further properties of the determinant are obtained in the following theorem.

Theorem 2.2 *The determinant satisfies the following properties.*

(1) *The determinant of a triangular matrix is the product of the diagonal entries.*

(2) *The matrix A is invertible if and only if* $\det A \neq 0$.

(3) *For any two $n \times n$ matrices A and B,* $\det(AB) = \det A \det B$.

(4) $\det A^T = \det A$.

Proof: **(1)** If A is a diagonal matrix, then it is clear that $\det A = a_{11} \cdots a_{nn}$ by (1) of Theorem 2.1 and rule $(\mathbf{R}_1)$. Suppose that A is a lower triangular matrix. Then a forward elimination, which does not change the determinant, produces a zero row if A has a zero diagonal entry, or makes A row equivalent to the diagonal matrix D whose diagonal entries are exactly those of A if the diagonal entries are all nonzero. Thus, in the former case, $\det A = 0$ and the product of the diagonal entries is also zero. In the latter case, $\det A = \det D = a_{11} \cdots a_{nn}$. Similar arguments apply when A is an upper triangular matrix.

(2) Note again that a forward elimination reduces a square matrix A to an upper triangular matrix, which has a zero row if A is singular and has no zero row if A is nonsingular (see Theorem 1.8).

(3) If A is not invertible, then AB is not invertible, and so $\det(AB) = 0 = \det A \det B$. By the properties of the elementary matrices, it is clear that for any elementary matrix E, $\det(EB) = \det E \det B$. If A is invertible, it can be written as a product of elementary matrices, say $A = E_1E_2\cdots E_k$. Then by induction on k, we get

$$\begin{aligned}\det(AB) &= \det(E_1E_2\cdots E_kB)\\ &= \det E_1\ \det E_2\ \cdots\ \det E_k\ \det B\\ &= \det(E_1E_2\cdots E_k)\det B\\ &= \det A\det B.\end{aligned}$$

(4) Clearly, A is not invertible if and only if A^T is not. Thus for a singular matrix A we have $\det A^T = 0 = \det A$. If A is invertible, then there is a factorization $PA = LDU$ for a permutation matrix P. By (3), we get

$$\det P\ \det A = \det L\ \det D\ \det U.$$

Note that the transpose of $PA = LDU$ is $A^TP^T = U^TD^TL^T$ and that for any triangular matrix B, $\det B = \det B^T$ by (1). In particular, since L, U, L^T, and U^T are triangular with 1's on the diagonal, their determinants are all equal to 1. Therefore, we have

$$\begin{aligned}\det A^T\det P^T &= \det U^T\det D^T\det L^T\\ &= \det L\ \det D\ \det U\ =\ \det A\ \det P.\end{aligned}$$

By the definition, a permutation matrix P is obtained from the identity matrix by a sequence of row interchanges: that is, $P = E_k\cdots E_1I_n$ for some k, where each E_i is an elementary matrix obtained from the identity matrix by interchanging two rows. Thus, $\det E_i = -1$ for each $i = 1,\ldots,k$, and clearly $E_i^T = E_i = E_i^{-1}$. Therefore, $\det P = (-1)^k = \det P^T$ by (3), so $\det A = \det A^T$. □

Remark: From the equality $\det A = \det A^T$, we could define the determinant in terms of columns instead of rows in Definition 2.2, and Theorem 2.1 is also true with "columns" instead of "rows".

Example 2.2 Evaluate the determinant of the following matrix A:

$$A = \begin{bmatrix} 2 & -4 & 0 & 0\\ 1 & -3 & 0 & 1\\ 1 & 0 & -1 & 2\\ 3 & -4 & 3 & -1\end{bmatrix}.$$

Solution: By using forward elimination, A can be transformed to an upper triangular matrix U. Since the forward elimination does not change the determinant, the determinant of A is simply the product of the diagonal entries of U:

$$\begin{aligned}\det A = \det U &= \det\begin{bmatrix} 2 & -4 & 0 & 0 \\ 0 & -1 & 0 & 1 \\ 0 & 0 & -1 & 4 \\ 0 & 0 & 0 & 13 \end{bmatrix} \\ &= 2\cdot(-1)^2\cdot 13 = 26.\end{aligned}$$ □

Problem 2.3 Prove that if A is invertible, then $\det A^{-1} = 1/\det A$.

Problem 2.4 Evaluate the determinant of each of the following matrices:

$$(1)\begin{bmatrix} 1 & 4 & 2 \\ 3 & 1 & 1 \\ -2 & 2 & 3 \end{bmatrix},\ (2)\begin{bmatrix} 11 & 12 & 13 & 14 \\ 21 & 22 & 23 & 24 \\ 31 & 32 & 33 & 34 \\ 41 & 42 & 43 & 44 \end{bmatrix},\ (3)\begin{bmatrix} 1 & x & x^2 & x^3 \\ x^3 & 1 & x & x^2 \\ x^2 & x^3 & 1 & x \\ x & x^2 & x^3 & 1 \end{bmatrix}.$$

2.2 Existence and uniqueness

Recall that $\det A = ad - bc$ defined in the previous section satisfies the three rules of Definition 2.2. Conversely, the following lemma shows that any function of $M_{2\times2}(\mathbb{R})$ into $\mathbb{R}$ satisfying the three rules $(\mathbf{R}_1)$ - $(\mathbf{R}_3)$ of Definition 2.2 must be det, which implies the uniqueness of the determinant function on $M_{2\times2}(\mathbb{R})$.

Lemma 2.3 *If $f : M_{2\times2}(\mathbb{R}) \to \mathbb{R}$ satisfies the three rules in* Definition 2.2, *then $f(A) = ad - bc$.*

Proof: First, note that $f\begin{bmatrix} 0 & 1 \\ 1 & 0 \end{bmatrix} = -1$ by the rules $(\mathbf{R}_1)$ and $(\mathbf{R}_2)$.

$$\begin{aligned} f(A) &= f\begin{bmatrix} a & b \\ c & d \end{bmatrix} = f\begin{bmatrix} a+0 & 0+b \\ c & d \end{bmatrix} \\ &= f\begin{bmatrix} a & 0 \\ c & d \end{bmatrix} + f\begin{bmatrix} 0 & b \\ c & d \end{bmatrix} \\ &= f\begin{bmatrix} a & 0 \\ 0 & d \end{bmatrix} + f\begin{bmatrix} a & 0 \\ c & 0 \end{bmatrix} + f\begin{bmatrix} 0 & b \\ 0 & d \end{bmatrix} + f\begin{bmatrix} 0 & b \\ c & 0 \end{bmatrix} \\ &= ad + 0 + 0 - bc. \end{aligned}$$ □

Therefore, when $n = 2$ there is only one function f on $M_{2\times 2}(\mathbb{R})$ which satisfies the three rules: *i.e.*, $f = \det$.

Now for $n = 3$, the same calculation as in the case of $n = 2$ can be applied. That is, by repeated use of the three rules $(\mathbf{R}_1)$ - $(\mathbf{R}_3)$ as in the proof of Lemma 2.3, we can obtain the explicit formula for the determinant function on $M_{3\times 3}(\mathbb{R})$ as follows:

$$\begin{aligned}
&\det \begin{bmatrix} a_{11} & a_{12} & a_{13} \\ a_{21} & a_{22} & a_{23} \\ a_{31} & a_{32} & a_{33} \end{bmatrix} \\
&= \det \begin{bmatrix} a_{11} & 0 & 0 \\ 0 & a_{22} & 0 \\ 0 & 0 & a_{33} \end{bmatrix} + \det \begin{bmatrix} 0 & a_{12} & 0 \\ 0 & 0 & a_{23} \\ a_{31} & 0 & 0 \end{bmatrix} + \det \begin{bmatrix} 0 & 0 & a_{13} \\ a_{21} & 0 & 0 \\ 0 & a_{32} & 0 \end{bmatrix} \\
&\quad + \det \begin{bmatrix} a_{11} & 0 & 0 \\ 0 & 0 & a_{23} \\ 0 & a_{32} & 0 \end{bmatrix} + \det \begin{bmatrix} 0 & a_{12} & 0 \\ a_{21} & 0 & 0 \\ 0 & 0 & a_{33} \end{bmatrix} + \det \begin{bmatrix} 0 & 0 & a_{13} \\ 0 & a_{22} & 0 \\ a_{31} & 0 & 0 \end{bmatrix} \\
&= a_{11}a_{22}a_{33} + a_{12}a_{23}a_{31} + a_{13}a_{21}a_{32} - a_{11}a_{23}a_{32} - a_{12}a_{21}a_{33} - a_{13}a_{22}a_{31}.
\end{aligned}$$

This expression of $\det A$ for a matrix $A \in M_{3\times 3}(\mathbb{R})$ satisfies the three rules. Therefore, for $n = 3$, it shows both the uniqueness and the existence of the determinant function on $M_{3\times 3}(\mathbb{R})$.

Problem 2.5 Show that the above formula of the determinant for 3×3 matrices satisfies the three rules in Definition 2.2.

To get the formula of the determinant for matrices of order $n > 3$, the same computational process can be repeated using the three rules again, but the computation is going to be more complicated as the order gets higher. To derive the explicit formula for $\det A$ of order $n > 3$, we examine the above case in detail. In the process of deriving the explicit formula for $\det A$ of a 3×3 matrix A, we can observe the following three steps:

(**1st**) By using the linearity of the determinant function in each row, $\det A$ of a 3×3 matrix A is expanded as the sum of the determinants of $3^3 = 27$ matrices. Except for exactly six matrices, all of them have zero columns so that their determinants are zero (see the proof of Lemma 2.3).

(**2nd**) In each of these remaining six matrices, all entries are zero except for exactly three entries that came from the given matrix A. Indeed, no two of the three entries came from the same column or from the same row of A.

In other words, in each row there is only one entry that came from A and at the same time in each column there is only one entry that came from A.

Actually, in each of the six matrices, the three entries from A, say a_{ij}, $a_{k\ell}$, and a_{pq}, are chosen as follows: If the first entry a_{ij} is chosen from the first row and the third column of A, say a_{13}, then the other entries $a_{k\ell}$ and a_{pq} in the product should be chosen from the second or the third row and the first or the second column. Thus, if the second entry $a_{k\ell}$ is taken from the second row, the column it belongs to must be either the first or the second, *i.e.*, either a_{21} or a_{22}. If a_{21} is taken, then the third entry a_{pq} must be, without option, a_{32}. Thus, the entries from A in the chosen matrix are a_{13}, a_{21}, and a_{32}. Therefore, the three entries in each of the six remaining matrices are determined as follows: when the row indices (the first indices i of a_{ij}) are arranged in the order 1, 2, 3, the assignment of the column indices 1, 2, 3 (the second indices j of a_{ij}) to each of the row indices is simply a re-arrangement of 1, 2, 3 without repetitions or omissions. In this way, one can recognize that the number $6 = 3!$ is simply the number of ways in which the three column indices 1, 2, 3 are rearranged.

(3rd) The determinant of each of the six matrices may be computed by converting it into a diagonal matrix using suitable "column interchanges" (see Theorem 2.2 (1)), so its determinant becomes $\pm a_{ij}a_{k\ell}a_{pq}$, where the sign depends on the number of column interchanges.

For example, for the matrix having entries a_{13}, a_{22}, and a_{31} from A, one can convert this matrix into a diagonal matrix in a couple of ways: for instance, one can take just one interchanging of the first and the third columns or take three interchanges: the first and the second, and then the second and the third, and then the first and the second. In any case,

$$\det \begin{bmatrix} 0 & 0 & a_{13} \\ 0 & a_{22} & 0 \\ a_{31} & 0 & 0 \end{bmatrix} = -\det \begin{bmatrix} a_{13} & 0 & 0 \\ 0 & a_{22} & 0 \\ 0 & 0 & a_{31} \end{bmatrix} = -a_{13}a_{22}a_{31}.$$

Note that an interchange of two columns is the same as an interchange of two corresponding column indices. As mentioned above, there may be several ways of column interchanges to convert the given matrix to a diagonal matrix. However, it is very interesting that, whatever ways of column interchanges we take, the parity of the number of column interchanges remains the same all the time.

In this example, the given arrangement of the column is expressed in the arrangement of column indices, which is 3, 2, 1. Thus, to arrive at the order 1, 2, 3, which represents the diagonal matrix, we can take either just one interchanging of 3 and 1, or three interchanges: 3 and 2, 3 and 1, and then 2 and 1. In either case, the parity is odd so that the "−" sign in the computation of determinant came from $(-1)^1 = (-1)^3$, where the exponents mean the numbers of interchanges of the column indices.

In summary, in the expansion of $\det A$ for $A \in M_{3\times3}(\mathbb{R})$, the number $6 = 3!$ of the determinants which contribute to the computation of $\det A$ is simply the number of ways in which the three numbers 1, 2, 3 are rearranged without repetitions or omissions. Moreover, the sign of each of the six determinants is determined by the parity (even or odd) of the number of column interchanges required to arrive at the order of 1, 2, 3 from the given arrangement of the column indices.

These observations can be used to derive the explicit formula of the determinant for matrices of order $n > 3$. We begin with the following definition.

Definition 2.3 A **permutation** of the set of integers $N_n = \{1,\ 2,\ \ldots,\ n\}$ is a one-to-one function from N_n onto itself.

Therefore, a permutation σ of N_n assigns a number $\sigma(i)$ in N_n to each number i in N_n, and this permutation σ is commonly denoted by

$$\sigma = \langle \sigma(1),\ \sigma(2),\ \ldots,\ \sigma(n) \rangle = \begin{pmatrix} 1 & 2 & \cdots & n \\ \sigma(1) & \sigma(2) & \cdots & \sigma(n) \end{pmatrix}.$$

Here, the first row is the usual lay-out of N_n as the domain set, and the second row is just an arrangement in a certain order without repetitions or omissions of the numbers in N_n as the image set. A permutation that interchanges only two numbers in N_n, leaving the rest of the numbers fixed, such as $\sigma = \langle 3, 2, 1, \ldots, n \rangle$, is called a **transposition**. Note that the composition of any two permutations is also a permutation. Moreover, the composition of a transposition to a permutation σ produces an interchanging of two numbers in the permutation σ. In particular, the composition of a transposition with itself always produces the identity permutation.

It is not hard to see that if S_n denotes the set of all permutations of N_n, then S_n has exactly $n!$ permutations.

Once we have listed all the permutations, the next step is to determine the sign of each permutation. A permutation $\sigma = \langle j_1,\ j_2,\ \ldots,\ j_n \rangle$ is said to have an **inversion** if $j_s > j_t$ for $s < t$ (*i.e.*, a larger number precedes a smaller number). For example, the permutation $\sigma = \langle 3, 1, 2 \rangle$ has two inversions since 3 precedes 1 and 2.

An inversion in a permutation can be eliminated by composing it with a suitable transposition: for example, if $\sigma = \langle 3, 2, 1\rangle$ with three inversions, then by multiplying a transposition $\langle 2, 1, 3\rangle$ to it, we get $\langle 2, 3, 1\rangle$ with two inversions, which is the same as interchanging the first two numbers 3, 2 in σ. Therefore, given a permutation $\sigma = \langle\sigma(1), \sigma(2), \ldots, \sigma(n)\rangle$ in S_n, one can convert it to the identity permutation $\langle 1, 2, \ldots, n\rangle$, which is the only one with no inversions, by composing it with certain number of transpositions. For example, by composing the three (which is the number of inversions in σ) transpositions $\langle 2, 1, 3\rangle$, $\langle 1, 3, 2\rangle$ and $\langle 2, 1, 3\rangle$ with $\sigma = \langle 3, 2, 1\rangle$, we get the identity permutation. However, the number of necessary transpositions to convert the given permutation into the identity permutation need not be unique as we have seen in the third step. Notice that even if the number of necessary transpositions is not unique the parity (even or odd) is always consistent with the number of inversions.

Recall that all we need in the computation of the determinant is just the parity (even or odd) of the number of column interchanges, which is the same as that of the number of inversions in the permutation of the column indices.

A permutation is said to be **even** if it has an even number of inversions, and it is said to be **odd** if it has an odd number of inversions. For example, when $n = 3$, the permutations $\langle 1, 2, 3\rangle$, $\langle 2, 3, 1\rangle$ and $\langle 3, 1, 2\rangle$ are even, while the permutations $\langle 1, 3, 2\rangle$, $\langle 2, 1, 3\rangle$ and $\langle 3, 2, 1\rangle$ are odd. In general, for a permutation σ in S_n, the **sign** of σ is defined as

$$\operatorname{sgn}(\sigma) = \begin{cases} 1 & \text{if } \sigma \text{ is an even permutation} \\ -1 & \text{if } \sigma \text{ is an odd permutation.} \end{cases}$$

It is not hard to see that the number of even permutations is equal to that of odd permutations, so it is $\frac{n!}{2}$. In the case $n = 3$, one can notice that there are 3 terms with $+$ sign and 3 terms with $-$ sign.

Problem 2.6 Show that the number of even permutations and the number of odd permutations in S_n are equal.

Now, we repeat the three steps to get an explicit formula for $\det A$ of a square matrix $A = [a_{ij}]$ of order n. First, the determinant $\det A$ can be expressed as the sum of determinants of $n!$ matrices, each of which has zero entries except the n entries $a_{1\sigma(1)}, a_{2\sigma(2)}, \cdots, a_{n\sigma(n)}$ taken from A, where σ is a permutation of the set $\{1, 2, \ldots, n\}$ of column indices. The n entries $a_{1\sigma(1)}, a_{2\sigma(2)}, \cdots, a_{n\sigma(n)}$ are chosen from A in such a way that no two of them come from the same row or the same column. Such a matrix can be converted to a diagonal matrix. Hence, its determinant is equal to $\pm a_{1\sigma(1)}a_{2\sigma(2)}\cdots a_{n\sigma(n)}$, where the sign $\pm$ is determined by the parity of the number of column interchanges to convert the matrix to a diagonal matrix, which is equal to that of inversions in σ: $\mathrm{sgn}(\sigma)$. Therefore, the determinant of the matrix whose entries are all zero except for $a_{i\sigma(i)}$'s is equal to

$$\mathrm{sgn}(\sigma)a_{1\sigma(1)}a_{2\sigma(2)}\cdots a_{n\sigma(n)},$$

which is called a **signed elementary product** of A. Now, our discussions can be summarized as follows:

Theorem 2.4 *For an $n \times n$ matrix A,*

$$\det A = \sum_{\sigma \in S_n} \mathrm{sgn}(\sigma)a_{1\sigma(1)}a_{2\sigma(2)}\cdots a_{n\sigma(n)}.$$

That is, $\det A$ *is the sum of all signed elementary products of* A.

It is not difficult to see that this explicit formula for $\det A$ satisfies the three rules in the definition of the determinant. Therefore, we have both *existence* and *uniqueness* for the determinant function of square matrices of any order $n \geq 1$.

Example 2.3 Consider a permutation $\sigma = \langle 3, 4, 2, 5, 1\rangle \in S_5$: *i.e.*, $\sigma(1) = 3$, $\sigma(2) = 4$, $\ldots$, $\sigma(5) = 1$. Then σ has total $2 + 4 = 6$ inversions: two inversions caused by the position of $\sigma(1) = 3$, which precedes 1 and 2, and four inversions in the permutation $\tau = \langle 4, 2, 5, 1\rangle$, which is a permutation of the set $\{1, 2, 4, 5\}$. Thus,

$$\mathrm{sgn}(\sigma) = (-1)^{2+4} = (-1)^2\mathrm{sgn}(\tau).$$

Note that the permutation τ can be considered as a permutation of N_4 by replacing the numbers 4 and 5 by 3 and 4, respectively.

Moreover, $\sigma = \langle 3, 4, 2, 5, 1\rangle$ can be converted to $\langle 1, 3, 4, 2, 5\rangle$ by shifting the number 1 by four transpositions, and then $\langle 1, 3, 4, 2, 5\rangle$ can be converted to the identity permutation $\langle 1, 2, 3, 4, 5\rangle$ by two transpositions. Hence, σ can be converted to the identity permutation by six transpositions. □

In general, for a fixed j, $1 \leq j \leq n$, there are $(n-1)!$ permutations σ's in S_n such that $\sigma(1) = j$. Each σ of those permutations has $j - 1$ inversions (j

precedes $j-1$ smaller numbers) and as many inversions as in the permutation $\tau = \langle\sigma(2), \ldots, \sigma(n)\rangle$. Therefore,

$$\operatorname{sgn}(\sigma) = (-1)^{j-1}\operatorname{sgn}(\tau).$$

Also, the permutation $\tau = \langle\sigma(2), \ldots, \sigma(n)\rangle$ can be considered as a permutation of N_{n-1} by replacing $\{j+1, \ldots, n\}$ by $\{j, \ldots, n-1\}$. Thus we have the following lemma.

Lemma 2.5 *For any permutation σ in S_n, if $\sigma(1) = j$, then*

$$\operatorname{sgn}(\sigma) = (-1)^{j-1}\operatorname{sgn}(\langle\sigma(2), \ldots, \sigma(n)\rangle),$$

where $\langle\sigma(2), \ldots, \sigma(n)\rangle$ is a permutation of $n-1$ numbers $N_n - \{j = \sigma(1)\}$.

Problem 2.7 Let $A = [\mathbf{c}_1 \cdots \mathbf{c}_n]$ be an $n \times n$ matrix with the column vectors $\mathbf{c}_j$'s. Show that $\det[\mathbf{c}_j\, \mathbf{c}_1 \cdots \mathbf{c}_{j-1}\, \mathbf{c}_{j+1} \cdots \mathbf{c}_n] = (-1)^{j-1}\det[\mathbf{c}_1 \cdots \mathbf{c}_j \cdots \mathbf{c}_n]$. Note that the same kind of equality holds when A is written in row vectors.

Problem 2.8 Compute the determinant of the matrix $\begin{bmatrix} 1 & 4 & 2 \\ 3 & 1 & 1 \\ -2 & 2 & 3 \end{bmatrix}$.

2.3 Cofactor expansion

Recall that the determinant of an $n \times n$ matrix A is the sum of all signed elementary products of A, and

$$\det A = \sum_{\sigma \in S_n} \operatorname{sgn}(\sigma) a_{1\sigma(1)} a_{2\sigma(2)} \cdots a_{n\sigma(n)}.$$

The first factor $a_{1\sigma(1)}$ in each term can be any one of $a_{11}, a_{12}, \ldots, a_{1n}$ in the first row of A. Among the $n!$ terms in this sum, there are precisely $(n-1)!$ permutations such that $a_{1\sigma(1)} = a_{11}$, *i.e.*, $\sigma(1) = 1$. The sum of those terms such that $\sigma(1) = 1$ can be written as $a_{11}A_{11}$, where

$$A_{11} = (-1)^0 \sum_{\tau} \operatorname{sgn}(\tau) a_{2\tau(2)} \cdots a_{n\tau(n)},$$

summing over all permutations τ of the numbers $\{2, 3, \ldots, n\}$. The term $(-1)^0$ means that there is no extra inversion other than that of τ if $\sigma(1) = 1$ is at the first place. Note that each term in A_{11} contains no entries from the first row or from the first column of A. Hence, all the terms of the sum in A_{11} are the signed elementary products of the submatrix M_{11} of A obtained by deleting the first row and the first column of A. Thus $A_{11} = (-1)^0 \det M_{11}$.

Similarly, if $a_{1\sigma(1)}$ is chosen to be a_{1j} with $1 \le j \le n$, then all $(n-1)!$ terms such that $\sigma(1) = j$ in the expression of $\det A$ add up to $a_{1j}A_{1j}$ with

$$A_{1j} = \sum_{\sigma \in S_n,\ \sigma(1)=j} \operatorname{sgn}(\sigma) a_{2\sigma(2)} a_{3\sigma(3)} \cdots a_{n\sigma(n)} = (-1)^{j-1} \det M_{1j},$$

where M_{1j} is the submatrix of A obtained by deleting the row and the column containing a_{1j}, and the sign $(-1)^{j-1}$ means the extra inversion numbers caused by placing $\sigma(1) = j$ at the first place as shown in Lemma 2.5.

By grouping $a_{1j}A_{1j}$ for all $j = 1, \ldots, n$ in the expression of $\det A$, we can get an expansion of $\det A$ with respect to the first row:

$$\det A = a_{11}A_{11} + a_{12}A_{12} + \cdots + a_{1n}A_{1n},$$

where $A_{1j} = (-1)^{j-1} \det M_{1j}$ and M_{1j} is the submatrix of A obtained by deleting the row and the column containing a_{1j}.

There is a similar expansion with respect to any other row, say the i-th row. To show this, first construct a new matrix $\bar{A}$ by using the i-th row of A as the first row and then shifting each of the preceding $i-1$ rows one row down. Then it is easy to see that $\det A = (-1)^{i-1} \det \bar{A}$ by Lemma 2.5. Now, with the expansion of $\det \bar{A}$ with respect to the first row $[a_{i1} \ \cdots \ a_{in}]$, we get

$$\det A = a_{i1}A_{i1} + a_{i2}A_{i2} + \cdots + a_{in}A_{in},$$

where $A_{ij} = (-1)^{i+j} \det M_{ij}$ and M_{ij} is the submatrix of A obtained by deleting the row and the column containing a_{ij}.

Also, we can do the same with the column vectors because $\det A^T = \det A$. This gives the following theorem:

Theorem 2.6 *Let A be an $n \times n$ matrix. Then,*

(1) *for each $1 \le i \le n$,*

$$\det A = a_{i1}A_{i1} + a_{i2}A_{i2} + \cdots + a_{in}A_{in},$$

called the **cofactor expansion** *of $\det A$ along the i-th row.*

(2) *For each $1 \le j \le n$,*

$$\det A = a_{1j}A_{1j} + a_{2j}A_{2j} + \cdots + a_{nj}A_{nj},$$

called the **cofactor expansion** *of $\det A$ along the j-th column.*

The submatrix M_{ij} is called the **minor** of the entry a_{ij} and the number $A_{ij} = (-1)^{i+j} \det M_{ij}$ is called the **cofactor** of the entry a_{ij}. Therefore, the determinant of an $n \times n$ matrix A can be computed by multiplying the entries in any one row by their cofactors and adding the resulting products. As a matter of fact, the determinant could be defined inductively by these explicit formulas.

Example 2.4 Let

$$A = \begin{bmatrix} 1 & 2 & 3 \\ 4 & 5 & 6 \\ 7 & 8 & 9 \end{bmatrix}.$$

Then the cofactors of a_{11}, a_{12} and a_{13} are

$$\begin{aligned} A_{11} &= (-1)^{1+1} \det \begin{bmatrix} 5 & 6 \\ 8 & 9 \end{bmatrix} &= 5 \cdot 9 - 8 \cdot 6 = -3, \\ A_{12} &= (-1)^{1+2} \det \begin{bmatrix} 4 & 6 \\ 7 & 9 \end{bmatrix} &= (-1)(4 \cdot 9 - 7 \cdot 6) = 6, \\ A_{13} &= (-1)^{1+3} \det \begin{bmatrix} 4 & 5 \\ 7 & 8 \end{bmatrix} &= 4 \cdot 8 - 7 \cdot 5 = -3, \end{aligned}$$

respectively. Hence the expansion of $\det A$ along the first column is

$$\det A = a_{11}A_{11} + a_{12}A_{12} + a_{13}A_{13} = 1 \cdot (-3) + 2 \cdot 6 + 3 \cdot (-3) = 0. \qquad \square$$

For a 3×3 matrix A, the cofactor expansion of A along the second column has the following form:

$$\begin{aligned} &\det \begin{bmatrix} a_{11} & a_{12} & a_{13} \\ a_{21} & a_{22} & a_{23} \\ a_{31} & a_{32} & a_{33} \end{bmatrix} = a_{12}A_{12} + a_{22}A_{22} + a_{32}A_{32} \\ &= -a_{12} \det \begin{bmatrix} a_{21} & a_{23} \\ a_{31} & a_{33} \end{bmatrix} + a_{22} \det \begin{bmatrix} a_{11} & a_{13} \\ a_{31} & a_{33} \end{bmatrix} - a_{32} \det \begin{bmatrix} a_{11} & a_{13} \\ a_{21} & a_{23} \end{bmatrix}. \end{aligned}$$

As this formula suggests, in the cofactor expansion of $\det A$ along a row or a column, the evaluation of A_{ij} can be avoided whenever $a_{ij} = 0$, because the product $a_{ij}A_{ij}$ is zero regardless of the value A_{ij}. Therefore it is beneficial to make the cofactor expansion along a row or a column that contains as many zero entries as possible. Moreover, by using the elementary operations, a matrix A may be simplified into another one having more zero entries in a row or in a column. This kind of simplification can be done by the elementary row (or column) operations, and generally gives the most efficient way to evaluate the determinant of a matrix. The next examples illustrate this method for an evaluation of the determinant.

Example 2.5 Evaluate the determinant of

$$A = \begin{bmatrix} 1 & -1 & 2 & -1 \\ -3 & 4 & 1 & -1 \\ 2 & -5 & -3 & 8 \\ -2 & 6 & -4 & 1 \end{bmatrix}.$$

Solution: Apply the elementary operations:

$$\begin{aligned} 3 \times \text{ row } 1 &+ \text{ row } 2, \\ (-2) \times \text{ row } 1 &+ \text{ row } 3, \\ 2 \times \text{ row } 1 &+ \text{ row } 4 \end{aligned}$$

to A. Then

$$\det A = \det \begin{bmatrix} 1 & -1 & 2 & -1 \\ 0 & 1 & 7 & -4 \\ 0 & -3 & -7 & 10 \\ 0 & 4 & 0 & -1 \end{bmatrix} = \det \begin{bmatrix} 1 & 7 & -4 \\ -3 & -7 & 10 \\ 4 & 0 & -1 \end{bmatrix}.$$

Now apply the operation: row 1 + row 2, to the matrix on the right side, then

$$\begin{aligned} \det \begin{bmatrix} 1 & 7 & -4 \\ -3 & -7 & 10 \\ 4 & 0 & -1 \end{bmatrix} &= \det \begin{bmatrix} 1 & 7 & -4 \\ -2 & 0 & 6 \\ 4 & 0 & -1 \end{bmatrix} \\ &= (-1)^{1+2} \cdot 7 \cdot \det \begin{bmatrix} -2 & 6 \\ 4 & -1 \end{bmatrix} \\ &= -7(2-24) = 154. \end{aligned}$$

Thus $\det A = 154$. □

Problem 2.9 Use cofactor expansions along a row or a column to evaluate the determinants of the following matrices:

$$(1)\ A = \begin{bmatrix} 0 & 1 & 1 & 1 \\ 2 & 0 & 1 & 1 \\ 2 & 2 & 0 & 1 \\ 2 & 2 & 2 & 0 \end{bmatrix}, \quad (2)\ A = \begin{bmatrix} 1 & -2 & 1 & 1 \\ -1 & 3 & 0 & 2 \\ 0 & 1 & 1 & 3 \\ 1 & 2 & 5 & 0 \end{bmatrix}.$$

Example 2.6 Show that $\det A = (x-y)(x-z)(x-w)(y-z)(y-w)(z-w)$ for

$$A = \begin{bmatrix} 1 & x & x^2 & x^3 \\ 1 & y & y^2 & y^3 \\ 1 & z & z^2 & z^3 \\ 1 & w & w^2 & w^3 \end{bmatrix}.$$

Solution: Use Gaussian elimination. To begin with, add $(-1)\times$ row 1 to rows 2, 3, and 4 of A:

$$\begin{aligned}
\det A &= \det \begin{bmatrix} 1 & x & x^2 & x^3 \\ 0 & y-x & y^2-x^2 & y^3-x^3 \\ 0 & z-x & z^2-x^2 & z^3-x^3 \\ 0 & w-x & w^2-x^2 & w^3-x^3 \end{bmatrix} \\
&= \det \begin{bmatrix} y-x & y^2-x^2 & y^3-x^3 \\ z-x & z^2-x^2 & z^3-x^3 \\ w-x & w^2-x^2 & w^3-x^3 \end{bmatrix} \\
&= (y-x)(z-x)(w-x)\det \begin{bmatrix} 1 & y+x & y^2+xy+x^2 \\ 1 & z+x & z^2+xz+x^2 \\ 1 & w+x & w^2+xw+x^2 \end{bmatrix} \\
&= (x-y)(x-z)(w-x)\det \begin{bmatrix} 1 & y+x & y^2+xy+x^2 \\ 0 & z-y & (z-y)(z+y+x) \\ 0 & w-y & (w-y)(w+y+x) \end{bmatrix} \\
&= (x-y)(x-z)(w-x)\det \begin{bmatrix} z-y & (z-y)(z+y+x) \\ w-y & (w-y)(w+y+x) \end{bmatrix} \\
&= (x-y)(x-z)(x-w)(y-z)(w-y)\det \begin{bmatrix} 1 & z+y+x \\ 1 & w+y+x \end{bmatrix} \\
&= (x-y)(x-z)(x-w)(y-z)(y-w)(z-w).
\end{aligned}$$

□

Problem 2.10 Let A be the Vandermonde matrix of order n:

$$A = \begin{bmatrix} 1 & x_1 & x_1^2 & \cdots & x_1^{n-1} \\ 1 & x_2 & x_2^2 & \cdots & x_2^{n-1} \\ \vdots & \vdots & \vdots & & \vdots \\ 1 & x_n & x_n^2 & \cdots & x_n^{n-1} \end{bmatrix}.$$

Show that

$$\det A = \prod_{1\le i<j\le n} (x_j - x_i).$$

2.4 Cramer's rule

The cofactor expansion of the determinant gives a method for computing the inverse of an invertible matrix A. For $i \neq j$, let A^* be the matrix A with the j-th row replaced by the i-th row. Then the determinant of A^* must be zero, because the entries of the i-th and j-th rows are the same. Moreover, the cofactors of A^* with respect to the j-th row are the same as those of A: that is, $A^*_{jk} = A_{jk}$ for all $k = 1, \ldots, n$. Therefore, we have

$$\begin{aligned} 0 = \det A^* &= a_{i1}A^*_{j1} + a_{i2}A^*_{j2} + \cdots + a_{in}A^*_{jn} \\ &= a_{i1}A_{j1} + a_{i2}A_{j2} + \cdots + a_{in}A_{jn}. \end{aligned}$$

This proves the following lemma.

Lemma 2.7

$$a_{i1}A_{j1} + a_{i2}A_{j2} + \cdots + a_{in}A_{jn} = \begin{cases} \det A & \textit{if } i = j \\ 0 & \textit{if } i \neq j. \end{cases}$$

Definition 2.4 If A is an $n \times n$ matrix and A_{ij} is the cofactor of a_{ij}, then the new matrix

$$\begin{bmatrix} A_{11} & A_{12} & \cdots & A_{1n} \\ A_{21} & A_{22} & \cdots & A_{2n} \\ \vdots & \vdots & \ddots & \vdots \\ A_{n1} & A_{n2} & \cdots & A_{nn} \end{bmatrix}$$

is called the **matrix of cofactors** of A. The transpose of this matrix is called the **adjoint** of A and is denoted by $\mathrm{adj}A$.

It follows from Lemma 2.7 that

$$A \cdot \mathrm{adj}A = \begin{bmatrix} \det A & 0 & \cdots & 0 \\ 0 & \det A & \cdots & 0 \\ \vdots & \vdots & \ddots & \vdots \\ 0 & 0 & \cdots & \det A \end{bmatrix} = (\det A)I.$$

If A is invertible, then $\det A \neq 0$ and we may write $A\left(\frac{1}{\det A}\,\mathrm{adj}A\right) = I$. Thus

$$A^{-1} = \frac{1}{\det A}\,\mathrm{adj}A, \quad \text{and} \quad A = (\det A)\,\mathrm{adj}(A^{-1})$$

by replacing A with A^{-1}.

Example 2.7 For a matrix $A = \begin{bmatrix} a & b \\ c & d \end{bmatrix}$, $\text{adj}A = \begin{bmatrix} d & -b \\ -c & a \end{bmatrix}$, and if $\det A = ad - bc \neq 0$, then

$$A^{-1} = \frac{1}{ad - bc} \begin{bmatrix} d & -b \\ -c & a \end{bmatrix}.$$

Problem 2.11 Compute $\text{adj}A$ and A^{-1} for $A = \begin{bmatrix} 1 & 3 & 1 \\ 2 & 1 & 1 \\ 2 & -2 & 1 \end{bmatrix}$.

Problem 2.12 Show that A is invertible if and only if $\text{adj}A$ is invertible, and that if A is invertible, then

$$(\text{adj}A)^{-1} = \frac{A}{\det A} = \text{adj}(A^{-1}).$$

Problem 2.13 Let A be an $n \times n$ matrix with $n > 1$. Show that
(1) $\det(\text{adj}A) = (\det A)^{n-1}$;
(2) $\text{adj}(\text{adj}A) = (\det A)^{n-2}A$, if A is invertible.

The next theorem establishes a formula for the solution of a system of n equations in n unknowns. It is not useful as a practical method but can be used to study properties of the solution without solving the system.

Theorem 2.8 (Cramer's rule) *Let $A\mathbf{x} = \mathbf{b}$ be a system of n linear equations in n unknowns such that $\det A \neq 0$. Then the system has the unique solution given by*

$$x_j = \frac{\det C_j}{\det A}, \qquad j = 1,\ 2,\ \ldots,\ n,$$

where C_j is the matrix obtained from A by replacing the j-th column with the column matrix $\mathbf{b} = [b_1\ b_2\ \cdots\ b_n]^T$.

Proof: If $\det A \neq 0$, then A is invertible and $\mathbf{x} = A^{-1}\mathbf{b}$ is the unique solution of $A\mathbf{x} = \mathbf{b}$. Since

$$\mathbf{x} = A^{-1}\mathbf{b} = \frac{1}{\det A}(\text{adj}A)\mathbf{b},$$

it follows that

$$x_j = \frac{b_1 A_{1j} + b_2 A_{2j} + \cdots + b_n A_{nj}}{\det A} = \frac{\det C_j}{\det A}.$$ □

Example 2.8 Use Cramer's rule to solve

$$\begin{cases} x_1 + 2x_2 + x_3 = 50 \\ 2x_1 + 2x_2 + x_3 = 60 \\ x_1 + 2x_2 + 3x_3 = 90. \end{cases}$$

Solution:

$$A = \begin{bmatrix} 1 & 2 & 1 \\ 2 & 2 & 1 \\ 1 & 2 & 3 \end{bmatrix}, \quad C_1 = \begin{bmatrix} 50 & 2 & 1 \\ 60 & 2 & 1 \\ 90 & 2 & 3 \end{bmatrix},$$

$$C_2 = \begin{bmatrix} 1 & 50 & 1 \\ 2 & 60 & 1 \\ 1 & 90 & 3 \end{bmatrix}, \quad C_3 = \begin{bmatrix} 1 & 2 & 50 \\ 2 & 2 & 60 \\ 1 & 2 & 90 \end{bmatrix}.$$

Therefore,

$$x_1 = \frac{\det C_1}{\det A} = 10, \quad x_2 = \frac{\det C_2}{\det A} = 10, \quad x_3 = \frac{\det C_3}{\det A} = 20. \qquad \square$$

Cramer's rule provides a convenient method for writing down the solution of a system of n linear equations in n unknowns in terms of determinants. To find the solution, however, one must evaluate $n+1$ determinants of order n. Evaluating even two of these determinants generally involves more computations than solving the system by using Gauss-Jordan elimination.

Problem 2.14 Use Cramer's rule to solve the systems

$$(1) \begin{cases} 4x_2 + 3x_3 = -2 \\ 3x_1 + 4x_2 + 5x_3 = 6 \\ -2x_1 + 5x_2 - 2x_3 = 1. \end{cases}$$

$$(2) \begin{cases} \frac{2}{x} - \frac{3}{y} + \frac{5}{z} = 3 \\ -\frac{4}{x} + \frac{7}{y} + \frac{2}{z} = 0 \\ \frac{2}{y} - \frac{1}{z} = 2. \end{cases}$$

Problem 2.15 Let A be the matrix obtained from the identity matrix I_n with i-th column replaced by the column vector $\mathbf{x} = [x_1 \ \cdots \ x_n]^T$. Compute $\det A$.

Problem 2.16 Prove that if A_{ij} is the cofactor of a_{ij} in $A = [a_{ij}]$, and if $n > 1$, then

$$\det \begin{bmatrix} 0 & x_1 & x_2 & \cdots & x_n \\ x_1 & a_{11} & a_{12} & \cdots & a_{1n} \\ x_2 & a_{21} & a_{22} & \cdots & a_{2n} \\ \vdots & \vdots & \vdots & & \vdots \\ x_n & a_{n1} & a_{n2} & \cdots & a_{nn} \end{bmatrix} = -\sum_{i=1}^{n}\sum_{j=1}^{n} A_{ij}x_ix_j.$$

2.5 Application: Area and Volume

In this section, we restrict our attention to the case of $n = 2$ or 3 in order to visualize the geometric figures conveniently, even if the same argument can be applied for $n > 3$.

For an $n \times n$ square matrix A, the row vectors $\mathbf{r}_i = [a_{i1} \ \cdots \ a_{in}]$, $i = 1, \ldots, n$, of A can be considered as elements in

$$\mathbb{R}^n = \{(a_1, \ldots, a_n) \ : \ a_i \in \mathbb{R}, i = 1, \ldots, n\}.$$

The set

$$\mathcal{P}(A) = \left\{ \sum_{i=1}^{n} t_i \mathbf{r}_i \ : \ 0 \le t_i \le 1, \ i = 1, \ldots, n \right\}$$

is called a **parallelogram** if $n = 2$, or a **parallelepiped** if $n \ge 3$. Note that the row vectors of A form the edges of $\mathcal{P}(A)$, and a different order of the row vectors does not alter the shape of $\mathcal{P}(A)$.

A geometrical meaning of the determinant is that it represents the volume (or area for $n = 2$) of the parallelepiped $\mathcal{P}(A)$.

Theorem 2.9 *The determinant* $\det A$ *of an* $n \times n$ *matrix* A *is the volume of* $\mathcal{P}(A)$ *up to sign. In fact, the volume of* $\mathcal{P}(A)$ *is equal to* $|\det A\,|$.

Proof: We present here a geometrical sketch since this way seems more intuitive and more convincing. We give only the proof of the case $n = 2$, and leave the case $n = 3$ to the readers. Let

$$A = \begin{bmatrix} a & b \\ c & d \end{bmatrix} = \begin{bmatrix} \mathbf{r}_1 \\ \mathbf{r}_2 \end{bmatrix},$$

where $\mathbf{r}_1$, $\mathbf{r}_2$ are the row vectors of A. Let $Area(A) = Area(\mathbf{r}_1, \ \mathbf{r}_2)$ denote the area of the parallelogram $\mathcal{P}(\mathbf{r}_1, \ \mathbf{r}_2)$ (see the figure below).

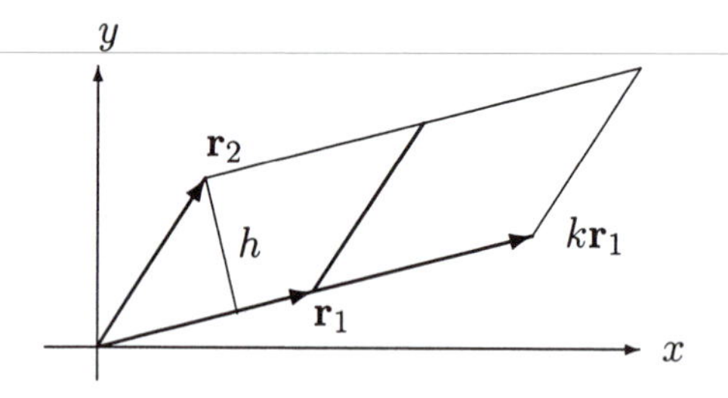

(1) It is quite clear that if $A = I_2$, then $Area \begin{bmatrix} 1 & 0 \\ 0 & 1 \end{bmatrix} = 1$.

(2) Since the shape of the parallelogram $\mathcal{P}(A)$ does not depend on the order of placing the row vectors: *i.e.*, $\mathcal{P}(\mathbf{r}_1,\ \mathbf{r}_2) = \mathcal{P}(\mathbf{r}_2,\ \mathbf{r}_1)$, we have $Area(\mathbf{r}_1, \mathbf{r}_2) = Area(\mathbf{r}_2, \mathbf{r}_1)$. On the other hand, $\det(\mathbf{r}_1, \mathbf{r}_2) = -\det(\mathbf{r}_2, \mathbf{r}_1)$. Thus

$$\det(\mathbf{r}_1,\ \mathbf{r}_2) = \pm Area(\mathbf{r}_1,\ \mathbf{r}_2),$$

which explains why we say "up to sign".

(3) From the figure above, if we replace $\mathbf{r}_1$ by $k\mathbf{r}_1$ in A, then the bottom edge $\mathbf{r}_1$ of $\mathcal{P}(A)$ is elongated by $|k|$ while the height h remains unchanged. Thus

$$Area(k\mathbf{r}_1,\ \mathbf{r}_2) = |k| Area(\mathbf{r}_1,\ \mathbf{r}_2).$$

(4) The additivity in the first row is a trivial consequence of examining the following figure: That is, if we replace $\mathbf{r}_1$ by $\mathbf{r}_1 + \mathbf{r}_1'$ while fixing $\mathbf{r}_2$, then, as the following figure shows, we have

$$Area(\mathbf{r}_1 + \mathbf{r}_1',\ \mathbf{r}_2) = Area(\mathbf{r}_1,\ \mathbf{r}_2) + Area(\mathbf{r}_1',\ \mathbf{r}_2).$$

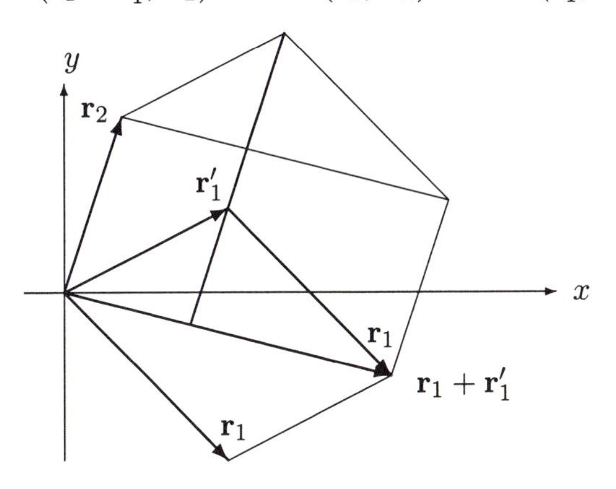

(5) Thus the area function $Area$ on $M_{2\times 2}(\mathbb{R})$ satisfies the rules $(\mathbf{R}_1)$ and $(\mathbf{R}_3)$ of the determinant except for the rule $(\mathbf{R}_2)$. Therefore, by uniqueness, $\det = \pm Area$. □

Remark: (1) Note that if we have constructed the parallelepiped $\mathcal{P}(A)$ using the column vectors of A, then the shape of the parallelepiped is totally different from the one constructed using the row vectors. However, $\det A = \det A^T$ means their volumes are the same, which is a totally nontrivial fact.

(2) For $n \geq 3$, the volume of $\mathcal{P}(A)$ can be defined by induction on n, and exactly the same argument in the proof can be applied to show that the volume is the determinant. However, there is another way of looking at this fact. Let $\{\mathbf{c}_1, \ldots, \mathbf{c}_n\}$ be n column vectors of an $m \times n$ matrix A. They constitute an n-dimensional parallelepiped in $\mathbb{R}^m$ such that

$$\mathcal{P}(A) = \left\{ \sum_{i=1}^{n} t_i \mathbf{c}_i \ : \ 0 \leq t_i \leq 1, \ i = 1, \ldots, n \right\}.$$

A formula for the volume of this parallelepiped may be expressed as follows: We first consider a two-dimensional parallelepiped (a parallelogram) determined by two column vectors $\mathbf{c}_1$ and $\mathbf{c}_2$ of $A = [\mathbf{c}_1 \ \mathbf{c}_2]$ in $\mathbb{R}^3$.

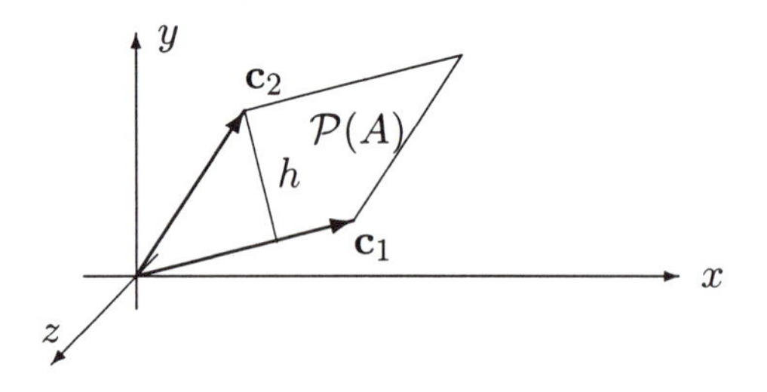

The area of this parallelogram is simply $Area(\mathcal{P}(A)) = \|\mathbf{c}_1\| h$, where $h = \|\mathbf{c}_2\| \sin\theta$ and θ is the angle between $\mathbf{c}_1$ and $\mathbf{c}_2$. Therefore, we have

$$\begin{aligned}
Area(\mathcal{P}(A))^2 &= \|\mathbf{c}_1\|^2 \|\mathbf{c}_2\|^2 \sin^2\theta \ = \ \|\mathbf{c}_1\|^2 \|\mathbf{c}_2\|^2 (1 - \cos^2\theta) \\
&= (\mathbf{c}_1 \cdot \mathbf{c}_1)(\mathbf{c}_2 \cdot \mathbf{c}_2) \left(1 - \frac{(\mathbf{c}_1 \cdot \mathbf{c}_2)^2}{(\mathbf{c}_1 \cdot \mathbf{c}_1)(\mathbf{c}_2 \cdot \mathbf{c}_2)} \right) \\
&= (\mathbf{c}_1 \cdot \mathbf{c}_1)(\mathbf{c}_2 \cdot \mathbf{c}_2) - (\mathbf{c}_1 \cdot \mathbf{r}_2)^2 \\
&= \det \begin{bmatrix} \mathbf{c}_1 \cdot \mathbf{c}_1 & \mathbf{c}_1 \cdot \mathbf{c}_2 \\ \mathbf{c}_2 \cdot \mathbf{c}_1 & \mathbf{c}_2 \cdot \mathbf{c}_2 \end{bmatrix} \\
&= \det \left(\begin{bmatrix} \mathbf{c}_1^T \\ \mathbf{c}_2^T \end{bmatrix} \begin{bmatrix} \mathbf{c}_1 & \mathbf{c}_2 \end{bmatrix} \right) \ = \ \det(A^T A).
\end{aligned}$$

In general, let $\mathbf{c}_1, \ldots, \mathbf{c}_n$ be n column vectors of an $m \times n$ matrix A. Then one can show (for a proof see Exercise **5.16**) that the volume of the n-dimensional parallelepiped $\mathcal{P}(A)$ determined by those n column vectors $\mathbf{c}_j$'s in $\mathbb{R}^m$ is

$$\mathrm{vol}(\mathcal{P}(A)) = \sqrt{\det(A^T A)}.$$

In particular, if A is an $m \times m$ square matrix, then

$$\mathrm{vol}(\mathcal{P}(A)) = \sqrt{\det(A^T A)} = \sqrt{\det(A^T)\det(A)} = |\det(A)|,$$

as expected.

Problem 2.17 Show that the area of a triangle ABC in the plane $\mathbb{R}^2$, where $A = (x_1,\ y_1)$, $B = (x_2,\ y_2)$, $C = (x_3,\ y_3)$, is equal to the absolute value of

$$\frac{1}{2}\det\begin{bmatrix} x_1 & y_1 & 1 \\ x_2 & y_2 & 1 \\ x_3 & y_3 & 1 \end{bmatrix}.$$

2.6 Exercises

2.1. Determine the values of k for which $\det\begin{bmatrix} k & k \\ 4 & 2k \end{bmatrix} = 0$.

2.2. Evaluate $\det(A^2BA^{-1})$ and $\det(B^{-1}A^3)$ for the following matrices:

$$A = \begin{bmatrix} 1 & -2 & 3 \\ -2 & 3 & 1 \\ 0 & 1 & 0 \end{bmatrix}, \quad B = \begin{bmatrix} 1 & 0 & 2 \\ 3 & -2 & 5 \\ 2 & 1 & 3 \end{bmatrix}.$$

2.3. Evaluate the determinant of

$$A = \begin{bmatrix} 3 & -2 & -5 & 4 \\ -5 & 2 & 8 & -5 \\ -3 & 4 & 7 & -3 \\ 2 & -3 & -5 & 8 \end{bmatrix}.$$

2.4. Evaluate $\det A$ for an $n \times n$ matrix $A = [a_{ij}]$ when

(1) $a_{ij} = \begin{cases} 1 & i \neq j \\ 0 & i = j, \end{cases}$ or (2) $a_{ij} = i + j$.

-A)$isapolynomialinxof c_1 x + c_0$.

2.5. Find all solutions of the equation $\det(AB) = 0$ for

$$A = \begin{bmatrix} x+2 & 3x \\ 3 & x+2 \end{bmatrix}, \qquad B = \begin{bmatrix} x & 0 \\ 5 & x+2 \end{bmatrix}.$$

2.6. Prove that if A is an $n \times n$ skew-symmetric matrix and n is odd, then $\det A = 0$. Give an example of 4×4 skew-symmetric matrix A with $\det A \neq 0$.

2.7. Use the determinant function to find

(1) the area of the parallelogram with edges determined by $(4,\ 3)$ and $(7,\ 5)$,

(2) the volume of the parallelepiped with edges determined by the vectors $(1,\ 0,\ 4)$, $(0,\ -2,\ 2)$ and $(3,\ 1,\ -1)$.

2.8. Use Cramer's rule to solve each system.

(1) $\begin{cases} x_1 + x_2 = 3 \\ x_1 - x_2 = -1 . \end{cases}$ (2) $\begin{cases} x_1 + x_2 + x_3 = 2 \\ x_1 + 2x_2 + x_3 = 2 \\ x_1 + 3x_2 - x_3 = -4 . \end{cases}$

(3) $\begin{cases} - x_2 + x_4 = -1 \\ x_1 + x_3 = 3 \\ x_1 - x_2 - x_3 - x_4 = 2 \\ x_1 + x_2 + x_3 + x_4 = 0 . \end{cases}$

2.9. Use Cramer's rule to solve the given system:

(1) $\begin{bmatrix} 1 & 2 \\ 4 & 3 \end{bmatrix} \mathbf{x} = \begin{bmatrix} 1 \\ 2 \end{bmatrix}$ (2) $\begin{bmatrix} 1 & 2 & -1 \\ 2 & 3 & 4 \\ 0 & 1 & 5 \end{bmatrix} \mathbf{x} = \begin{bmatrix} -1 \\ 2 \\ 0 \end{bmatrix}$.

2.10. Find a constant k so that the system of linear equations

$$\begin{cases} kx - 2y - z = 0 \\ (k+1)y + 4z = 0 \\ (k-1)z = 0 . \end{cases}$$

has more than one solution. (Is it possible to apply Cramer's rule here?)

2.11. Solve the following system of linear equations by using Cramer's rule and by using Gaussian elimination:

$$\begin{bmatrix} 1 & 1 & 1 & 1 \\ 1 & 2 & 1 & 1 \\ 1 & 1 & 2 & 1 \\ 1 & 1 & 1 & 2 \end{bmatrix} \mathbf{x} = \begin{bmatrix} 1 \\ 2 \\ 3 \\ 4 \end{bmatrix}.$$

2.12. Solve the following system of equations by using Cramer's rule:

$$\begin{cases} 3x + 2y = 3z + 1 \\ 3x + 2z = 8 - 5y \\ 3z - 1 = x - 2y . \end{cases}$$

2.13. Calculate the cofactors A_{11}, A_{12}, A_{13} and A_{33} for the matrix A:

(1) $A = \begin{bmatrix} 1 & 2 & 1 \\ 0 & 1 & 3 \\ 2 & 1 & 1 \end{bmatrix}$, (2) $A = \begin{bmatrix} 1 & 4 & 0 \\ 1 & 0 & 2 \\ 3 & 1 & 2 \end{bmatrix}$, (3) $A = \begin{bmatrix} 2 & -1 & 3 \\ -1 & 2 & 2 \\ 3 & 2 & 1 \end{bmatrix}$.

2.14. Let A be the $n \times n$ matrix whose entries are all 1. Show that

(1) $\det(A - nI_n) = 0$.

(2) $(A - nI_n)_{ij} = (-1)^{n-1}n^{n-2}$ for all $i,\ j$, where $(A - nI_n)_{ij}$ denotes the cofactor of the $(i,\ j)$-entry of $A - nI_n$.

2.15. Show that if A is symmetric, then so is $\mathrm{adj}A$. Moreover, if it is invertible, then the inverse of A is also symmetric.

2.16. Use the adjoint formula to compute the inverse of the each of the following matrices:

$$A = \begin{bmatrix} -2 & 3 & 2 \\ 6 & 0 & 3 \\ 4 & 1 & -1 \end{bmatrix}, \qquad B = \begin{bmatrix} \cos\theta & 0 & -\sin\theta \\ 0 & 1 & 0 \\ \sin\theta & 0 & \cos\theta \end{bmatrix}.$$

2.17. Compute $\mathrm{adj}A$, $\det A$, $\det(\mathrm{adj}A)$, A^{-1}, and verify $A \cdot \mathrm{adj}A = (\det A)I$ for

$$(1)\ A = \begin{bmatrix} 2 & 1 & 3 \\ -1 & 2 & 0 \\ 3 & -2 & 1 \end{bmatrix}, \quad (2)\ A = \begin{bmatrix} 1 & 2 & 3 \\ 2 & 3 & 4 \\ 1 & 5 & 7 \end{bmatrix}.$$

2.18. Let A, B be invertible matrices. Show that $\mathrm{adj}(AB) = \mathrm{adj}B\ \mathrm{adj}A$. (The reader may also try to prove this equality for noninvertible matrices.)

2.19. For an $m \times n$ matrix A and $n \times m$ matrix B, show that

$$\det \begin{bmatrix} 0 & A \\ -B & I \end{bmatrix} = \det(AB).$$

2.20. Find the area of the triangle with vertices at $(0,0)$, $(1,3)$ and $(3,1)$ in $\mathbb{R}^2$.

2.21. Find the area of the triangle with vertices at $(0,0,0)$, $(1,1,2)$ and $(2,2,1)$ in $\mathbb{R}^3$.

2.22. For $A, B, C, D \in M_{n\times n}(\mathbb{R})$, show that $\det \begin{bmatrix} A & B \\ 0 & D \end{bmatrix} = \det A \det D$. But, in general, $\det \begin{bmatrix} A & B \\ C & D \end{bmatrix} \neq \det A \det D - \det B \det C$.

2.23. Determine whether or not the following statements are true in general, and justify your answers.

(1) For any square matrices A and B of the same size, $\det(A+B) = \det A + \det B$.

(2) For any square matrices A and B of the same size, $\det(AB) = \det(BA)$.

(3) If A is an $n \times n$ square matrix, then for any scalar c, $\det(cI_n - A) = c^n - \det A$.

(4) If A is an $n \times n$ square matrix, then for any scalar c, $\det(cI_n - A^T) = \det(cI_n - A)$.

(5) If E is an elementary matrix, then $\det E = \pm 1$.

(6) There is no matrix A of order 3 such that $A^2 = -I_3$.

(7) Let A be a nilpotent matrix, *i.e.*, $A^k = \mathbf{0}$ for some natural number k. Then $\det A = 0$.

(8) $\det(kA) = k \det A$ for any square matrix A.

(9) Any system $A\mathbf{x} = \mathbf{b}$ has a solution if and only if $\det A \neq 0$.

(10) For any $n \times 1$, $n \geq 2$, column vectors $\mathbf{u}$ and $\mathbf{v}$, $\det(\mathbf{u}\mathbf{v}^T) = 0$.

(11) If A is a square matrix with $\det A = 1$, then $\mathrm{adj}(\mathrm{adj}A) = A$.

(12) If the entries of A are all integers and $\det A = 1$ or -1, then the entries of A^{-1} are also integers.
(13) If the entries of A are 0's or 1's, then $\det A = 1$, 0, or -1.
(14) Every system of n linear equations in n unknowns can be solved by Cramer's rule.
(15) If A is a permutation matrix, then $A^T = A$.

Chapter 3

Vector Spaces

3.1 Vector spaces and subspaces

We discussed how to solve a system $A\mathbf{x} = \mathbf{b}$ of linear equations, and we saw that the basic questions of the existence or uniqueness of the solution were much easier to answer after Gaussian-elimination. In this chapter, we introduce the notion of a vector space, which is an abstraction of the usual algebraic structures of the 3-space $\mathbb{R}^3$ and then elaborate our study of a system of linear equations to this framework.

Usually, many physical quantities, such as length, area, mass, temperature are described by real numbers as magnitudes. Other physical quantities like force or velocity have directions as well as magnitudes. Such quantities with direction are called **vectors**, while the numbers are called **scalars**. For instance, an element (or a point) $\mathbf{x}$ in the 3-space $\mathbb{R}^3$ is usually represented as a triple of real numbers:

$$\mathbf{x} = (x_1,\ x_2,\ x_3),$$

where $x_i \in \mathbb{R}$, $i = 1,\ 2,\ 3$, are called the **coordinates** of $\mathbf{x}$. This expression provides a rectangular coordinate system in a natural way. On the other hand, pictorially such a point in the 3-space $\mathbb{R}^3$ can also be represented by an arrow from the origin to $\mathbf{x}$. In this way, a point in the 3-space $\mathbb{R}^3$ can be understood as a vector. The direction of the arrow specifies the direction of the vector, and the length of the arrow describes its magnitude.

In order to have a more general definition of vectors, we extract the most basic properties of those arrows in $\mathbb{R}^3$. Note that for all vectors (or points) in $\mathbb{R}^3$, there are two algebraic operations: the addition of any two vectors and scalar multiplication of a vector by a scalar. That is, for two vectors $\mathbf{x} = (x_1, x_2, x_3)$, $\mathbf{y} = (y_1, y_2, y_3)$ in $\mathbb{R}^3$ and k a scalar, we define

$$\begin{aligned} \mathbf{x}+\mathbf{y} &= (x_1+y_1, x_2+y_2, x_3+y_3), \\ k\mathbf{x} &= (kx_1, kx_2, kx_3). \end{aligned}$$

The addition of vectors and scalar multiplication of a vector in the 3-space $\mathbb{R}^3$ are illustrated as follows:

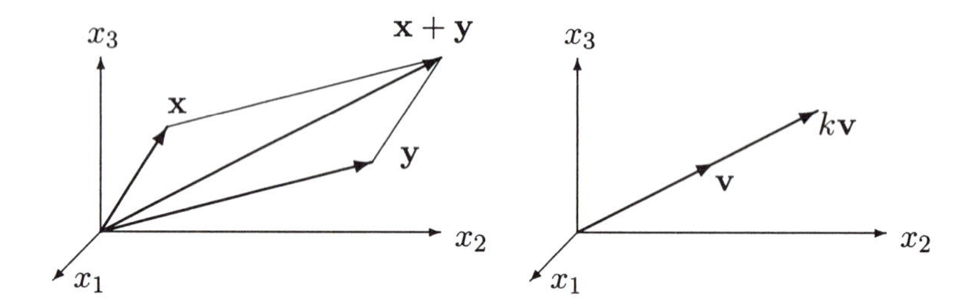

Even though our geometric visualization of vectors does not go beyond the 3-space $\mathbb{R}^3$, it is possible to extend the above algebraic operations of vectors in the 3-space $\mathbb{R}^3$ to the general **n-space** $\mathbb{R}^n$ for any positive integer n. It is defined to be the set of all ordered n-tuples $(a_1, a_2, \ldots, a_n)$ of real numbers, called *vectors*: *i.e.*,

$$\mathbb{R}^n = \{(a_1, a_2, \ldots, a_n) : a_i \in \mathbb{R}, i = 1, 2, \ldots, n\}.$$

For any two vectors $\mathbf{x} = (x_1, x_2, \ldots, x_n)$ and $\mathbf{y} = (y_1, y_2, \ldots, y_n)$ in the n-space $\mathbb{R}^n$, and a scalar k, the *sum* $\mathbf{x}+\mathbf{y}$ and the *scalar multiplication* $k\mathbf{x}$ of them are vectors in $\mathbb{R}^n$ defined by

$$\begin{aligned} \mathbf{x}+\mathbf{y} &= (x_1+y_1, x_2+y_2, \ldots, x_n+y_n), \\ k\mathbf{x} &= (kx_1, kx_2, \ldots, kx_n). \end{aligned}$$

It is easy to verify the following list of arithmetical rules of the operations:

Theorem 3.1 *For any scalars k and ℓ, and vectors $\mathbf{x} = (x_1, x_2, \ldots, x_n)$, $\mathbf{y} = (y_1, y_2, \ldots, y_n)$, and $\mathbf{z} = (z_1, z_2, \ldots, z_n)$ in the n-space $\mathbb{R}^n$, the following rules hold:*

(1) $\mathbf{x}+\mathbf{y} = \mathbf{y}+\mathbf{x}$,

(2) $\mathbf{x}+(\mathbf{y}+\mathbf{z}) = (\mathbf{x}+\mathbf{y})+\mathbf{z}$,

(3) $\mathbf{x}+\mathbf{0} = \mathbf{x} = \mathbf{0}+\mathbf{x}$,

(4) $\mathbf{x} + (-1)\mathbf{x} = \mathbf{0}$,

(5) $k(\mathbf{x} + \mathbf{y}) = k\mathbf{x} + k\mathbf{y}$,

(6) $(k + \ell)\mathbf{x} = k\mathbf{x} + \ell\mathbf{x}$,

(7) $k(\ell\mathbf{x}) = (k\ell)\mathbf{x}$,

(8) $1\mathbf{x} = \mathbf{x}$,

where $\mathbf{0} = (0, 0, \ldots, 0)$ *is the* **zero** *vector.*

We usually write a vector $(a_1, a_2, \ldots, a_n)$ in the n-space $\mathbb{R}^n$ as an $n \times 1$ column matrix

$$(a_1, a_2, \ldots, a_n) = \begin{bmatrix} a_1 \\ a_2 \\ \vdots \\ a_n \end{bmatrix} = [a_1 \ a_2 \ \cdots \ a_n]^T,$$

also called a *column vector*. Then the two operations of the matrix sum and the scalar multiplication of column matrices coincide with those of vectors in $\mathbb{R}^n$, and the above theorem is just Theorem 1.2.

These rules of arithmetic of vectors are the most important ones because they are the only rules that we need to manipulate vectors in the n-space $\mathbb{R}^n$. Hence, an (abstract) vector space can be defined with respect to these rules of operations of vectors in the n-space $\mathbb{R}^n$ so that $\mathbb{R}^n$ itself becomes a vector space. In general, a *vector space* is defined to be a set with two operations: an addition and a scalar multiplication which satisfy the above rules of operations in $\mathbb{R}^n$.

Definition 3.1 A (real) **vector space** is a nonempty set V of elements, called **vectors**, with two algebraic operations that satisfy the following rules.

(A) There is an operation called *vector addition* that associates to every pair $\mathbf{x}$ and $\mathbf{y}$ of vectors in V a unique vector $\mathbf{x} + \mathbf{y}$ in V, called the the **sum** of $\mathbf{x}$ and $\mathbf{y}$, so that the following rules hold for all vectors $\mathbf{x}, \mathbf{y}, \mathbf{z}$ in V:

(1) $\mathbf{x} + \mathbf{y} = \mathbf{y} + \mathbf{x}$ (commutativity in addition),

(2) $\mathbf{x} + (\mathbf{y}+\mathbf{z}) = (\mathbf{x}+\mathbf{y}) + \mathbf{z} (= \mathbf{x} + \mathbf{y} + \mathbf{z})$ (associativity in addition),

(3) there is a unique vector $\mathbf{0}$ in V such that $\mathbf{x} + \mathbf{0} = \mathbf{x} = \mathbf{0} + \mathbf{x}$ for all $\mathbf{x} \in V$ (it is called the **zero vector**),

(4) for any $\mathbf{x} \in V$, there is a vector $-\mathbf{x} \in V$, called the **negative** of $\mathbf{x}$, such that $\mathbf{x} + (-\mathbf{x}) = (-\mathbf{x}) + \mathbf{x} = \mathbf{0}$.

(B) There is an operation called **scalar multiplication** that associates to each vector $\mathbf{x}$ in V and each scalar k a unique vector $k\mathbf{x}$ in V called the **multiplication** of $\mathbf{x}$ by a (real) scalar k, so that the following rules hold for all vectors $\mathbf{x}$, $\mathbf{y}$, $\mathbf{z}$ in V and all scalars k, ℓ:

(5) $k(\mathbf{x}+\mathbf{y}) = k\mathbf{x}+k\mathbf{y}$ (distributivity with respect to vector addition),

(6) $(k+\ell)\mathbf{x} = k\mathbf{x}+\ell\mathbf{x}$ (distributivity with respect to scalar addition),

(7) $k(\ell\mathbf{x}) = (k\ell)\mathbf{x}$ (associativity in scalar multiplication),

(8) $1\mathbf{x} = \mathbf{x}$.

Clearly, the n-space $\mathbb{R}^n$ is a vector space by Theorem 3.1. A **complex vector space** is obtained if, instead of real numbers, we take complex numbers for scalars. For example, the set $\mathbb{C}^n$ of all ordered n-tuples of complex numbers is a complex vector space. In Chapter 7 we shall discuss complex vector spaces, but until then we will discuss only real vector spaces unless otherwise stated.

Example 3.1 **(1)** For any positive integer m and n, the set $M_{m\times n}(\mathbb{R})$ of all $m \times n$ matrices forms a vector space under the matrix sum and scalar multiplication defined in Section 1.3. The zero vector in this space is the zero matrix $\mathbf{0}_{m\times n}$, and $-A$ is the negative of a matrix A.

(2) Let $C(\mathbb{R})$ denote the set of real-valued continuous functions defined on the real line $\mathbb{R}$. For two functions $f(x)$ and $g(x)$, and a real number k, the sum $f+g$ and the scalar multiplication kf of them are defined by

$$\begin{aligned}(f+g)(x) &= f(x)+g(x),\\ (kf)(x) &= kf(x).\end{aligned}$$

Then one can easily verify, as an exercise, that the set $C(\mathbb{R})$ is a vector space under these operations. The zero vector in this space is the constant function whose value at each point is zero.

(3) Let A be an $m \times n$ matrix. Then it is easy to show that the set of solutions of the homogeneous system $A\mathbf{x} = \mathbf{0}$ is a vector space (under the sum and scalar multiplication of matrices).

Theorem 3.2 *Let V be a vector space and let $\mathbf{x}$, $\mathbf{y}$ be vectors in V. Then*

(1) $\mathbf{x}+\mathbf{y} = \mathbf{y}$ *implies* $\mathbf{x} = \mathbf{0}$,

(2) $0\mathbf{x} = \mathbf{0}$,

(3) $k\mathbf{0} = \mathbf{0}$ *for any* $k \in \mathbb{R}$,

(4) $-\mathbf{x}$ *is unique and* $-\mathbf{x} = (-1)\mathbf{x}$,

(5) *if* $k\mathbf{x} = \mathbf{0}$, *then* $k = 0$ *or* $\mathbf{x} = \mathbf{0}$.

Proof: (1) By adding $-\mathbf{y}$ to both sides of $\mathbf{x} + \mathbf{y} = \mathbf{y}$, we have

$$\mathbf{x} = \mathbf{x} + \mathbf{0} = \mathbf{x} + \mathbf{y} + (-\mathbf{y}) = \mathbf{y} + (-\mathbf{y}) = \mathbf{0}.$$

(2) $0\mathbf{x} = (0 + 0)\mathbf{x} = 0\mathbf{x} + 0\mathbf{x}$ implies $0\mathbf{x} = \mathbf{0}$ by (1).

(3) This is an easy exercise.

(4) The uniqueness of the negative $-\mathbf{x}$ of $\mathbf{x}$ can be shown by a simple modification of Lemma 1.6. In fact, if $\bar{\mathbf{x}}$ is another negative of $\mathbf{x}$ such that $\mathbf{x} + \bar{\mathbf{x}} = \mathbf{0}$, then

$$-\mathbf{x} = -\mathbf{x} + \mathbf{0} = -\mathbf{x} + (\mathbf{x} + \bar{\mathbf{x}}) = (-\mathbf{x} + \mathbf{x}) + \bar{\mathbf{x}} = \mathbf{0} + \bar{\mathbf{x}} = \bar{\mathbf{x}}.$$

On the other hand, the equation

$$\mathbf{x} + (-1)\mathbf{x} = 1\mathbf{x} + (-1)\mathbf{x} = (1 - 1)\mathbf{x} = 0\mathbf{x} = \mathbf{0}$$

shows that $(-1)\mathbf{x}$ is another negative of $\mathbf{x}$, and hence $-\mathbf{x} = (-1)\mathbf{x}$ by the uniqueness of $-\mathbf{x}$.

(5) Suppose $k\mathbf{x} = \mathbf{0}$ and $k \neq 0$. Then $\mathbf{x} = 1\mathbf{x} = \frac{1}{k}(k\mathbf{x}) = \frac{1}{k}\mathbf{0} = \mathbf{0}$. □

Problem 3.1 Let V be the set of all pairs (x, y) of real numbers. Suppose that an addition and scalar multiplication of pairs are defined by

$$(x, y) + (u, v) = (x + 2u, y + 2v), \quad k(x, y) = (kx, ky).$$

Is the set V a vector space under those operations? Justify your answer.

A subset W of a vector space V is called a **subspace** of V if W is itself a vector space under the addition and scalar multiplication defined in V. Usually, in order to show that a subset W is a subspace, it is not necessary to verify all the rules of the definition of a vector space, because certain rules satisfied in the larger space V are automatically satisfied in every subset, if vector addition and scalar multiplication are closed in subset.

Theorem 3.3 *A nonempty subset* W *of a vector space* V *is a subspace if and only if* $\mathbf{x} + \mathbf{y}$ *and* $k\mathbf{x}$ *are contained in* W *(or equivalently,* $\mathbf{x} + k\mathbf{y} \in W$*) for any vectors* $\mathbf{x}$ *and* $\mathbf{y}$ *in* W *and any scalar* $k \in \mathbb{R}$.

Proof: We need only to prove the sufficiency. Assume both conditions hold and let $\mathbf{x}$ be any vector in W. Since W is closed under scalar multiplication, $\mathbf{0} = 0\mathbf{x}$ and $-\mathbf{x} = (-1)\mathbf{x}$ are in W, so rules (3) and (4) for a vector space hold. All the other rules for a vector space are clear. □

A vector space V itself and the zero vector $\{\mathbf{0}\}$ are trivially subspaces. Some nontrivial subspaces are given in the following examples.

Example 3.2 Let $W = \{(x,\ y,\ z) \in \mathbb{R}^3\ :\ ax + by + cz = 0\}$, where a, b, c are constants. If $\mathbf{x} = (x_1,\ x_2,\ x_3)$, $\mathbf{y} = (y_1,\ y_2,\ y_3)$ are points in W, then clearly $\mathbf{x} + \mathbf{y} = (x_1 + y_1,\ x_2 + y_2,\ x_3 + y_3)$ is also a point in W, because it satisfies the equation in W. Similarly, $k\mathbf{x}$ also lies in W for any scalar k. Hence, W is a subspace of $\mathbb{R}^3$ and is a plane passing through the origin in $\mathbb{R}^3$.

Example 3.3 Let A be an $m \times n$ matrix. Then, as we have seen in Example 3.1 (3), the set

$$W = \{\mathbf{x} \in \mathbb{R}^n\ :\ A\mathbf{x} = \mathbf{0}\}$$

of solutions of the homogeneous system $A\mathbf{x} = \mathbf{0}$ is a vector space. Moreover, since the operations in W and in $\mathbb{R}^n$ coincide, W is a subspace of $\mathbb{R}^n$.

Example 3.4 For a nonnegative integer n, let $P_n(\mathbb{R})$ denote the set of all real polynomials in x with degree $\leq n$. Then $P_n(\mathbb{R})$ is a subspace of the vector space $C(\mathbb{R})$ of all continuous functions on $\mathbb{R}$.

Example 3.5 Let W be the set of all $n \times n$ real symmetric matrices. Then W is a subspace of the vector space $M_{n\times n}(\mathbb{R})$ of all $n \times n$ matrices, because the sum of two symmetric matrices is symmetric and a scalar multiplication of a symmetric matrix is also symmetric. Similarly, the set of all $n \times n$ skew-symmetric matrices is also a subspace of $M_{n\times n}(\mathbb{R})$.

Problem 3.2 Which of the following sets are subspaces of the 3-space $\mathbb{R}^3$? Justify your answer.

(1) $W = \{(x,\ y,\ z) \in \mathbb{R}^3\ :\ xyz = 0\}$,
(2) $W = \{(2t,\ 3t,\ 4t) \in \mathbb{R}^3\ :\ t \in \mathbb{R}\}$,
(3) $W = \{(x,\ y,\ z) \in \mathbb{R}^3\ :\ x^2 + y^2 - z^2 = 0\}$,
(4) $W = \{\mathbf{x} \in \mathbb{R}^3\ :\ \mathbf{x}^T\mathbf{u} = \mathbf{0} = \mathbf{x}^T\mathbf{v}\}$, where $\mathbf{u}$ and $\mathbf{v}$ are any two fixed nonzero vectors in $\mathbb{R}^3$.

Can you describe all subspaces of the 3-space $\mathbb{R}^3$?

Problem 3.3 Let $V = C(\mathbb{R})$ be the vector space of all continuous functions on $\mathbb{R}$. Which of the following sets W are subspaces of V? Justify your answer.

(1) W is the set of all differentiable functions on $\mathbb{R}$.

(2) W is the set of all bounded continuous functions on $\mathbb{R}$.

(3) W is the set of all continuous nonnegative-valued functions on $\mathbb{R}$, *i.e.*, $f(x) \geq 0$ for any $x \in \mathbb{R}$.

(4) W is the set of all continuous odd functions on $\mathbb{R}$, *i.e.*, $f(-x) = -f(x)$ for any $x \in \mathbb{R}$.

(5) W is the set of all polynomials with integer coefficients.

3.2 Bases

Recall that any vector in the 3-space $\mathbb{R}^3$ is of the form $(x_1,\ x_2,\ x_3)$ which can alsobe written as

$$(x_1,\ x_2,\ x_3) = x_1(1,\ 0,\ 0) + x_2(0,\ 1,\ 0) + x_3(0,\ 0,\ 1).$$

That is, any vector in $\mathbb{R}^3$ can be expressed as the sum of scalar multiples of $\mathbf{e}_1 = (1,\ 0,\ 0)$, $\mathbf{e}_2 = (0,\ 1,\ 0)$ and $\mathbf{e}_3 = (0,\ 0,\ 1)$, which are also denoted by **i**, **j** and **k**, respectively. The following definition gives a name to such expressions.

Definition 3.2 Let V be a vector space, and let $\{\mathbf{x}_1,\ \mathbf{x}_2,\ \ldots,\ \mathbf{x}_m\}$ be a set of vectors in V. Then a vector $\mathbf{y}$ in V of the form

$$\mathbf{y} = a_1\mathbf{x}_1 + a_2\mathbf{x}_2 + \cdots + a_m\mathbf{x}_m,$$

where $a_1,\ \ldots,\ a_m$ are scalars, is called a **linear combination** of the vectors $\mathbf{x}_1,\ \mathbf{x}_2,\ \ldots,\ \mathbf{x}_m$.

The next theorem shows that the set of all linear combinations of a finite set of vectors in a vector space forms a subspace.

Theorem 3.4 *Let* $\mathbf{x}_1,\ \mathbf{x}_2,\ \ldots,\ \mathbf{x}_m$ *be vectors in a vector space* V. *Then the set* $W = \{a_1\mathbf{x}_1 + a_2\mathbf{x}_2 + \cdots + a_m\mathbf{x}_m\ :\ a_i \in \mathbb{R}\}$ *of all linear combinations of* $\mathbf{x}_1,\ \mathbf{x}_2,\ \ldots,\ \mathbf{x}_m$ *is a subspace of* V *called the* **subspace of V spanned** *by* $\mathbf{x}_1,\ \mathbf{x}_2,\ \ldots,\ \mathbf{x}_m$.

Proof: We want to show that W is closed under addition and scalar multiplication. Let $\mathbf{u}$ and $\mathbf{w}$ be any two vectors in W. Then

$$\begin{array}{rcl} \mathbf{u} & = & a_1\mathbf{x}_1 + a_2\mathbf{x}_2 + \cdots + a_m\mathbf{x}_m, \\ \mathbf{w} & = & b_1\mathbf{x}_1 + b_2\mathbf{x}_2 + \cdots + b_m\mathbf{x}_m \end{array}$$

for some scalars a_i's and b_i's. Therefore,

$$\mathbf{u} + \mathbf{w} = (a_1 + b_1)\mathbf{x}_1 + (a_2 + b_2)\mathbf{x}_2 + \cdots + (a_m + b_m)\mathbf{x}_m$$

and, for any scalar k,

$$k\mathbf{u} = (ka_1)\mathbf{x}_1 + (ka_2)\mathbf{x}_2 + \cdots + (ka_m)\mathbf{x}_m.$$

Thus, $\mathbf{u} + \mathbf{w}$ and $k\mathbf{u}$ are linear combinations of $\mathbf{x}_1$, $\mathbf{x}_2$, $\ldots$, $\mathbf{x}_m$ and consequently contained in W. Therefore, W is a subspace of V. □

Suppose that $\{\mathbf{x}_1, \mathbf{x}_2, \ldots, \mathbf{x}_m\}$ is any set of m vectors in a vector space V. If any vector in V can be written as a linear combination of these vector $\mathbf{x}_i$'s, we say that it is a **spanning set** of V.

Example 3.6 (1) For a nonzero vector $\mathbf{v}$ in a vector space V, linear combinations of $\mathbf{v}$ are simply scalar multiples of $\mathbf{v}$. Thus the subspace W of V spanned by $\mathbf{v}$ is $W = \{k\mathbf{v} : k \in \mathbb{R}\}$.

(2) Consider three vectors $\mathbf{v}_1 = (1, 1, 1)$, $\mathbf{v}_2 = (1, -1, 1)$ and $\mathbf{v}_3 = (1, 0, 1)$ in $\mathbb{R}^3$. The subspace W_1 spanned by $\mathbf{v}_1$ and $\mathbf{v}_2$ is written as

$$W_1 = \{a_1\mathbf{v}_1 + a_2\mathbf{v}_2 = (a_1 + a_2, a_1 - a_2, a_1 + a_2) : a_i \in \mathbb{R}\},$$

and the subspace W_2 spanned by $\mathbf{v}_1$, $\mathbf{v}_2$ and $\mathbf{v}_3$ is written as

$$W_2 = \{a_1\mathbf{v}_1 + a_2\mathbf{v}_2 + a_3\mathbf{v}_3 = (a_1 + a_2 + a_3, a_1 - a_2, a_1 + a_2 + a_3) : a_i \in \mathbb{R}\}.$$

Then $a_1\mathbf{v}_1 + a_2\mathbf{v}_2 = a_1\mathbf{v}_1 + a_2\mathbf{v}_2 + 0\mathbf{v}_3$ implies $W_1 \subseteq W_2$. On the other hand, any vector in W_2 is of the form $a_1\mathbf{v}_1 + a_2\mathbf{v}_2 + a_3\mathbf{v}_3$. But, since $\mathbf{v}_3 = \frac{1}{2}(\mathbf{v}_1 + \mathbf{v}_2)$, this can be rewritten as $c_1\mathbf{v}_1 + c_2\mathbf{v}_2$. This means that $W_2 \subseteq W_1$, thus $W_1 = W_2$ which is a plane in $\mathbb{R}^3$ containing the vectors $\mathbf{v}_1$, $\mathbf{v}_2$ and $\mathbf{v}_3$. In general, a subspace in a vector space can have many different spanning sets. □

Example 3.7 Let

$$\begin{aligned} \mathbf{e}_1 &= (1,\ 0,\ 0,\ \ldots,\ 0), \\ \mathbf{e}_2 &= (0,\ 1,\ 0,\ \ldots,\ 0), \\ &\vdots \\ \mathbf{e}_n &= (0,\ 0,\ 0,\ \ldots,\ 1) \end{aligned}$$

be n vectors in the n-space $\mathbb{R}^n$ ($n \geq 3$). Then a linear combination of $\mathbf{e}_1,\ \mathbf{e}_2,\ \mathbf{e}_3$ is of the form

$$a_1\mathbf{e}_1 + a_2\mathbf{e}_2 + a_3\mathbf{e}_3 = (a_1,\ a_2,\ a_3,\ 0,\ \ldots,\ 0).$$

Hence, the set

$$W = \{(a_1,\ a_2,\ a_3,\ 0,\ \ldots,\ 0) \in \mathbb{R}^n\ :\ a_1,\ a_2,\ a_3 \in \mathbb{R}\}$$

is the subspace of the n-space $\mathbb{R}^n$ spanned by the vectors $\mathbf{e}_1,\ \mathbf{e}_2,\ \mathbf{e}_3$. Note that the subspace W can be identified with the 3-space $\mathbb{R}^3$ through the identification

$$(a_1,\ a_2,\ a_3,\ 0,\ \ldots,\ 0) \equiv (a_1,\ a_2,\ a_3)$$

with $a_i \in \mathbb{R}$. In general, for $m \leq n$, the m-space $\mathbb{R}^m$ can be identified as a subspace of the n-space $\mathbb{R}^n$. □

Example 3.8 Let $A = [\mathbf{a}_1\ \mathbf{a}_2\ \cdots\ \mathbf{a}_n]$ be an $m \times n$ matrix. Then the column vectors $\mathbf{a}_i$'s are in $\mathbb{R}^m$, and the matrix product $A\mathbf{x}$ represents the linear combination of the column vector $\mathbf{a}_i$'s whose coefficients are the components of $\mathbf{x} \in \mathbb{R}^n$, *i.e.*, $A\mathbf{x} = x_1\mathbf{a}_1 + x_2\mathbf{a}_2 + \cdots + x_n\mathbf{a}_n$. Therefore, the set

$$W = \{A\mathbf{x} \in \mathbb{R}^m\ :\ \mathbf{x} \in \mathbb{R}^n\}$$

of all linear combinations of the column vectors of A is a subspace of $\mathbb{R}^m$ called the **column space** of A. Therefore, $A\mathbf{x} = \mathbf{b}$ *has a solution* $(x_1, \cdots, x_n)$ *in* $\mathbb{R}^n$ *if and only if the vector* $\mathbf{b}$ *belongs to the subspace* W *spanned by the column vectors of* A. □

Problem 3.4 Let $\mathbf{x}_1,\ \mathbf{x}_2,\ \ldots,\ \mathbf{x}_m$ be vectors in a vector space V and let W be the subspace spanned by $\mathbf{x}_1,\ \mathbf{x}_2,\ \ldots,\ \mathbf{x}_m$. Show that W is the smallest subspace of V containing $\mathbf{x}_1,\ \mathbf{x}_2,\ \ldots,\ \mathbf{x}_m$, *i.e.*, if U is a subspace of V containing $\mathbf{x}_1,\ \mathbf{x}_2,\ \ldots,\ \mathbf{x}_m$, then $W \subseteq U$.

Problem 3.5 Show that the set of all matrices of the form $AB - BA$ does not span the vector space $M_{n\times n}(\mathbb{R})$.

As we saw above, any nonempty subset of a vector space V spans a subspace through the linear combinations of the vectors, and a subspace can have many spanning sets with a different number of vectors. This means that a vector can be written as linear combinations in various ways. If one can find a finite number of vectors in V such that any vector in V can be expressed in a unique way as a linear combination of them, then the study of the vector space V might be easier and the computations of vectors may be simplified. Thus, for a fixed set of vectors $\mathbf{x}_1, \mathbf{x}_2, \ldots, \mathbf{x}_m$ in a vector space V, we look at their linear combinations $c_1\mathbf{x}_1 + c_2\mathbf{x}_2 + \cdots + c_m\mathbf{x}_m$ and see whether any vector in V can be written in this form in exactly one way. This problem can be rephrased as to whether or not a nontrivial linear combination produces the zero vector, while the trivial combination, with all scalars $c_i = 0$, obviously produces the zero vector.

Definition 3.3 A set of vectors $\{\mathbf{x}_1, \mathbf{x}_2, \ldots, \mathbf{x}_m\}$ in a vector space V is said to be **linearly independent** if the vector equation, called the **linear dependence** of $\mathbf{x}_i$'s,

$$c_1\mathbf{x}_1 + c_2\mathbf{x}_2 + \cdots + c_m\mathbf{x}_m = \mathbf{0}$$

has only the trivial solution $c_1 = c_2 = \cdots = c_m = 0$. Otherwise, it is said to be **linearly dependent**.

Therefore, a set of vectors $\{\mathbf{x}_1, \mathbf{x}_2, \ldots, \mathbf{x}_m\}$ is linearly dependent if and only if there is a linear dependence

$$c_1\mathbf{x}_1 + c_2\mathbf{x}_2 + \cdots + c_m\mathbf{x}_m = \mathbf{0}$$

with a nontrivial solution $(c_1, c_2, \ldots, c_m)$. In this case, we may assume that $c_m \neq 0$. Then the equation can be rewritten as

$$\mathbf{x}_m = -\frac{c_1}{c_m}\mathbf{x}_1 - \frac{c_2}{c_m}\mathbf{x}_2 - \cdots - \frac{c_{m-1}}{c_m}\mathbf{x}_{m-1}.$$

That is, *a set of vectors is linearly dependent if and only if at least one of the vectors in the set can be written as a linear combination of the others.*

Example 3.9 Let $\mathbf{x} = (1, 2, 3)$ and $\mathbf{y} = (3, 2, 1)$ be two vectors in the 3-space $\mathbb{R}^3$. Then clearly $\mathbf{y} \neq \lambda\mathbf{x}$ for any $\lambda \in \mathbb{R}$ (or $a\mathbf{x} + b\mathbf{y} = \mathbf{0}$ only when $a = b = 0$). This means that $\{\mathbf{x}, \mathbf{y}\}$ is linearly independent in $\mathbb{R}^3$. If $\mathbf{w} = (3, 6, 9)$, then $\{\mathbf{x}, \mathbf{w}\}$ is linearly dependent since $\mathbf{w} - 3\mathbf{x} = \mathbf{0}$. In general, if $\mathbf{x}, \mathbf{y}$ are noncollinear vectors in the 3-space $\mathbb{R}^3$, the set of all linear combinations of $\mathbf{x}$ and $\mathbf{y}$ determines a plane W through the origin in $\mathbb{R}^3$, *i.e.*, $W = \{a\mathbf{x} + b\mathbf{y} : a, b \in \mathbb{R}\}$. Let $\mathbf{z}$ be another nonzero vector in the 3-space $\mathbb{R}^3$. If $\mathbf{z} \in W$, then there are some numbers a, $b \in \mathbb{R}$, not all of them are zero, such that $\mathbf{z} = a\mathbf{x} + b\mathbf{y}$, that is, the set $\{\mathbf{x}, \ \mathbf{y}, \ \mathbf{z}\}$ is linearly dependent. If $\mathbf{z} \notin W$, then $a\mathbf{x} + b\mathbf{y} + c\mathbf{z} = \mathbf{0}$ is possible only when $a = b = c = 0$ (prove it). Therefore, the set $\{\mathbf{x}, \ \mathbf{y}, \ \mathbf{z}\}$ is linearly independent if and only if $\mathbf{z}$ does not lie in W. □

By abuse of language, it is sometimes convenient to say that "the vectors $\mathbf{x}_1, \mathbf{x}_2, \ldots, \mathbf{x}_m$ are linearly independent," although this is really a property of a set.

Example 3.10 The columns of the matrix

$$A = \begin{bmatrix} 1 & -2 & -1 & 0 \\ 4 & 2 & 6 & 8 \\ 2 & -1 & 1 & 3 \end{bmatrix}$$

are linearly dependent in the 3-space $\mathbb{R}^3$, since the third column is the sum of the first and the second.

As this example shows, the concept of linear dependence can be applied to the row or column vectors of any matrix.

Example 3.11 Consider an upper triangular matrix

$$A = \begin{bmatrix} 2 & 3 & 5 \\ 0 & 1 & 6 \\ 0 & 0 & 4 \end{bmatrix}.$$

The linear dependence of the column vectors of A may be written as

$$c_1 \begin{bmatrix} 2 \\ 0 \\ 0 \end{bmatrix} + c_2 \begin{bmatrix} 3 \\ 1 \\ 0 \end{bmatrix} + c_3 \begin{bmatrix} 5 \\ 6 \\ 4 \end{bmatrix} = \begin{bmatrix} 0 \\ 0 \\ 0 \end{bmatrix},$$

which, in matrix notation, may be written as a homogeneous system:

$$\begin{bmatrix} 2 & 3 & 5 \\ 0 & 1 & 6 \\ 0 & 0 & 4 \end{bmatrix} \begin{bmatrix} c_1 \\ c_2 \\ c_3 \end{bmatrix} = \begin{bmatrix} 0 \\ 0 \\ 0 \end{bmatrix}.$$

From the third row, $c_3 = 0$, from the second row $c_2 = 0$, and substitution of them into the first row forces $c_1 = 0$, *i.e.*, the homogeneous system has only the trivial solution, so that the column vectors are linearly independent. □

The following theorem can be proven by the same argument.

Theorem 3.5 *The nonzero rows of a matrix of a row-echelon form are linearly independent, and so are the columns that contain leading 1's.*

In particular, the rows of any triangular matrix with nonzero diagonals are linearly independent, and so are the columns.

In general, if $V = \mathbb{R}^m$ and $\mathbf{v}_1, \mathbf{v}_2, \ldots, \mathbf{v}_n$ are n vectors in $\mathbb{R}^m$, then they form an $m \times n$ matrix $A = [\mathbf{v}_1 \ \mathbf{v}_2 \ \cdots \ \mathbf{v}_n]$. On the other hand, Example 3.8 shows that the linear dependence $c_1\mathbf{v}_1 + c_2\mathbf{v}_2 + \cdots + c_n\mathbf{v}_n = \mathbf{0}$ of $\mathbf{v}_i$'s is nothing but the homogeneous equation $A\mathbf{x} = \mathbf{0}$, where $\mathbf{x} = (c_1, c_2, \cdots, c_n)$. Thus, *the column vectors $\mathbf{v}_i$'s of A are linearly independent in $\mathbb{R}^m$ if and only if the homogeneous system $A\mathbf{x} = \mathbf{0}$ has only the trivial solution,* and *they are linearly dependent if and only if $A\mathbf{x} = \mathbf{0}$ has a nontrivial solution.*

If U is the reduced row-echelon form of A, then we know that $A\mathbf{x} = \mathbf{0}$ and $U\mathbf{x} = \mathbf{0}$ have the same set of solutions. Moreover, a homogeneous system $A\mathbf{x} = \mathbf{0}$ with unknowns more than the number of equations always has a nontrivial solution (see the remark on page 11). This proves the following lemma.

Lemma 3.6 **(1)** *Any set of n vectors in the m-space $\mathbb{R}^m$ is linearly dependent if $n > m$.*

(2) *If U is the reduced row-echelon form of A, then the columns of U are linearly independent if and only if the columns of A are linearly independent.*

Example 3.12 Consider the vectors $\mathbf{e}_1 = (1, 0, 0)$, $\mathbf{e}_2 = (0, 1, 0)$ and $\mathbf{e}_3 = (0, 0, 1)$ in the 3-space $\mathbb{R}^3$. The vector equation $c_1\mathbf{e}_1 + c_2\mathbf{e}_2 + c_3\mathbf{e}_3 = \mathbf{0}$ becomes

$$c_1(1, 0, 0) + c_2(0, 1, 0) + c_3(0, 0, 1) = (0, 0, 0)$$

or, equivalently, $(c_1, c_2, c_3) = (0, 0, 0)$. Thus, $c_1 = c_2 = c_3 = 0$, so the set of vectors $\{\mathbf{e}_1, \mathbf{e}_2, \mathbf{e}_3\}$ is linearly independent and also spans $\mathbb{R}^3$.

Example 3.13 In general, it is quite clear that the vectors $\mathbf{e}_1, \mathbf{e}_2, \ldots, \mathbf{e}_n$ in $\mathbb{R}^n$ are linearly independent. Moreover, they span the n-space $\mathbb{R}^n$: In fact, when we write a vector $\mathbf{x} \in \mathbb{R}^n$ as $(x_1, \ldots, x_n)$, it means just the linear combination of the vector $\mathbf{e}_i$'s:

$$\mathbf{x} = (x_1, \ldots, x_n) = x_1\mathbf{e}_1 + \cdots + x_n\mathbf{e}_n.$$

However, if any one of the $\mathbf{e}_i$'s is missed, then they cannot span $\mathbb{R}^n$. Thus, this kind of vector plays a special role in the vector space.

Definition 3.4 Let V be a vector space. A **basis** for V is a set of linearly independent vectors that spans V.

For example, as we saw in Example 3.13, the set $\{\mathbf{e}_1, \mathbf{e}_2, \ldots, \mathbf{e}_n\}$ forms a basis, called the **standard basis** for the n-space $\mathbb{R}^n$. Of course, there are many other bases for $\mathbb{R}^n$.

Example 3.14 **(1)** The set of vectors $(1,1,0)$, $(0,-1,1)$, and $(1,0,1)$ is not a basis for the 3-space $\mathbb{R}^3$, since this set is linearly dependent (the third is the sum of the first two vectors), and cannot span $\mathbb{R}^3$. (The vector $(1,0,0)$ cannot be obtained as a linear combination of them (prove it).) This set does not have enough vectors spanning $\mathbb{R}^3$.

(2) The set of vectors $(1,0,0)$, $(0,1,1)$, $(1,0,1)$ and $(0,1,0)$ is not a basis either, since they are not linearly independent (the sum of the first two minus the third makes the fourth) even though they span $\mathbb{R}^3$. This set of vectors has some redundant vectors spanning $\mathbb{R}^3$.

(3) The set of vectors $(1,1,1)$, $(0,1,1)$, and $(0,0,1)$ is linearly independent and also spans $\mathbb{R}^3$. That is, it is a basis for $\mathbb{R}^3$, different from the standard basis. This set has the proper number of vectors spanning $\mathbb{R}^3$, since the set cannot be reduced to a smaller set nor does it need any additional vector spanning $\mathbb{R}^3$. □

By definition, in order to show that a set of vectors in a vector space is a basis, one needs to show two things: *it is linearly independent, and it spans the whole space.* The following theorem shows that a basis for a vector space represents a coordinate system just like the rectangular coordinate system by the standard basis for $\mathbb{R}^n$.

Theorem 3.7 *Let $\alpha = \{\mathbf{v}_1, \mathbf{v}_2, \ldots, \mathbf{v}_n\}$ be a basis for a vector space V. Then each vector $\mathbf{x}$ in V can be uniquely expressed as a linear combination of $\mathbf{v}_1, \mathbf{v}_2, \ldots, \mathbf{v}_n$, i.e., there are unique scalars a_i's, $i = 1, \ldots, n$, such that*

$$\mathbf{x} = a_1\mathbf{v}_1 + a_2\mathbf{v}_2 + \cdots + a_n\mathbf{v}_n.$$

In this case, the column vector $[a_1\ a_2\ \cdots\ a_n]^T$ is called the **coordinate vector** of $\mathbf{x}$ with respect to the basis α, and it is denoted $[\mathbf{x}]_\alpha$.

Proof: If $\mathbf{x}$ can be also expressed as $\mathbf{x} = b_1\mathbf{v}_1 + b_2\mathbf{v}_2 + \cdots + b_n\mathbf{v}_n$, then we have $\mathbf{0} = (a_1 - b_1)\mathbf{v}_1 + (a_2 - b_2)\mathbf{v}_2 + \cdots + (a_n - b_n)\mathbf{v}_n$. By the linear independence of $\mathbf{x}_i$'s, $a_i = b_i$ for all $i = 1, \ldots, n$. □

Example 3.15 Let $\alpha = \{\mathbf{e}_1, \mathbf{e}_2, \mathbf{e}_3\}$ be the standard basis for $\mathbb{R}^3$, and let $\beta = \{\mathbf{v}_1, \mathbf{v}_2, \mathbf{v}_3\}$ with $\mathbf{v}_1 = (1, 1, 1) = \mathbf{e}_1 + \mathbf{e}_2 + \mathbf{e}_3$, $\mathbf{v}_2 = (0, 1, 1) = \mathbf{e}_2 + \mathbf{e}_3$, $\mathbf{v}_3 = (0, 0, 1) = \mathbf{e}_3$. Then

$$[\mathbf{v}_1]_\alpha = \begin{bmatrix} 1 \\ 1 \\ 1 \end{bmatrix}, \quad [\mathbf{v}_2]_\alpha = \begin{bmatrix} 0 \\ 1 \\ 1 \end{bmatrix}, \quad [\mathbf{v}_3]_\alpha = \begin{bmatrix} 0 \\ 0 \\ 1 \end{bmatrix},$$

while $[\mathbf{v}_1]_\beta = [1\ 0\ 0]^T$, $[\mathbf{v}_2]_\beta = [0\ 1\ 0]^T$, $[\mathbf{v}_3]_\beta = [0\ 0\ 1]^T$.

Problem 3.6 Show that the vectors $\mathbf{v}_1 = (1,\ 2,\ 1)$, $\mathbf{v}_2 = (2,\ 9,\ 0)$ and $\mathbf{v}_3 = (3,\ 3,\ 4)$ in the 3-space $\mathbb{R}^3$ form a basis.

Problem 3.7 Show that the set $\{1,\ x,\ x^2,\ \ldots,\ x^n\}$ is a basis for $P_n(\mathbb{R})$, the vector space of all polynomials of degree $\leq n$ with real coefficients.

Problem 3.8 In the n-space $\mathbb{R}^n$, determine whether or not the set

$$\{\mathbf{e}_1 - \mathbf{e}_2,\ \mathbf{e}_2 - \mathbf{e}_3,\ \ldots,\ \mathbf{e}_{n-1} - \mathbf{e}_n,\ \mathbf{e}_n - \mathbf{e}_1\}$$

is linearly dependent.

Problem 3.9 Let $\mathbf{x}_k$ denote the vector in $\mathbb{R}^n$ whose first $k-1$ coordinates are zero and whose last $n-k+1$ coordinates are 1. Show that the set $\{\mathbf{x}_1,\ \mathbf{x}_2,\ \ldots,\ \mathbf{x}_n\}$ is a basis for $\mathbb{R}^n$.

3.3 Dimensions

We often say that the line $\mathbb{R}^1$ is one-dimensional, the plane $\mathbb{R}^2$ is two-dimensional and the space $\mathbb{R}^3$ is three-dimensional, etc. This is mostly due to the fact that the freedom in choosing coordinates for each element in the space is 1, 2 or 3, respectively. This means that the concept of *dimension* is closely related to the concept of bases. Note that for a vector space in general there is no unique way to choose a basis. However, there is something common to all bases, and this is related to the notion of dimension. We first need the following important lemma from which one can define the dimension of a vector space.

Lemma 3.8 *Let V be a vector space and let $\alpha = \{\mathbf{x}_1,\ \mathbf{x}_2,\ \ldots,\ \mathbf{x}_m\}$ be a set of m-vectors in V.*

(1) *If α spans V, then every set of vectors with more than m vectors cannot be linearly independent.*

(2) *If α is linearly independent, then any set of vectors with fewer than m vectors cannot span V.*

Proof: Since (2) follows from (1) directly, we prove only (1). Let $\beta = \{\mathbf{y}_1, \mathbf{y}_2, \ldots, \mathbf{y}_n\}$ be a set of n-vectors in V with $n > m$. We will show that β is linearly dependent. Indeed, since each vector $\mathbf{y}_j$ is a linear combination of the vectors in the spanning set α, *i.e.*, for $j = 1, \ldots, n$,

$$\mathbf{y}_j = a_{1j}\mathbf{x}_1 + a_{2j}\mathbf{x}_2 + \cdots + a_{mj}\mathbf{x}_m = \sum_{i=1}^{m} a_{ij}\mathbf{x}_i,$$

we have

$$\begin{aligned}
c_1\mathbf{y}_1 + c_2\mathbf{y}_2 + \cdots + c_n\mathbf{y}_n &= c_1(a_{11}\mathbf{x}_1 + a_{21}\mathbf{x}_2 + \cdots + a_{m1}\mathbf{x}_m) \\
&\quad + c_2(a_{12}\mathbf{x}_1 + a_{22}\mathbf{x}_2 + \cdots + a_{m2}\mathbf{x}_m) \\
&\quad\ \vdots \\
&\quad + c_n(a_{1n}\mathbf{x}_1 + a_{2n}\mathbf{x}_2 + \cdots + a_{mn}\mathbf{x}_m) \\
&= (a_{11}c_1 + a_{12}c_2 + \cdots + a_{1n}c_n)\mathbf{x}_1 \\
&\quad + (a_{21}c_1 + a_{22}c_2 + \cdots + a_{2n}c_n)\mathbf{x}_2 \\
&\quad\ \vdots \\
&\quad + (a_{m1}c_1 + a_{m2}c_2 + \cdots + a_{mn}c_n)\mathbf{x}_m.
\end{aligned}$$

Thus, β is linearly dependent if and only if the system of linear equations

$$c_1\mathbf{y}_1 + c_2\mathbf{y}_2 + \cdots + c_n\mathbf{y}_n = \mathbf{0}$$

has a nontrivial solution $(c_1, c_2, \ldots, c_n) \neq (0, 0, \cdots, 0)$. This is true if all the coefficients of $\mathbf{x}_i$'s are zero but not all of c_i's are zero. This means that the homogeneous system of linear equations in c_i's,

$$\begin{bmatrix} a_{11} & a_{12} & \cdots & a_{1n} \\ a_{21} & a_{22} & \cdots & a_{2n} \\ \vdots & \vdots & & \vdots \\ a_{m1} & a_{m2} & \cdots & a_{mn} \end{bmatrix} \begin{bmatrix} c_1 \\ c_2 \\ \vdots \\ c_n \end{bmatrix} = \begin{bmatrix} 0 \\ 0 \\ \vdots \\ 0 \end{bmatrix},$$

has a nontrivial solution. This is guaranteed by Lemma 3.6, since A is an $m \times n$ matrix with $m < n$. □

It is clear by Lemma 3.8 that if a set $\alpha = \{\mathbf{x}_1, \mathbf{x}_2, \ldots, \mathbf{x}_n\}$ of n vectors is a basis for a vector space V, then no other set $\beta = \{\mathbf{y}_1, \mathbf{y}_2, \ldots, \mathbf{y}_r\}$ of r vectors can be a basis for V if $r \neq n$. This means that all bases for a vector space V have the same number of vectors, even if there are many different bases for a vector space. Therefore, we obtain the following important result:

Theorem 3.9 *If a basis for a vector space V consists of n vectors, then so does every other basis.*

Definition 3.5 The **dimension** of a vector space V is the number, say n, of vectors in a basis for V, denoted by $\dim V = n$. When V has a basis of a finite number of vectors, V is said to be **finite dimensional**.

Example 3.16 The following can be easily verified:

(1) If V has only the zero vector: $V = \{\mathbf{0}\}$, then $\dim V = 0$.

(2) If $V = \mathbb{R}^n$, then $\dim \mathbb{R}^n = n$, since V has the standard basis $\{\mathbf{e}_1, \mathbf{e}_2, \ldots, \mathbf{e}_n\}$.

(3) If $V = P_n(\mathbb{R})$ of all polynomials of degree less than or equal to n, then $\dim P_n(\mathbb{R}) = n + 1$ since $\{1, x, x^2, \ldots, x^n\}$ is a basis for V.

(4) If $V = M_{m\times n}(\mathbb{R})$ of all $m \times n$ matrices, then $\dim M_{m\times n}(\mathbb{R}) = mn$ since $\{E_{ij} : i = 1, \ldots, m,\ j = 1, \ldots, n\}$ is a basis for V, where E_{ij} is the $m \times n$ matrix whose (i, j)-th entry is 1 and all others are zero. □

If $V = C(\mathbb{R})$ of all real-valued continuous functions defined on the real line, then one can show that V is not finite dimensional. A vector space V is **infinite dimensional** if it is not finite dimensional. In this book, we are concerned only with finite dimensional vector spaces unless otherwise stated.

Theorem 3.10 *Let V be a finite dimensional vector space.*

(1) *Any linearly independent set in V can be extended to a basis by adding more vectors if necessary.*

(2) *Any set of vectors that spans V can be reduced to a basis by discarding vectors if necessary.*

Proof: We prove **(1)** only and leave **(2)** as an exercise. Let $\alpha = \{\mathbf{x}_1, \ldots, \mathbf{x}_k\}$ be a linearly independent set in V. If α spans V, then α is a basis. If α does not span V, then there exists a vector, say $\mathbf{x}_{k+1}$, in V that is not contained in the subspace spanned by the vectors in α. Now $\{\mathbf{x}_1, \ldots, \mathbf{x}_k, \mathbf{x}_{k+1}\}$ is linearly independent (check why). If $\{\mathbf{x}_1, \ldots, \mathbf{x}_k, \mathbf{x}_{k+1}\}$ spans V, then this is a basis for V. If it does not span V, then the same procedure can be repeated, yielding a linearly independent set that spans V, *i.e.*, a basis for V. This procedure must stop in a finite step because of Lemma 3.8 for a finite dimensional vector space V. □

Theorem 3.10 shows that a basis for a vector space V is a set of vectors in V which is *maximally* independent and *minimally* spanning in the above sense. In particular, if W is a subspace of V, then any basis for W is linearly independent also in V, and can be extended to a basis for V. Thus $\dim W \leq \dim V$.

Corollary 3.11 *Let V be a vector space of dimension n. Then*
(1) *any set of n vectors that spans V is a basis for V, and*
(2) *any set of n linearly independent vectors is a basis for V.*

Proof: Again we prove **(1)** only. If a spanning set of n vectors were not linearly independent, then the set would be reduced to a basis that has a smaller number of vectors than n vectors. □

Corollary 3.11 means that if it is known that $\dim V = n$ and if a set of n vectors either is linearly independent or spans V, then it is already a basis for the space V.

Example 3.17 Let W be the subspace of $\mathbb{R}^4$ spanned by the vectors

$$\mathbf{x}_1 = (1,\ -2,\ 5,\ -3),\ \ \mathbf{x}_2 = (0,\ 1,\ 1,\ 4),\ \ \mathbf{x}_3 = (1,\ 0,\ 1,\ 0).$$

Find a basis for W and extend it to a basis for $\mathbb{R}^4$.

Solution: Note that $\dim W \leq 3$ since W is spanned by three vectors $\mathbf{x}_i$'s. Let A be the 3×4 matrix whose rows are $\mathbf{x}_1$, $\mathbf{x}_2$ and $\mathbf{x}_3$:

$$A = \begin{bmatrix} 1 & -2 & 5 & -3 \\ 0 & 1 & 1 & 4 \\ 1 & 0 & 1 & 0 \end{bmatrix}.$$

Reduce A to a row-echelon form:

$$U = \begin{bmatrix} 1 & -2 & 5 & -3 \\ 0 & 1 & 1 & 4 \\ 0 & 0 & 1 & \frac{5}{6} \end{bmatrix}.$$

The three nonzero row vectors of U are clearly linearly independent, and they also span W because the vectors $\mathbf{x}_1$, $\mathbf{x}_2$ and $\mathbf{x}_3$ can be expressed as a linear combination of these three nonzero row vectors of U. Hence, U provides a basis for W. (Note that this implies $\dim W = 3$ and hence $\mathbf{x}_1$, $\mathbf{x}_2$, $\mathbf{x}_3$ is also a basis for W by Corollary 3.11. The linear independence of $\mathbf{x}_i$'s is a by-product of this fact).

To extend this basis, just add any nonzero vector of the form $\mathbf{x}_4 = (0,\ 0,\ 0,\ t)$ to the rows of U to get a basis for the space $\mathbb{R}^4$. □

Problem 3.10 Let W be a subspace of a vector space V. Show that if $\dim W = \dim V$, then $W = V$.

Problem 3.11 Find a basis and the dimension of each of the following subspaces of $M_{n\times n}(\mathbb{R})$ of all $n \times n$ matrices:

(1) the space of all $n \times n$ diagonal matrices whose traces are zero;

(2) the space of all $n \times n$ symmetric matrices;

(3) the space of all $n \times n$ skew-symmetric matrices.

Now consider two subspaces U and W of a vector space V. The **sum** of these subspaces U and W is defined by

$$U + W = \{\mathbf{u} + \mathbf{w} \ : \ \mathbf{u} \in U,\ \mathbf{w} \in W\}.$$

It is not hard to see that this is a subspace of V.

Problem 3.12 Let U and W be subspaces of a vector space V.

(1) Show that $U + W$ is the smallest subspace of V containing U and W.

(2) Prove that $U \cap W$ is also a subspace of V. Is $U \cup W$ a subspace of V? Justify your answer.

Definition 3.6 A vector space V is called the **direct sum** of two subspaces U and W, written $V = U \oplus W$, if $V = U + W$ and $U \cap W = \{\mathbf{0}\}$.

For example, one can easily show that $\mathbb{R}^3 = \mathbb{R}^1 \oplus \mathbb{R}^2 = \mathbb{R}^1 \oplus \mathbb{R}^1 \oplus \mathbb{R}^1$.

Theorem 3.12 *A vector space V is the direct sum of subspaces U and W, i.e., $V = U \oplus W$, if and only if for any $\mathbf{v} \in V$ there exist unique $\mathbf{u} \in U$ and $\mathbf{w} \in W$ such that $\mathbf{v} = \mathbf{u} + \mathbf{w}$.*

Proof: Suppose that $V = U \oplus W$. Then, for any $\mathbf{v} \in V$, there exist vectors $\mathbf{u} \in U$ and $\mathbf{w} \in W$ such that $\mathbf{v} = \mathbf{u} + \mathbf{w}$, since $V = U + W$. To show the uniqueness, suppose that $\mathbf{v}$ is also expressed as a sum $\mathbf{u}' + \mathbf{w}'$ for $\mathbf{u}' \in U$ and $\mathbf{w}' \in W$. Then $\mathbf{u} + \mathbf{w} = \mathbf{u}' + \mathbf{w}'$ implies

$$\mathbf{u} - \mathbf{u}' = \mathbf{w}' - \mathbf{w} \in U \cap W = \{\mathbf{0}\}.$$

Hence, $\mathbf{u} = \mathbf{u}'$ and $\mathbf{w} = \mathbf{w}'$.

Conversely, if there exists a nonzero vector $\mathbf{v}$ in $U \cap W$, then $\mathbf{v}$ can be written as sum of vectors in U and W in many different ways:

$$\mathbf{v} = \mathbf{v} + \mathbf{0} = \mathbf{0} + \mathbf{v} = \frac{1}{2}\mathbf{v} + \frac{1}{2}\mathbf{v} = \frac{1}{3}\mathbf{v} + \frac{2}{3}\mathbf{v} \in U + W. \qquad \square$$

Example 3.18 Consider the three vectors $\mathbf{e}_1$, $\mathbf{e}_2$ and $\mathbf{e}_3$ in $\mathbb{R}^3$. Let $U = \{a_1\mathbf{e}_1 + b_3\mathbf{e}_3 : a_1, b_3 \in \mathbb{R}\}$ be the subspace spanned by $\mathbf{e}_1$ and $\mathbf{e}_3$ (xz-plane), and let $W = \{a_2\mathbf{e}_2 + c_2\mathbf{e}_3 : a_2, c_3 \in \mathbb{R}\}$ be the subspace of $\mathbb{R}^3$ spanned by $\mathbf{e}_2$ and $\mathbf{e}_3$ (yz-plane). Then a vector in $U + W$ is of the form

$$(a_1\mathbf{e}_1 + b_3\mathbf{e}_3) + (a_2\mathbf{e}_2 + c_2\mathbf{e}_3) = a_1\mathbf{e}_1 + a_2\mathbf{e}_2 + (b_3 + c_2)\mathbf{e}_3 = a_1\mathbf{e}_1 + a_2\mathbf{e}_2 + a_3\mathbf{e}_3$$

where $a_3 = b_3 + c_3$ and a_1, a_2, a_3 are arbitrary numbers. Thus $U + W = \mathbb{R}^3$. However, $\mathbb{R}^3 \neq U \oplus W$ since clearly $\mathbf{e}_3 \in U \cap W \neq \{\mathbf{0}\}$. In fact, the vector $\mathbf{e}_3 \in \mathbb{R}^3$ can be written as many linear combinations of vectors in U and W:

$$\mathbf{e}_3 = \frac{1}{2}\mathbf{e}_3 + \frac{1}{2}\mathbf{e}_3 = \frac{1}{3}\mathbf{e}_3 + \frac{2}{3}\mathbf{e}_3 \in U + W.$$

Note that if we had taken W to be the subspace spanned by $\mathbf{e}_2$ alone, then it would be easy to see that $\mathbb{R}^3 = U \oplus W$. Note also that there are many choices for W. $\square$

As a direct consequence of Theorem 3.10 and the definition of the direct sum, one can show the following.

Corollary 3.13 *If U is a subspace of V, then there is a subspace W in V such that $V = U \oplus W$.*

Proof: Choose a basis $\{\mathbf{u}_1, \ldots, \mathbf{u}_k\}$ for U, and extend it to a basis $\{\mathbf{u}_1, \ldots, \mathbf{u}_k, \mathbf{u}_{k+1}, \ldots, \mathbf{u}_n\}$ for V. Then the subspace W spanned by $\{\mathbf{v}_{k+1}, \ldots, \mathbf{v}_n\}$ satisfies the requirement. $\square$

Problem 3.13 Let U and W be the subspaces of the vector space $M_{n\times n}(\mathbb{R})$ consisting of all symmetric matrices and all skew-symmetric matrices, respectively. Show that $M_{n\times n}(\mathbb{R}) = U \oplus W$. Therefore, the decomposition of a square matrix A given in (3) of Problem 1.10 is unique.

Problem 3.14 Let $\{\mathbf{v}_1, \mathbf{v}_2, \ldots, \mathbf{v}_n\}$ be a basis for a vector space V and let $W_i = \{r\mathbf{v}_i : r \in \mathbb{R}\}$ be the subspace of V spanned by $\mathbf{v}_i$. Show that $V = W_1 \oplus W_2 \oplus \cdots \oplus W_n$.

3.4 Row and column spaces

In this section, we go back to systems of linear equations and study them in terms of the concepts introduced in the previous sections. Note that an $m \times n$ matrix A can be abbreviated by the row vectors or column vectors as follows:

$$\begin{aligned} A &= \begin{bmatrix} a_{11} & a_{12} & \cdots & a_{1n} \\ a_{21} & a_{22} & \cdots & a_{2n} \\ \vdots & \vdots & & \vdots \\ a_{m1} & a_{m2} & \cdots & a_{mn} \end{bmatrix} = \begin{bmatrix} \mathbf{r}_1 \\ \mathbf{r}_2 \\ \vdots \\ \mathbf{r}_m \end{bmatrix} \\ &= \begin{bmatrix} \mathbf{c}_1 & \mathbf{c}_2 & \cdots & \mathbf{c}_n \end{bmatrix}, \end{aligned}$$

where the $\mathbf{r}_i$'s are the row vectors of A that are in $\mathbb{R}^n$, and the $\mathbf{c}_j$'s are the column vectors of A that are in $\mathbb{R}^m$.

Definition 3.7 Let A be an $m \times n$ matrix with row vectors $\{\mathbf{r}_1, \ldots, \mathbf{r}_m\}$ and column vectors $\{\mathbf{c}_1, \ldots, \mathbf{c}_n\}$.

(1) The **row space** of A is the subspace in $\mathbb{R}^n$ spanned by the row vectors $\{\mathbf{r}_1, \ldots, \mathbf{r}_m\}$, denoted by $\mathcal{R}(A)$.

(2) The **column space** of A is the subspace in $\mathbb{R}^m$ spanned by the column vectors $\{\mathbf{c}_1, \ldots, \mathbf{c}_n\}$, denoted by $\mathcal{C}(A)$.

(3) The solution set of the homogeneous equation $A\mathbf{x} = \mathbf{0}$ is called the **null space** of A, denoted by $\mathcal{N}(A)$.

Note that the null space $\mathcal{N}(A)$ is a subspace of the n-space $\mathbb{R}^n$, whose dimension is called the **nullity** of A. Since the row vectors of A are just the column vectors of its transpose A^T, and the column vectors of A are the row vectors of A^T, the row space of A is the column space of A^T; that is,

$$\mathcal{R}(A) = \mathcal{C}(A^T) \quad \text{and} \quad \mathcal{C}(A) = \mathcal{R}(A^T).$$

Since $A\mathbf{x} = x_1\mathbf{c}_1 + x_2\mathbf{c}_2 + \ldots x_n\mathbf{c}_n$ for any vector $\mathbf{x} = (x_1, x_2, \ldots, x_n) \in \mathbb{R}^n$, we get

$$\mathcal{C}(A) = \{A\mathbf{x} \,:\, \mathbf{x} \in \mathbb{R}^n\}.$$

Thus, for a vector $\mathbf{b} \in \mathbb{R}^m$, *the system* $A\mathbf{x} = \mathbf{b}$ *has a solution if and only if* $\mathbf{b} \in \mathcal{C}(A) \subseteq \mathbb{R}^m$. Thus, the column space $\mathcal{C}(A)$ is the set of vectors $\mathbf{b} \in \mathbb{R}^m$ for which $A\mathbf{x} = \mathbf{b}$ has a solution.

It is quite natural to ask what the dimensions of those subspaces are, and how one can find bases for them. This will help us to understand the structure of all the solutions of the equation $A\mathbf{x} = \mathbf{b}$. Since the set of the row vectors and the set of the column vectors of A are spanning sets for the row space and the column space, respectively, a minimally spanning subset of each of them will be a basis for each of them.

This is not difficult for a matrix of a (reduced) row-echelon form.

Example 3.19 Let U be in a reduced row-echelon form given as

$$U = \begin{bmatrix} 1 & 0 & 0 & 2 & 2 \\ 0 & 1 & 0 & -1 & 3 \\ 0 & 0 & 1 & 4 & -1 \\ 0 & 0 & 0 & 0 & 0 \end{bmatrix}.$$

Clearly, the first three nonzero row vectors containing leading 1's are linearly independent and they form a basis for the row space $\mathcal{R}(U)$, so that $\dim \mathcal{R}(U) = 3$. On the other hand, note that the first three columns containing leading 1's are linearly independent (see Theorem 3.5), and that the last two column vectors can be expressed as linear combinations of them. Hence, they form a basis for $\mathcal{C}(U)$, and $\dim \mathcal{C}(U) = 3$. To find a basis for the null space $\mathcal{N}(U)$, we first solve the system $U\mathbf{x} = \mathbf{0}$ with arbitrary values s and t for the free variables x_4 and x_5, and get the solution

$$\begin{bmatrix} x_1 \\ x_2 \\ x_3 \\ x_4 \\ x_5 \end{bmatrix} = \begin{bmatrix} -2s & - & 2t \\ s & - & 3t \\ -4s & + & t \\ s & & \\ & & t \end{bmatrix} = s\begin{bmatrix} -2 \\ 1 \\ -4 \\ 1 \\ 0 \end{bmatrix} + t\begin{bmatrix} -2 \\ -3 \\ 1 \\ 0 \\ 1 \end{bmatrix} = s\mathbf{n}_s + t\mathbf{n}_t,$$

where $\mathbf{n}_s = (-2,\ 1,\ -4,\ 1,\ 0)$, $\mathbf{n}_t = (-2,\ -3,\ 1,\ 0,\ 1)$. It shows that these two vectors $\mathbf{n}_s$ and $\mathbf{n}_t$ span the null space $\mathcal{N}(U)$, and they are clearly linearly independent. Hence, the set $\{\mathbf{n}_s,\ \mathbf{n}_t\}$ is a basis for the null space $\mathcal{N}(U)$. □

In the following, the row, the column or the null space of a matrix A will be discussed in relation to the corresponding space of its (reduced) row-echelon form. We first investigate the row space $\mathcal{R}(A)$ and the null space $\mathcal{N}(A)$ of A by comparing them with those of the reduced row-echelon form U of A. Since $A\mathbf{x} = \mathbf{0}$ and $U\mathbf{x} = \mathbf{0}$ have the same solution set by Theorem 1.1, we have $\mathcal{N}(A) = \mathcal{N}(U)$.

Let $A = \begin{bmatrix} \mathbf{r}_1 \\ \vdots \\ \mathbf{r}_m \end{bmatrix}$ be an $m \times n$ matrix, where $\mathbf{r}_i$'s are the row vectors of A. The three elementary row operations change A into the following three types:

$$A_1 = \begin{bmatrix} \mathbf{r}_1 \\ \vdots \\ k\mathbf{r}_i \\ \vdots \\ \mathbf{r}_m \end{bmatrix} \text{ for } k \neq 0, \quad A_2 = \begin{bmatrix} \vdots \\ \mathbf{r}_j \\ \vdots \\ \mathbf{r}_i \\ \vdots \end{bmatrix} \text{ for } i < j, \quad A_3 = \begin{bmatrix} \mathbf{r}_1 \\ \vdots \\ \mathbf{r}_i + k\mathbf{r}_j \\ \vdots \\ \mathbf{r}_m \end{bmatrix}.$$

It is clear that the row vectors of the three matrices A_1, A_2 and A_3 are linear combinations of the row vectors of A. On the other hand, by the inverse elementary row operations, these matrices can be changed into A. Thus, the row vectors of A can also be written as linear combinations of those of A_i's. This means that if matrices A and B are row equivalent, then their row spaces must be equal, *i.e.*, $\mathcal{R}(A) = \mathcal{R}(B)$.

Now the nonzero row vectors in the reduced row-echelon form U are always linearly independent and span the row space of U (see Theorem 3.5). Thus they form a basis for the row space $\mathcal{R}(A)$ of A. We have the following theorem.

Theorem 3.14 *Let U be a (reduced) row-echelon form of a matrix A. Then*

$$\mathcal{R}(A) = \mathcal{R}(U) \quad \textit{and} \quad \mathcal{N}(A) = \mathcal{N}(U).$$

Moreover, if U has r nonzero row vectors containing leading 1's, then they form a basis for the row space $\mathcal{R}(A)$, so that the dimension of $\mathcal{R}(A)$ is r.

The following example shows how to find bases for the row and the null spaces, and at the same time how to find a basis for the column space $\mathcal{C}(A)$.

Example 3.20 Let A be a matrix given as

$$A = \begin{bmatrix} 1 & 2 & 0 & 2 & 5 \\ -2 & -5 & 1 & -1 & -8 \\ 0 & -3 & 3 & 4 & 1 \\ 3 & 6 & 0 & -7 & 2 \end{bmatrix} = \begin{bmatrix} \mathbf{r}_1 \\ \mathbf{r}_2 \\ \mathbf{r}_3 \\ \mathbf{r}_4 \end{bmatrix}.$$

Find bases for the row space $\mathcal{R}(A)$, the null space $\mathcal{N}(A)$, and the column space $\mathcal{C}(A)$ of A.

Solution: (1) Find a basis for $\mathcal{R}(A)$: By Gauss-Jordan elimination on A, we get the reduced row-echelon form U:

$$U = \begin{bmatrix} 1 & 0 & 2 & 0 & 1 \\ 0 & 1 & -1 & 0 & 1 \\ 0 & 0 & 0 & 1 & 1 \\ 0 & 0 & 0 & 0 & 0 \end{bmatrix}.$$

Since the three nonzero row vectors

$$\begin{aligned} \mathbf{v}_1 &= (1, \ 0, \ 2, \ 0, \ 1), \\ \mathbf{v}_2 &= (0, \ 1, \ -1, \ 0, \ 1), \\ \mathbf{v}_3 &= (0, \ 0, \ 0, \ 1, \ 1) \end{aligned}$$

of U are linearly independent, they form a basis for the row space $\mathcal{R}(U) = \mathcal{R}(A)$, so $\dim \mathcal{R}(A) = 3$. (Note that in the process of Gaussian elimination, we did not use a permutation matrix. This means that the three nonzero rows of U were obtained from the first three row vectors $\mathbf{r}_1$, $\mathbf{r}_2$, $\mathbf{r}_3$ of A and the fourth row $\mathbf{r}_4$ of A turned out to be a linear combination of them. Thus the first three row vectors of A also form a basis for the row space.)

(2) Find a basis for $\mathcal{N}(A)$. It is enough to solve the homogeneous system $U\mathbf{x} = \mathbf{0}$, since $\mathcal{N}(A) = \mathcal{N}(U)$. That is, neglecting the fourth zero equation, the equation $U\mathbf{x} = \mathbf{0}$ takes the following system of equations:

$$\left\{ \begin{array}{ccccccccc} x_1 & & + & 2x_3 & & + & x_5 & = & 0 \\ & x_2 & - & x_3 & & + & x_5 & = & 0 \\ & & & & x_4 & + & x_5 & = & 0. \end{array} \right.$$

Since the first, the second and the fourth columns of U contain the leading 1's, we see that the basic variables are x_1, x_2, x_4, and the free variables are

x_3, x_5. By assigning arbitrary values s and t to the free variables x_3 and x_5, we find the solution $\mathbf{x}$ of $U\mathbf{x} = \mathbf{0}$ as

$$\mathbf{x} = \begin{bmatrix} x_1 \\ x_2 \\ x_3 \\ x_4 \\ x_5 \end{bmatrix} = \begin{bmatrix} -2s & - & t \\ s & - & t \\ s & & \\ & - & t \\ & & t \end{bmatrix} = s \begin{bmatrix} -2 \\ 1 \\ 1 \\ 0 \\ 0 \end{bmatrix} + t \begin{bmatrix} -1 \\ -1 \\ 0 \\ -1 \\ 1 \end{bmatrix} = s\mathbf{n}_s + t\mathbf{n}_t,$$

where $\mathbf{n}_s = (-2,\ 1,\ 1,\ 0,\ 0)$ and $\mathbf{n}_t = (-1,\ -1,\ 0,\ -1,\ 1)$. In fact, the two vectors $\mathbf{n}_s$ and $\mathbf{n}_t$ are the solutions when the values of $(x_3, x_5) = (s, t)$ are $(1, 0)$ and those of $(x_3, x_5) = (s, t)$ are $(0, 1)$, respectively. They must be linearly independent, since $(1,\ 0)$ and $(0,\ 1)$, as the (x_3, x_5)-coordinates of $\mathbf{n}_s$ and $\mathbf{n}_t$ respectively, are clearly linearly independent. Since any solution of $U\mathbf{x} = \mathbf{0}$ is a linear combination of them, the set $\{\mathbf{n}_s,\ \mathbf{n}_t\}$ is a basis for the null space $\mathcal{N}(U) = \mathcal{N}(A)$. Thus $\dim \mathcal{N}(A) = 2 =$ *the number of free variables in* $U\mathbf{x} = \mathbf{0}$.

(3) Find a basis for $\mathcal{C}(A)$. Let $\mathbf{c}_1,\ \mathbf{c}_2,\ \mathbf{c}_3,\ \mathbf{c}_4,\ \mathbf{c}_5$ denote the column vectors of A in the given order. Since these column vectors of A span $\mathcal{C}(A)$, we only need to discard some of the columns that can be expressed as linear combinations of other column vectors. But, the linear dependence

$$x_1\mathbf{c}_1 + x_2\mathbf{c}_2 + x_3\mathbf{c}_3 + x_4\mathbf{c}_4 + x_5\mathbf{c}_5 = \mathbf{0}, \quad \textit{i.e.}, \ A\mathbf{x} = \mathbf{0},$$

holds if and only if $\mathbf{x} = (x_1,\ \cdots,\ x_5) \in \mathcal{N}(A)$. By taking $\mathbf{x} = \mathbf{n}_s = (-2,\ 1,\ 1,\ 0,\ 0)$ or $\mathbf{x} = \mathbf{n}_t = (-1,\ -1,\ 0,\ -1,\ 1)$, the basis vectors of $\mathcal{N}(A)$ given in (2), we obtain two nontrivial linear dependencies of $\mathbf{c}_i$'s:

$$\begin{aligned} -2\mathbf{c}_1 + \mathbf{c}_2 + \mathbf{c}_3 &= \mathbf{0}, \\ -\mathbf{c}_1 - \mathbf{c}_2 - \mathbf{c}_4 + \mathbf{c}_5 &= \mathbf{0}, \end{aligned}$$

respectively. Hence, the column vectors $\mathbf{c}_3$ and $\mathbf{c}_5$ corresponding to the free variables in $A\mathbf{x} = \mathbf{0}$ can be written as

$$\begin{aligned} \mathbf{c}_3 &= 2\mathbf{c}_1 - \mathbf{c}_2, \\ \mathbf{c}_5 &= \mathbf{c}_1 + \mathbf{c}_2 + \mathbf{c}_4. \end{aligned}$$

That is, the column vectors $\mathbf{c}_3, \mathbf{c}_5$ of A are linear combinations of the column vectors $\mathbf{c}_1, \mathbf{c}_2, \mathbf{c}_4$, which correspond to the basic variables in $A\mathbf{x} = \mathbf{0}$. Hence, $\{\mathbf{c}_1,\ \mathbf{c}_2,\ \mathbf{c}_4\}$ spans the column space $\mathcal{C}(A)$.

We claim that $\{\mathbf{c}_1, \mathbf{c}_2, \mathbf{c}_4\}$ is linearly independent. Let $\tilde{A} = [\mathbf{c}_1\ \mathbf{c}_2\ \mathbf{c}_4]$ and $\tilde{U} = [\mathbf{u}_1\ \mathbf{u}_2\ \mathbf{u}_4]$ be submatrices of A and U, respectively, where $\mathbf{u}_j$ is the j-th column vector of the reduced row-echelon form U of A obtained in (1). Then clearly $\tilde{U}$ is the reduced row-echelon form of $\tilde{A}$ so that $\mathcal{N}(\tilde{A}) = \mathcal{N}(\tilde{U})$. Since the vectors $\mathbf{u}_1, \mathbf{u}_2, \mathbf{u}_4$ are just the columns of U containing leading 1's, they are linearly independent, by Theorem 3.5, and $\tilde{U}\mathbf{x} = \mathbf{0}$ has only a trivial solution. This means that $\tilde{A}\mathbf{x} = \mathbf{0}$ has also only a trivial solution, so $\{\mathbf{c}_1, \mathbf{c}_2, \mathbf{c}_4\}$ is linearly independent. Therefore, it is a basis for the column space $\mathcal{C}(A)$ and $\dim \mathcal{C}(A) = 3 =$ *the number of basic variables*. That is, *the column vectors of A corresponding to the basic variables in $U\mathbf{x} = \mathbf{0}$ form a basis for the column space $\mathcal{C}(A)$.* □

In summary, given a matrix A, we first find the (reduced) row-echelon form U of A by Gauss-Jordan elimination. Then a basis for $\mathcal{R}(A) = \mathcal{R}(U)$ is the set of nonzero rows vectors of U, and a basis for $\mathcal{N}(A) = \mathcal{N}(U)$ can be found by solving $U\mathbf{x} = \mathbf{0}$, which is easy. On the other hand, one has to be careful for $\mathcal{C}(U) \neq \mathcal{C}(A)$ in general, since the column space of A is not preserved by Gauss-Jordan elimination. However, we have $\dim \mathcal{C}(A) = \dim \mathcal{C}(U)$, and a basis for $\mathcal{C}(A)$ can be selected from the column vectors in A, not in U, as those corresponding to the basic variables (or the leading 1's in U). To show that those column vectors indeed form a basis for $\mathcal{C}(A)$, we used a basis for the null space $\mathcal{N}(A)$ to eliminate the redundant columns.

Note that a basis for the column space $\mathcal{C}(A)$ can be also found with the elementary column operations, which is the same as finding a basis for the row space $\mathcal{R}(A^T)$ of A^T.

Problem 3.15 Let A be the matrix given in Example 3.20. Find a relation of a, b, c, d so that the vector $\mathbf{x} = (a, b, c, d)$ belongs to $\mathcal{C}(A)$.

Problem 3.16 Find bases for $\mathcal{R}(A)$ and $\mathcal{N}(A)$ of the matrix

$$A = \begin{bmatrix} 1 & -2 & 0 & 0 & 3 \\ 2 & -5 & -3 & -2 & 6 \\ 0 & 5 & 15 & 10 & 0 \\ 2 & 6 & 18 & 8 & 6 \end{bmatrix}.$$

Also find a basis for $\mathcal{C}(A)$ by finding a basis for $\mathcal{R}(A^T)$.

Problem 3.17 Let A and B be two $n \times n$ matrices. Show that $AB = \mathbf{0}$ if and only if the column space of B is a subspace of the nullspace of A.

Problem 3.18 Find an example of a matrix A and its row-echelon form U such that $\mathcal{C}(A) \neq \mathcal{C}(U)$.

3.5 Rank and nullity

The argument in Example 3.20 is so general that it can be used to prove the following theorem, which is one of the most fundamental results in linear algebra. The proof given here is just a repetition of the argument in Example 3.20 in a general form, and so may be skipped at the reader's discretion.

Theorem 3.15 (The first fundamental theorem) *For any $m \times n$ matrix A, the row space and the column space of A have the same dimension; that is,* $\dim \mathcal{R}(A) = \dim \mathcal{C}(A)$.

Proof: Let $\dim \mathcal{R}(A) = r$ and let U be the reduced row-echelon form of A. Then r is the number of the nonzero row (or column) vectors of U containing leading 1's, which is equal to the number of basic variables in $U\mathbf{x} = \mathbf{0}$ or $A\mathbf{x} = \mathbf{0}$. We shall prove that the r columns of A corresponding to the r leading 1's (or basic variables) form a basis for $\mathcal{C}(A)$, so that $\dim \mathcal{C}(A) = r = \dim \mathcal{R}(A)$.

(1) They are linearly independent: Let $\tilde{A}$ denote the submatrix of A whose columns are those of A corresponding to the r basic variables (or leading 1's) in U, and let $\tilde{U}$ denote the submatrix of U containing r leading 1's. Then, it is quite clear that $\tilde{U}$ is the reduced row-echelon form of $\tilde{A}$, so that $\tilde{A}\mathbf{x} = \mathbf{0}$ if and only if $\tilde{U}\mathbf{x} = \mathbf{0}$. However, $\tilde{U}\mathbf{x} = \mathbf{0}$ has only a trivial solution since the columns of U containing the leading 1's are linearly independent by Theorem 3.5. Therefore, $\tilde{A}\mathbf{x} = \mathbf{0}$ also has only the trivial solution, so the columns of $\tilde{A}$ are linearly independent.

(2) They span $\mathcal{C}(A)$: Note that the columns A corresponding to the free variables are not contained in $\tilde{A}$, and each of these column vector of A can be written as a linear combination of the column vectors of $\tilde{A}$ (see Example 3.20). In fact, if $\{x_{i_1}, x_{i_2}, \ldots, x_{i_k}\}$ is the set of free variables whose corresponding columns are not in $\tilde{A}$, then, for an assignment of value 1 to x_{i_j} and 0 to all the other free variables, one can always find a nontrivial solution of

$$A\mathbf{x} = x_1\mathbf{c}_1 + x_2\mathbf{c}_2 + \ldots + x_n\mathbf{c}_n = \mathbf{0}.$$

When the solution is substituted into this equation, one can see that the column $\mathbf{c}_{i_j}$ of A corresponding to $x_{i_j} = 1$ is written as a linear combination of the columns of $\tilde{A}$. This can be done for each $j = 1, \ldots, k$, so the columns of A corresponding to those free variables are redundant in the spanning set of $\mathcal{C}(A)$. □

Remark: In the proof of Theorem 3.15, once we have shown that the columns in $\tilde{A}$ are linearly independent as in (1), we may replace step (2) by the following argument: One can easily see that $\dim \mathcal{C}(A) \geq \dim \mathcal{R}(A)$ by Theorem 3.10. On the other hand, since this inequality holds for arbitrary matrices, in particular for A^T, we get $\dim \mathcal{C}(A^T) \geq \dim \mathcal{R}(A^T)$. Moreover, $\mathcal{C}(A^T) = \mathcal{R}(A)$ and $\mathcal{R}(A^T) = \mathcal{C}(A)$ implies $\dim \mathcal{C}(A) \leq \dim \mathcal{R}(A)$, which means $\dim \mathcal{C}(A) = \dim \mathcal{R}(A)$. This also means that the column vectors of $\tilde{A}$ span $\mathcal{C}(A)$, and so form a basis.

In summary, the following equalities are now clear from Theorem 3.14 and 3.15:

$$\begin{aligned}
\dim \mathcal{R}(A) &= \dim \mathcal{R}(U) \\
&= \text{the number of nonzero row vectors of } U \\
&= \text{the maximal number of linearly independent} \\
&\ \text{row vectors of } A \\
&= \text{the number of basic variables in } U\mathbf{x} = \mathbf{0}. \\
&= \text{the maximal number of linearly independent} \\
&\ \text{column vectors of } A \\
&= \dim \mathcal{C}(A).
\end{aligned}$$

$$\begin{aligned}
\dim \mathcal{N}(A) &= \dim \mathcal{N}(U) \\
&= \text{the number of free variables in } U\mathbf{x} = \mathbf{0}.
\end{aligned}$$

Definition 3.8 For an $m \times n$ matrix A, the **rank** of A is defined to be the dimension of the row space (or the column space), denoted by rank A.

Clearly, rank $I_n = n$ and rank $A =$ rank A^T. And for an $m \times n$ matrix A, rank $A = \dim \mathcal{R}(A) = \dim \mathcal{C}(A)$. Since $\dim \mathcal{R}(A) \leq m$ and $\dim \mathcal{C}(A) \leq n$, we have the following corollary:

Corollary 3.16 *If A is an $m \times n$ matrix, then* rank $A \leq \min\{m, n\}$.

Since $\dim \mathcal{R}(A) = \dim \mathcal{C}(A) =$ rank A is the number of basic variables in $A\mathbf{x} = \mathbf{0}$, and $\dim \mathcal{N}(A) =$ nullity of A is the number of free variables $A\mathbf{x} = \mathbf{0}$, we have the following corollary.

Corollary 3.17 *For any $m \times n$ matrix A,*

$$\begin{aligned}
\dim \mathcal{R}(A) + \dim \mathcal{N}(A) &= \text{rank } A + \text{nullity of } A &= n, \\
\dim \mathcal{C}(A) + \dim \mathcal{N}(A^T) &= \text{rank } A + \text{nullity of } A^T &= m.
\end{aligned}$$

If $\dim \mathcal{N}(A) = 0$ (or $\mathcal{N}(A) = \{\mathbf{0}\}$), then $\dim \mathcal{R}(A) = n$ (or $\mathcal{R}(A) = \mathbb{R}^n$), which means that A has exactly n linearly independent rows and n linearly independent columns. In particular, if A is a square matrix of order n, then the row vectors are linearly independent if and only if the column vectors are linearly independent. Therefore, by Theorem 1.8, we get the following corollary.

Corollary 3.18 *Let A be an $n \times n$ square matrix. Then A is invertible if and only if* rank $A = n$.

Example 3.21 For a 4×5 matrix

$$A = \begin{bmatrix} 1 & 2 & 0 & 2 & 1 \\ -1 & -2 & 1 & 1 & 0 \\ 1 & 2 & -3 & -7 & 2 \\ 1 & 2 & -2 & -4 & 3 \end{bmatrix},$$

by Gaussian elimination, we get

$$U = \begin{bmatrix} 1 & 2 & 0 & 2 & 1 \\ 0 & 0 & 1 & 3 & 1 \\ 0 & 0 & 0 & 0 & 1 \\ 0 & 0 & 0 & 0 & 0 \end{bmatrix}.$$

The first three nonzero rows containing leading 1's in U form a basis for $\mathcal{R}(U) = \mathcal{R}(A)$. Note that x_1, x_3 and x_5 are the basic variables in $U\mathbf{x} = \mathbf{0}$, since the first, third and fifth columns of U contain leading 1's. Thus the three columns $\mathbf{c}_1 = (1, -1, 1, 1)$, $\mathbf{c}_3 = (0, 1, -3, -2)$ and $\mathbf{c}_5 = (1, 0, 2, 3)$ of A, not the three columns in U, corresponding to those basic variables x_1, x_3 and x_5 form a basis for $\mathcal{C}(A)$. Therefore, rank $A = \dim \mathcal{R}(A) = \dim \mathcal{C}(A) = 3$, the nullity of $A = \dim \mathcal{N}(A) = 2$, and $\dim \mathcal{N}(A^T) = 1$. □

Problem 3.19 Find the nullity and the rank of each of the following matrices:

$$(1)\, A = \begin{bmatrix} 1 & 3 & 1 & 7 \\ 2 & 3 & -1 & 9 \\ -1 & -2 & 0 & -5 \end{bmatrix}, \quad (2)\, A = \begin{bmatrix} 1 & 2 & 1 & 2 \\ 1 & 1 & 2 & 0 \\ 2 & 1 & 5 & -2 \end{bmatrix}.$$

For each of the matrices, show that $\dim \mathcal{R}(A) = \dim \mathcal{C}(A)$ directly by finding their bases.

Problem 3.20 Show that a system of linear equations $A\mathbf{x} = \mathbf{b}$ has a solution if and only if rank A = rank $[A\ \mathbf{b}]$, where $[A\ \mathbf{b}]$ denotes the augmented matrix of $A\mathbf{x} = \mathbf{b}$.

Theorem 3.19 *For any two matrices A and B for which AB can be defined,*

(1) $\mathcal{N}(AB) \supseteq \mathcal{N}(B)$,

(2) $\mathcal{N}((AB)^T) \supseteq \mathcal{N}(A^T)$,

(3) $\mathcal{C}(AB) \subseteq \mathcal{C}(A)$,

(4) $\mathcal{R}(AB) \subseteq \mathcal{R}(B)$.

Proof: **(1)** and **(2)** are clear, since $B\mathbf{x} = \mathbf{0}$ implies $(AB)\mathbf{x} = A(B\mathbf{x}) = \mathbf{0}$.

(3) For an $m \times n$ matrix A and an $n \times p$ matrix B,

$$\begin{aligned} \mathcal{C}(AB) &= \{AB\mathbf{x} \ : \ \mathbf{x} \in \mathbb{R}^p\} \\ &\subseteq \{A\mathbf{y} \ : \ \mathbf{y} \in \mathbb{R}^n\} = \mathcal{C}(A), \end{aligned}$$

because $B\mathbf{x} \in \mathbb{R}^n$ for any $\mathbf{x} \in \mathbb{R}^p$.

(4) $\mathcal{R}(AB) = \mathcal{C}((AB)^T) = \mathcal{C}(B^T A^T) \subseteq \mathcal{C}(B^T) = \mathcal{R}(B)$. □

Corollary 3.20 rank$(AB) \leq \min\{\text{rank } A, \ \text{rank } B\}$.

In some particular cases, the equality holds. In fact, it will be shown later in Theorem 5.23 that for any square matrix A, rank$(A^T A) =$ rank $A =$ rank(AA^T). The following problem illustrates another such case.

Problem 3.21 Let A be an invertible square matrix. Show that, for any matrix B, rank$(AB) =$ rank $B =$ rank(BA).

Theorem 3.21 *Let A be an $m \times n$ matrix of rank r. Then*

(1) *for every submatrix C of A,* rank $C \leq r$, *and*

(2) *the matrix A has at least one $r \times r$ submatrix of rank r, that is, A has an invertible submatrix of order r.*

Proof: **(1)** We consider an intermediate matrix B which is obtained from A by removing the rows that are not wanted in C. Then clearly $\mathcal{R}(B) \subseteq \mathcal{R}(A)$ and hence rank $B \leq$ rank A. Moreover, since the columns of C are taken from those of B, $\mathcal{C}(C) \subseteq \mathcal{C}(B)$ and rank $C \leq$ rank B.

(2) Note that we can find r linearly independent row vectors of A, which form a basis for the row space of A. Let B be the matrix whose row vectors consist of these vectors. Then rank $B = r$ and the column space of B must be of dimension r. By taking r linearly independent column vectors of B, one can find an $r \times r$ submatrix C of A with rank r. □

Problem 3.22 Prove that the rank of a matrix is equal to the largest order of its invertible submatrices.

Problem 3.23 For each of the matrices given in Problem 3.19, find an invertible submatrix of the largest order.

3.6 Bases for subspaces

In this section, we discuss how to find bases for $V+W$ and $V\cap W$ of two subspaces V and W of the n-space $\mathbb{R}^n$, and then derive an important relationship between the dimensions of those subspaces in terms of the dimensions of V and W.

Let $\alpha = \{\mathbf{v}_1, \dots, \mathbf{v}_k\}$ and $\beta = \{\mathbf{w}_1, \dots, \mathbf{w}_\ell\}$ be bases for V and W, respectively. Let Q be the $n \times (k+\ell)$ matrix whose columns are those bases vectors:

$$Q = [\mathbf{v}_1 \ \cdots \ \mathbf{v}_k \ \mathbf{w}_1 \ \cdots \ \mathbf{w}_\ell]_{n\times(k+\ell)}.$$

Then it is quite clear that $\mathcal{C}(Q) = V + W$, so that a basis for $\mathcal{C}(Q)$ is a basis for $V + W$. On the other hand, one can show that $\mathcal{N}(Q)$ can be identified with $V \cap W$.

In fact, if $\mathbf{x} = (a_1, \dots, a_k, b_1, \dots, b_\ell) \in \mathcal{N}(Q) \subseteq \mathbb{R}^{k+\ell}$, then

$$Q\mathbf{x} = a_1\mathbf{v}_1 + \cdots + a_k\mathbf{v}_k + b_1\mathbf{w}_1 + \cdots + b_\ell\mathbf{w}_\ell = \mathbf{0}.$$

This means that corresponding to $\mathbf{x}$ there is a vector

$$\mathbf{y} = a_1\mathbf{v}_1 + \cdots + a_k\mathbf{v}_k = -(b_1\mathbf{w}_1 + \cdots + b_\ell\mathbf{w}_\ell)$$

that belongs to $V\cap W$, since the middle part is in V as a linear combination of the basis vectors in α and the right side is in W as a linear combination of the basis vectors in β. On the other hand, if $\mathbf{y} \in V \cap W$, $\mathbf{y}$ can be written as linear combinations of both bases for V and W:

$$\begin{aligned} \mathbf{y} &= a_1\mathbf{v}_1 + \cdots + a_k\mathbf{v}_k \\ &= b_1\mathbf{w}_1 + \cdots + b_\ell\mathbf{w}_\ell, \end{aligned}$$

for some $a_1, \dots, a_k$ and $b_1, \dots, b_\ell$. Let $\mathbf{x} = (a_1, \dots, a_k, -b_1, \dots, -b_\ell)$. Then it is quite clear that $Q\mathbf{x} = \mathbf{0}$, *i.e.*, $\mathbf{x} \in \mathcal{N}(Q)$. That is, for each $\mathbf{x} \in \mathcal{N}(Q)$, there corresponds a vector $\mathbf{y} \in V \cap W$, and *vice versa*. Moreover, if $\mathbf{x}_i$, $i = 1, 2$, correspond to $\mathbf{y}_i$, then one can easily check that $\mathbf{x}_1 + \mathbf{x}_2$ corresponds to $\mathbf{y}_1 + \mathbf{y}_2$, and $k\mathbf{x}_1$ corresponds to $k\mathbf{y}_1$. This means that the two vector spaces $\mathcal{N}(Q)$ and $V \cap W$ can be identified as vector spaces. In particular, for a basis for $\mathcal{N}(Q)$, the corresponding set in $V \cap W$ is also a basis, that is, if the set of vectors

$$\begin{cases} \mathbf{x}_1 &= (a_{11}, \ldots, a_{1k}, b_{11}, \ldots, b_{1\ell}), \\ \quad\vdots & \\ \mathbf{x}_s &= (a_{s1}, \ldots, a_{sk}, b_{s1}, \ldots, b_{s\ell}), \end{cases}$$

is a basis for $\mathcal{N}(Q)$, then the set

$$\begin{cases} \mathbf{y}_1 &= a_{11}\mathbf{v}_1 + \cdots + a_{1k}\mathbf{v}_k, \\ \quad\vdots & \\ \mathbf{y}_s &= a_{s1}\mathbf{v}_1 + \cdots + a_{sk}\mathbf{v}_k, \end{cases} \quad \text{or} \quad \begin{cases} \mathbf{y}_1 &= -(b_{11}\mathbf{w}_1 + \cdots + b_{1\ell}\mathbf{w}_\ell), \\ \quad\vdots & \\ \mathbf{y}_s &= -(b_{s1}\mathbf{w}_1 + \cdots + b_{s\ell}\mathbf{w}_\ell) \end{cases}$$

is also a basis for $V \cap W$, and *vice versa*. This means that

$$\dim \mathcal{N}(Q) = \dim V \cap W.$$

Note that $\dim(V + W) \neq \dim V + \dim W$, in general. The following corollary gives a relation of them.

Corollary 3.22 *For any subspaces V and W of the n-space $\mathbb{R}^n$,*

$$\dim(V + W) + \dim(V \cap W) = \dim V + \dim W.$$

Proof: Let $\dim V = k$ and $\dim W = \ell$. Recall that rank $A+$ nullity $A =$ the number of the columns of a matrix A. Thus, for the matrix Q above, we have

$$\dim \mathcal{C}(Q) + \dim \mathcal{N}(Q) = k + \ell.$$

However, we have $\dim \mathcal{C}(Q) = \dim(V + W)$, $\dim \mathcal{N}(Q) = \dim(V \cap W)$. $\dim V = k$ and $\dim W = \ell$. $\square$

Example 3.22 Let V and W be two subspaces of $\mathbb{R}^5$ with bases

$$\begin{cases} \mathbf{v}_1 = (1, & 3, & -2, & 2, & 3), \\ \mathbf{v}_2 = (1, & 4, & -3, & 4, & 2), \\ \mathbf{v}_3 = (1, & 3, & 0, & 2, & 3), \end{cases} \qquad \begin{cases} \mathbf{w}_1 = (2, & 3, & -1, & -2, & 9), \\ \mathbf{w}_2 = (1, & 5, & -6, & 6, & 1), \\ \mathbf{w}_3 = (2, & 4, & 4, & 2, & 8), \end{cases}$$

respectively. Then the matrix Q takes the following form:

$$Q = [\mathbf{v}_1\ \mathbf{v}_2\ \mathbf{v}_3\ \mathbf{w}_1\ \mathbf{w}_2\ \mathbf{w}_3\ ] = \begin{bmatrix} 1 & 1 & 1 & 2 & 1 & 2 \\ 3 & 4 & 3 & 3 & 5 & 4 \\ -2 & -3 & 0 & -1 & -6 & 4 \\ 2 & 4 & 2 & -2 & 6 & 2 \\ 3 & 2 & 3 & 9 & 1 & 8 \end{bmatrix}.$$

After Gauss-Jordan elimination, we get

$$U = \begin{bmatrix} 1 & 0 & 0 & 5 & 0 & 0 \\ 0 & 1 & 0 & -3 & 2 & 0 \\ 0 & 0 & 1 & 0 & -1 & 0 \\ 0 & 0 & 0 & 0 & 0 & 1 \end{bmatrix}.$$

From this, one can directly see that $\dim(V + W) = 4$. The columns $\mathbf{v}_1, \mathbf{v}_2, \mathbf{v}_3, \mathbf{w}_3$ corresponding to the basic variables in $Q\mathbf{x} = \mathbf{0}$ form a basis for $\mathcal{C}(Q) = V + W$. Moreover, $\dim \mathcal{N}(Q) = \dim(V \cap W) = 2$, since there are two free variables x_4 and x_5 in $Q\mathbf{x} = \mathbf{0}$.

To find a basis for $V \cap W$, we solve $U\mathbf{x} = \mathbf{0}$ for $(x_1, x_2, x_3, 1, 0, x_5)$ and $(x_1, x_2, x_3, 0, 1, x_5)$. After a simple computation, we obtain a basis for $\mathcal{N}(Q)$:

$$\mathbf{x}_1 = (-5, 3, 0, 1, 0, 0) \text{ and } \mathbf{x}_2 = (0, -2, 1, 0, 1, 0).$$

From $Q\mathbf{x}_i = \mathbf{0}$, we obtain two equations:

$$\begin{array}{rcrcrcl} -5\mathbf{v}_1 & + & 3\mathbf{v}_2 & + & \mathbf{w}_1 & = & \mathbf{0}, \\ -2\mathbf{v}_2 & + & \mathbf{v}_3 & + & \mathbf{w}_2 & = & \mathbf{0}. \end{array}$$

Therefore, $\{\mathbf{y}_1, \mathbf{y}_2\}$ is a basis for $V \cap W$, where

$$\mathbf{y}_1 = 5\mathbf{v}_1 - 3\mathbf{v}_2 = \begin{bmatrix} 2 \\ 3 \\ -1 \\ -2 \\ 9 \end{bmatrix} = \mathbf{w}_1, \quad \mathbf{y}_2 = 2\mathbf{v}_2 - \mathbf{v}_3 = \begin{bmatrix} 1 \\ 5 \\ -6 \\ 6 \\ 1 \end{bmatrix} = \mathbf{w}_2.$$

Clearly, the equality

$$\dim(V + W) + \dim(V \cap W) = 4 + 2 = 3 + 3 = \dim V + \dim W$$

holds in this example. □

Remark: In Example 3.22, we showed a method for finding bases for $V+W$ and $V \cap W$ for given subspaces V and W of $\mathbb{R}^n$ by constructing a matrix Q whose columns are basis vectors for V and basis vectors for W. There is another method for finding their bases by constructing a matrix Q whose rows are basis vectors for V and basis vectors for W.

If Q is the matrix whose row vectors are basis vectors for V and basis vectors for W in order, then clearly $V + W = \mathcal{R}(Q)$. By finding a basis for the row space $\mathcal{R}(Q)$, we can get a basis for $V + W$.

On the other hand, a basis for $V \cap W$ can be found as follows: Let A be the $k \times n$ matrix whose rows are basis vectors for V, and B the $\ell \times n$ matrix whose rows are basis vectors for W. Then, $V = \mathcal{R}(A)$ and $W = \mathcal{R}(B)$. Let $\bar{A}$ denote the matrix A with an unknown vector $\mathbf{x} = (x_1, \ldots, x_n) \in \mathbb{R}^n$ attached at the bottom row, *i.e.*,

$$\bar{A} = \begin{bmatrix} A \\ \mathbf{x} \end{bmatrix},$$

and the matrix $\bar{B}$ is defined similarly. Then it is clear that $\mathcal{R}(A) = \mathcal{R}(\bar{A})$ and $\mathcal{R}(B) = \mathcal{R}(\bar{B})$ if and only if $\mathbf{x} \in V \cap W = \mathcal{R}(A) \cap \mathcal{R}(B)$. This means that the row-echelon form of A and that of $\bar{A}$ should be the same via the same Gaussian elimination. Thus, by comparing the row vectors of the row-echelon form of A with those of $\bar{A}$, we can obtain a system of linear equations for $\mathbf{x} = (x_1, \ldots, x_n)$. By the same argument applied to B and $\bar{B}$, we get another system of linear equations for the same $\mathbf{x} = (x_1, \ldots, x_n)$. Solutions to these two systems together will provide us with a basis for $V \cap W$.

The following example illustrates how one can apply this argument to find bases for $V + W$ and $V \cap W$.

Example 3.23 Let V be the subspace of $\mathbb{R}^5$ spanned by

$$\begin{aligned} \mathbf{v}_1 &= (1, \; 3, \; -2, \; 2, \; 3), \\ \mathbf{v}_2 &= (1, \; 4, \; -3, \; 4, \; 2), \\ \mathbf{v}_3 &= (2, \; 3, \; -1, \; -2, \; 10), \end{aligned}$$

and W the subspace spanned by

$$\begin{aligned} \mathbf{w}_1 &= (1, \; 3, \; 0, \; 2, \; 1), \\ \mathbf{w}_2 &= (1, \; 5, \; -6, \; 6, \; 3), \\ \mathbf{w}_3 &= (2, \; 5, \; 3, \; 2, \; 1). \end{aligned}$$

Find a basis for $V + W$ and for $V \cap W$.

Solution: Note that the matrix A whose row vectors are $\mathbf{v}_i$'s is reduced to a row-echelon form

$$\begin{bmatrix} 1 & 3 & -2 & 2 & 3 \\ 0 & 1 & -1 & 2 & -1 \\ 0 & 0 & 0 & 0 & 1 \end{bmatrix},$$

so that $\dim V = 3$. Similarly, the matrix B whose row vectors are $\mathbf{w}_j$'s is reduced to a row-echelon form

$$\begin{bmatrix} 1 & 3 & 0 & 2 & 1 \\ 0 & 2 & -6 & 4 & 2 \\ 0 & 0 & 0 & 0 & 0 \end{bmatrix},$$

so that $\dim W = 2$.

Now, if Q denotes the 6×5 matrix whose row vectors are $\mathbf{v}_i$'s and $\mathbf{w}_j$'s, then $V + W = \mathcal{R}(Q)$. By Gaussian elimination, Q is reduced to a row-echelon form, excluding zero rows:

$$\begin{bmatrix} 1 & 3 & -2 & 2 & 3 \\ 0 & 1 & -1 & 2 & -1 \\ 0 & 0 & 1 & 0 & -1 \\ 0 & 0 & 0 & 0 & 1 \end{bmatrix}.$$

Thus, the four nonzero row vectors

$$(1,\ 3,\ -2,\ 2,\ 3),\ (0,\ 1,\ -1,\ 2,\ -1),\ (0,\ 0,\ 1,\ 0,\ -1),\ (0,\ 0,\ 0,\ 0,\ 1)$$

form a basis for $V + W$, so that $\dim(V + W) = 4$.

We now find a basis for $V \cap W$. A vector $\mathbf{x} = (x_1,\ x_2,\ x_3,\ x_4,\ x_5) \in \mathbb{R}^5$ is contained in $V \cap W$ if and only if $\mathbf{x}$ is contained in both the row space of A and that of B.

Let $\bar{A}$ be A with $\mathbf{x}$ attached at the last row:

$$\bar{A} = \begin{bmatrix} 1 & 3 & -2 & 2 & 3 \\ 1 & 4 & -3 & 4 & 2 \\ 2 & 3 & -1 & -2 & 10 \\ x_1 & x_2 & x_3 & x_4 & x_5 \end{bmatrix}.$$

Then by the same Gaussian elimination $\bar{A}$ is reduced to

$$\begin{bmatrix} 1 & 3 & -2 & 2 & 3 \\ 0 & 1 & -1 & 2 & -1 \\ 0 & 0 & 0 & 0 & 1 \\ 0 & 0 & -x_1+x_2+x_3 & 4x_1-2x_2+x_4 & 0 \end{bmatrix}.$$

Therefore, $\mathbf{x} \in \mathcal{R}(A) = V$ if and only if $\mathcal{R}(A) = \mathcal{R}(\bar{A})$. By comparing the row vectors of the row-echelon form of A with those of $\bar{A}$, it gives that $\mathbf{x} \in \mathcal{R}(A)$ if and only if the last row vector of the row-echelon form of $\bar{A}$ is the zero vector, that is, $\mathbf{x}$ is a solution of the homogeneous system of equations

$$\begin{cases} -x_1 + x_2 + x_3 = 0 \\ 4x_1 - 2x_2 + x_4 = 0. \end{cases}$$

We do the same calculation with $\bar{B}$, and obtain another homogeneous system of linear equations for $\mathbf{x}$:

$$\begin{cases} -9x_1 + 3x_2 + x_3 = 0 \\ 4x_1 - 2x_2 + x_4 = 0 \\ 2x_1 - x_2 + x_5 = 0. \end{cases}$$

Solving these two homogeneous systems together yields

$$V \cap W = \{t(1,\ 4,\ -3,\ 4,\ 2) : t \in \mathbb{R}\}.$$

Hence, $\{(1,\ 4,\ -3,\ 4,\ 2)\}$ is a basis for $V \cap W$ and $\dim(V \cap W) = 1$. □

Problem 3.24 Let V and W be the subspaces of the vector space $P_3(\mathbb{R})$ spanned by

$$\begin{cases} v_1(x) = 3 - x + 4x^2 + x^3, \\ v_2(x) = 5 + 5x^2 + x^3, \\ v_3(x) = 5 - 5x + 10x^2 + 3x^3, \end{cases}$$

and

$$\begin{cases} w_1(x) = 9 - 3x + 3x^2 + 2x^3, \\ w_2(x) = 5 - x + 2x^2 + x^3, \\ w_3(x) = 6 + 4x^2 + x^3, \end{cases}$$

respectively. Find the dimensions and bases for $V + W$ and $V \cap W$.

Problem 3.25 Let

$$\begin{aligned} V &= \{(x,\ y,\ z,\ u) \in \mathbb{R}^4 : y + z + u = 0\}, \\ W &= \{(x,\ y,\ z,\ u) \in \mathbb{R}^4 : x + y = 0,\ z = 2u\} \end{aligned}$$

be two subspaces of $\mathbb{R}^4$. Find bases for V, W, $V + W$, and $V \cap W$.

3.7 Invertibility

We now can have the following existence and uniqueness theorems for a solution of a system of linear equations $A\mathbf{x} = \mathbf{b}$ for an $m \times n$ matrix A and a vector $\mathbf{b} \in \mathbb{R}^m$.

Theorem 3.23 (Existence) *Let A be an $m \times n$ matrix. Then the following statements are equivalent.*

(1) *For each $\mathbf{b} \in \mathbb{R}^m$, $A\mathbf{x} = \mathbf{b}$ has* at least *one solution $\mathbf{x}$ in $\mathbb{R}^n$.*

(2) *The column vectors of A span $\mathbb{R}^m$, i.e., $\mathcal{C}(A) = \mathbb{R}^m$.*

(3) rank $A = m$, *and hence $m \leq n$.*

(4) *There exists an $n \times m$ right inverse B of A such that $AB = I_m$.*

Proof: **(1)** $\Leftrightarrow$ **(2)**: Note that $\mathcal{C}(A) \subseteq \mathbb{R}^m$ in general. For any $\mathbf{b} \in \mathbb{R}^m$, there is a solution $\mathbf{x} \in \mathbb{R}^n$ of $A\mathbf{x} = \mathbf{b}$ if and only if $\mathbf{b}$ is a linear combination of the column vectors of A. This is equivalent to saying that $\mathbb{R}^m = \mathcal{C}(A)$.

(2) $\Leftrightarrow$ **(3)**: Since $\dim \mathcal{C}(A) = \text{rank } A = \dim \mathcal{R}(A) \leq \min\{m, n\}$, $\mathcal{C}(A) = \mathbb{R}^m$ if and only if $\dim \mathcal{C}(A) = m \leq n$ (see Problem 3.10).

(1) $\Rightarrow$ **(4)** : Let $\mathbf{e}_1, \mathbf{e}_2, \ldots, \mathbf{e}_m$ be the standard basis for $\mathbb{R}^m$. Then for each $i = 1, 2, \ldots, m$ we can find at least one solution $\mathbf{x}_i \in \mathbb{R}^n$ such that $A\mathbf{x}_i = \mathbf{e}_i$ by the condition. If B is the $n \times m$ matrix whose columns are these solutions, *i.e.*, $B = [\mathbf{x}_1 \ \mathbf{x}_2 \ \cdots \ \mathbf{x}_m]$, then it follows by matrix multiplication that

$$AB = A\,[\mathbf{x}_1 \ \mathbf{x}_2 \ \cdots \ \mathbf{x}_m] = [\mathbf{e}_1 \ \mathbf{e}_2 \ \cdots \ \mathbf{e}_m] = I_m.$$

Hence, the matrix B is a required right inverse.

(4) $\Rightarrow$ **(1)** : If B is a right inverse of A, then for any $\mathbf{b} \in \mathbb{R}^m$, $\mathbf{x} = B\mathbf{b}$ is a solution of $A\mathbf{x} = \mathbf{b}$. □

Condition (2) means that A has m linearly independent column vectors, and condition (3) implies that there exist m linearly independent row vectors of A, since rank $A = m = \dim \mathcal{R}(A)$.

Note that if $\mathcal{C}(A) \subsetneq \mathbb{R}^m$, then $A\mathbf{x} = \mathbf{b}$ has no solution for $\mathbf{b} \notin \mathcal{C}(A)$.

Theorem 3.24 (Uniqueness) *Let A be an $m \times n$ matrix. Then the following statements are equivalent.*

(1) *For each $\mathbf{b} \in \mathbb{R}^m$, $A\mathbf{x} = \mathbf{b}$ has* at most *one solution $\mathbf{x}$ in $\mathbb{R}^n$.*

(2) *The column vectors of A are linearly independent.*

(3) $\dim \mathcal{C}(A) = \text{rank } A = n$, *and hence* $n \le m$.

(4) $\mathcal{R}(A) = \mathbb{R}^n$.

(5) $\mathcal{N}(A) = \{\mathbf{0}\}$.

(6) *There exists an* $n \times m$ *left inverse* C *of* A *such that* $CA = I_n$.

Proof: **(1)** $\Rightarrow$ **(2)** : Note that the column vectors of A are linearly independent if and only if the homogeneous equation $A\mathbf{x} = \mathbf{0}$ has only a trivial solution. However, $A\mathbf{x} = \mathbf{0}$ has always a trivial solution $\mathbf{x} = \mathbf{0}$ and (1) means that it is the only one.

(2) $\Leftrightarrow$ **(3)** : Clear, because all the column vectors are linearly independent if and only if they form a basis for $\mathcal{C}(A)$, or $\dim \mathcal{C}(A) = n \le m$.

(3) $\Leftrightarrow$ **(4)**: Clear, because $\dim \mathcal{R}(A) = \text{rank } A = \dim \mathcal{C}(A) = n$ if and only if $\mathcal{R}(A) = \mathbb{R}^n$ (see Problem 3.10).

(4) $\Leftrightarrow$ **(5)**: Clear, since $\dim \mathcal{R}(A) + \dim \mathcal{N}(A) = n$.

(2) $\Rightarrow$ **(6)** : Suppose that the columns of A are linearly independent so that rank $A = n$. Extend these column vectors of A to a basis for $\mathbb{R}^m$ by adding $m - n$ more independent vectors to them. Construct an $m \times m$ matrix S with those vectors in columns. Then the matrix S has rank m and is hence invertible. Let C be the $n \times m$ matrix obtained from S^{-1} by throwing away the last $m - n$ rows. Since the first n columns of S constitute the matrix A, we have $CA = I_n$.

(6) $\Rightarrow$ **(1)** : Let C be a left inverse of A. If $A\mathbf{x} = \mathbf{b}$ has no solution, then we are done. Suppose that $A\mathbf{x} = \mathbf{b}$ has two solutions, say $\mathbf{x}_1$ and $\mathbf{x}_2$. Then

$$\mathbf{x}_1 = CA\mathbf{x}_1 = C\mathbf{b} = CA\mathbf{x}_2 = \mathbf{x}_2.$$

Hence, the system can have at most one solution. □

Remark: **(1)** We have proved that an $m\times n$ matrix A has a right inverse if and only if rank $A = m$, and A has a left inverse if and only if rank $A = n$. In the first case $A\mathbf{x} = \mathbf{b}$ always has a solution, and in the second case the solution (if it exists) is unique. Therefore, if $m \ne n$, A cannot have both left and right inverses.

(2) For a practical way of finding a right or a left inverse of an $m\times n$ matrix A, we will show later (see Corollary 5.24) that if rank $A = m$, then $(AA^T)^{-1}$ exists and $A^T(AA^T)^{-1}$ is a right inverse of A, and if rank $A = n$, then $(A^TA)^{-1}$ exists and $(A^TA)^{-1}A^T$ is a left inverse of A.

(3) Note that if $m = n$ so that A is a square matrix, then A has a right inverse (and a left inverse) if and only if rank $A = m = n$. Moreover, in this case the inverses are the same (see Theorem 1.8). Therefore, a square matrix A has rank n if and only if A is invertible. This means that for a square matrix "Existence = Uniqueness", and the ten conditions in the above two theorems are all equivalent. In particular, for the invertibility of a square matrix it is enough to show the existence of a one-side inverse.

Problem 3.26 For each of the following matrices, discuss the number of possible solutions to the system of linear equations $A\mathbf{x} = \mathbf{b}$ for any $\mathbf{b}$:

$$(1)\ A = \begin{bmatrix} 1 & 3 & -2 & 5 & 4 \\ 1 & 4 & 1 & 3 & 5 \\ 2 & 7 & -3 & 6 & 13 \end{bmatrix}, \qquad (2)\ A = \begin{bmatrix} 2 & 3 \\ 3 & -7 \\ -6 & 1 \end{bmatrix},$$

$$(3)\ A = \begin{bmatrix} 1 & 2 & -3 & -2 & -3 \\ 1 & 3 & -2 & 0 & -4 \\ 3 & 8 & -7 & -2 & -11 \\ 2 & 1 & -9 & -10 & -3 \end{bmatrix}, \qquad (4)\ A = \begin{bmatrix} 1 & 1 & 2 \\ 4 & 5 & 5 \\ 1 & 2 & -2 \end{bmatrix}.$$

The following theorem is a collection of the results proved in Theorems 1.8, 3.23, 3.24, and the Remark before Definition 4.3.

Theorem 3.25 *For a square matrix A of order n, the following statements are equivalent.*

(1) *A is invertible.*

(2) $\det A \neq 0$.

(3) *A is row equivalent to I_n.*

(4) *A is a product of elementary matrices.*

(5) *Elimination can be completed: $PA = LDU$, with all $d_i \neq 0$.*

(6) *$A\mathbf{x} = \mathbf{b}$ has a solution for every $\mathbf{b} \in \mathbb{R}^n$.*

(7) *$A\mathbf{x} = \mathbf{0}$ has only a trivial solution, i.e., $\mathcal{N}(A) = \{\mathbf{0}\}$.*

(8) *The columns of A are linearly independent.*

(9) *The columns of A span $\mathbb{R}^n$, i.e., $\mathcal{C}(A) = \mathbb{R}^n$.*

(10) *A has a left inverse.*

(11) rank $A = n$.

(12) *The rows of A are linearly independent.*

(13) *The rows of A span $\mathbb{R}^n$,* i.e., $\mathcal{R}(A) = \mathbb{R}^n$.

(14) *A has a right inverse.*

(15)* *The linear transformation* $A : \mathbb{R}^n \to \mathbb{R}^n$ *via* $A(\mathbf{x}) = A\mathbf{x}$ *is injective.*

(16)* *The linear transformation* $A : \mathbb{R}^n \to \mathbb{R}^n$ *is surjective.*

(17)* *Zero is not an eigenvalue of* A.

Proof: Exercise: where have we proved which claim? Prove any not covered. The numbers with asterisks will be explained in the following places: (15) and (16) in the Remark on page 141 and (17) in Theorem 6.1. □

3.8 Application: Interpolation

In many scientific experiments, a scientist wants to find the precise functional relationship between input data and output data. That is, in his experiment, he puts various input values into his experimental device and obtains output values corresponding to those input values. After his experiment, what he has is a table of inputs and outputs. The precise functional relationship might be very complicated, and sometimes it might be very hard or almost impossible to find the precise function. In this case, one thing he can do is to find a polynomial whose graph passes through each of the data points and comes very close to the function he wanted to find. That is, he is looking for a polynomial that approximates the precise function. Such a polynomial is called an **interpolating polynomial**. This problem is closely related to systems of linear equations.

Let us begin with a set of given data: Suppose that for $n+1$ distinct experimental input values $x_0, x_1, \ldots, x_n$, we obtained $n+1$ output values $y_0 = f(x_0), y_1 = f(x_1), \ldots, y_n = f(x_n)$. The output values are supposed to be related to the inputs by a certain function f. We wish to construct a polynomial $p(x)$ of degree less than or equal to n which interpolates $f(x)$ at $x_0, x_1, \ldots, x_n$: *i.e.*, $p(x_i) = y_i = f(x_i)$ for $i = 0, 1, \ldots, n$.

Note that if there is such a polynomial, it must be unique. Indeed, if $q(x)$ is another such polynomial, then $h(x) = p(x) - q(x)$ is also a polynomial of degree less than or equal to n vanishing at $n+1$ distinct points $x_0, x_1, \ldots, x_n$. Hence $h(x)$ must be the identically zero polynomial so that $p(x) = q(x)$ for all $x \in \mathbb{R}$.

In fact, the unique polynomial $p(x)$ can be found by solving a system of linear equations: If we write $p(x) = a_0 + a_1x + \cdots + a_nx^n$, then we are supposed to determine the coefficients a_i's. The set of equations

$$p(x_i) = a_0 + a_1 x_i + \cdots + a_n x_i^n = y_i = f(x_i),$$

for $i = 0, 1, \ldots, n$, constitutes a system of $n+1$ linear equations in $n+1$ unknowns a_i's:

$$\begin{bmatrix} 1 & x_0 & \cdots & x_0^n \\ 1 & x_1 & \cdots & x_1^n \\ \vdots & \vdots & & \vdots \\ 1 & x_n & \cdots & x_n^n \end{bmatrix} \begin{bmatrix} a_0 \\ a_1 \\ \vdots \\ a_n \end{bmatrix} = \begin{bmatrix} y_0 \\ y_1 \\ \vdots \\ y_n \end{bmatrix}.$$

The coefficient matrix A is a square matrix of order $n+1$, known as **Vandermonde's matrix** (see Problem 2.10), whose determinant is

$$\det A = \prod_{0 \le i < j \le n} (x_j - x_i).$$

Since the x_i's are all distinct, $\det A \neq 0$. It follows that A is nonsingular, and hence $A\mathbf{x} = \mathbf{b}$ always has a unique solution, which determines the unique polynomial $p(x)$ of degree $\le n$ passing through the given $n+1$ points (x_0, y_0), (x_1, y_1), $\cdots$, (x_n, y_n) in the plane $\mathbb{R}^2$.

Example 3.24 Given four points

$$(0, 3),\ (1, 0),\ (-1, 2),\ (3, 6)$$

in the plane $\mathbb{R}^2$, let $p(x) = a_0 + a_1 x + a_2 x^2 + a_3 x^3$ be the polynomial passing through the given four points. Then, we have a system of equations

$$\left\{ \begin{array}{lllllllll} a_0 & & & & & & & = & 3 \\ a_0 & + & a_1 & + & a_2 & + & a_3 & = & 0 \\ a_0 & - & a_1 & + & a_2 & - & a_3 & = & 2 \\ a_0 & + & 3a_1 & + & 9a_2 & + & 27a_3 & = & 6. \end{array} \right.$$

Solving this system, we find that $a_0 = 3$, $a_1 = -2$, $a_2 = -2$, $a_3 = 1$ is the unique solution, and the unique polynomial is $p(x) = 3 - 2x - 2x^2 + x^3$. $\square$

Problem 3.27 Let $f(x) = \sin x$. Then at $x = 0, \frac{\pi}{4}, \frac{\pi}{3}, \frac{3\pi}{4}, \pi$, the values of f are $y = 0, \frac{1}{\sqrt{2}}, \frac{\sqrt{3}}{2}, \frac{1}{\sqrt{2}}, 0$. Find the polynomial $p(x)$ of degree ≤ 4 that passes through these five points. (One may need to use a computer due to messy computation).

Problem 3.28 Find a polynomial $p(x) = a + bx + cx^2 + dx^3$ that satisfies $p(0) = 1$, $p'(0) = 2$, $p(1) = 4$, $p'(1) = 4$.

Problem 3.29 Find the equation of a circle that passes through the three points $(2, -2)$, $(3, 5)$, and $(-4, 6)$ in the plane $\mathbb{R}^2$.

Remark: (1) It is suggested that the readers think about the differences between this interpolation and the Taylor polynomial approximation to a differentiable function.

(2) Note again that the interpolating polynomial $p(x)$ of degree $\leq n$ is uniquely determined when we have the correct data, *i.e.*, when we are given precisely $n+1$ values of y at precisely $n+1$ distinct points $x_0, x_1, \ldots, x_n$.

However, if we are given fewer data, then the polynomial is under-determined: *i.e.*, if we have m values of y with $m < n+1$ at m distinct points $x_1, x_2, \ldots, x_m$, then there are as many interpolating polynomials as the null space of A since in this case A is an $m \times (n+1)$ matrix with $m < n+1$.

On the other hand, if we are given more than $n+1$ data, then the polynomial is over-determined: *i.e.*, if we have m values of y with $m > n+1$ at m distinct points $x_1, x_2, \ldots, x_m$, then there need not be any interpolating polynomial since the system could be inconsistent. In this case, the best we can do is to find a polynomial of degree $\leq n$ to which the data is closest. We will review this statement again in Section 5.8.

3.9 Application: The Wronskian

Let $\mathbf{y}_1, \mathbf{y}_2, \ldots, \mathbf{y}_n$ be n vectors in an m-dimensional vector space V. To check the independence of the vectors $\mathbf{y}_i$'s, consider its linear dependence:

$$c_1\mathbf{y}_1 + c_2\mathbf{y}_2 + \cdots + c_n\mathbf{y}_n = \mathbf{0}.$$

Let $\alpha = \{\mathbf{x}_1, \mathbf{x}_2, \ldots, \mathbf{x}_m\}$ be a basis for V. By expressing each $\mathbf{y}_i$ as a linear combination of the basis vectors $\mathbf{x}_i$'s, the linear dependence of $\mathbf{y}_i$'s can be written as a linear combination of the basis vectors $\mathbf{x}_i$'s, so that all of the coefficients (which are also linear combinations of c_i's) must be zero. It gives a homogeneous system of linear equations in c_i's, say $A\mathbf{c} = \mathbf{0}$ with an $m \times n$ matrix A, as in the proof of Lemma 3.8. Recall that the vectors $\mathbf{y}_i$'s are linearly independent if and only if the system $A\mathbf{c} = \mathbf{0}$ has only a trivial solution. Hence, the linear independence of a set of vectors in a finite dimensional vector space can be tested by solving a homogeneous system of linear equations. But, if V is not finite dimensional, this test for the linear independence of a set of vectors cannot be applied.

In this section, we introduce a test for the linear independence of a set of functions. For our purpose, let V be the vector space of all functions on $\mathbb{R}$ which are differentiable infinitely many times. Then one can easily see that V is not finite dimensional.

Let $f_1(x)$, $f_2(x)$, $\cdots$, $f_n(x)$ be n functions in V. The n functions are linearly independent in V if the linear equation

$$c_1 f_1(x) + c_2 f_2(x) + \cdots + c_n f_n(x) = 0$$

for all $x \in \mathbb{R}$ implies that all $c_i = 0$. By taking the differentiation $n-1$ times, we obtain n equations:

$$c_1 f_1^{(i)}(x) + c_2 f_2^{(i)}(x) + \cdots + c_n f_n^{(i)}(x) = 0, \quad 0 \le i \le n-1,$$

for all $x \in \mathbb{R}$. Or, in a matrix form:

$$\begin{bmatrix} f_1(x) & f_2(x) & \cdots & f_n(x) \\ f_1'(x) & f_2'(x) & \cdots & f_n'(x) \\ \vdots & & & \vdots \\ f_1^{(n-1)}(x) & f_2^{(n-1)}(x) & \cdots & f_n^{(n-1)}(x) \end{bmatrix} \begin{bmatrix} c_1 \\ c_2 \\ \vdots \\ c_n \end{bmatrix} = \begin{bmatrix} 0 \\ 0 \\ \vdots \\ 0 \end{bmatrix}.$$

The determinant of the coefficient matrix is called the **Wronskian** for $\{f_1(x), f_2(x), \cdots, f_n(x)\}$ and denoted by $W(x)$. Therefore, if there is a point $x_0 \in \mathbb{R}$ such that $W(x) \neq 0$, then the coefficient matrix is nonsingular at $x = x_0$, and so all $c_i = 0$. Therefore, if the Wronskian is nonzero at a point in $\mathbb{R}$, then $\{f_1(x), f_2(x), \cdots, f_n(x)\}$ are linearly independent.

Example 3.25 For the sets of functions $F_1 = \{x, \cos x, \sin x\}$ and $F_2 = \{x, e^x, e^{-x}\}$, the Wronskians are

$$W_1(x) = \det \begin{bmatrix} x & \cos x & \sin x \\ 1 & -\sin x & \cos x \\ 0 & -\cos x & -\sin x \end{bmatrix} = x$$

and

$$W_2(x) = \det \begin{bmatrix} x & e^x & e^{-x} \\ 1 & e^x & -e^{-x} \\ 0 & e^x & e^{-x} \end{bmatrix} = 2x.$$

Since $W_i(x) \neq 0$ for $x \neq 0$, both F_i are linearly independent. □

Problem 3.30 Show that $1, x, x^2, \cdots, x^n$ are linearly independent in the vector space $C(\mathbb{R})$ of continuous functions.

3.10 Exercises

3.1. Let V be the set of all pairs $(x,\ y)$ of real numbers. Define

$$\begin{aligned}(x,\ y)+(x_1,\ y_1) &= (x+x_1,\ y+y_1)\\ c(x,\ y) &= (cx,\ y).\end{aligned}$$

Is V a vector space with these operations?

3.2. For $\mathbf{x},\ \mathbf{y} \in \mathbb{R}^n$ and $k \in \mathbb{R}$, define two operations as

$$\mathbf{x} \oplus \mathbf{y} = \mathbf{x} - \mathbf{y}, \qquad k \cdot \mathbf{x} = -k\mathbf{x}.$$

The operations on the right sides are the usual ones. Which of the rules in the definition of a vector space are satisfied for $(\mathbb{R}^n,\ \oplus,\ \cdot)$?

3.3. Determine whether the given set is a vector space with the usual addition and scalar multiplication of functions.

(1) The set of all functions f defined on the interval $[-1,\ 1]$ such that $f(0) = 0$.
(2) The set of all functions f defined on $\mathbb{R}$ such that $\lim_{x\to\infty} f(x) = 0$.
(3) The set of all twice differentiable functions f defined on $\mathbb{R}$ such that $f''(x) + f(x) = 0$.

3.4. Let $C^2[-1,\ 1]$ be the vector space of all functions with continuous second derivatives on the domain $[-1,\ 1]$. Which of the following subsets is a subspace of $C^2[-1,\ 1]$?

(1) $W = \{f(x) \in C^2[-1,\ 1] : f''(x) + f(x) = 0,\ -1 \le x \le 1\}$.
(2) $W = \{f(x) \in C^2[-1,\ 1] : f''(x) + f(x) = x^2,\ -1 \le x \le 1\}$.

3.5. Which of the following subsets of $C[-1,\ 1]$ is a subspace of the vector space $C[-1,\ 1]$ of continuous functions on $[-1,\ 1]$?

(1) $W = \{f(x) \in C[-1,\ 1] : f(-1) = -f(1)\}$.
(2) $W = \{f(x) \in C[-1,\ 1] : f(x) \ge 0 \text{ for all } x \text{ in } [-1,\ 1]\}$.
(3) $W = \{f(x) \in C[-1,\ 1] : f(-1) = -2 \text{ and } f(1) = 2\}$.
(4) $W = \{f(x) \in C[-1,\ 1] : f(\frac{1}{2}) = 0\}$.

3.6. Does the vector $(3,\ -1,\ 0,\ -1)$ belong to the subspace of $\mathbb{R}^4$ spanned by the vectors $(2,\ -1,\ 3,\ 2)$, $(-1,\ 1,\ 1,\ -3)$ and $(1,\ 1,\ 9,\ -5)$?

3.7. Express the given function as a linear combination of functions in the given set Q.

(1) $p(x) = -1 - 3x + 3x^2$ and $Q = \{p_1(x),\ p_2(x),\ p_3(x)\}$, where $p_1(x) = 1 + 2x + x^2$, $p_2(x) = 2 + 5x$, $p_3(x) = 3 + 8x - 2x^2$.
(2) $p(x) = -2 - 4x + x^2$ and $Q = \{p_1(x),\ p_2(x),\ p_3(x),\ p_4(x)\}$, where $p_1(x) = 1 + 2x^2 + x^3$, $p_2(x) = 1 + x + 2x^3$, $p_3(x) = -1 - 3x - 4x^3$, $p_4(x) = 1 + 2x - x^2 + x^3$.

3.8. Is $\{\cos^2 x,\ \sin^2 x,\ 1,\ e^x\}$ linearly independent in the vector space $C(\mathbb{R})$?

3.9. Show that the given sets of functions are linearly independent in the vector space $C[-\pi,\ \pi]$.

(1) $\{1,\ x,\ x^2,\ x^3,\ x^4\}$

(2) $\{1,\ e^x,\ e^{2x},\ e^{3x}\}$

(3) $\{1,\ \sin x,\ \cos x,\ \ldots,\ \sin kx,\ \cos kx\}$

3.10. Are the vectors

$$\mathbf{v}_1 = (1,\ 1,\ 2,\ 4), \qquad \mathbf{v}_2 = (2,\ -1,\ -5,\ 2),$$
$$\mathbf{v}_3 = (1,\ -1,\ -4,\ 0), \quad \mathbf{v}_4 = (2,\ 1,\ 1,\ 6)$$

linearly independent in the 4-space $\mathbb{R}^4$?

3.11. In the 3-space $\mathbb{R}^3$, let W be the set of all vectors $(x_1,\ x_2,\ x_3)$ that satisfy the equation $x_1 - x_2 - x_3 = 0$. Prove that W is a subspace of $\mathbb{R}^3$. Find a basis for the subspace W.

3.12. With respect to the basis $\alpha = \{1,\ x,\ x^2\}$ for the vector space $P_2(\mathbb{R})$, find the coordinate vector of the following polynomials:

$$(1)\ f(x) = x^2 - x + 1, \quad (2)\ f(x) = x^2 + 4x - 1, \quad (3)\ f(x) = 2x + 5.$$

3.13. Let W be the subspace of $C[-\pi,\ \pi]$ consisting of functions of the form $f(x) = a\sin x + b\cos x$. Determine the dimension of W.

3.14. Let V denote the set of all infinite sequences of real numbers:

$$V = \{\mathbf{x} : \mathbf{x} = \{x_i\}_{i=1}^{\infty}, x_i \in \mathbb{R}\}.$$

If $\mathbf{x} = \{x_i\}$ and $\mathbf{y} = \{y_i\}$ are in V, then $\mathbf{x} + \mathbf{y}$ is the sequence $\{x_i + y_i\}_{i=1}^{\infty}$. If c is a real number, then $c\mathbf{x}$ is the sequence $\{cx_i\}_{i=1}^{\infty}$.

(1) Prove that V is a vector space.

(2) Prove that V is not finite dimensional.

3.15. For two matrices A and B for which AB can be defined, prove the following statements:

(1) If both A and B have linearly independent column vectors, then the column vectors of AB are also linearly independent.

(2) If both A and B have linearly independent row vectors, then the row vectors of AB are also linearly independent.

(3) If the column vectors of B are linearly dependent, then the column vectors of AB are also linearly dependent.

(4) If the row vectors of A are linearly dependent, then the row vectors of AB are also linearly dependent.

3.16. Let $U = \{(x,\ y,\ z)\ :\ 2x + 3y + z = 0\}$ and $V = \{(x,\ y,\ z)\ :\ x + 2y - z = 0\}$ be subspaces of $\mathbb{R}^3$.

(1) Find a basis for $U \cap V$.
(2) Determine the dimension of $U + V$.
(3) Describe U, V, $U \cap V$ and $U + V$ geometrically.

3.17. How many 5×5 permutation matrices are there? Are they linearly independent? Do they span the vector space $M_{5\times5}(\mathbb{R})$?

3.18. Find bases for the row space, the column space, and the null space for each of the following matrices.

$$(1)\ A = \begin{bmatrix} 1 & 2 & 1 & 5 \\ 2 & 4 & -3 & 0 \\ 1 & 2 & -1 & 1 \end{bmatrix}, \quad (2)\ B = \begin{bmatrix} 0 & 2 & 1 & -5 \\ 1 & 1 & -2 & 2 \\ 1 & 5 & 0 & 0 \end{bmatrix},$$

$$(3)\ C = \begin{bmatrix} 1 & 3 & 2 \\ 2 & 6 & 4 \\ 3 & 9 & 6 \end{bmatrix}, \quad (4)\ D = \begin{bmatrix} 0 & 1 & -1 & -2 & 1 \\ 1 & 1 & -1 & 3 & 1 \\ 2 & 1 & -1 & 8 & 3 \\ 0 & 0 & -2 & 2 & 1 \\ 3 & 5 & -5 & 5 & 10 \end{bmatrix}.$$

3.19. Find the rank of A as a function of x: $A = \begin{bmatrix} 2 & 2 & -6 & 8 \\ 3 & 3 & -9 & 8 \\ 1 & 1 & x & 4 \end{bmatrix}$.

3.20. Find the rank and the largest invertible submatrix of each of the following matrices.

$$(1) \begin{bmatrix} 0 & 0 & 0 & 1 \\ 0 & 0 & 1 & 0 \\ 0 & 1 & 0 & 0 \\ 0 & 0 & 0 & 0 \end{bmatrix}, \quad (2) \begin{bmatrix} 1 & 1 & 0 & 1 \\ 2 & 1 & 1 & 2 \\ 1 & 1 & 1 & 4 \end{bmatrix}, \quad (3) \begin{bmatrix} 1 & 2 & 3 & 1 \\ 1 & 4 & 0 & 1 \\ 0 & 2 & 3 & 0 \\ 1 & 0 & 0 & 0 \end{bmatrix}.$$

3.21. For any nonzero column vectors $\mathbf{u}$, $\mathbf{v}$, show that the matrix $A = \mathbf{u}\mathbf{v}^T$ has rank 1. Conversely, every matrix of rank 1 can be written as $\mathbf{u}\mathbf{v}^T$ for some $\mathbf{u}$, $\mathbf{v}$.

3.22. Determine whether the following statements are true or false, and justify your answers.

(1) The set of all $n \times n$ matrices A such that $A^T = A^{-1}$ is a subspace of the vector space $M_{n\times n}(\mathbb{R})$.
(2) If α and β are linearly independent subsets of a vector space V, then so is their union $\alpha \cup \beta$.
(3) If U and W are subspaces of a vector space V with bases α and β respectively, then the intersection $\alpha \cap \beta$ is a basis for $U \cap W$.
(4) Let U be the row-echelon form of a square matrix A. If the first r columns of U are linearly independent, then so are the first r columns of A.
(5) Any two row-equivalent matrices have the same column space.

(6) Let A be an $m \times n$ matrix with rank m. Then the column vectors of A span $\mathbb{R}^m$.

(7) Let A be an $m \times n$ matrix with rank n. Then $A\mathbf{x} = \mathbf{b}$ has at most one solution.

(8) If U is a subspace of V and $\mathbf{x}$, $\mathbf{y}$ are vectors in V such that $\mathbf{x} + \mathbf{y}$ is contained in U, then $\mathbf{x} \in U$ and $\mathbf{y} \in U$.

(9) Let U and V are vector spaces. Then U is a subspace of V if and only if $\dim U \leq \dim V$.

(10) For any $m \times n$ matrix A, $\dim \mathcal{C}(A) + \dim \mathcal{N}(A^T) = m$.

Chapter 4

Linear Transformations

4.1 Introduction

As we saw in Chapter 3, there are many vector spaces. Naturally, one can ask whether or not two vector spaces are the same. To say two vector spaces are the same or not, one has to compare them first as sets, and then see whether or not their arithmetical rules are preserved. A usual way of comparing two sets is defining a **function** between them. Recall that a function from a set X into another set Y is a rule which assigns a unique element y in Y to each element x in X. Such a function is denoted as $f : X \to Y$ and sometimes referred to as a **transformation** or a **mapping**. We say that f transforms (or maps) X into Y. When given sets are vector spaces, one can compare their arithmetical rules also by a transformation f if f preserves the arithmetical rules, that is, $f(\mathbf{x}+\mathbf{y}) = f(\mathbf{x}) + f(\mathbf{y})$ and $f(k\mathbf{x}) = kf(\mathbf{x})$ for any vectors $\mathbf{x}, \mathbf{y}$ and any scalar k. In this chapter, we discuss this kind of transformations between vector spaces via the linear equation $A\mathbf{x} = \mathbf{b}$.

For an $m \times n$ matrix A, the equation $A\mathbf{x} = \mathbf{b}$ means that to every vector $\mathbf{x} = [x_1\ x_2\ \cdots\ x_n]^T$ in $\mathbb{R}^n$ the matrix multiplication $A\mathbf{x}$ assigns a vector $\mathbf{b}$ $(= A\mathbf{x})$ in $\mathbb{R}^m$. That is, the matrix A transforms every vector $\mathbf{x}$ in $\mathbb{R}^n$ into a vector $\mathbf{b}$ in $\mathbb{R}^m$ by the matrix multiplication $A\mathbf{x} = \mathbf{b}$. Moreover, the distributive law $A(\mathbf{x} + k\mathbf{y}) = A\mathbf{x} + kA\mathbf{y}$, for $k \in \mathbb{R}$ and $\mathbf{x}, \mathbf{y} \in \mathbb{R}^n$, of matrix multiplication means that A preserves the sum of vectors and scalar multiplication.

Definition 4.1 Let V and W be vector spaces. A function $T : V \to W$ is called a **linear transformation** from V to W if for all $\mathbf{x},\ \mathbf{y} \in V$ and scalar k the following conditions hold:

(1) $T(\mathbf{x}+\mathbf{y}) = T(\mathbf{x}) + T(\mathbf{y})$,
(2) $T(k\mathbf{x}) = kT(\mathbf{x})$.

We often call T simply **linear**. It is not hard to see that the two conditions for a linear transformation can be combined into a single requirement

$$T(\mathbf{x}+k\mathbf{y}) = T(\mathbf{x}) + kT(\mathbf{y}).$$

Geometrically, this is just the requirement for a straight line to be transformed into a straight line, since $\mathbf{x}+k\mathbf{y}$ represents a straight line through $\mathbf{x}$ in the direction $\mathbf{y}$ in V, and its image $T(\mathbf{x})+kT(\mathbf{y})$ also represents a straight line through $T(\mathbf{x})$ in the direction of $T(\mathbf{y})$ in W. The following theorem is a direct consequence of the definition, and the proof is left for an exercise.

Theorem 4.1 *Let $T : V \to W$ be a linear transformation. Then*
(1) $T(\mathbf{0}) = \mathbf{0}$.
(2) *For any $\mathbf{x}_1, \mathbf{x}_2, \ldots, \mathbf{x}_n \in V$ and scalars $k_1, k_2, \ldots, k_n$,*

$$T(k_1\mathbf{x}_1 + k_2\mathbf{x}_2 + \cdots + k_n\mathbf{x}_n) = k_1T(\mathbf{x}_1) + k_2T(\mathbf{x}_2) + \cdots + k_nT(\mathbf{x}_n).$$

Example 4.1 Consider the following functions:
(1) $f : \mathbb{R} \to \mathbb{R}$ defined by $f(x) = 2x$;
(2) $g : \mathbb{R} \to \mathbb{R}$ defined by $g(x) = x^2 - x$;
(3) $h : \mathbb{R}^2 \to \mathbb{R}^2$ defined by $h(x, y) = (x - y, 2x)$;
(4) $k : \mathbb{R}^2 \to \mathbb{R}^2$ defined by $k(x, y) = (xy, x^2 + 1)$.
One can easily see that g and k are not linear, while f and h are linear.

Example 4.2 **(1)** For an $m \times n$ matrix A, the transformation $T : \mathbb{R}^n \to \mathbb{R}^m$ defined by the matrix multiplication

$$T(\mathbf{x}) = A\mathbf{x}$$

is a linear transformation by the distributive law $A(\mathbf{x}+k\mathbf{y}) = A\mathbf{x}+kA\mathbf{y}$ for any $\mathbf{x}, \mathbf{y} \in \mathbb{R}^n$ and for any scalar $k \in \mathbb{R}$. Therefore, a matrix A, identified with T, may be considered to be a linear transformation of $\mathbb{R}^n$ to $\mathbb{R}^m$.

(2) For a vector space V, the **identity transformation** $Id : V \to V$ is defined by $Id(\mathbf{x}) = \mathbf{x}$ for all $\mathbf{x} \in V$. If W is another vector space, the **zero transformation** $T_0 : V \to W$ is defined by $T_0(\mathbf{x}) = \mathbf{0}$ (the zero vector) for all $\mathbf{x} \in V$. Clearly, both transformations are linear. □

Nontrivial important examples of linear transformations are the rotations, reflections, and projections in geometry defined in the following example.

Example 4.3 **(1)** Let θ denote the angle between the x-axis and a fixed vector in $\mathbb{R}^2$. Then the matrix

$$R_\theta = \begin{bmatrix} \cos\theta & -\sin\theta \\ \sin\theta & \cos\theta \end{bmatrix}$$

defines a linear transformation on $\mathbb{R}^2$ that rotates any vector in $\mathbb{R}^2$ through the angle θ about the origin, and is called a **rotation** by the angle θ.

(2) The **projection** on the x-axis is the linear transformation $T : \mathbb{R}^2 \to \mathbb{R}^2$ defined by, for $\mathbf{x} = (x, y) \in \mathbb{R}^2$,

$$T(\mathbf{x}) = \begin{bmatrix} 1 & 0 \\ 0 & 0 \end{bmatrix} \begin{bmatrix} x \\ y \end{bmatrix} = \begin{bmatrix} x \\ 0 \end{bmatrix}.$$

(3) The linear transformation $T : \mathbb{R}^2 \to \mathbb{R}^2$ defined by, for $\mathbf{x} = (x, y)$,

$$T(\mathbf{x}) = \begin{bmatrix} 1 & 0 \\ 0 & -1 \end{bmatrix} \begin{bmatrix} x \\ y \end{bmatrix} = \begin{bmatrix} x \\ -y \end{bmatrix}$$

is called the **reflection** about the x-axis. □

Problem 4.1 Find the matrix of reflection about the line $y = x$ in the plane $\mathbb{R}^2$.

Example 4.4 The transformation $\mathrm{tr} : M_{n\times n}(\mathbb{R}) \to \mathbb{R}$ defined as the sum of diagonal entries

$$\mathrm{tr}(A) = a_{11} + a_{22} + \cdots + a_{nn} = \sum_{i=1}^{n} a_{ii},$$

for $A = [a_{ij}] \in M_{n\times n}(\mathbb{R})$, is called the **trace**. It is easy to show that

$$\mathrm{tr}(A + B) = \mathrm{tr}(A) + \mathrm{tr}(B) \quad \text{and} \quad \mathrm{tr}(kA) = k\ \mathrm{tr}(A)$$

for any matrices A and B in $M_{n\times n}(\mathbb{R})$, which means that "tr" is a linear transformation. In particular, one can easily show that the set of all $n\times n$ matrices with trace 0 is a subspace of $M_{n\times n}(\mathbb{R})$. □

Problem 4.2 Let $W = \{A \in M_{n\times n}(\mathbb{R}) : \mathrm{tr}(A) = 0\}$. Show that W is a subspace, and then find a basis for W.

Problem 4.3 Show that, for any matrices A and B in $M_{n\times n}(\mathbb{R})$, $\text{tr}\,(AB) = \text{tr}\,(BA)$.

Example 4.5 From the calculus, it is well known that two transformations

$$D : P_n(\mathbb{R}) \to P_{n-1}(\mathbb{R}), \quad \mathcal{I} : P_{n-1}(\mathbb{R}) \to P_n(\mathbb{R})$$

defined by differentiation and integration,

$$D(f)(x) = f'(x), \quad \mathcal{I}(f)(x) = \int_0^x f(t)dt,$$

satisfy linearity, and so they are linear transformations. Many problems related with differential and integral equations may be reformulated in terms of linear transformations. □

Definition 4.2 Let V and W be two vector spaces, and let $T : V \to W$ be a linear transformation from V into W.

(1) $\text{Ker}(T) = \{\mathbf{v} \in V : T(\mathbf{v}) = \mathbf{0}\} \subseteq V$ is called the **kernel** of T.

(2) $\text{Im}(T) = \{T(\mathbf{v}) \in W : \mathbf{v} \in V\} = T(V) \subseteq W$ is called the **image** of T.

Example 4.6 Let V and W be vector spaces and let $Id : V \to V$ and $T_0 : V \to W$ be the identity and the zero transformations, respectively. Then it is easy to see that $\text{Ker}(Id) = \{\mathbf{0}\}$, $\text{Im}(Id) = V$, $\text{Ker}(T_0) = V$, and $\text{Im}(T_0) = \{\mathbf{0}\}$. □

Theorem 4.2 *Let $T : V \to W$ be a linear transformation from a vector space V to a vector space W. Then the kernel* $\text{Ker}(T)$ *and the image* $\text{Im}(T)$ *are subspaces of V and W, respectively.*

Proof: Since $T(\mathbf{0}) = \mathbf{0}$, each of $\text{Ker}(T)$ and $\text{Im}(T)$ is nonempty having $\mathbf{0}$. multiplication.

(1) For any $\mathbf{x}, \ \mathbf{y} \in \text{Ker}(T)$ and for any scalar k,

$$T(\mathbf{x} + k\mathbf{y}) = T(\mathbf{x}) + kT(\mathbf{y}) = \mathbf{0} + k\mathbf{0} = \mathbf{0}.$$

Hence $\mathbf{x} + k\mathbf{y} \in \text{Ker}(T)$ so that $\text{Ker}(T)$ is a subspace of V.

(2) If $\mathbf{v}, \ \mathbf{w} \in \text{Im}(T)$, then there exist $\mathbf{x}$ and $\mathbf{y}$ in V such that $T(\mathbf{x}) = \mathbf{v}$ and $T(\mathbf{y}) = \mathbf{w}$. Thus, for any scalar k,

$$\mathbf{v} + k\mathbf{w} = T(\mathbf{x}) + kT(\mathbf{y}) = T(\mathbf{x} + k\mathbf{y}).$$

Thus $\mathbf{v} + k\mathbf{w} \in \text{Im}(T)$, so that $\text{Im}(T)$ is a subspace of W. □

Example 4.7 Let $A : \mathbb{R}^n \to \mathbb{R}^m$ be the linear transformation defined by an $m \times n$ matrix A as in Example 4.2 (1). The kernel $\mathrm{Ker}(A)$ of A consists of all solution vectors $\mathbf{x}$ of the homogeneous system $A\mathbf{x} = \mathbf{0}$. Therefore, the kernel $\mathrm{Ker}(A)$ of A is nothing but the null space $\mathcal{N}(A)$ of the matrix A, and the image $\mathrm{Im}(A)$ of A is just the column space $\mathcal{C}(A) = \mathrm{Im}(A) = A(\mathbb{R}^n) \subseteq \mathbb{R}^m$ of the matrix A. Recall that $A\mathbf{x}$ is a linear combination of the column vectors of A. □

One of the most important properties of linear transformations is that they are completely determined by their values on a basis.

Theorem 4.3 *Let V and W be vector spaces. Let $\{\mathbf{v}_1, \ldots, \mathbf{v}_n\}$ be a basis for V and let $\mathbf{w}_1, \ldots, \mathbf{w}_n$ be any vectors (possibly repeated) in W. Then there exists a unique linear transformation $T : V \to W$ such that $T(\mathbf{v}_i) = \mathbf{w}_i$ for $i = 1, \ldots, n$.*

Proof: Let $\mathbf{x} \in V$. Then it has a unique expression: $\mathbf{x} = \sum_{i=1}^n a_i\mathbf{v}_i$ for some scalars $a_1, \ldots, a_n$. Define

$$T : V \to W \quad \text{by} \quad T(\mathbf{x}) = \sum_{i=1}^n a_i\mathbf{w}_i.$$

In particular, $T(\mathbf{v}_i) = \mathbf{w}_i$ for $i = 1, 2, \ldots, n$.

Linearity: For $\mathbf{x} = \sum_{i=1}^n a_i\mathbf{v}_i$, $\mathbf{y} = \sum_{i=1}^n b_i\mathbf{v}_i \in V$ and k a scalar, we have $\mathbf{x} + k\mathbf{y} = \sum_{i=1}^n (a_i + kb_i)\mathbf{v}_i$. Then

$$T(\mathbf{x} + k\mathbf{y}) = \sum_{i=1}^n (a_i + kb_i)\mathbf{w}_i = \sum_{i=1}^n a_i\mathbf{w}_i + k\sum_{i=1}^n b_i\mathbf{w}_i = T(\mathbf{x}) + kT(\mathbf{y}).$$

Uniqueness: Suppose that $S : V \to W$ is linear and $S(\mathbf{v}_i) = \mathbf{w}_i$ for $i = 1, \ldots, n$. Then for any $\mathbf{x} \in V$ with $\mathbf{x} = \sum_{i=1}^n a_i\mathbf{v}_i$, we have

$$S(\mathbf{x}) = \sum_{i=1}^n a_iS(\mathbf{v}_i) = \sum_{i=1}^n a_i\mathbf{w}_i = T(\mathbf{x}).$$

Hence, we have $S = T$. □

Therefore, from an assignment $T(\mathbf{v}_i) = \mathbf{w}_i$ of an arbitrary vector in $W$to each vector $\mathbf{v}_i$ in a basis for V, one can extend it uniquely to a linear transformation T from a vector space V into W. The uniqueness in the above theorem may be rephrased as the following corollary.

Corollary 4.4 *Let V and W be vector spaces, and let $\{\mathbf{v}_1, \ldots, \mathbf{v}_n\}$ be a basis for V. If $S, T : V \to W$ are linear transformations and $S(\mathbf{v}_i) = T(\mathbf{v}_i)$ for $i = 1, \ldots, n$, then $S = T$, i.e., $S(\mathbf{x}) = T(\mathbf{x})$ for all $\mathbf{x} \in V$.*

Example 4.8 Let $\mathbf{w}_1 = (1, 0)$, $\mathbf{w}_2 = (2, -1)$, $\mathbf{w}_3 = (4, 3)$ be three vectors in $\mathbb{R}^2$.

(1) Let $\alpha = \{\mathbf{e}_1, \mathbf{e}_2, \mathbf{e}_3\}$ be the standard basis for the 3-space $\mathbb{R}^3$, and let $T : \mathbb{R}^3 \to \mathbb{R}^2$ be the linear transformation defined by

$$T(\mathbf{e}_1) = \mathbf{w}_1, \qquad T(\mathbf{e}_2) = \mathbf{w}_2, \qquad T(\mathbf{e}_3) = \mathbf{w}_3.$$

Find a formula for $T(x_1, x_2, x_3)$, and then use it to compute $T(2, -3, 5)$.

(2) Let $\beta = \{\mathbf{v}_1, \mathbf{v}_2, \mathbf{v}_3\}$ be another basis for $\mathbb{R}^3$, where $\mathbf{v}_1 = (1, 1, 1)$, $\mathbf{v}_2 = (1, 1, 0)$, $\mathbf{v}_3 = (1, 0, 0)$, and let $T : \mathbb{R}^3 \to \mathbb{R}^2$ be the linear transformation defined by

$$T(\mathbf{v}_1) = \mathbf{w}_1, \qquad T(\mathbf{v}_2) = \mathbf{w}_2, \qquad T(\mathbf{v}_3) = \mathbf{w}_3.$$

Find a formula for $T(x_1, x_2, x_3)$, and then use it to compute $T(2, -3, 5)$.

Solution: **(1)** For $\mathbf{x} = (x_1, x_2, x_3) = x_1\mathbf{e}_1 + x_2\mathbf{e}_2 + x_3\mathbf{e}_3 \in \mathbb{R}^3$,

$$\begin{aligned} T(\mathbf{x}) &= \sum_{i=1}^{3} x_i T(\mathbf{e}_i) = \sum_{i=1}^{3} x_i \mathbf{w}_i \\ &= x_1(1, 0) + x_2(2, -1) + x_3(4, 3) \\ &= (x_1 + 2x_2 + 4x_3, -x_2 + 3x_3). \end{aligned}$$

Thus, $T(2, -3, 5) = (16, 18)$. In matrix notation, this can be written as

$$\begin{bmatrix} 1 & 2 & 4 \\ 0 & -1 & 3 \end{bmatrix} \begin{bmatrix} x_1 \\ x_2 \\ x_3 \end{bmatrix} = \begin{bmatrix} x_1 + 2x_2 + 4x_3 \\ -x_2 + 3x_3 \end{bmatrix}.$$

(2) In this case, we need to express $\mathbf{x} = (x_1, x_2, x_3)$ as a linear combination of $\mathbf{v}_1, \mathbf{v}_2, \mathbf{v}_3$, *i.e.*,

$$\begin{aligned} (x_1, x_2, x_3) = \sum_{i=1}^{3} k_i \mathbf{v}_i &= k_1(1, 1, 1) + k_2(1, 1, 0) + k_3(1, 0, 0) \\ &= (k_1 + k_2 + k_3)\mathbf{e}_1 + (k_1 + k_2)\mathbf{e}_2 + k_1\mathbf{e}_3. \end{aligned}$$

By equating corresponding components we obtain a system of equations

$$\begin{cases} k_1 + k_2 + k_3 = x_1 \\ k_1 + k_2 = x_2 \\ k_1 = x_3 . \end{cases}$$

The solution is $k_1 = x_3$, $k_2 = x_2 - x_3$, $k_3 = x_1 - x_2$. Therefore,

$$\begin{aligned} (x_1, x_2, x_3) &= x_3\mathbf{v}_1 + (x_2 - x_3)\mathbf{v}_2 + (x_1 - x_2)\mathbf{v}_3, \quad \text{and} \\ T(x_1, x_2, x_3) &= x_3T(\mathbf{v}_1) + (x_2 - x_3)T(\mathbf{v}_2) + (x_1 - x_2)T(\mathbf{v}_3) \\ &= x_3(1, 0) + (x_2 - x_3)(2, -1) + (x_1 - x_2)(4, 3) \\ &= (4x_1 - 2x_2 - x_3, 3x_1 - 4x_2 + x_3). \end{aligned}$$

From this formula we obtain $T(2, -3, 5) = (9, 23)$. In matrix notation, it can be written as

$$\begin{bmatrix} 4 & -2 & -1 \\ 3 & -4 & 1 \end{bmatrix} \begin{bmatrix} x_1 \\ x_2 \\ x_3 \end{bmatrix} = \begin{bmatrix} 4x_1 - 2x_2 - x_3 \\ 3x_1 - 4x_2 + x_3 \end{bmatrix}.$$ □

Problem 4.4 Is there a linear transformation $T : \mathbb{R}^3 \to \mathbb{R}^2$ such that $T(3, 1, 0) = (1, 1)$ and $T(-6, -2, 0) = (2, 1)$? If yes, can you find an expression of $T(\mathbf{x})$ for $\mathbf{x} = (x_1, x_2, x_3)$ in $\mathbb{R}^3$?

Problem 4.5 Let V and W be vector spaces and $T : V \to W$ be linear. Let $\{\mathbf{w}_1, \mathbf{w}_2, \ldots, \mathbf{w}_k\}$ be a linearly independent subset of the image $\mathrm{Im}(T) \subseteq W$. Suppose that $\alpha = \{\mathbf{v}_1, \mathbf{v}_2, \ldots, \mathbf{v}_k\}$ is chosen so that $T(\mathbf{v}_i) = \mathbf{w}_i$ for $i = 1, 2, \ldots, k$. Prove that α is linearly independent.

4.2 Invertible linear transformations

Note that a function f from a set X to a set Y is said to be **invertible** if there is a function g, which is called the **inverse** function of f and denoted by $g = f^{-1}$, from Y to X such that their compositions satisfy $g \circ f = Id$ and $f \circ g = Id$. One can notice that if there exists an invertible function from a set X into another set Y, then it gives a one-to-one correspondence between these two sets so that they can be identified as sets. A useful criterion for a function between two given sets to be invertible is that it is one-to-one and onto. Recall that a function $f : X \to Y$ is said to be **one-to-one** (or

injective) if $f(u) = f(v)$ in Y implies $u = v$ in X, and said to be **onto** (or **surjective**) if for each element y in Y there is an element x in X such that $f(x) = y$. A function is said to be **bijective** if it is both one-to-one and onto, that is, if for each element y in Y there is a *unique* element x in X such that $f(x) = y$.

Lemma 4.5 *A function $f : X \to Y$ is invertible if and only if it is bijective (or one-to-one and onto).*

Proof: Suppose $f : X \to Y$ is invertible, and let $g : Y \to X$ be its inverse. If $f(u) = f(v)$, then $u = g(f(u)) = g(f(v)) = v$. Thus f is one-to-one. For each $y \in Y$, $g(y) = x \in X$. Then $f(x) = f(g(y)) = y$. Thus it is onto.

Conversely, suppose f is bijective. Then, for each $y \in Y$, there is unique $x \in X$ such that $f(x) = y$. Now for each $y \in Y$ define $g(y) = x$. Then one can easily check that $g : Y \to X$ is a well-defined function such that $f \circ g = Id$ and $g \circ f = Id$, *i.e.*, g is the inverse of f. □

The following lemma shows that if a given function is an *invertible linear* transformation from a *vector space* into another, then the linearity is also preserved by the inversion.

Lemma 4.6 *Let V and W be vector spaces. If $T : V \to W$ is an invertible linear transformation, then its inverse $T^{-1} : W \to V$ is also linear.*

Proof: Let $\mathbf{w}_1$, $\mathbf{w}_2 \in W$, and let k be any scalar. Since T is invertible, it is one-to-one and onto, so there exist unique vectors $\mathbf{v}_1$ and $\mathbf{v}_2$ in V such that $T(\mathbf{v}_1) = \mathbf{w}_1$ and $T(\mathbf{v}_2) = \mathbf{w}_2$. Then

$$\begin{aligned} T^{-1}(\mathbf{w}_1 + k\mathbf{w}_2) &= T^{-1}\left(T(\mathbf{v}_1) + kT(\mathbf{v}_2)\right) \\ &= T^{-1}\left(T(\mathbf{v}_1 + k\mathbf{v}_2)\right) \\ &= \mathbf{v}_1 + k\mathbf{v}_2 \\ &= T^{-1}(\mathbf{w}_1) + kT^{-1}(\mathbf{w}_2). \end{aligned}$$ □

Definition 4.3 A linear transformation $T : V \to W$ from a vector space V to a vector space W is called an **isomorphism** if it is invertible (or one-to-one and onto). In this case, we say V and W are **isomorphic** to each other.

Lemma 4.6 shows that if T is an isomorphism, then its inverse T^{-1} is also an isomorphism with $(T^{-1})^{-1} = T$. Therefore, if V and W are isomorphic to each other, then it means that they look the same as vector spaces.

If $T : V \to W$ and $S : W \to Z$ are linear transformations, then it is quite easy to show that their composition $(S \circ T)(\mathbf{v}) = S(T(\mathbf{v}))$ is also a linear transformation from V to Z. In particular, if two linear transformations are given by matrices $A : \mathbb{R}^n \to \mathbb{R}^m$ and $B : \mathbb{R}^m \to \mathbb{R}^k$, then their composition is nothing but the matrix multiplication BA of them, *i.e.*, $(B \circ A)(\mathbf{x}) = B(A\mathbf{x}) = (BA)\mathbf{x}$. Hence, if a linear transformation is given by an invertible $n \times n$ square matrix $A : \mathbb{R}^n \to \mathbb{R}^n$, then the inverse matrix A^{-1} plays the inverse linear transformation, so that it is an isomorphism of $\mathbb{R}^n$. That is, a linear transformation given by an $n \times n$ square matrix $A : \mathbb{R}^n \to \mathbb{R}^n$ is an isomorphism if and only if rank $A = n$.

Problem 4.6 Suppose that S and T are linear transformations whose composition $S \circ T$ is well-defined. Prove that

(1) if $S \circ T$ is one-to-one, so is T,
(2) if $S \circ T$ is onto, so is S,
(3) if S and T are isomorphisms, then so is $S \circ T$,
(4) if A and B are two $n \times n$ matrices of rank n, then so is AB.

Theorem 4.7 *Two vector spaces V and W are isomorphic if and only if* $\dim V = \dim W$.

Proof: Let $T : V \to W$ be an isomorphism, and let $\{\mathbf{v}_1, \ldots, \mathbf{v}_n\}$ be a basis for V. Then we show that the set $\{T(\mathbf{v}_1), \ldots, T(\mathbf{v}_n)\}$ is a basis for W so that $\dim W = n = \dim V$.

(1) *It is linearly independent:* Since T is one-to-one, the equation

$$\mathbf{0} = c_1 T(\mathbf{v}_1) + \cdots + c_n T(\mathbf{v}_n) = T(c_1\mathbf{v}_1 + \cdots + c_n\mathbf{v}_n)$$

implies that $\mathbf{0} = c_1\mathbf{v}_1 + \cdots + c_n\mathbf{v}_n$. Since the $\mathbf{v}_i$'s are linearly independent, we have $c_i = 0$ for all $i = 1, \ldots, n$.

(2) *It spans W:* Since T is onto, for any $\mathbf{y} \in W$ there exists an $\mathbf{x} \in V$ such that $T(\mathbf{x}) = \mathbf{y}$. Write $\mathbf{x} = \sum_{i=1}^n a_i\mathbf{v}_i$. Then

$$\mathbf{y} = T(\mathbf{x}) = T(a_1\mathbf{v}_1 + \cdots + a_n\mathbf{v}_n) = a_1 T(\mathbf{v}_1) + \cdots + a_n T(\mathbf{v}_n),$$

i.e., $\mathbf{y}$ is a linear combination of $T(\mathbf{v}_1), \cdots, T(\mathbf{v}_n)$.

Conversely, suppose that $\dim V = \dim W$. Then one can choose any bases $\{\mathbf{v}_1, \ldots, \mathbf{v}_n\}$ and $\{\mathbf{w}_1, \ldots, \mathbf{w}_n\}$ for V and W, respectively. By Theorem 4.3 there exists a linear transformation $T : V \to W$ such that $T(\mathbf{v}_i) = \mathbf{w}_i$ for $i = 1, \cdots, n$. It is not hard to show that T is invertible so that T is an isomorphism. Hence V and W are isomorphic. □

Problem 4.7 Let $T : V \to W$ be a linear transformation. Prove that
(1) T is one-to-one if and only if $\text{Ker}(T) = \{\mathbf{0}\}$,
(2) if $V = W$, then T is one-to-one if and only if T is onto.

Corollary 4.8 *Any n-dimensional vector space V is isomorphic to the n-space $\mathbb{R}^n$.*

An **ordered basis** for a vector space is a basis endowed with a specific order. Let V be a vector space of dimension n with an ordered basis $\alpha = \{\mathbf{v}_1, \ldots, \mathbf{v}_n\}$. Let $\beta = \{\mathbf{e}_1, \ldots, \mathbf{e}_n\}$ be the standard basis for $\mathbb{R}^n$ in this order. Then clearly the linear transformation Φ defined by $\Phi(\mathbf{v}_i) = \mathbf{e}_i$ is an isomorphism from V to $\mathbb{R}^n$, called the **natural isomorphism** with respect to the basis α. Now for any $\mathbf{x} = \sum_{i=1}^n a_i\mathbf{v}_i \in V$, the image of $\mathbf{x}$ under this natural isomorphism is written as

$$\Phi(\mathbf{x}) = \sum_{i=1}^n a_i\Phi(\mathbf{v}_i) = \sum_{i=1}^n a_i\mathbf{e}_i = (a_1, \ldots, a_n) = \begin{bmatrix} a_1 \\ \vdots \\ a_n \end{bmatrix} \in \mathbb{R}^n,$$

which is called the **coordinate vector** of $\mathbf{x}$ with respect to the basis α, and is denoted by $[\mathbf{x}]_\alpha (= \Phi(\mathbf{x}))$. Clearly $[\mathbf{v}_i]_\alpha = \mathbf{e}_i$.

Example 4.9 Recall that, from Example 4.3, the rotation by the angle θ of $\mathbb{R}^2$ is given by the matrix

$$R_\theta = \begin{bmatrix} \cos\theta & -\sin\theta \\ \sin\theta & \cos\theta \end{bmatrix}.$$

Clearly, it is invertible and hence an isomorphism of $\mathbb{R}^2$. In fact, one can easily check that the inverse R_θ^{-1} is simply $R_{-\theta}$.

Let $\alpha = \{\mathbf{e}_1, \mathbf{e}_2\}$ be the standard basis, and let $\beta = \{\mathbf{v}_1, \mathbf{v}_2\}$, where $\mathbf{v}_i = R_\theta\mathbf{e}_i$, $i = 1, 2$. Then β is also a basis for $\mathbb{R}^2$. The coordinate vectors of $\mathbf{v}_i$ with respect to α are themselves

$$[\mathbf{v}_1]_\alpha = \begin{bmatrix} \cos\theta \\ \sin\theta \end{bmatrix}, \quad [\mathbf{v}_2]_\alpha = \begin{bmatrix} -\sin\theta \\ \cos\theta \end{bmatrix},$$

while

$$[\mathbf{v}_1]_\beta = \begin{bmatrix} 1 \\ 0 \end{bmatrix}, \ [\mathbf{v}_2]_\beta = \begin{bmatrix} 0 \\ 1 \end{bmatrix}.$$

□

Example 4.10 In Problem 4.1, one can notice that the reflection about the line $y = x$ may be obtained by the compositions of rotation by $-\frac{\pi}{4}$ of the plane, reflection about the x-axis, and rotation by $\frac{\pi}{4}$. Actually, it is multiplication of the matrices given in (1) and (3) of Example 4.3 with $\theta = \frac{\pi}{4}$: that is, if we denote rotation by $\frac{\pi}{4}$ by

$$R_{\frac{\pi}{4}} = \begin{bmatrix} \cos\frac{\pi}{4} & -\sin\frac{\pi}{4} \\ \sin\frac{\pi}{4} & \cos\frac{\pi}{4} \end{bmatrix} = \begin{bmatrix} \frac{1}{\sqrt{2}} & -\frac{1}{\sqrt{2}} \\ \frac{1}{\sqrt{2}} & \frac{1}{\sqrt{2}} \end{bmatrix},$$

and reflection about the x-axis by $\begin{bmatrix} 1 & 0 \\ 0 & -1 \end{bmatrix}$, then $R_{-\frac{\pi}{4}} = R_{\frac{\pi}{4}}^{-1}$, and the matrix we want is

$$R_{\frac{\pi}{4}} \begin{bmatrix} 1 & 0 \\ 0 & -1 \end{bmatrix} R_{\frac{\pi}{4}}^{-1} = \begin{bmatrix} \frac{1}{\sqrt{2}} & -\frac{1}{\sqrt{2}} \\ \frac{1}{\sqrt{2}} & \frac{1}{\sqrt{2}} \end{bmatrix} \begin{bmatrix} 1 & 0 \\ 0 & -1 \end{bmatrix} \begin{bmatrix} \frac{1}{\sqrt{2}} & \frac{1}{\sqrt{2}} \\ -\frac{1}{\sqrt{2}} & \frac{1}{\sqrt{2}} \end{bmatrix} = \begin{bmatrix} 0 & 1 \\ 1 & 0 \end{bmatrix}.$$

The reflection about any line ℓ in the plane can be obtained in this way:

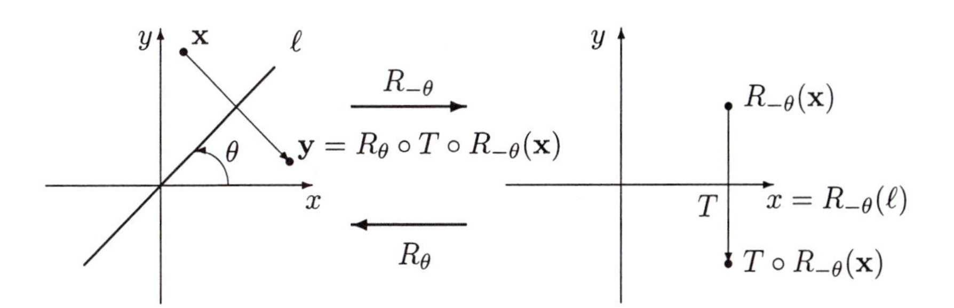

where T is the reflection about the x-axis. □

Problem 4.8 Find the matrix of reflection about the line $y = \sqrt{3}x$ in $\mathbb{R}^2$.

Problem 4.9 Find the coordinate vector of $5 + 2x + 3x^2$ with respect to the given ordered basis α for $P_2(\mathbb{R})$:

(1) $\alpha = \{1, x, x^2\}$; (2) $\alpha = \{1 + x, 1 + x^2, x + x^2\}$.

Example 4.11 Let A be an $n \times n$ matrix. It is a linear transformation on the n-space $\mathbb{R}^n$ defined by the matrix multiplication $A\mathbf{x}$ for any $\mathbf{x} \in \mathbb{R}^n$. Suppose that $\mathbf{r}_1, \ldots, \mathbf{r}_n$ are linearly independent vectors in $\mathbb{R}^n$ constituting a parallelepiped (see Remark (2) on page 70). Then A transforms this parallelepiped into another parallelepiped determined by $A\mathbf{r}_1, \ldots, A\mathbf{r}_n$. Hence, if we denote the $n \times n$ matrix whose j-th column is $\mathbf{r}_j$ by B, and the $n \times n$ matrix whose j-th column is $A\mathbf{r}_j$ by C, then clearly, $C = AB$, so

$$\mathrm{vol}(\mathcal{P}(C)) = |\det(AB)| = |\det\ A||\det\ B| = |\det\ A|\mathrm{vol}(\mathcal{P}(B)).$$

This means that, for a square matrix A considered as a linear transformation, the absolute value of the determinant of A is the ratio between the volumes of a parallelepiped $\mathcal{P}(B)$ and its image parallelepiped $\mathcal{P}(C)$ under the transformation by A. If $\det A = 0$, then the image $\mathcal{P}(C)$ is a parallelepiped in a subspace of dimension less than n. □

Problem 4.10 Let $T : \mathbb{R}^3 \to \mathbb{R}^3$ be the linear transformation given by $T(x, y, z) = (x + y, y + z, x + z)$. Let C denote the unit cube determined by the standard basis $\mathbf{e}_1, \mathbf{e}_2, \mathbf{e}_3$. Find the volume of the image parallelepiped $T(C)$ of C under T.

4.3 Application: Computer graphics

One of the simple applications of a linear transformation is to animations or graphical display of pictures on a computer screen. For a simple display of the idea, let us consider a picture in 2-plane $\mathbb{R}^2$. Note that a picture or an image on a screen usually consists of a number of points, lines or curves connecting some of them, and information about how to fill the regions bounded by the lines and curves. Assuming that the computer has information about how to connect the points and curves, a figure can be defined by a list of points. For example, consider the capital letters "LA" below:

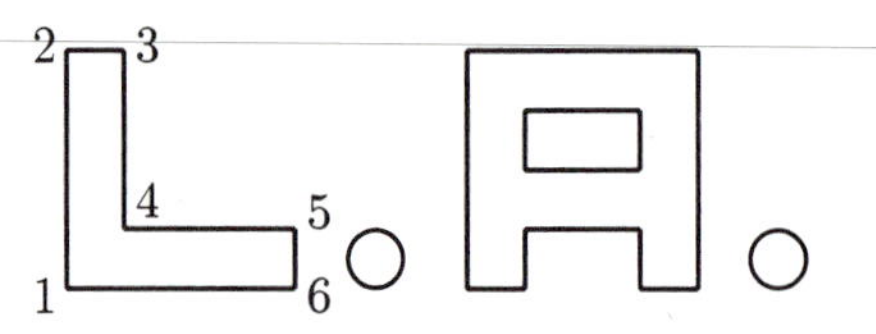

They can be represented by a matrix with coordinates of the vertices. For the sake of brevity we write it just for "L" as follows: The coordinates of the 6 vertices form a matrix:

$$\begin{array}{r} \textit{vertices} \\ \text{x}-\textit{coordinate} \\ \text{y}-\textit{coordinate} \end{array} \begin{array}{c} \begin{array}{cccccc} 1 & 2 & 3 & 4 & 5 & 6 \end{array} \\ \left[\begin{array}{cccccc} 0 & 0 & 0.5 & 0.5 & 2.0 & 2.0 \\ 0 & 2.0 & 2.0 & 0.5 & 0.5 & 0.0 \end{array}\right] \end{array} = A.$$

Of course, we assume that the computer knows which vertices are connected to which by lines via some other algorithm. We know that line segments are transformed to other line segments by a matrix, considered as a linear transformation. Thus, by multiplying a matrix to A, the vertices are transformed to the other set of vertices, and the line segments connecting the vertices are preserved. For example, the matrix $B = \begin{bmatrix} 1 & 0.25 \\ 0 & 1 \end{bmatrix}$ transforms the matrix A to the following form, which represents new coordinates of the vertices:

$$\begin{array}{r} \textit{vertices} \\ BA \;= \end{array} \begin{array}{c} \begin{array}{cccccc} 1 & 2 & 3 & 4 & 5 & 6 \end{array} \\ \left[\begin{array}{cccccc} 0 & 0.5 & 1.0 & 0.625 & 2.125 & 2.0 \\ 0 & 2.0 & 2.0 & 0.5 & 0.5 & 0.0 \end{array}\right] \end{array}.$$

Now, the computer connects these vertices properly by lines according to the given algorithm and displays on the screen the changed figure as the left side of the following:

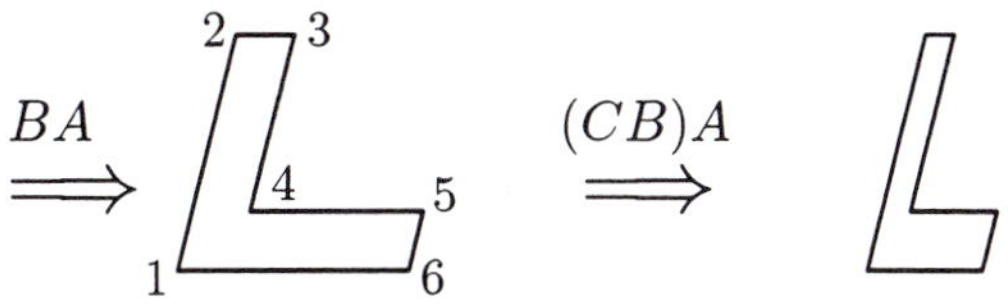

The multiplication of the matrix $C = \begin{bmatrix} 0.5 & 0 \\ 0 & 1 \end{bmatrix}$ to BA shrinks the width of BA by half, the right side of the above figure. Thus, changes in the shape of a figure may be obtained by compositions of appropriate linear transformations. Now, it is suggested that the readers try to find various matrices such as reflections, rotations, or any other linear transformations, and multiply them to A to see how the shape of the figure changes.

Remark: Incidentally, one can see that the composition of a rotation by π followed by a reflection about an axis is the same as the composition of the reflection followed by the rotation. In general, a rotation and a reflection are not commutative, neither are a reflection and another reflection.

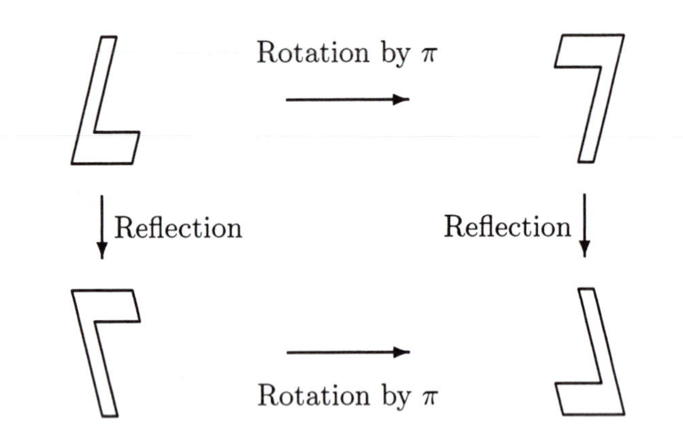

The above argument generally applies to figures in any dimension. For instance, a 3×3 matrix may be used to convert a figure in $\mathbb{R}^3$ since each point has 3 components.

Example 4.12 It is easy to see that the matrices

$$R_{(x,\alpha)} = \begin{bmatrix} 1 & 0 & 0 \\ 0 & \cos\alpha & -\sin\alpha \\ 0 & \sin\alpha & \cos\alpha \end{bmatrix}, \quad R_{(y,\beta)} = \begin{bmatrix} \cos\beta & 0 & -\sin\beta \\ 0 & 1 & 0 \\ \sin\beta & 0 & \cos\beta \end{bmatrix},$$

$$R_{(z,\gamma)} = \begin{bmatrix} \cos\gamma & -\sin\gamma & 0 \\ \sin\gamma & \cos\gamma & 0 \\ 0 & 0 & 1 \end{bmatrix}$$

are the rotations about the x, y, z-axes by the angles α, β and γ, respectively.

In general, the matrix that rotates $\mathbb{R}^3$ with respect to a given axis is useful in many applications. One can easily express such a general rotation as a composition of basic rotations such as $R_{(x,\alpha)}$, $R_{(y,\beta)}$ and $R_{(z,\gamma)}$.

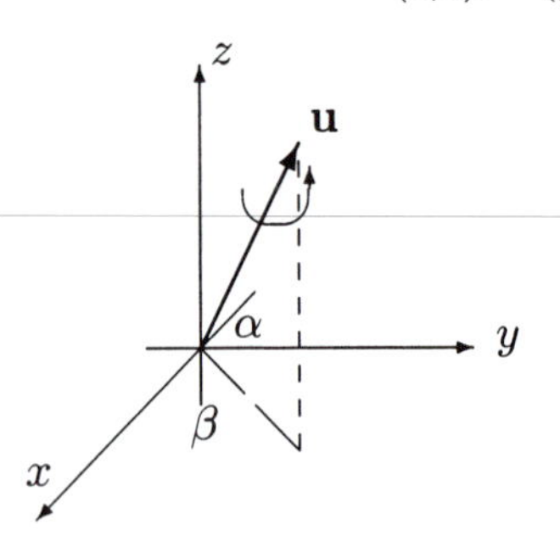

Suppose that the axis of a rotation is the line determined by the vector $\mathbf{u} = (\cos\alpha\cos\beta,\ \cos\alpha\sin\beta,\ \sin\alpha)$, $-\frac{\pi}{2} \le \alpha \le \frac{\pi}{2}$, $0 \le \beta \le 2\pi$, in spherical coordinates, and we want to find the matrix $R_{(\mathbf{u},\theta)}$ of the rotation about the $\mathbf{u}$-axis by θ: For this, we first rotate the $\mathbf{u}$-axis about the z-axis into the xz-plane by $R_{(z,-\beta)}$ and then about the y-axis into the x-axis by $R_{(y,-\alpha)}$. The rotation about the $\mathbf{u}$-axis is the same as the rotation about the x-axis, *i.e.*, one can use the rotation $R_{(x,\theta)}$ about the x-axis. After this, we get back to the rotation about the $\mathbf{u}$-axis via $R_{(y,\alpha)}$ and $R_{(z,\beta)}$. In summary,

$$R_{(\mathbf{u},\theta)} = R_{(z,\beta)}R_{(y,\alpha)}R_{(x,\theta)}R_{(y,-\alpha)}R_{(z,-\beta)}. \qquad \square$$

Problem 4.11 Find the matrix $R_{(\mathbf{u},\frac{\pi}{4})}$ for the rotation about the line determined by $\mathbf{u} = (1, 1, 1)$ by $\dfrac{\pi}{4}$.

4.4 Matrices of linear transformations

We saw that multiplication of an $m\times n$ matrix A with an $n\times 1$ column matrix $\mathbf{x}$ gives rise to a linear transformation from $\mathbb{R}^n$ to $\mathbb{R}^m$. In this section, we show that for any vector spaces V and W (not necessarily the n-spaces), a linear transformation $T : V \to W$ can be represented by a matrix.

Recall that, for any n-dimensional vector space V with an ordered basis, there is a natural isomorphism from V to the n-space $\mathbb{R}^n$, which depends on the choice of a basis for V. Let $T : V \to W$ be a linear transformation from an n-dimensional vector space V to an m-dimensional vector space W. Take ordered bases $\alpha = \{\mathbf{v}_1, \ldots, \mathbf{v}_n\}$ for V and $\beta = \{\mathbf{w}_1, \ldots, \mathbf{w}_m\}$ for W, and fix them in the following discussion. Then each vector $T(\mathbf{v}_j)$ in W is expressed uniquely as a linear combination of the vectors $\mathbf{w}_1, \ldots, \mathbf{w}_m$ in the basis β for W, say

$$\left\{\begin{array}{lcccccccc} T(\mathbf{v}_1) & = & a_{11}\mathbf{w}_1 & + & a_{21}\mathbf{w}_2 & + & \cdots & + & a_{m1}\mathbf{w}_m \\ T(\mathbf{v}_2) & = & a_{12}\mathbf{w}_1 & + & a_{22}\mathbf{w}_2 & + & \cdots & + & a_{m2}\mathbf{w}_m \\ & \vdots & & & & & & \vdots & \\ T(\mathbf{v}_n) & = & a_{1n}\mathbf{w}_1 & + & a_{2n}\mathbf{w}_2 & + & \cdots & + & a_{mn}\mathbf{w}_m, \end{array}\right.$$

or, in a short form,

$$T(\mathbf{v}_j) = \sum_{i=1}^{m} a_{ij}\mathbf{w}_i \quad \text{for } 1 \le j \le n,$$

for some scalars a_{ij} $(i = 1, \ldots, m;\ j = 1, \ldots, n)$. Notice the indexing order of a_{ij} in this expression: The coordinate vector $[T(\mathbf{v}_j)]_\beta$ of $T(\mathbf{v}_j)$ with respect to the basis β can be written as a column vector

$$[T(\mathbf{v}_j)]_\beta = \begin{bmatrix} a_{1j} \\ \vdots \\ a_{mj} \end{bmatrix}.$$

Now for any vector $\mathbf{x} = \sum_{j=1}^n x_j \mathbf{v}_j \in V$,

$$\begin{aligned} T(\mathbf{x}) &= \sum_{j=1}^n x_j T(\mathbf{v}_j) = \sum_{j=1}^n x_j \sum_{i=1}^m a_{ij}\mathbf{w}_i \\ &= \sum_{i=1}^m \left(\sum_{j=1}^n x_j a_{ij}\right)\mathbf{w}_i = \sum_{i=1}^m \left(\sum_{j=1}^n a_{ij} x_j\right)\mathbf{w}_i. \end{aligned}$$

Therefore, the coordinate vector of $T(\mathbf{x})$ with respect to the basis β is

$$[T(\mathbf{x})]_\beta = \begin{bmatrix} \sum_{j=1}^n a_{1j}x_j \\ \vdots \\ \sum_{j=1}^n a_{mj}x_j \end{bmatrix} = \begin{bmatrix} a_{11} & \cdots & a_{1n} \\ \vdots & & \vdots \\ a_{m1} & \cdots & a_{mn} \end{bmatrix} \begin{bmatrix} x_1 \\ \vdots \\ x_n \end{bmatrix} = A[\mathbf{x}]_\alpha,$$

where $[\mathbf{x}]_\alpha = [x_1 \ \cdots \ x_n]^T$ is the coordinate vector of $\mathbf{x}$ with respect to the basis α in V. In this sense, we say that matrix multiplication by A represents the transformation T. *Note that $A = [a_{ij}]$ is the matrix whose column vectors are just the coordinate vectors $[T(\mathbf{v}_j)]_\beta$ of $T(\mathbf{v}_j)$ with respect to the basis β.* Moreover, for the fixed bases α for V and β for W, the matrix A associated with the linear transformation T with respect to these bases is unique, because the coordinate expression of a vector with respect to a basis is unique. Thus, the assignment of the matrix A to a linear transformation T is well-defined.

Definition 4.4 The matrix A is called the **associated matrix** for T (or **matrix representation** of T) with respect to the bases α and β, and denoted by $A = [T]_\alpha^\beta$.

Now, the above argument can be summarized in the following theorem.

Theorem 4.9 *Let $T : V \to W$ be a linear transformation from an n-dimensional vector space V to an m-dimensional vector space W. For fixed* *ordered bases α for V and β for W, the coordinate vector $[T(\mathbf{x})]_\beta$ of $T(\mathbf{x})$ with respect to β is given as a matrix product of the associated matrix $[T]_\alpha^\beta$ of T and $[\mathbf{x}]_\alpha$, i.e.,*

$$[T(\mathbf{x})]_\beta = [T]_\alpha^\beta [\mathbf{x}]_\alpha.$$

The associated matrix $[T]_\alpha^\beta$ is given as

$$[T]_\alpha^\beta = \left[\, [T(\mathbf{v}_1)]_\beta \; [T(\mathbf{v}_2)]_\beta \; \cdots \; [T(\mathbf{v}_n)]_\beta \,\right].$$

This situation can be incorporated in the following commutative diagram:

$$\begin{array}{ccccc} V & & \xrightarrow{\quad T \quad} & & W \\ & \mathbf{x} & \longmapsto & T(\mathbf{x}) & \\ \Phi \big\downarrow & \downarrow & & \downarrow & \big\downarrow \Psi \\ & [\mathbf{x}]_\alpha & \longmapsto & [T(\mathbf{x})]_\beta & \\ \mathbb{R}^n & & \xrightarrow[A = [T]_\alpha^\beta]{} & & \mathbb{R}^m, \end{array}$$

where Φ and Ψ denote the natural isomorphisms, defined in Section 4.2, from V to $\mathbb{R}^n$ with respect to α, and from W to $\mathbb{R}^m$ with respect to β, respectively. Note that the commutativity of the above diagram means that $A \circ \Phi = \Psi \circ T$. When $V = W$ and $\alpha = \beta$, we simply write $[T]_\alpha$ for $[T]_\alpha^\alpha$.

Remark: (1) Note that an $m \times n$ matrix A is the matrix representation of A itself with respect to the standard bases α for $\mathbb{R}^n$ and γ for $\mathbb{R}^m$, *i.e.*, $A = [A]_\alpha^\gamma$. In particular, if A is an invertible $n \times n$ square matrix, then the column vectors $\mathbf{c}_1, \ldots, \mathbf{c}_n$ form another basis β for $\mathbb{R}^n$. Thus, A is simply the linear transformation on $\mathbb{R}^n$ that takes the standard basis α to β, in fact,

$$A\mathbf{e}_j = \begin{bmatrix} a_{1j} \\ \vdots \\ a_{nj} \end{bmatrix} = \mathbf{c}_j = \sum_{i=1}^{n} a_{ij}\mathbf{e}_i,$$

the j-th column of A.

(2) Let V and W be vector spaces with bases α and β, respectively, and let $T : V \to W$ be a linear transformation with the matrix representation $[T]_\alpha^\beta = A$. Then it is quite clear that $\mathrm{Ker}(T)$ and $\mathrm{Im}(T)$ are isomorphic to $\mathcal{N}(A)$ and $\mathcal{C}(A)$, respectively, via the natural isomorphisms. In particular, if $V = \mathbb{R}^n$ and $W = \mathbb{R}^m$ with the standard bases, then $\mathrm{Ker}(T) = \mathcal{N}(A)$, and $\mathrm{Im}(T) = \mathcal{C}(A)$. Therefore, from Corollary 3.17, we have

$$\dim(\mathrm{Ker}(T)) + \dim(\mathrm{Im}(T)) = \dim V.$$

The following examples illustrate the computation of matrices associated with linear transformations.

Example 4.13 Let $Id : V \to V$ be the identity transformation on a vector space V. Then for any ordered basis α for V, the matrix $[Id]_\alpha = I$, the identity matrix.

Example 4.14 Let $T : P_1(\mathbb{R}) \to P_2(\mathbb{R})$ be the linear transformation defined by

$$(T(p))(x) = xp(x).$$

Then, with the bases $\alpha = \{1,\ x\}$ and $\beta = \{1,\ x,\ x^2\}$ for $P_1(\mathbb{R})$ and $P_2(\mathbb{R})$, respectively, the associated matrix for T is $[T]_\alpha^\beta = \begin{bmatrix} 0 & 0 \\ 1 & 0 \\ 0 & 1 \end{bmatrix}$. □

Example 4.15 Let $T : \mathbb{R}^2 \to \mathbb{R}^3$ be the linear transformation defined by $T(x,\ y) = (x + 2y,\ 0,\ 2x + 3y)$ with respect to the standard bases α and β for $\mathbb{R}^2$ and $\mathbb{R}^3$, respectively. Then

$$\begin{aligned} T(\mathbf{e}_1) &= T(1,\ 0) = (1,\ 0,\ 2) = 1\mathbf{e}_1 + 0\mathbf{e}_2 + 2\mathbf{e}_3, \\ T(\mathbf{e}_2) &= T(0,\ 1) = (2,\ 0,\ 3) = 2\mathbf{e}_1 + 0\mathbf{e}_2 + 3\mathbf{e}_3. \end{aligned}$$

Hence, $[T]_\alpha^\beta = \begin{bmatrix} 1 & 2 \\ 0 & 0 \\ 2 & 3 \end{bmatrix}$. If $\beta' = \{\mathbf{e}_3,\ \mathbf{e}_2,\ \mathbf{e}_1\}$, then $[T]_\alpha^{\beta'} = \begin{bmatrix} 2 & 3 \\ 0 & 0 \\ 1 & 2 \end{bmatrix}$. □

Example 4.16 Let $T : \mathbb{R}^2 \to \mathbb{R}^2$ be a linear transformation given by $T(1,\ 1) = (0,\ 1)$ and $T(-1,\ 1) = (2,\ 3)$. Find the matrix representation $[T]_\alpha$ of T with respect to the standard basis $\alpha = \{\mathbf{e}_1,\ \mathbf{e}_2\}$.

Solution: Note that $(a,\ b) = a\mathbf{e}_1 + b\mathbf{e}_2$ for any $(a,\ b) \in \mathbb{R}^2$. Thus the definition of T shows

$$\begin{aligned} T(\mathbf{e}_1) + T(\mathbf{e}_2) &= T(\mathbf{e}_1 + \mathbf{e}_2) = T(1,\ 1) = (0,\ 1) = \mathbf{e}_2, \\ -T(\mathbf{e}_1) + T(\mathbf{e}_2) &= T(-\mathbf{e}_1 + \mathbf{e}_2) = T(-1,\ 1) = (2,\ 3) = 2\mathbf{e}_1 + 3\mathbf{e}_2. \end{aligned}$$

By solving these equations, we obtain

$$\begin{aligned} T(\mathbf{e}_1) &= -\mathbf{e}_1 - \mathbf{e}_2, \\ T(\mathbf{e}_2) &= \mathbf{e}_1 + 2\mathbf{e}_2. \end{aligned}$$

Therefore, $[T]_\alpha = \begin{bmatrix} -1 & 1 \\ -1 & 2 \end{bmatrix}$. □

Example 4.17 Let T be the linear transformation in Example 4.16. Find $[T]_\beta$ for a basis $\beta = \{\mathbf{v}_1, \mathbf{v}_2\}$, where $\mathbf{v}_1 = (0, 1)$ and $\mathbf{v}_2 = (2, 3)$.

Solution: From Example 4.16,

$$\begin{aligned} T(\mathbf{v}_1) &= \begin{bmatrix} -1 & 1 \\ -1 & 2 \end{bmatrix} \begin{bmatrix} 0 \\ 1 \end{bmatrix} = \begin{bmatrix} 1 \\ 2 \end{bmatrix} = [T(\mathbf{v}_1)]_\alpha, \\ T(\mathbf{v}_2) &= \begin{bmatrix} -1 & 1 \\ -1 & 2 \end{bmatrix} \begin{bmatrix} 2 \\ 3 \end{bmatrix} = \begin{bmatrix} 1 \\ 4 \end{bmatrix} = [T(\mathbf{v}_2)]_\alpha. \end{aligned}$$

Writing these vectors with respect to β, we get

$$\begin{bmatrix} 1 \\ 2 \end{bmatrix} = a\mathbf{v}_1 + b\mathbf{v}_2 = \begin{bmatrix} 2b \\ a + 3b \end{bmatrix}, \quad \begin{bmatrix} 1 \\ 4 \end{bmatrix} = c\mathbf{v}_1 + d\mathbf{v}_2 = \begin{bmatrix} 2d \\ c + 3d \end{bmatrix}.$$

Solving for a, b, c and d, we obtain

$$[T(\mathbf{v}_1)]_\beta = \begin{bmatrix} a \\ b \end{bmatrix} = \frac{1}{2}\begin{bmatrix} 1 \\ 1 \end{bmatrix}, \text{ and } [T(\mathbf{v}_2)]_\beta = \begin{bmatrix} c \\ d \end{bmatrix} = \frac{1}{2}\begin{bmatrix} 5 \\ 1 \end{bmatrix}.$$

Therefore, $[T]_\beta = \frac{1}{2}\begin{bmatrix} 1 & 5 \\ 1 & 1 \end{bmatrix}$. □

Problem 4.12 Find the matrix representation of each of the following linear transformations T of $\mathbb{R}^3$ with respect to the standard basis $\alpha = \{\mathbf{e}_1, \mathbf{e}_2, \mathbf{e}_3\}$, and $\beta = \{\mathbf{e}_3, \mathbf{e}_2, \mathbf{e}_1\}$:
(1) $T(x, y, z) = (2x - 3y + 4z,\ 5x - y + 2z,\ 4x + 7y)$,
(2) $T(x, y, z) = (2y + z,\ x - 4y,\ 3x)$.

Problem 4.13 Let $T : \mathbb{R}^4 \to \mathbb{R}^3$ be the linear transformation defined by

$$T(x, y, z, u) = (x + 2y,\ x - 3z + u,\ 2y + 3z + 4u).$$

Let α and β be the standard bases for $\mathbb{R}^4$ and $\mathbb{R}^3$, respectively. Find $[T]_\alpha^\beta$.

Problem 4.14 Let $Id : \mathbb{R}^n \to \mathbb{R}^n$ be the identity transformation. Let $\mathbf{x}_k$ denote the vector in $\mathbb{R}^n$ whose first $k-1$ coordinates are zero and the last $n-k+1$ coordinates are 1. Then clearly $\beta = \{\mathbf{x}_1, \ldots, \mathbf{x}_n\}$ is a basis for $\mathbb{R}^n$ (see Problem 3.9). Let $\alpha = \{\mathbf{e}_1, \ldots, \mathbf{e}_n\}$ be the standard basis for $\mathbb{R}^n$. Find the matrix representations $[Id]_\alpha^\beta$ and $[Id]_\beta^\alpha$.

4.5 Vector spaces of linear transformations

Let V and W be two vector spaces. Let $\mathcal{L}(V;W)$ denote the set of all linear transformations from V to W, *i.e.*,

$$\mathcal{L}(V;W) = \{T \ : \ T \text{ is a linear transformation from } V \text{ into } W\}.$$

For $S, T \in \mathcal{L}(V;W)$ and $\lambda \in \mathbb{R}$, define the **sum** $S+T$ and the **scalar multiplication** λS by

$$(S+T)(\mathbf{v}) = S(\mathbf{v}) + T(\mathbf{v}), \qquad \text{and} \qquad (\lambda S)(\mathbf{v}) = \lambda(S(\mathbf{v}))$$

for any $\mathbf{v} \in V$. Then clearly $S+T$ and λS belong to $\mathcal{L}(V;W)$, so that $\mathcal{L}(V;W)$ becomes a vector space. In particular, if $V = \mathbb{R}^n$ and $W = \mathbb{R}^m$, then the set $M_{m\times n}(\mathbb{R})$ is precisely the vector space of the linear transformations of $\mathbb{R}^n$ into $\mathbb{R}^m$ with respect to the standard bases. Hence, by fixing the standard bases, we have identified $\mathcal{L}(\mathbb{R}^n;\mathbb{R}^m) = M_{m\times n}(\mathbb{R})$ via the matrix representation.

In general, for any vector spaces V and W of dimensions n and m with ordered bases α and β, respectively, there is a one-to-one correspondence between $\mathcal{L}(V;W)$ and $M_{m\times n}(\mathbb{R})$ via the matrix representation.

Let us first define a transformation $\phi : \mathcal{L}(V;W) \to M_{m\times n}(\mathbb{R})$ as

$$\phi(T) = [T]_\alpha^\beta \in M_{m\times n}(\mathbb{R})$$

for any $T \in \mathcal{L}(V;W)$ (see Section 4.4). If $[S]_\alpha^\beta = [T]_\alpha^\beta$ for S and $T \in \mathcal{L}(V;W)$, then we have $S = T$ by Corollary 4.4. This means that ϕ is one-to-one.

On the other hand, an $m \times n$ matrix A, considered as a linear transformation from $\mathbb{R}^n$ to $\mathbb{R}^m$, gives rise to a linear transformation T from V to W via the composition of A with the natural isomorphisms Φ and Ψ, *i.e.*, $T = \Psi^{-1} \circ A \circ \Phi$, which satisfies $[T]_\alpha^\beta = A$. This means that ϕ is onto.

Therefore, ϕ gives an one-to-one correspondence between $\mathcal{L}(V;W)$ and $M_{m\times n}(\mathbb{R})$. Furthermore, the following theorem shows that ϕ is linear, so that it is in fact an isomorphism from $\mathcal{L}(V;W)$ to $M_{m\times n}(\mathbb{R})$.

Theorem 4.10 *Let V and W be vector spaces with ordered bases α and β, respectively, and let S, $T : V \to W$ be linear. Then we have*

$$[S+T]_\alpha^\beta = [S]_\alpha^\beta + [T]_\alpha^\beta \qquad \textit{and} \qquad [kS]_\alpha^\beta = k[S]_\alpha^\beta.$$

Proof: Let $\alpha = \{\mathbf{v}_1, \ldots, \mathbf{v}_n\}$ and $\beta = \{\mathbf{w}_1, \ldots, \mathbf{w}_m\}$. Then we have unique expressions $S(\mathbf{v}_j) = \sum_{i=1}^m a_{ij}\mathbf{w}_i$ and $T(\mathbf{v}_j) = \sum_{i=1}^m b_{ij}\mathbf{w}_i$ for each $1 \le j \le n$. Hence

$$(S+T)(\mathbf{v}_j) = \sum_{i=1}^m a_{ij}\mathbf{w}_i + \sum_{i=1}^m b_{ij}\mathbf{w}_i = \sum_{i=1}^m (a_{ij}+b_{ij})\mathbf{w}_i.$$

Thus

$$[S+T]_\alpha^\beta = [S]_\alpha^\beta + [T]_\alpha^\beta.$$

The proof of the second equality $[kS]_\alpha^\beta = k[S]_\alpha^\beta$ is left as an exercise. □

In summary, for vector spaces V of dimension n and W of dimension m with fixed ordered bases α and β respectively, the vector space $\mathcal{L}(V;W)$ of all linear transformations from V to W can be identified with the vector space $M_{m\times n}(\mathbb{R})$ of all $m \times n$ matrices so that

$$\dim \mathcal{L}(V;W) = \dim M_{m\times n}(\mathbb{R}) = mn = \dim V \dim W.$$

Remark: (1) Let $A\mathbf{x} = \mathbf{b}$ be a system of linear equations for an $m\times n$ matrix A. By considering the coefficient matrix A as a linear transformation, one can have other equivalent conditions to those in Theorems 3.23 and 3.24: The conditions in Theorem 3.23 (*e.g.*, rank $A = m$) are equivalent to the condition that A is *surjective*, and those in Theorem 3.24 (*e.g.*, rank $A = n$) are equivalent to the condition that A is *one-to-one*. This observation gives the proof of (15)-(16) in Theorem 3.25.

(2) With the identification of vector spaces $\mathcal{L}(V;W)$ and $M_{m\times n}(\mathbb{R})$ as above, we can have, by Theorem 3.25, the following equivalent conditions for a linear transformation T on a vector space V:

(i) *T is an isomorphism,*

(ii) *T is one-to-one,*

(iii) *T is surjective.*

(One can also prove them directly by using the definition of a basis for V.)

The next theorem shows that the one-to-one correspondence between $\mathcal{L}(V;W)$ and $M_{m\times n}(\mathbb{R})$ preserves not only the linear structure but also the compositions of linear transformations. Let V, W and Z be vector spaces. Suppose that $S: V \to W$ and $T: W \to Z$ are linear transformations. Then clearly the composition $T \circ S: V \to Z$ is also linear. Often we refer this composition to the product operation of linear transformations.

Theorem 4.11 *Let V, W and Z be vector spaces with ordered bases α, β, and γ, respectively. Suppose that $S : V \to W$ and $T : W \to Z$ are linear transformations. Then*

$$[T \circ S]_\alpha^\gamma = [T]_\beta^\gamma [S]_\alpha^\beta.$$

Proof: Let $\alpha = \{\mathbf{v}_1, \ldots, \mathbf{v}_n\}$, $\beta = \{\mathbf{w}_1, \ldots, \mathbf{w}_m\}$ and $\gamma = \{\mathbf{z}_1, \ldots, \mathbf{z}_\ell\}$. Let $[T]_\beta^\gamma = [a_{ij}]$ and $[S]_\alpha^\beta = [b_{pq}]$. Then, for $1 \le i \le n$

$$\begin{aligned}(T \circ S)(\mathbf{v}_i) &= T(S(\mathbf{v}_i)) = T\left(\sum_{k=1}^m b_{ki}\mathbf{w}_k\right) = \sum_{k=1}^m b_{ki}T(\mathbf{w}_k) \\ &= \sum_{k=1}^m b_{ki}\left(\sum_{j=1}^\ell a_{jk}\mathbf{z}_j\right) = \sum_{j=1}^\ell\left(\sum_{k=1}^m a_{jk}b_{ki}\right)\mathbf{z}_j.\end{aligned}$$

It shows that $[T \circ S]_\alpha^\gamma = [T]_\beta^\gamma [S]_\alpha^\beta$. □

Problem 4.15 Let α be the standard basis for $\mathbb{R}^3$, and let S, $T : \mathbb{R}^3 \to \mathbb{R}^3$ be two linear transformations given by

$$S(\mathbf{e}_1) = (2,\ 2,\ 1), \quad S(\mathbf{e}_2) = (0,\ 1,\ 2), \quad S(\mathbf{e}_3) = (-1,\ 2,\ 1),$$
$$T(\mathbf{e}_1) = (1,\ 0,\ 1), \quad T(\mathbf{e}_2) = (0,\ 1,\ 1), \quad T(\mathbf{e}_3) = (1,\ 1,\ 2).$$

Compute $[S+T]_\alpha$, $[2T-S]_\alpha$ and $[T \circ S]_\alpha$.

Problem 4.16 Let $T : P_2(\mathbb{R}) \to P_2(\mathbb{R})$ be the linear transformation defined by $T(f) = (3+x)f' + 2f$, and $S : P_2(\mathbb{R}) \to \mathbb{R}^3$ defined by $S(a+bx+cx^2) = (a-b,\ a+b,\ c)$. For a basis $\alpha = \{1,\ x,\ x^2\}$ for $P_2(\mathbb{R})$ and the standard basis $\beta = \{\mathbf{e}_1,\ \mathbf{e}_2,\ \mathbf{e}_3\}$ for $\mathbb{R}^3$, compute $[S]_\alpha^\beta$, $[T]_\alpha$, and $[S \circ T]_\alpha^\beta$.

Theorem 4.12 *Let V and W be vector spaces with ordered bases α and β, respectively, and let $T : V \to W$ be an isomorphism. Then*

$$[T^{-1}]_\beta^\alpha = ([T]_\alpha^\beta)^{-1}.$$

Proof: Since T is invertible, $\dim V = \dim W$, and the matrices $[T]_\alpha^\beta$ and $[T^{-1}]_\beta^\alpha$ are square and of the same size. Thus,

$$[T]_\alpha^\beta [T^{-1}]_\beta^\alpha = [T \circ T^{-1}]_\beta = [Id]_\beta$$

is the identity matrix. Hence, $[T^{-1}]_\beta^\alpha = ([T]_\alpha^\beta)^{-1}$. □

In particular, if a linear transformation $T : V \to W$ is an isomorphism, then $[T]_\alpha^\beta$ is an invertible matrix for any bases α for V and β for W.

Problem 4.17 For the vector spaces $P_1(\mathbb{R})$ and $\mathbb{R}^2$, choose the bases $\alpha = \{1,\ x\}$ for $P_1(\mathbb{R})$ and $\beta = \{\mathbf{e}_1,\ \mathbf{e}_2\}$ for $\mathbb{R}^2$, respectively. Let $T : P_1(\mathbb{R}) \to \mathbb{R}^2$ be the linear transformation defined by $T(a + bx) = (a,\ a + b)$.

(1) Show that T is invertible. (2) Find $[T]_\alpha^\beta$ and $[T^{-1}]_\beta^\alpha$.

4.6 Change of bases

In Section 4.2, we saw that any vector space V of dimension n with an ordered basis α is isomorphic to the n-space $\mathbb{R}^n$ via the natural isomorphism Φ. It assigns the coordinate vector in $\mathbb{R}^n$ to each $\mathbf{x} \in V$, *i.e.*, $\Phi(\mathbf{x}) = [\mathbf{x}]_\alpha$. Of course, we can get a different isomorphism if we take another basis β instead of α: That is, the coordinate expression $[\mathbf{x}]_\beta$ of $\mathbf{x}$ with respect to β may be different from $[\mathbf{x}]_\alpha$. Thus, one may naturally ask what the relation between $[\mathbf{x}]_\alpha$ and $[\mathbf{x}]_\beta$ is for the two different bases. In this section, we discuss this question. One of the fundamental problems in linear algebra is to find bases for which the matrix representation of a linear transformation is as simple as possible.

Let us begin with an example in the plane $\mathbb{R}^2$. The coordinate expression of $\mathbf{x} = (x,\ y) \in \mathbb{R}^2$ with respect to the standard basis $\alpha = \{\mathbf{e}_1,\ \mathbf{e}_2\}$ is $\mathbf{x} = x\mathbf{e}_1 + y\mathbf{e}_2$, so that $[\mathbf{x}]_\alpha = \begin{bmatrix} x \\ y \end{bmatrix}$.

Now let $\beta = \{\mathbf{e}'_1,\ \mathbf{e}'_2\}$ be another basis for $\mathbb{R}^2$ obtained by rotating α counterclockwise through an angle θ.

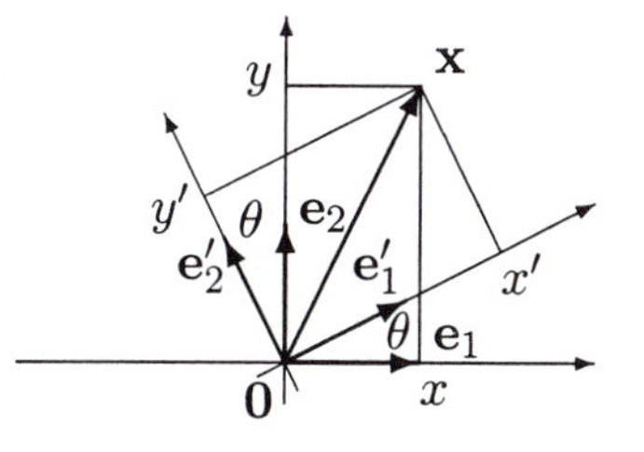

Then the coordinate expression of $\mathbf{x} \in \mathbb{R}^2$ with respect to β is written as $\mathbf{x} = x'\mathbf{e}'_1 + y'\mathbf{e}'_2$, or $[\mathbf{x}]_\beta = \begin{bmatrix} x' \\ y' \end{bmatrix}$.

In particular, the expression of the vectors in β with respect to α are

$$\begin{array}{rcrcrcr} \mathbf{e}_1' & = & Id(\mathbf{e}_1') & = & \cos\theta\,\mathbf{e}_1 & + & \sin\theta\,\mathbf{e}_2 \\ \mathbf{e}_2' & = & Id(\mathbf{e}_2') & = & -\sin\theta\,\mathbf{e}_1 & + & \cos\theta\,\mathbf{e}_2, \end{array}$$

so

$$[\mathbf{e}_1']_\alpha = \begin{bmatrix} \cos\theta \\ \sin\theta \end{bmatrix}, \quad [\mathbf{e}_2']_\alpha = \begin{bmatrix} -\sin\theta \\ \cos\theta \end{bmatrix}.$$

Therefore,

$$\begin{aligned} \mathbf{x} &= x'\mathbf{e}_1' + y'\mathbf{e}_2' = (x'\cos\theta - y'\sin\theta)\mathbf{e}_1 + (x'\sin\theta + y'\cos\theta)\mathbf{e}_2 \\ &= x\mathbf{e}_1 + y\mathbf{e}_2. \end{aligned}$$

This can be written as the following matrix equation:

$$\begin{bmatrix} x \\ y \end{bmatrix} = \begin{bmatrix} \cos\theta & -\sin\theta \\ \sin\theta & \cos\theta \end{bmatrix} \begin{bmatrix} x' \\ y' \end{bmatrix}, \quad \text{or} \quad [\mathbf{x}]_\alpha = [Id]_\beta^\alpha [\mathbf{x}]_\beta,$$

where

$$[Id]_\beta^\alpha = [[\mathbf{e}_1']_\alpha \ [\mathbf{e}_2']_\alpha] = \begin{bmatrix} \cos\theta & -\sin\theta \\ \sin\theta & \cos\theta \end{bmatrix}.$$

Note that $[Id]_\alpha^\beta = ([Id]_\beta^\alpha)^{-1} = \begin{bmatrix} \cos\theta & \sin\theta \\ -\sin\theta & \cos\theta \end{bmatrix}$ by Theorem 4.12.

In general, let $\alpha = \{\mathbf{v}_1, \mathbf{v}_2, \ldots, \mathbf{v}_n\}$ and $\beta = \{\mathbf{w}_1, \mathbf{w}_2, \ldots, \mathbf{w}_n\}$ be two ordered bases for V. Then any vector $\mathbf{x} \in V$ has two expressions:

$$\mathbf{x} = \sum_{i=1}^{n} x_i\mathbf{v}_i = \sum_{j=1}^{n} y_j\mathbf{w}_j.$$

Now, each vector in β is expressed as a linear combination of the vectors in α: $\mathbf{w}_j = Id(\mathbf{w}_j) = \sum_{i=1}^{n} q_{ij}\mathbf{v}_i$ for $j = 1, 2, \ldots, n$, so that

$$[\mathbf{w}_j]_\alpha = [Id(\mathbf{w}_j)]_\alpha = \begin{bmatrix} q_{1j} \\ \vdots \\ q_{nj} \end{bmatrix}.$$

Then for any $\mathbf{x} \in V$,

$$\mathbf{x} = \sum_{i=1}^{n} x_i\mathbf{v}_i = \sum_{j=1}^{n} y_j\mathbf{w}_j = \sum_{j=1}^{n} y_j \sum_{i=1}^{n} q_{ij}\mathbf{v}_i = \sum_{i=1}^{n} \left(\sum_{j=1}^{n} q_{ij}y_j \right) \mathbf{v}_i.$$

This is equivalent to the following matrix equation:

$$\begin{bmatrix} x_1 \\ \vdots \\ x_n \end{bmatrix} = \begin{bmatrix} q_{11} & \cdots & q_{1n} \\ & \ddots & \\ q_{n1} & \cdots & q_{nn} \end{bmatrix} \begin{bmatrix} y_1 \\ \vdots \\ y_n \end{bmatrix},$$

or

$$[\mathbf{x}]_\alpha = [Id]_\beta^\alpha [\mathbf{x}]_\beta$$

$$\begin{array}{ccccc} V & & \xrightarrow{\quad Id \quad} & & V \\ & \mathbf{x} & \longmapsto & \mathbf{x} & \\ \Phi' \downarrow & \downarrow & & \downarrow & \downarrow \Phi \\ & [\mathbf{x}]_\beta & \longmapsto & [\mathbf{x}]_\alpha & \\ \mathbb{R}^n & & \xrightarrow[Q = [Id]_\beta^\alpha]{} & & \mathbb{R}^n, \end{array}$$

where

$$[Id]_\beta^\alpha = \begin{bmatrix} q_{11} & \cdots & q_{1n} \\ & \ddots & \\ q_{n1} & \cdots & q_{nn} \end{bmatrix} = [\ [\mathbf{w}_1]_\alpha \ \cdots \ [\mathbf{w}_n]_\alpha].$$

Definition 4.5 The matrix representation $[Id]_\beta^\alpha$ of the identity transformation $Id : V \to V$ with respect to any two bases α and β is called the **transition matrix** or the **coordinate change matrix** from β to α.

Since the identity transformation $Id : V \to V$ is invertible, the transition matrix $Q = [Id]_\beta^\alpha$ is also invertible by Theorem 4.12. If we had taken the expressions of the vectors in the basis α with respect to the basis β: $\mathbf{v}_j = Id(\mathbf{v}_j) = \sum_{i=1}^n p_{ij}\mathbf{w}_i$ for $j = 1, 2, \ldots, n$, then we would have $[p_{ij}] = [Id]_\alpha^\beta = Q^{-1}$ and

$$[\mathbf{x}]_\beta = [Id]_\alpha^\beta [\mathbf{x}]_\alpha = Q^{-1}[\mathbf{x}]_\alpha.$$

Example 4.18 Let the 3-space $\mathbb{R}^3$ be equipped with the standard xyz-coordinate system, *i.e.*, with the standard basis $\alpha = \{\mathbf{e}_1, \mathbf{e}_2, \mathbf{e}_3\}$. Take a new $x'y'z'$-coordinate system by rotating the xyz-system around its z-axis counterclockwise through an angle θ, *i.e.*, we take a new basis $\beta = \{\mathbf{e}_1', \mathbf{e}_2', \mathbf{e}_3'\}$ by rotating the basis α about z axis through θ. Then we get

$$[\mathbf{e}_1']_\alpha = \begin{bmatrix} \cos\theta \\ \sin\theta \\ 0 \end{bmatrix}, \quad [\mathbf{e}_2']_\alpha = \begin{bmatrix} -\sin\theta \\ \cos\theta \\ 0 \end{bmatrix}, \quad [\mathbf{e}_3']_\alpha = \begin{bmatrix} 0 \\ 0 \\ 1 \end{bmatrix}.$$

Hence, the transition matrix from β to α is

$$Q = [Id]_\beta^\alpha = \begin{bmatrix} \cos\theta & -\sin\theta & 0 \\ \sin\theta & \cos\theta & 0 \\ 0 & 0 & 1 \end{bmatrix},$$

so

$$[\mathbf{x}]_\alpha = \begin{bmatrix} x \\ y \\ z \end{bmatrix} = \begin{bmatrix} \cos\theta & -\sin\theta & 0 \\ \sin\theta & \cos\theta & 0 \\ 0 & 0 & 1 \end{bmatrix} \begin{bmatrix} x' \\ y' \\ z' \end{bmatrix} = Q[\mathbf{x}]_\beta.$$

Moreover, $Q = [Id]_\beta^\alpha$ is invertible and the transition matrix from α to β is

$$Q^{-1} = [Id]_\alpha^\beta = \begin{bmatrix} \cos\theta & \sin\theta & 0 \\ -\sin\theta & \cos\theta & 0 \\ 0 & 0 & 1 \end{bmatrix},$$

so that

$$\begin{bmatrix} x' \\ y' \\ z' \end{bmatrix} = \begin{bmatrix} \cos\theta & \sin\theta & 0 \\ -\sin\theta & \cos\theta & 0 \\ 0 & 0 & 1 \end{bmatrix} \begin{bmatrix} x \\ y \\ z \end{bmatrix}.$$

□

Problem 4.18 Find the transition matrix from a basis α to another basis β for the 3-space $\mathbb{R}^3$, where

$$\alpha = \{(1, 0, 1), (1, 1, 0), (0, 1, 1)\}, \quad \beta = \{(2, 3, 1), (1, 2, 0), (2, 0, 3)\}.$$

4.7 Similarity

The coordinate expression of a vector in a vector space V depends on the choice of an ordered basis. Hence, the matrix representation of a linear transformation is also dependent on the choice of bases.

Let V and W be two vector spaces of dimensions n and m with two ordered bases α and β, respectively, and let $T : V \to W$ be a linear transformation. In Section 4.4, we discussed how to find $[T]_\alpha^\beta$. If we have different bases α' and β' for V and W, respectively, then we get another matrix representation $[T]_{\alpha'}^{\beta'}$ of T. We, in fact, have two different expressions

$$\begin{array}{ll} [\mathbf{x}]_\alpha \text{ and } [\mathbf{x}]_{\alpha'} \text{ in } \mathbb{R}^n & \text{for each } \mathbf{x} \in V, \\ [T(\mathbf{x})]_\beta \text{ and } [T(\mathbf{x})]_{\beta'} \text{ in } \mathbb{R}^m & \text{for } T(\mathbf{x}) \in W. \end{array}$$

They are related by the transition matrices in the following equations:

$$[\mathbf{x}]_{\alpha'} = [Id_V]_{\alpha}^{\alpha'}[\mathbf{x}]_{\alpha}, \quad \text{and} \quad [T(\mathbf{x})]_{\beta'} = [Id_W]_{\beta}^{\beta'}[T(\mathbf{x})]_{\beta}.$$

On the other hand, by Theorem 4.9, we have

$$[T(\mathbf{x})]_{\beta} = [T]_{\alpha}^{\beta}[\mathbf{x}]_{\alpha}, \quad \text{and} \quad [T(\mathbf{x})]_{\beta'} = [T]_{\alpha'}^{\beta'}[\mathbf{x}]_{\alpha'}.$$

Therefore, we get

$$\begin{aligned}[T]_{\alpha'}^{\beta'}[\mathbf{x}]_{\alpha'} &= [T(\mathbf{x})]_{\beta'} = [Id_W]_{\beta}^{\beta'}[T(\mathbf{x})]_{\beta} = [Id_W]_{\beta}^{\beta'}[T]_{\alpha}^{\beta}[\mathbf{x}]_{\alpha} \\ &= [Id_W]_{\beta}^{\beta'}[T]_{\alpha}^{\beta}[Id_V]_{\alpha'}^{\alpha}[\mathbf{x}]_{\alpha'}.\end{aligned}$$

Actually, from Theorem 4.11, this relation can be obtained directly as

$$[T]_{\alpha'}^{\beta'} = [Id_W \circ T \circ Id_V]_{\alpha'}^{\beta'} = [Id_W]_{\beta}^{\beta'}[T]_{\alpha}^{\beta}[Id_V]_{\alpha'}^{\alpha},$$

since $T = Id_W \circ T \circ Id_V$. Note that $[T]_{\alpha}^{\beta}$ and $[T]_{\alpha'}^{\beta'}$ are $m \times n$ matrices, $[Id_V]_{\alpha'}^{\alpha}$ is an $n \times n$ matrix and $[Id_W]_{\beta}^{\beta'}$ is an $m \times m$ matrix.

The relation can also be incorporated in the following commutative diagrams:

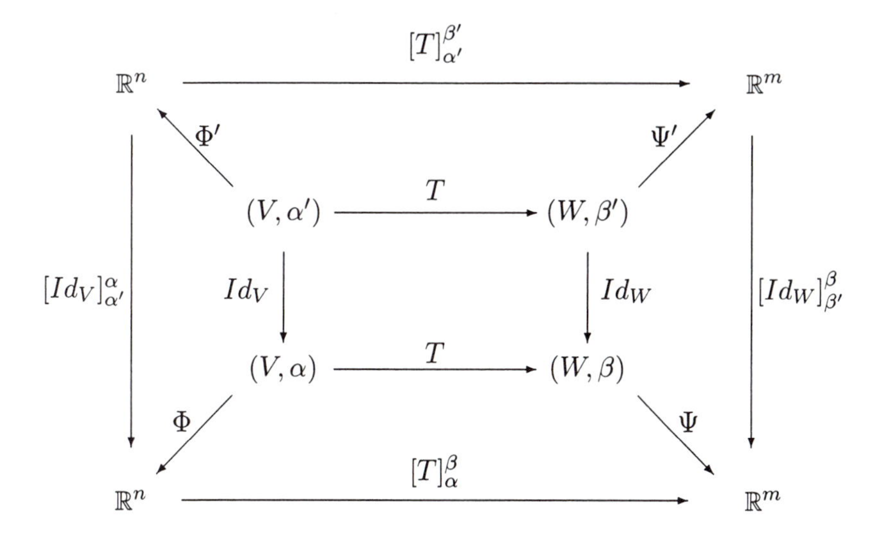

The following theorem summarizes the above argument.

Theorem 4.13 *Let $T : V \to W$ be a linear transformation on a vector space V with bases α and α' to another vector space W with bases β and β'. Then*

$$[T]_{\alpha'}^{\beta'} = P^{-1}[T]_{\alpha}^{\beta}Q,$$

where $Q = [Id_V]_{\alpha'}^{\alpha}$ and $P = [Id_W]_{\beta'}^{\beta}$ are the transition matrices.

In particular, if we take $W = V$, $\alpha = \beta$ and $\alpha' = \beta'$, then $P = Q$ and we get to the following corollary.

Corollary 4.14 *Let $T : V \to V$ be a linear transformation on a vector space V, and let α and β be ordered bases for V. Let $Q = [Id]_{\beta}^{\alpha}$ be the transition matrix from β to α. Then*

(1) *Q is invertible, and $Q^{-1} = [Id]_{\alpha}^{\beta}$.*

(2) *For any $\mathbf{x} \in V$, $[\mathbf{x}]_{\alpha} = Q[\mathbf{x}]_{\beta}$.*

(3) *$[T]_{\beta} = Q^{-1}[T]_{\alpha}Q$.*

Relation (3) of $[T]_{\beta}$ and $[T]_{\alpha}$ in Corollary 4.14 is called a *similarity*. In general, we have the following definition.

Definition 4.6 For any square matrices A and B, A is said to be **similar** to B if there exists a nonsingular matrix Q such that $B = Q^{-1}AQ$.

Note that if A is similar to B, then B is also similar to A. Thus we simply say that A and B are similar matrices. We saw in Theorem 4.14 that if A and B are $n \times n$ matrices representing the same linear transformation T, then A and B are similar.

Example 4.19 Let $\beta = \{\mathbf{v}_1, \mathbf{v}_2, \mathbf{v}_3\}$ be a basis for $\mathbb{R}^3$ consisting of $\mathbf{v}_1 = (1, 1, 0)$, $\mathbf{v}_2 = (1, 0, 1)$ and $\mathbf{v}_3 = (0, 1, 1)$. Let T be the linear transformation on $\mathbb{R}^3$ given by the matrix

$$[T]_{\beta} = \begin{bmatrix} 2 & 1 & -1 \\ 1 & 2 & 3 \\ -1 & 1 & 1 \end{bmatrix}.$$

Let $\alpha = \{\mathbf{e}_1, \mathbf{e}_2, \mathbf{e}_3\}$ be the standard basis. Find the transition matrix $[Id]_{\alpha}^{\beta}$ and $[T]_{\alpha}$.

Solution: Since $\mathbf{v}_1 = \mathbf{e}_1 + \mathbf{e}_2$, $\mathbf{v}_2 = \mathbf{e}_1 + \mathbf{e}_3$, $\mathbf{v}_3 = \mathbf{e}_2 + \mathbf{e}_3$, we have

$$[Id]_\beta^\alpha = \begin{bmatrix} 1 & 1 & 0 \\ 1 & 0 & 1 \\ 0 & 1 & 1 \end{bmatrix}, \quad \text{and} \quad [Id]_\alpha^\beta = ([Id]_\beta^\alpha)^{-1} = \frac{1}{2}\begin{bmatrix} 1 & 1 & -1 \\ 1 & -1 & 1 \\ -1 & 1 & 1 \end{bmatrix}.$$

Therefore,

$$[T]_\alpha = [Id]_\beta^\alpha [T]_\beta [Id]_\alpha^\beta = \frac{1}{2}\begin{bmatrix} 4 & 2 & 2 \\ 3 & -1 & 1 \\ -1 & 1 & 7 \end{bmatrix}.$$

□

Example 4.20 Let $T : \mathbb{R}^3 \to \mathbb{R}^3$ be the linear transformation defined by

$$T(x_1,\ x_2,\ x_3) = (2x_1 + x_2,\ x_1 + x_2 + 3x_3,\ -x_2).$$

Let $\alpha = \{\mathbf{e}_1,\ \mathbf{e}_2,\ \mathbf{e}_3\}$ be the standard ordered basis. Then we clearly have

$$[T]_\alpha = \begin{bmatrix} 2 & 1 & 0 \\ 1 & 1 & 3 \\ 0 & -1 & 0 \end{bmatrix}.$$

Let $\beta = \{\mathbf{v}_1,\ \mathbf{v}_2,\ \mathbf{v}_3\}$ be another ordered basis for $\mathbb{R}^3$ consisting of $\mathbf{v}_1 = (-1,\ 0,\ 0)$, $\mathbf{v}_2 = (2,\ 1,\ 0)$, and $\mathbf{v}_3 = (1,\ 1,\ 1)$. Let $Q = [Id]_\beta^\alpha$ be the transition matrix from β to α. Since α is the standard ordered basis for $\mathbb{R}^3$, the columns of Q are simply the vectors in β written in the same order, with an easily calculated inverse. Thus

$$Q = \begin{bmatrix} -1 & 2 & 1 \\ 0 & 1 & 1 \\ 0 & 0 & 1 \end{bmatrix}, \qquad Q^{-1} = \begin{bmatrix} -1 & 2 & -1 \\ 0 & 1 & -1 \\ 0 & 0 & 1 \end{bmatrix}.$$

A straightforward multiplication shows that

$$[T]_\beta = Q^{-1}[T]_\alpha Q = \begin{bmatrix} 0 & 2 & 8 \\ -1 & 4 & 6 \\ 0 & -1 & -1 \end{bmatrix}.$$

To show that this is the correct matrix, we can verify that the image under T of the j-th vector of β is the linear combination of the vectors of β with the entries of the j-th column of $[T]_\beta$ as its coefficients. For example, for $j = 2$

we have $T(\mathbf{v}_2) = T(2, 1, 0) = (5, 3, -1)$. On the other hand, the coefficients of $[T(\mathbf{v}_2)]_\beta$ are just the entries of the second column of $[T]_\beta$. Therefore,

$$\begin{aligned} T(\mathbf{v}_2) &= 2\mathbf{v}_1 + 4\mathbf{v}_2 - \mathbf{v}_3 \\ &= 12\mathbf{e}_1 + 4(2\mathbf{e}_1 + \mathbf{e}_2) - (\mathbf{e}_1 + \mathbf{e}_2 + \mathbf{e}_3) \\ &= 5\mathbf{e}_1 + 3\mathbf{e}_2 - \mathbf{e}_3 \;=\; (5, 3, -1), \end{aligned}$$

as expected. □

The next theorem shows that two similar matrices are matrix representations of the same linear transformation.

Theorem 4.15 *Suppose that A represents a linear transformation $T : V \to V$ on a vector space V with respect to an ordered basis $\alpha = \{\mathbf{v}_1, \ldots, \mathbf{v}_n\}$, i.e., $[T]_\alpha = A$. If $B = Q^{-1}AQ$ for some nonsingular matrix Q, then there exists a basis β for V such that $B = [T]_\beta$, and $Q = [Id]_\beta^\alpha$.*

Proof: Let $Q = [q_{ij}]$ and let $\mathbf{w}_1, \ldots, \mathbf{w}_n$ be the vectors in V defined by

$$\left\{ \begin{aligned} \mathbf{w}_1 &= q_{11}\mathbf{v}_1 + q_{21}\mathbf{v}_2 + \cdots + q_{n1}\mathbf{v}_n \\ \mathbf{w}_2 &= q_{12}\mathbf{v}_1 + q_{22}\mathbf{v}_2 + \cdots + q_{n2}\mathbf{v}_n \\ &\;\vdots \\ \mathbf{w}_n &= q_{1n}\mathbf{v}_1 + q_{2n}\mathbf{v}_2 + \cdots + q_{nn}\mathbf{v}_n. \end{aligned} \right.$$

Then the nonsingularity of $Q = [q_{ij}]$ implies that $\beta = \{\mathbf{w}_1, \ldots, \mathbf{w}_n\}$ is an ordered basis for V, and Theorem 4.14 (3) shows that $[T]_\beta = Q^{-1}[T]_\alpha Q = Q^{-1}AQ = B$ with $Q = [Id]_\beta^\alpha$. □

Example 4.21 Let D be the differential operator on the vector space $P_2(\mathbb{R})$. Given two ordered bases $\alpha = \{1, x, x^2\}$ and $\beta = \{1, 2x, 4x^2 - 2\}$ for $P_2(\mathbb{R})$, we first note that

$$\begin{aligned} D(1) &= 0 \cdot 1 + 0 \cdot x + 0 \cdot x^2 \\ D(x) &= 1 \cdot 1 + 0 \cdot x + 0 \cdot x^2 \\ D(x^2) &= 0 \cdot 1 + 2 \cdot x + 0 \cdot x^2. \end{aligned}$$

Hence, the matrix representation of D with respect to α is given by

$$[D]_\alpha = \begin{bmatrix} 0 & 1 & 0 \\ 0 & 0 & 2 \\ 0 & 0 & 0 \end{bmatrix}.$$

Applying D to 1, $2x$ and $4x^2-2$, one obtains

$$\begin{aligned} D(1) &= 0\cdot 1+0\cdot 2x+0\cdot(4x^2-2) \\ D(2x) &= 2\cdot 1+0\cdot 2x+0\cdot(4x^2-2) \\ D(4x^2-2) &= 0\cdot 1+4\cdot 2x+0\cdot(4x^2-2). \end{aligned}$$

Thus,

$$[D]_\beta = \begin{bmatrix} 0 & 2 & 0 \\ 0 & 0 & 4 \\ 0 & 0 & 0 \end{bmatrix}.$$

The transition matrix Q from $\beta = \{1,\ 2x,\ 4x^2-2\}$ to $\alpha = \{1,\ x,\ x^2\}$ and its inverse are easily calculated as

$$Q = [Id]_\beta^\alpha = \begin{bmatrix} 1 & 0 & -2 \\ 0 & 2 & 0 \\ 0 & 0 & 4 \end{bmatrix}, \qquad Q^{-1} = [Id]_\alpha^\beta = \frac{1}{4}\begin{bmatrix} 4 & 0 & 2 \\ 0 & 2 & 0 \\ 0 & 0 & 1 \end{bmatrix}.$$

A simple computation shows that $[D]_\beta = Q^{-1}[D]_\alpha Q$. □

Problem 4.19 Let $T: \mathbb{R}^3 \to \mathbb{R}^3$ be the linear transformation defined by

$$T(x_1,\ x_2,\ x_3) = (x_1+2x_2+x_3,\ -x_2,\ x_1+4x_3).$$

Let α be the standard basis, and let $\beta = \{\mathbf{v}_1,\ \mathbf{v}_2,\ \mathbf{v}_3\}$ be another ordered basis consisting of $\mathbf{v}_1 = (1,\ 0,\ 0)$, $\mathbf{v}_2 = (1,\ 1,\ 0)$, and $\mathbf{v}_3 = (1,\ 1,\ 1)$ for $\mathbb{R}^3$. Find the associated matrix of T with respect to α and the associated matrix of T with respect to β. Are they similar?

Problem 4.20 Suppose that A and B are similar $n \times n$ matrices. Show that

(1) $\det A = \det B$,

(2) $\operatorname{tr} A = \operatorname{tr} B$,

(3) $\operatorname{rank} A = \operatorname{rank} B$.

Problem 4.21 Let A and B be $n \times n$ matrices. Show that if A is similar to B, then A^2 is similar to B^2.

4.8 Dual spaces

In this section, we are concerned exclusively with linear transformations from a vector space V to the one-dimensional vector space $\mathbb{R}^1$. Such a linear transformation is called a **linear functional** of V. The definite integrals of continuous functions is one of the most important examples of linear functionals in mathematics.

For a matrix A regarded as a linear transformation $A : \mathbb{R}^n \to \mathbb{R}^m$, we saw that the transpose A^T of A is another linear transformation $A^T : \mathbb{R}^m \to \mathbb{R}^n$. For a linear transformation $T : V \to W$ on a vector space V to W, one can naturally ask what its *transpose* is and what the definition is. This section will answer those questions.

Example 4.22 Let $C[a, b]$ be the vector space of all continuous real-valued functions on the interval $[a, b]$. The definite integral $\mathcal{I} : C[a, b] \to \mathbb{R}$ defined by

$$\mathcal{I}(f) = \int_a^b f(t)dt$$

is a linear functional of $C[a, b]$. In particular, if the interval is $[0, 2\pi]$ and n is an integer, then

$$\mathcal{F}_n(f) = \frac{1}{2\pi}\int_0^{2\pi} f(t)e^{-int}dt$$

is a linear functional, called the n-th **Fourier coefficient** of f.

Example 4.23 The trace function $\mathrm{tr} : M_{n\times n}(\mathbb{R}) \to \mathbb{R}$ is a linear functional of $M_{n\times n}(\mathbb{R})$.

Note that as we saw in Section 4.5, the set of all linear functionals of V is the vector space $\mathcal{L}(V; \mathbb{R}^1)$ whose dimension equals the dimension of V (see page 141).

Definition 4.7 For a vector space V, the vector space of all linear functionals of V is called the **dual space** of V and denoted by V^*.

Recall that such a linear transformation $T : V \to \mathbb{R}$ is completely determined by the values on a basis for V. Thus if $\alpha = \{\mathbf{v}_1, \mathbf{v}_2, \ldots, \mathbf{v}_n\}$ is a basis for a vector space V, then the functions $\mathbf{v}_i^* : V \to \mathbb{R}$ defined by $\mathbf{v}_i^*(\mathbf{v}_j) = \delta_{ij}$ for each $i, j = 1, \ldots, n$ are clearly linear functionals of V, called the i-th **coordinate function** with respect to the basis α. In particular, for any $\mathbf{x} = \sum a_i\mathbf{v}_i \in V$, $\mathbf{v}_i^*(\mathbf{x}) = a_i$, the i-th coordinate of $\mathbf{x}$ with respect to α.

Theorem 4.16 *The set $\alpha^* = \{\mathbf{v}_1^*,\ \mathbf{v}_2^*,\ \ldots,\ \mathbf{v}_n^*\}$ forms a basis for the dual space V^*, and for any $T \in V^*$ we have*

$$T = \sum_{i=1}^{n} T(\mathbf{v}_i)\mathbf{v}_i^*.$$

Proof: Clearly, the set $\alpha^* = \{\mathbf{v}_1^*,\ \mathbf{v}_2^*,\ \cdots,\ \mathbf{v}_n^*\}$ is linearly independent, since $\mathbf{0} = \sum_{i=1}^n c_i\mathbf{v}_i^*$ implies $0 = \sum_{i=1}^n c_i\mathbf{v}_i^*(\mathbf{v}_j) = c_j$ for each $j = 1,\ \ldots,\ n$. Moreover, the set α^* spans V^*; for any $T \in V^*$ and any $\mathbf{v}_j \in \alpha$, we have

$$\left(\sum_{i=1}^{n} T(\mathbf{v}_i)\mathbf{v}_i^*\right)(\mathbf{v}_j) = \sum_{i=1}^{n} T(\mathbf{v}_i)(\mathbf{v}_i^*(\mathbf{v}_j)) = T(\mathbf{v}_j).$$

Hence, by Corollary 4.4, we get $T = \sum_{i=1}^n T(\mathbf{v}_i)\mathbf{v}_i^*$. □

Definition 4.8 For a basis $\alpha = \{\mathbf{v}_1,\ \mathbf{v}_2,\ \ldots,\ \mathbf{v}_n\}$ for a vector space V, the basis α^* for V^* is called the **dual basis** of α.

This theorem says that, for a fixed basis $\alpha = \{\mathbf{v}_1,\ \ldots,\ \mathbf{v}_n\}$ for V, the transformation $* : V \to V^*$ given by $*(\mathbf{v}_i) = \mathbf{v}_i^*$ is an isomorphism between V and V^*. Therefore, we have the following corollary.

Corollary 4.17 *Any finite-dimensional vector space is isomorphic to its dual space.*

Example 4.24 Let $\alpha = \{(1,\ 2), (1,\ 3)\}$ be a basis for $\mathbb{R}^2$. To determine the dual basis $\alpha^* = \{f,\ g\}$ of α, we consider the equations

$$\begin{aligned} 1 &= f(1,\ 2) = f(\mathbf{e}_1) + 2f(\mathbf{e}_2) \\ 0 &= f(1,\ 3) = f(\mathbf{e}_1) + 3f(\mathbf{e}_2). \end{aligned}$$

Solving these equations, we obtain that $f(\mathbf{e}_1) = 3$ and $f(\mathbf{e}_2) = -1$, and $f(x, y) = 3x - y$. Similarly, it can be shown that $g(x, y) = -2x + y$. □

Example 4.25 Consider $V = \mathbb{R}^n$ with the standard basis $\alpha = \{\mathbf{e}_1, \ldots, \mathbf{e}_n\}$, and its dual basis $\alpha^* = \{\mathbf{e}_1^*, \ldots, \mathbf{e}_n^*\}$ for $\mathbb{R}^{n*}$. Then for a vector $\mathbf{a} = (a_1, \ldots, a_n) = a_1\mathbf{e}_1 + \cdots + a_n\mathbf{e}_n \in \mathbb{R}^n$, we have $\mathbf{e}_i^*(\mathbf{a}) = \mathbf{e}_i^*(a_1\mathbf{e}_1 + \cdots + a_n\mathbf{e}_n) = a_i$. That is,

$$\mathbf{a} = (a_1, \ldots, a_n) = (\mathbf{e}_1^*(\mathbf{a}), \ldots, \mathbf{e}_n^*(\mathbf{a})) = (\mathbf{e}_1^*, \ldots, \mathbf{e}_n^*)(\mathbf{a}).$$

On the other hand, when we write a vector in $\mathbb{R}^n$ as $\mathbf{x} = (x_1, \ldots, x_n)$ in coordinate functions (or unknowns) x_i, it means that given a point $\mathbf{a} = (a_1, \ldots, a_n) \in \mathbb{R}^n$ each x_i gives us the i-th coordinate of $\mathbf{a}$, that is,

$$(x_1, \ldots, x_n)(\mathbf{a}) = (x_1(\mathbf{a}), \ldots, x_n(\mathbf{a})) = (a_1, \ldots, a_n).$$

In this way, we have identified $\mathbf{e}_i^* = x_i$ for $i = 1, \ldots, n$, *i.e.*, $\mathbb{R}^{n*} = \mathbb{R}^n$. Thus, the actual meaning of the usual coordinate expression $(x_1, \ldots, x_n)$ of $\mathbf{x}$ is just a vector in $\mathbb{R}^{n*}$ such that $(x_1, \ldots, x_n)(\mathbf{a}) = (a_1, \ldots, a_n)$ for a point $\mathbf{a} \in \mathbb{R}^n$. □

Now, consider two vector spaces V and W with fixed bases α and β, respectively. Let $S : V \to W$ be a linear transformation from V to W. Then for any linear functional $g \in W^*$, *i.e.*, $g : W \to \mathbb{R}$, it is easy to see that the composition $g \circ S(\mathbf{x}) = g(S(\mathbf{x}))$ for $\mathbf{x} \in V$ defines a linear functional on V, *i.e.*, $g \circ S \in V^*$. Thus, we have a transformation $S^* : W^* \to V^*$ defined by $S^*(g) = g \circ S$ for $g \in W^*$.

Theorem 4.18 *The mapping $S^* : W^* \to V^*$ defined by $S^*(g) = g \circ S$ for $g \in W^*$ is a linear transformation and $[S^*]_{\beta^*}^{\alpha^*} = \left([S]_\alpha^\beta\right)^T$.*

Proof: The mapping S^* is clearly linear by the definition of a composition of functions. Let $\alpha = \{\mathbf{v}_1, \ldots, \mathbf{v}_n\}$ and $\beta = \{\mathbf{w}_1, \ldots, \mathbf{w}_m\}$ be bases for V and W with their dual bases $\alpha^* = \{\mathbf{v}_1^*, \ldots, \mathbf{v}_n^*\}$ and $\beta^* = \{\mathbf{w}_1^*, \ldots, \mathbf{w}_m^*\}$, respectively. Let $[S]_\alpha^\beta = [a_{ij}]$ and $[S^*]_{\beta^*}^{\alpha^*} = [b_{k\ell}]$. Then,

$$S(\mathbf{v}_i) = \sum_{k=1}^m a_{ki}\mathbf{w}_k \quad \text{and} \quad S^*(\mathbf{w}_j^*) = \sum_{i=1}^n b_{ij}\mathbf{v}_i^*,$$

for $1 \le i \le n$ and $1 \le j \le m$. Thus,

$$\begin{aligned} b_{ij} &= S^*(\mathbf{w}_j^*)(\mathbf{v}_i) = (\mathbf{w}_j^* \circ S)(\mathbf{v}_i) \\ &= \mathbf{w}_j^*(S(\mathbf{v}_i)) = \mathbf{w}_j^*\left(\sum_{k=1}^m a_{ki}\mathbf{w}_k\right) = \sum_{k=1}^m a_{ki}\mathbf{w}_j^*(\mathbf{w}_k) = a_{ji}. \end{aligned}$$

Hence, we get $[S^*]_{\beta^*}^{\alpha^*} = \left([S]_\alpha^\beta\right)^T$. □

Remark: Theorem 4.18 shows that the matrix representation of S^* is just the transpose of that of S. And hence, the linear transformation S^* is called the **transpose** (or *adjoint*) of S, denoted also by S^T.

Example 4.26 With the identification $\mathbb{R}^{n*} = \mathbb{R}^n$ in Example 4.25, the transpose A^T of a matrix A is actually A^*:

$$A^T = A^* : \mathbb{R}^{m*} \to \mathbb{R}^{n*}. \qquad \square$$

For two linear transformations $S : U \to V$ and $T : V \to W$, it is quite easy to show (the readers may try to) that

$$(T \circ S)^* = S^* \circ T^*.$$

Thus, if $S : V \to W$ is an isomorphism, then so is its transpose $S^* : W^* \to V^*$. In particular, since $* : V \to V^*$ is an isomorphism, so is its transpose $** : V^* \to V^{**}$. Note that even though the isomorphism $* : V \to V^*$ depends on a choice of a basis for V, there is an isomorphism between V and V^{**} that does not depend on a choice of bases for the two vector spaces: We first define, for each $\mathbf{x} \in V$, $\tilde{\mathbf{x}} : V^* \to \mathbb{R}$ by $\tilde{\mathbf{x}}(f) = f(\mathbf{x})$ for every $f \in V^*$. It is easy to verify that $\tilde{\mathbf{x}}$ is a linear functional on V^*, so $\tilde{\mathbf{x}} \in V^{**}$. We will show below that the mapping $\Phi : V \to V^{**}$ defined by $\Phi(\mathbf{x}) = \tilde{\mathbf{x}}$ is the desired isomorphism between V and V^{**}.

Lemma 4.19 *If $\tilde{\mathbf{x}}(f) = \mathbf{0}$ for all $f \in V^*$, i.e., $\tilde{\mathbf{x}} = \mathbf{0}$ in V^{**}, then $\mathbf{x} = \mathbf{0}$.*

Proof: Suppose that $\mathbf{x} \neq \mathbf{0}$. Choose a basis $\alpha = \{\mathbf{v}_1, \mathbf{v}_2, \ldots, \mathbf{v}_n\}$ for V with $\mathbf{v}_1 = \mathbf{x}$. Let $\alpha^* = \{\mathbf{v}_1^*, \mathbf{v}_2^*, \ldots, \mathbf{v}_n^*\}$ be the dual basis of α. Then

$$\tilde{\mathbf{x}}(\mathbf{v}_1^*) = \mathbf{v}_1^*(\mathbf{x}) = \mathbf{v}_1^*(\mathbf{v}_1) = 1,$$

which contradicts the hypothesis. $\square$

Theorem 4.20 *The mapping $\Phi : V \to V^{**}$ defined by $\Phi(\mathbf{x}) = \tilde{\mathbf{x}}$ is an isomorphism from V to V^{**}.*

Proof: To show the linearity of Φ, let $\mathbf{x}, \mathbf{y} \in V$ and k a scalar. Then, for any $f \in V^*$,

$$\begin{aligned} \Phi(\mathbf{x} + k\mathbf{y})(f) &= (\widetilde{\mathbf{x} + k\mathbf{y}})(f) = f(\mathbf{x} + k\mathbf{y}) \\ &= f(\mathbf{x}) + kf(\mathbf{y}) = \tilde{\mathbf{x}}(f) + k\tilde{\mathbf{y}}(f) \\ &= (\tilde{\mathbf{x}} + k\tilde{\mathbf{y}})(f) = (\Phi(\mathbf{x}) + k\Phi(\mathbf{y}))(f). \end{aligned}$$

Hence, $\Phi(\mathbf{x} + k\mathbf{y}) = \Phi(\mathbf{x}) + k\Phi(\mathbf{y})$. The injectivity of Φ comes from Lemma 4.19. Since $\dim V = \dim V^{**}$, Φ is an isomorphism. $\square$

Problem 4.22 Let $\alpha = \{(1,\ 0,\ 1),\ (1,\ 2,\ 1),\ (0,\ 0,\ 1)\}$ be a basis for $\mathbb{R}^3$. Find the dual basis α^*.

Problem 4.23 Let $V = \mathbb{R}^3$ and define $f_i \in V^*$ as follows:

$$f_1(x,\ y,\ z) = x - 2y, \quad f_2(x,\ y,\ z) = x + y + z, \quad f_3(x,\ y,\ z) = y - 3z.$$

Prove that $\{f_1,\ f_2,\ f_3\}$ is a basis for V^*, and then find a basis for V for which it is the dual.

4.9 Exercises

4.1. Which of the following functions T are linear transformations?

(1) $T(x,\ y) = (x^2 - y^2,\ x^2 + y^2)$.
(2) $T(x,\ y,\ z) = (x + y,\ 0,\ 2x + 4z)$.
(3) $T(x,\ y) = (\sin x,\ y)$.
(4) $T(x,\ y) = (x + 1,\ 2y,\ x + y)$.
(5) $T(x,\ y,\ z) = (|x|,\ 0)$.

4.2. Let $T : P_2(\mathbb{R}) \to P_3(\mathbb{R})$ be a linear transformation such that $T(1) = 1$, $T(x) = x^2$, and $T(x^2) = x^3 + x$. Find $T(ax^2 + bx + c)$.

4.3. Find $S \circ T$ and/or $T \circ S$ whenever it is defined.

(1) $T(x,\ y,\ z) = (x - y + z,\ x + z)$, $S(x,\ y) = (x,\ x - y,\ y)$;
(2) $T(x,\ y) = (x,\ 3y + x,\ 2x - 4y,\ y)$, $S(x,\ y,\ z) = (2x,\ y)$;

4.4. Let $S : C(\mathbb{R}) \to C(\mathbb{R})$ be the function on the vector space $C(\mathbb{R})$ defined by, for $f \in C(\mathbb{R})$,

$$S(f)(x) = f(x) - \int_1^x uf(u)du.$$

Show that S is a linear transformation on the vector space $C(\mathbb{R})$.

4.5. Let T be a linear transformation on a vector space V such that $T^2 = Id$ and $T \neq Id$. Let $U = \{\mathbf{v} \in V : T(\mathbf{v}) = \mathbf{v}\}$ and $W = \{\mathbf{v} \in V : T(\mathbf{v}) = -\mathbf{v}\}$. Show that

(1) at least one of U and W is a nonzero subspace of V;
(2) $U \cap W = \{\mathbf{0}\}$;
(3) $V = U + W$.

4.6. If $T : \mathbb{R}^3 \to \mathbb{R}^3$ is defined by $T(x,\ y,\ z) = (2x - z,\ 3x - 2y,\ x - 2y + z)$,

(1) determine the null space $\mathcal{N}(T)$ of T,
(2) determine whether T is one-to-one,
(3) find a basis for $\mathcal{N}(T)$.

4.7. Show that each of the following linear transformations T on $\mathbb{R}^3$ is invertible, and find a formula for T^{-1}:

(1) $T(x,\ y,\ z) = (3x,\ x - y,\ 2x + y + z)$.

(2) $T(x,\ y,\ z) = (2x,\ 4x - y,\ 2x + 3y - z)$.

4.8. Let $S, T : V \to V$ be linear transformations of a vector space V.

(1) Show that if $T \circ S$ is one-to-one, then T is an isomorphism.

(2) Show that if $T \circ S$ is onto, then T is an isomorphism.

(3) Show that if T^k is an isomorphism for some positive k, then T is an isomorphism.

4.9. Let T be a linear transformation from $\mathbb{R}^3$ to $\mathbb{R}^2$, and let S be a linear transformation from $\mathbb{R}^2$ to $\mathbb{R}^3$. Prove that the composition $S \circ T$ is not invertible.

4.10. Let T be a linear transformation on a vector space V satisfying $T - T^2 = Id$. Show that T is invertible.

4.11. Let $T : P_3(\mathbb{R}) \to P_3(\mathbb{R})$ be the linear transformation defined by

$$Tf(x) = f''(x) - 4f'(x) + f(x).$$

Find the matrix $[T]_\alpha$ for the basis $\alpha = \{x,\ 1+x,\ x+x^2,\ x^3\}$.

4.12. Let T be the linear transformation on $\mathbb{R}^2$ defined by $T(x,\ y) = (-y,\ x)$.

(1) What is the matrix of T with respect to an ordered basis $\alpha = \{\mathbf{v}_1,\ \mathbf{v}_2\}$, where $\mathbf{v}_1 = (1,\ 2)$, $\mathbf{v}_2 = (1,\ -1)$?

(2) Show that for every real number c the linear transformation $T - c\,Id$ is invertible.

4.13. Find the matrix representation of each of the following linear transformations T on $\mathbb{R}^2$ with respect to the standard basis $\{\mathbf{e}_1,\ \mathbf{e}_2\}$.

(1) $T(x,\ y) = (2y,\ 3x - y)$.

(2) $T(x,\ y) = (3x - 4y,\ x + 5y)$.

4.14. Let $M = \begin{bmatrix} 4 & 2 & 1 \\ 0 & 1 & 3 \end{bmatrix}$.

(1) Find the unique linear transformation $T : \mathbb{R}^3 \to \mathbb{R}^2$ so that M is the matrix of T with respect to the bases

$$\alpha_1 = \left\{ \begin{bmatrix} 1 \\ 0 \\ 0 \end{bmatrix}, \begin{bmatrix} 1 \\ 1 \\ 0 \end{bmatrix}, \begin{bmatrix} 1 \\ 1 \\ 1 \end{bmatrix} \right\}, \quad \alpha_2 = \left\{ \begin{bmatrix} 1 \\ 0 \end{bmatrix}, \begin{bmatrix} 1 \\ 1 \end{bmatrix} \right\}.$$

(2) Find $T(x,\ y,\ z)$.

4.15. Find the matrix representation of each of the following linear transformations T on $P_2(\mathbb{R})$ with respect to the basis $\{1,\ x,\ x^2\}$.

(1) $T : p(x) \to p(x+1)$.

(2) $T : p(x) \to p'(x)$.
(3) $T : p(x) \to p(0)x$.
(4) $T : p(x) \to \dfrac{p(x) - p(0)}{x}$.

4.16. Consider the following ordered bases of $\mathbb{R}^3$: $\alpha = \{\mathbf{e}_1, \mathbf{e}_2, \mathbf{e}_3\}$ the standard basis and $\beta = \{\mathbf{u}_1 = (1, 1, 1), \mathbf{u}_2 = (1, 1, 0), \mathbf{u}_3 = (1, 0, 0)\}$.
(1) Find the transition matrix P from α to β.
(2) Find the transition matrix Q from β to α.
(3) Verify that $Q = P^{-1}$.
(4) Show that $[\mathbf{v}]_\beta = P[\mathbf{v}]_\alpha$ for any vector $\mathbf{v} \in \mathbb{R}^3$.
(5) Show that $[T]_\beta = Q^{-1}[T]_\alpha Q$ for the linear transformation T defined by $T(x, y, z) = (2y + x, x - 4y, 3x)$.

4.17. There are no matrices A and B in $M_{n\times n}(\mathbb{R})$ such that $AB - BA = I_n$.

4.18. Let $T : \mathbb{R}^3 \to \mathbb{R}^2$ be the linear transformation defined by
$T(x, y, x) = (3x + 2y - 4z, x - 5y + 3z)$,
and let $\alpha = \{(1, 1, 1), (1, 1, 0), (1, 0, 0)\}$ and $\beta = \{(1, 3), (2, 5)\}$ be bases for $\mathbb{R}^3$ and $\mathbb{R}^2$, respectively.
(1) Find the associated matrix $[T]_\alpha^\beta$ for T.
(2) Verify $[T]_\alpha^\beta[\mathbf{v}]_\alpha = [T(\mathbf{v})]_\beta$ for any $\mathbf{v} \in \mathbb{R}^3$.

4.19. Find the transition matrix $[Id]_\alpha^\beta$ from α to β, when
(1) $\alpha = \{(2,3), (0,1)\}$, $\beta = \{(6,4), (4,8)\}$;
(2) $\alpha = \{(5,1), (1,2)\}$, $\beta = \{(1,0), (0,1)\}$;
(3) $\alpha = \{(1,1,1), (1,1,0), (1,0,0)\}$, $\beta = \{(2,0,3), (-1,4,1), (3,2,5)\}$;
(4) $\alpha = \{t, 1, t^2\}$, $\beta = \{3 + 2t + t^2, t^2 - 4, 2 + t\}$.

4.20. Show that all matrices of the form $\begin{bmatrix} \cos\theta & \sin\theta \\ \sin\theta & -\cos\theta \end{bmatrix}$ are similar.

4.21. Show that the matrix $A = \begin{bmatrix} 1 & 0 \\ 1 & 1 \end{bmatrix}$ cannot be similar to a diagonal matrix.

4.22. Are the matrices $\begin{bmatrix} 1 & 2 & 5 \\ 0 & 1 & 6 \\ 1 & 0 & 1 \end{bmatrix}$ and $\begin{bmatrix} -1 & 0 & 1 \\ 0 & 4 & 2 \\ 0 & 0 & 3 \end{bmatrix}$ similar?

4.23. For a linear transformation T on a vector space V, show that T is one-to-one if and only if its transpose T^* is one-to-one.

4.24. Let $T : \mathbb{R}^3 \to \mathbb{R}^3$ be the linear transformation defined by
$T(x, y, z) = (2y + z, -x + 4y + z, x + z)$.
Compute $[T]_\alpha$ and $[T^*]_{\alpha^*}$ for the standard basis $\alpha = \{\mathbf{e}_1, \mathbf{e}_2, \mathbf{e}_3\}$.

4.25. Let T be the linear transformation from $\mathbb{R}^3$ into $\mathbb{R}^2$ defined by
$T(x_1,\ x_2,\ x_3) = (x_1 + x_2,\ 2x_3 - x_1)$.

(1) For the standard ordered bases α and β for $\mathbb{R}^3$ and $\mathbb{R}^2$ respectively, find the associated matrix for T with respect to the bases α and β.

(2) Let $\alpha = \{\mathbf{x}_1,\ \mathbf{x}_2,\ \mathbf{x}_3\}$ and $\beta = \{\mathbf{y}_1,\ \mathbf{y}_2\}$, where $\mathbf{x}_1 = (1, 0, -1)$, $\mathbf{x}_2 = (1, 1, 1)$, $\mathbf{x}_3 = (1, 0, 0)$, and $\mathbf{y}_1 = (0, 1)$, $\mathbf{y}_2 = (1, 0)$. Find the associated matrices $[T]_\alpha^\beta$ and $[T^*]_{\beta^*}^{\alpha^*}$.

4.26. Let T be the linear transformation from $\mathbb{R}^3$ to $\mathbb{R}^4$ defined by
$T(x,\ y,\ z) = (2x + y + 4z,\ x + y + 2z,\ y + 2z,\ x + y + 3z)$.
Find the range and the kernel of T. What is the dimension of $\mathcal{C}(T)$? Find $[T]_\alpha^\beta$ and $[T^*]_{\alpha^*}^{\beta^*}$, where
$\alpha = \{(1, 0, 0),\ (0, 1, 0),\ (0, 0, 1)\}$
$\beta = \{(1, 0, 0, 0),\ (1, 1, 0, 0),\ (1, 1, 1, 0),\ (1, 1, 1, 1)\}$.

4.27. Let T be the linear transformation on $V = \mathbb{R}^3$, for which the associated matrix with respect to the standard ordered basis is

$$A = \begin{bmatrix} 1 & 2 & 1 \\ 0 & 1 & 1 \\ -1 & 3 & 4 \end{bmatrix}.$$

Find the bases for the range and the null space of the transpose T^* on V^*.

4.28. Define three linear functionals on the vector space $P_2(\mathbb{R})$ by
$f_1(p) = \int_0^1 p(x)dx, \quad f_2(p) = \int_0^2 p(x)dx, \quad f_3(p) = \int_0^{-1} p(x)dx.$
Show that $\{f_1,\ f_2,\ f_3\}$ is a basis for V^* by finding its dual basis for V.

4.29. Determine whether or not the following statements are true in general, and justify your answers.

(1) For a linear transformation $T : \mathbb{R}^n \to \mathbb{R}^m$, $\mathrm{Ker}(T) = \{\mathbf{0}\}$ if $m > n$.

(2) For a linear transformation $T : \mathbb{R}^n \to \mathbb{R}^m$, $\mathrm{Ker}(T) \neq \{\mathbf{0}\}$ if $m < n$.

(3) A linear transformation $T : \mathbb{R}^n \to \mathbb{R}^m$ is one-to-one if and only if the nullspace of $[T]_\alpha^\beta$ is $\{\mathbf{0}\}$, for any bases α and β of $\mathbb{R}^n$ and $\mathbb{R}^m$ respectively.

(4) For a linear transformation T on $\mathbb{R}^n$, the dimension of the image of T is equal to that of the row space of $[T]_\alpha$ for any basis α for $\mathbb{R}^n$.

(5) Any polynomial $p(x)$ is linear if and only if the degree of $p(x)$ is 1.

(6) Let $T : \mathbb{R}^3 \to \mathbb{R}^2$ be a function given as $T(\mathbf{x}) = (T_1(\mathbf{x}),\ T_2(\mathbf{x}))$ for any $\mathbf{x} \in \mathbb{R}^3$. Then T is linear if and only if their coordinate functions T_i, $i = 1,\ 2$, are linear.

(7) For a linear transformation $T : \mathbb{R}^n \to \mathbb{R}^n$, if $[T]_\alpha^\beta = I_n$ for some bases α and β of $\mathbb{R}^n$, then T must be the identity transformation.

(8) If a linear transformation $T : \mathbb{R}^n \to \mathbb{R}^n$ is one-to-one, then any matrix representation of T is nonsingular.

(9) Any $m \times n$ matrix A can be a matrix representation of a linear transformation $T : \mathbb{R}^n \to \mathbb{R}^m$.

Chapter 5

Inner Product Spaces

5.1 Inner products

In order to study the geometry of a vector space, we go back to the case of the Euclidean 3-space $\mathbb{R}^3$. Recall that the **dot** (or **Euclidean inner**) **product** of two vectors $\mathbf{x} = (x_1,\ x_2,\ x_3)$ and $\mathbf{y} = (y_1,\ y_2,\ y_3)$ in $\mathbb{R}^3$ is defined by the formula

$$\mathbf{x} \cdot \mathbf{y} = x_1y_1 + x_2y_2 + x_3y_3 = \mathbf{x}^T\mathbf{y},$$

where $\mathbf{x}^T\mathbf{y}$ is the matrix product of $\mathbf{x}^T$ and $\mathbf{y}$. Using the dot product, the **length** (or **magnitude**) of a vector $\mathbf{x} = (x_1,\ x_2,\ x_3)$ is defined by

$$\|\mathbf{x}\| = (\mathbf{x} \cdot \mathbf{x})^{\frac{1}{2}} = \sqrt{x_1^2 + x_2^2 + x_3^2},$$

and the **distance** of two vectors $\mathbf{x}$ and $\mathbf{y}$ in $\mathbb{R}^3$ is defined by

$$d(\mathbf{x}, \mathbf{y}) = \|\mathbf{x} - \mathbf{y}\|.$$

In this way, the dot product can be considered to be a ruler for measuring the length of a line segment in $\mathbb{R}^3$. Furthermore, it can also be used to measure the angle between two vectors: in fact, the **angle** θ between two vectors $\mathbf{x}$ and $\mathbf{y}$ in $\mathbb{R}^3$ is measured by the formula involving the dot product

$$\cos\theta = \frac{\mathbf{x} \cdot \mathbf{y}}{\|\mathbf{x}\|\|\mathbf{y}\|},\ \ 0 \le \theta \le \pi,$$

since the dot product satisfies the formula

$$\mathbf{x} \cdot \mathbf{y} = \|\mathbf{x}\|\|\mathbf{y}\| \cos\theta.$$

In particular, two vectors $\mathbf{x}$ and $\mathbf{y}$ are orthogonal (*i.e.*, they form a right angle $\theta = \pi/2$) if and only if the Pythagorean theorem holds:

$$\|\mathbf{x}\|^2 + \|\mathbf{y}\|^2 = \|\mathbf{x} + \mathbf{y}\|^2.$$

By rewriting this formula in terms of the dot product, we obtain another equivalent condition:

$$\mathbf{x} \cdot \mathbf{y} = x_1 y_1 + x_2 y_2 + x_3 y_3 = 0.$$

In fact, this dot product is one of the most important structures with which $\mathbb{R}^3$ is equipped. Euclidean geometry begins with the vector space $\mathbb{R}^3$ together with the dot product, because the Euclidean distance can be defined by this dot product.

The dot product has a direct extension to the n-space $\mathbb{R}^n$ for any positive integer n, *i.e.*, for vectors $\mathbf{x} = (x_1, \ldots, x_n)$ and $\mathbf{y} = (y_1, \ldots, y_n)$ in $\mathbb{R}^n$, the dot product, also called the **Euclidean inner product**, and the **length** (or **magnitude**) of a vector are defined similarly as

$$\begin{aligned} \mathbf{x} \cdot \mathbf{y} &= x_1 y_1 + \cdots + x_n y_n = \mathbf{x}^T \mathbf{y}, \\ \|\mathbf{x}\| &= (\mathbf{x} \cdot \mathbf{x})^{\frac{1}{2}} = \sqrt{x_1^2 + \cdots + x_n^2}. \end{aligned}$$

In order to extend this notion of dot product to vector spaces in general, we extract the most essential properties that the dot product in $\mathbb{R}^n$ satisfies and take these properties as axioms for an inner product of a vector space V. First of all, we note that it is a rule that assigns a real number $\mathbf{x} \cdot \mathbf{y}$ to each pair of vectors $\mathbf{x}$ and $\mathbf{y}$ in $\mathbb{R}^n$, and the essential rules it satisfies are those in the following definition.

Definition 5.1 An **inner product** on a real vector space V is a function that associates a real number $\langle \mathbf{x}, \mathbf{y} \rangle$ to each pair of vectors $\mathbf{x}$ and $\mathbf{y}$ in V in such a way that the following rules are satisfied for all vectors $\mathbf{x}$, $\mathbf{y}$ and $\mathbf{z}$ in V and all scalars k in $\mathbb{R}$:

(1) $\langle \mathbf{x}, \mathbf{y} \rangle = \langle \mathbf{y}, \mathbf{x} \rangle$ (symmetry),
(2) $\langle \mathbf{x} + \mathbf{y}, \mathbf{z} \rangle = \langle \mathbf{x}, \mathbf{z} \rangle + \langle \mathbf{y}, \mathbf{z} \rangle$ (additivity),
(3) $\langle k\mathbf{x}, \mathbf{y} \rangle = k\langle \mathbf{x}, \mathbf{y} \rangle$ (homogeneity),
(4) $\langle \mathbf{x}, \mathbf{x} \rangle \geq 0$, and $\langle \mathbf{x}, \mathbf{x} \rangle = 0 \Leftrightarrow \mathbf{x} = \mathbf{0}$ (positive definiteness).

A pair $(V, \langle\ ,\ \rangle)$ of a (real) vector space V and an inner product $\langle\ ,\ \rangle$ is called a **(real) inner product space**. In particular, the pair $(\mathbb{R}^n, \cdot)$ is called the **Euclidean n-space**.

Note that by symmetry (1), additivity (2) and homogeneity (3) also hold for the second variable: *i.e.*,

(2′) $\langle \mathbf{x}, \mathbf{y}+\mathbf{z}\rangle = \langle \mathbf{x}, \mathbf{y}\rangle + \langle \mathbf{x}, \mathbf{z}\rangle$,

(3′) $\langle \mathbf{x}, k\mathbf{y}\rangle = k\langle \mathbf{x}, \mathbf{y}\rangle$.

Now it is easy to show that $\langle \mathbf{0}, \mathbf{y}\rangle = 0\langle \mathbf{0}, \mathbf{y}\rangle = 0$, and $\langle \mathbf{x}, \mathbf{0}\rangle = 0$.

Example 5.1 For vectors $\mathbf{x} = (x_1,\ x_2)$ and $\mathbf{y} = (y_1,\ y_2)$ in $\mathbb{R}^2$, define

$$\langle \mathbf{x}, \mathbf{y}\rangle = ax_1y_1 + cx_1y_2 + cx_2y_1 + bx_2y_2,$$

where a, b and c are arbitrary real numbers. Then this function $\langle\ ,\ \rangle$ clearly satisfies the first three rules of the inner product. Moreover, if $a > 0$ and $ab - c^2 > 0$ hold, then it also satisfies rule (4), the positive definiteness of the inner product. (Hint: The equation $\langle \mathbf{x}, \mathbf{x}\rangle = ax_1^2 + 2cx_1x_2 + bx_2^2 \geq 0$ if and only if either $x_2 = 0$ or the discriminant of $\langle \mathbf{x}, \mathbf{x}\rangle / x_2^2$ is nonpositive.) Note that the equation can be written as matrix products:

$$\langle \mathbf{x}, \mathbf{y}\rangle = [x_1\ x_2]\begin{bmatrix} a & c \\ c & b \end{bmatrix}\begin{bmatrix} y_1 \\ y_2 \end{bmatrix} = \mathbf{x}^T A\mathbf{y}.$$

In the case of $c = 0$, this reduces to $\langle \mathbf{x}, \mathbf{y}\rangle = ax_1y_1 + bx_2y_2$. Notice also that $a = \langle \mathbf{e}_1, \mathbf{e}_1\rangle$, $b = \langle \mathbf{e}_2, \mathbf{e}_2\rangle$ and $c = \langle \mathbf{e}_1, \mathbf{e}_2\rangle = \langle \mathbf{e}_2, \mathbf{e}_1\rangle$. □

Example 5.2 Let $V = C\,[0,\ 1]$ be the vector space of all real-valued continuous functions on $[0,\ 1]$. For any two functions $f(x)$ and $g(x)$ in V, define

$$\langle f, g\rangle = \int_0^1 f(x)g(x)dx\,.$$

Then $\langle\ ,\ \rangle$ is an inner product on V (verify this). Let

$$f(x) = \begin{cases} 1-2x & \text{if } 0 \leq x \leq \frac{1}{2}, \\ 0 & \text{if } \frac{1}{2} \leq x \leq 1, \end{cases} \quad \text{and} \quad g(x) = \begin{cases} 0 & \text{if } 0 \leq x \leq \frac{1}{2}, \\ 2x-1 & \text{if } \frac{1}{2} \leq x \leq 1. \end{cases}$$

Then $f \neq \mathbf{0} \neq g$, but $\langle f, g\rangle = 0$. □

By a subspace W of an inner product space V, we mean a subspace of the vector space V together with the inner product that is the restriction of the inner product of V to W.

Example 5.3 The set $W = D^1[0,\ 1]$ of all real-valued *differentiable* functions on $[0,\ 1]$ is a subspace of $V = C[0,\ 1]$. The restriction to W of the inner product on V defined in Example 5.2 makes W an inner product subspace of V. However, suppose we define another inner product on W by the following formula: For any two functions $f(x)$ and $g(x)$ in W,

$$\langle\langle f, g\rangle\rangle = \int_0^1 f(x)g(x)dx + \int_0^1 f'(x)g'(x)dx\,.$$

Then $\langle\langle\ ,\ \rangle\rangle$ is also an inner product on W but is not defined on V. This means that this inner product is quite different from the restriction of the inner product of V to W, and hence W with this new inner product is not a subspace of the space V as an inner product space. □

5.2 The lengths and angles of vectors

The following inequality will enable us to define an angle between two vectors in an inner product space V.

Theorem 5.1 (Cauchy-Schwarz inequality) *If* $\mathbf{x}$ *and* $\mathbf{y}$ *are vectors in an inner product space* V, *then*

$$\langle \mathbf{x}, \mathbf{y}\rangle^2 \le \langle \mathbf{x}, \mathbf{x}\rangle\langle \mathbf{y}, \mathbf{y}\rangle.$$

Proof: If $\mathbf{x} = \mathbf{0}$, it is clear. Assume $\mathbf{x} \neq \mathbf{0}$. For any scalar t, we have

$$0 \le \langle t\mathbf{x} + \mathbf{y}, t\mathbf{x} + \mathbf{y}\rangle = \langle \mathbf{x}, \mathbf{x}\rangle t^2 + 2\langle \mathbf{x}, \mathbf{y}\rangle t + \langle \mathbf{y}, \mathbf{y}\rangle.$$

This inequality implies that the polynomial $\langle \mathbf{x}, \mathbf{x}\rangle t^2 + 2\langle \mathbf{x}, \mathbf{y}\rangle t + \langle \mathbf{y}, \mathbf{y}\rangle$ in t has either no real roots or a repeated real root. Therefore, its discriminant must be nonpositive:

$$\langle \mathbf{x}, \mathbf{y}\rangle^2 - \langle \mathbf{x}, \mathbf{x}\rangle\langle \mathbf{y}, \mathbf{y}\rangle \le 0,$$

which implies the inequality. □

Problem 5.1 Prove that equality in the Cauchy-Schwarz inequality holds if and only if the vectors $\mathbf{x}$ and $\mathbf{y}$ are linearly dependent.

The lengths and angles of vectors in an inner product space are defined similarly to the case of the Euclidean n-space.

Definition 5.2 Let V be an inner product space. Then the **magnitude** or the **length** of a vector $\mathbf{x}$, denoted by $\|\mathbf{x}\|$, is defined by

$$\|\mathbf{x}\| = \sqrt{\langle \mathbf{x}, \mathbf{x} \rangle}\,.$$

The **distance** between two vectors $\mathbf{x}$ and $\mathbf{y}$, denoted by $d(\mathbf{x}, \mathbf{y})$, is defined by

$$d(\mathbf{x}, \mathbf{y}) = \|\mathbf{x} - \mathbf{y}\|.$$

From the Cauchy-Schwarz inequality, we have

$$-1 \le \frac{\langle \mathbf{x}, \mathbf{y} \rangle}{\|\mathbf{x}\|\|\mathbf{y}\|} \le 1.$$

Hence, there is a unique number $\theta \in [0, \pi]$ such that $\cos\theta = \frac{\langle \mathbf{x}, \mathbf{y} \rangle}{\|\mathbf{x}\|\|\mathbf{y}\|}$.

Definition 5.3 The real number θ in the interval $[0,\ \pi]$ that satisfies

$$\cos\theta = \frac{\langle \mathbf{x}, \mathbf{y} \rangle}{\|\mathbf{x}\|\|\mathbf{y}\|}, \quad \text{or} \quad \langle \mathbf{x}, \mathbf{y} \rangle = \|\mathbf{x}\|\|\mathbf{y}\|\cos\theta,$$

is called the **angle** between $\mathbf{x}$ and $\mathbf{y}$.

Example 5.4 In $\mathbb{R}^2$ equipped with an inner product $\langle \mathbf{x}, \mathbf{y} \rangle = 2x_1y_1 + 3x_2y_2$, the angle between $\mathbf{x} = (1, 2)$ and $\mathbf{y} = (1, 0)$ is computed as

$$\cos\theta = \frac{\langle \mathbf{x}, \mathbf{y} \rangle}{\|\mathbf{x}\|\|\mathbf{y}\|} = \frac{2}{\sqrt{14 \cdot 2}}.$$

Thus $\theta = \cos^{-1}(\frac{1}{\sqrt{7}})$. □

Problem 5.2 Prove the following properties of length in an inner product space V: For any vectors $\mathbf{x},\ \mathbf{y} \in V$,

(1) $\|\mathbf{x}\| \ge 0$,
(2) $\|\mathbf{x}\| = 0$ if and only if $\mathbf{x} = \mathbf{0}$,
(3) $\|k\mathbf{x}\| = |k|\,\|\mathbf{x}\|$,
(4) $\|\mathbf{x} + \mathbf{y}\| \le \|\mathbf{x}\| + \|\mathbf{y}\|$ (triangular inequality).

Problem 5.3 Let V be an inner product space. Show that for any vectors $\mathbf{x},\ \mathbf{y}$ and $\mathbf{z}$ in V,

(1) $d(\mathbf{x}, \mathbf{y}) \geq 0$,
(2) $d(\mathbf{x}, \mathbf{y}) = 0$ if and only if $\mathbf{x} = \mathbf{y}$,
(3) $d(\mathbf{x}, \mathbf{y}) = d(\mathbf{y}, \mathbf{x})$,
(4) $d(\mathbf{x}, \mathbf{y}) \leq d(\mathbf{x}, \mathbf{z}) + d(\mathbf{z}, \mathbf{y})$ (triangular inequality).

Therefore, an inner product in the 3-space $\mathbb{R}^3$ may play the roles of a ruler and a protractor in our physical world.

Definition 5.4 Two vectors $\mathbf{x}$ and $\mathbf{y}$ in an inner product space are said to be **orthogonal** (or **perpendicular**) if $\langle \mathbf{x}, \mathbf{y} \rangle = 0$.

Note that for nonzero vectors $\mathbf{x}$ and $\mathbf{y}$, $\langle \mathbf{x}, \mathbf{y} \rangle = 0$ if and only if $\theta = \pi/2$.

Lemma 5.2 *Let V be an inner product space and let $\mathbf{x} \in V$. Then the vector $\mathbf{x}$ is orthogonal to every vector $\mathbf{y}$ in V (i.e., $\langle \mathbf{x}, \mathbf{y} \rangle = 0$ for all $\mathbf{y}$ in V) if and only if $\mathbf{x} = \mathbf{0}$.*

Proof: If $\mathbf{x} = \mathbf{0}$, clearly $\langle \mathbf{x}, \mathbf{y} \rangle = 0$ for all $\mathbf{y}$ in V. Suppose that $\langle \mathbf{x}, \mathbf{y} \rangle = 0$ for all $\mathbf{y}$ in V. Then $\langle \mathbf{x}, \mathbf{x} \rangle = 0$ in particular. The positive definiteness of the inner product implies that $\mathbf{x} = \mathbf{0}$. □

Corollary 5.3 *Let V be an inner product space, and let $\alpha = \{\mathbf{v}_1, \ldots, \mathbf{v}_n\}$ be a basis for V. Then a vector $\mathbf{x}$ in V is orthogonal to every basis vector $\mathbf{v}_i$ in α if and only if $\mathbf{x} = \mathbf{0}$.*

Proof: If $\langle \mathbf{x}, \mathbf{v}_i \rangle = 0$ for $i = 1, \ldots, n$, then $\langle \mathbf{x}, \mathbf{y} \rangle = \sum_{i=1}^n y_i \langle \mathbf{x}, \mathbf{v}_i \rangle = 0$ for any $\mathbf{y} = \sum_{i=1}^n y_i \mathbf{v}_i \in V$. □

Example 5.5 (Pythagorean theorem) Let V be an inner product space, and let $\mathbf{x}$ and $\mathbf{y}$ be any two vectors in V with the angle θ. Then, $\langle \mathbf{x}, \mathbf{y} \rangle = \|\mathbf{x}\| \|\mathbf{y}\| \cos\theta$ gives the equality

$$\|\mathbf{x} + \mathbf{y}\|^2 = \|\mathbf{x}\|^2 + \|\mathbf{y}\|^2 + 2\|\mathbf{x}\| \|\mathbf{y}\| \cos\theta.$$

Moreover, it deduces the Pythagorean theorem: $\|\mathbf{x} + \mathbf{y}\|^2 = \|\mathbf{x}\|^2 + \|\mathbf{y}\|^2$ for any orthogonal vector $\mathbf{x}$ and $\mathbf{y}$. □

Theorem 5.4 *If* $\{\mathbf{x}_1,\ \mathbf{x}_2,\ \ldots,\ \mathbf{x}_k\}$ *nonzero vectors in an inner product space* V *are mutually orthogonal (i.e., each vector is orthogonal to every other vector), then they are linearly independent.*

Proof: Suppose $c_1\mathbf{x}_1 + c_2\mathbf{x}_2 + \cdots + c_k\mathbf{x}_k = \mathbf{0}$. Then for each $i = 1,\ \ldots,\ k$,

$$\begin{aligned} 0 &= \langle \mathbf{0}, \mathbf{x}_i \rangle = \langle c_1\mathbf{x}_1 + \cdots + c_k\mathbf{x}_k, \mathbf{x}_i \rangle \\ &= c_1\langle \mathbf{x}_1, \mathbf{x}_i \rangle + \cdots + c_i\langle \mathbf{x}_i, \mathbf{x}_i \rangle + \cdots + c_k\langle \mathbf{x}_k, \mathbf{x}_i \rangle \\ &= c_i\|\mathbf{x}_i\|^2, \end{aligned}$$

because $\mathbf{x}_1,\ \ldots,\ \mathbf{x}_k$ are mutually orthogonal. Since each $\mathbf{x}_i$ is not the zero vector, $\|\mathbf{x}_i\| \neq 0$; so $c_i = 0$ for $i = 1,\ \ldots,\ k$. □

Problem 5.4 Let $f(x)$ and $g(x)$ be continuous real-valued functions on $[0,\ 1]$. Prove

(1) $\left[\int_0^1 f(x)g(x)dx\right]^2 \leq \left[\int_0^1 f^2(x)dx\right]\left[\int_0^1 g^2(x)dx\right]$,

(2) $\left[\int_0^1 (f(x)+g(x))^2dx\right]^{\frac{1}{2}} \leq \left[\int_0^1 f^2(x)dx\right]^{\frac{1}{2}} + \left[\int_0^1 g^2(x)dx\right]^{\frac{1}{2}}$.

5.3 Matrix representations of inner products

As we saw at the end of Example 5.1, the inner product on an inner product space $(V,\ \langle , \rangle)$ can be expressed in terms of a symmetric matrix. In fact, let $\alpha = \{\mathbf{v}_1,\ \ldots,\ \mathbf{v}_n\}$ be a fixed ordered basis for V. Then for any $\mathbf{x} = \sum_{i=1}^n x_i\mathbf{v}_i$ and $\mathbf{y} = \sum_{j=1}^n y_j\mathbf{v}_j$ in V,

$$\langle \mathbf{x}, \mathbf{y} \rangle = \sum_{i=1}^n \sum_{j=1}^n x_i y_j \langle \mathbf{v}_i, \mathbf{v}_j \rangle$$

holds. If we set $a_{ij} = \langle \mathbf{v}_i, \mathbf{v}_j \rangle$ for $i, j = 1,\ \ldots,\ n$, then these numbers constitute a symmetric matrix $A = [a_{ij}]$, since $\langle \mathbf{v}_i, \mathbf{v}_j \rangle = \langle \mathbf{v}_j, \mathbf{v}_i \rangle$. Thus, in matrix notation, the inner product may be written as

$$\langle \mathbf{x}, \mathbf{y} \rangle = \sum_{i=1}^n \sum_{j=1}^n x_i y_j a_{ij} = [\mathbf{x}]_\alpha^T A [\mathbf{y}]_\alpha.$$

The matrix A is called the **matrix representation** of the inner product with respect to α.

Example 5.6 **(1)** With respect to the standard basis $\{\mathbf{e}_1, \mathbf{e}_2, \ldots, \mathbf{e}_n\}$ of the Euclidean n-space $\mathbb{R}^n$, the matrix representation of the dot product is the identity matrix, since $\mathbf{e}_i \cdot \mathbf{e}_j = \delta_{ij}$. Thus for $\mathbf{x} = \sum x_i\mathbf{e}_i$, $\mathbf{y} = \sum y_j\mathbf{e}_j \in \mathbb{R}^n$ the dot product is the matrix product $\mathbf{x}^T\mathbf{y}$:

$$\mathbf{x}\cdot\mathbf{y} = [x_1 \cdots x_n]\begin{bmatrix} 1 & & 0 \\ & \ddots & \\ 0 & & 1 \end{bmatrix}\begin{bmatrix} y_1 \\ \vdots \\ y_n \end{bmatrix} = [x_1 \cdots x_n]\begin{bmatrix} y_1 \\ \vdots \\ y_n \end{bmatrix} = \mathbf{x}^T\mathbf{y}.$$

(2) Let $V = P_2(\mathbb{R})$, and define an inner product of V as

$$\langle f, g\rangle = \int_0^1 f(x)g(x)dx.$$

Then for a basis $\alpha = \{f_1(x) = 1,\ f_2(x) = x,\ f_3(x) = x^2\}$ for V, one can easily find $A = [a_{ij}]$: for instance,

$$a_{23} = \langle f_2, f_3\rangle = \int_0^1 f_2(x)f_3(x)dx = \int_0^1 x\cdot x^2 dx = \frac{1}{4}. \qquad \square$$

The expression of the dot product as a matrix product is very useful in stating and proving theorems in the Euclidean space.

On the other hand, for any symmetric matrix A and for a fixed basis α, the formula $\langle\mathbf{x}, \mathbf{y}\rangle = [\mathbf{x}]_\alpha^T A[\mathbf{y}]_\alpha$ seems to give rise to an inner product on V. In fact, the formula clearly satisfies the first three rules in the definition of the inner product, but not necessarily the fourth rule, positive definiteness. The following theorem gives a necessary condition for a symmetric matrix A to give rise to an inner product. Some necessary and sufficient conditions will be discussed in Chapter 8.

Theorem 5.5 *The matrix representation A of an inner product (with respect to any basis) on a vector space V is invertible.*

Proof: It is enough to show that the column vectors of A are linearly independent. Let $\alpha = \{\mathbf{v}_1, \ldots, \mathbf{v}_n\}$ be a basis for an inner product space V. We denote the column vectors of $A = [a_{ij}] = [\langle\mathbf{v}_i, \mathbf{v}_j\rangle]$ by $\mathbf{a}_j$ for $j = 1, \cdots, n$. Consider the linear dependence of the column vectors of A: for $c_1, \cdots, c_n \in \mathbb{R}$,

$$\mathbf{0} = c_1\mathbf{a}_1 + \cdots + c_n\mathbf{a}_n,$$

Let $\mathbf{c} = \sum_{i=1}^{n} c_i \mathbf{v}_i \in V$ so that $[\mathbf{c}]_\alpha = [c_1 \ \cdots \ c_n]^T$. Then this equation becomes a homogeneous system $\mathbf{0} = A[\mathbf{c}]_\alpha$ of n linear equations in n unknowns:

$$\begin{cases} 0 = a_{11}c_1 + \cdots + a_{1n}c_n = \sum_{j=1}^{n} a_{1j}c_j = [\mathbf{v}_1]_\alpha^T A[\mathbf{c}]_\alpha = \langle \mathbf{v}_1, \mathbf{c} \rangle, \\ \quad \vdots \\ 0 = a_{n1}c_1 + \cdots + a_{nn}c_n = \sum_{j=1}^{n} a_{nj}c_j = [\mathbf{v}_n]_\alpha^T A[\mathbf{c}]_\alpha = \langle \mathbf{v}_n, \mathbf{c} \rangle, \end{cases}$$

where we used $[\mathbf{v}_i]_\alpha = \mathbf{e}_i$. Thus, by Corollary 5.3, we get $\mathbf{c} = \sum_{i=1}^{n} c_i \mathbf{v}_i = \mathbf{0}$, and the columns of A are linearly independent. □

Recall that the conditions $a > 0$ and $ab - c^2 > 0$ in (2) of Example 5.1 are sufficient for A to give rise to an inner product on $\mathbb{R}^2$.

The standard basis of the Euclidean n-space $\mathbb{R}^n$ has a special property: The basis vectors are mutually orthogonal and are of length 1. In this sense, it is called the **rectangular coordinate system** for $\mathbb{R}^n$. In an inner product space, a vector with length 1 is called a **unit vector**. If $\mathbf{x}$ is a nonzero vector in an inner product space V, the vector $\frac{1}{\|\mathbf{x}\|}\mathbf{x}$ is a unit vector. The process of obtaining a unit vector from a nonzero vector by multiplying by the inverse of its length is called a **normalization**. Thus, if there is a set of vectors (or a basis) in an inner product space consisting of mutually orthogonal vectors, then the vectors can be converted to unit vectors by normalizing them without losing their mutual orthogonality.

Problem 5.5 Normalize each of the following vectors in the Euclidean space $\mathbb{R}^3$:
(1) $\mathbf{u} = (2,\ 1,\ -1)$, (2) $\mathbf{v} = (1/2,\ 1/3,\ -1/4)$.

Definition 5.5 A set of vectors $\mathbf{x}_1, \mathbf{x}_2, \ldots, \mathbf{x}_k$ in an inner product space V is said to be **orthonormal** if

$$\langle \mathbf{x}_i, \mathbf{x}_j \rangle = \delta_{ij} = \begin{cases} 0 & \text{if } i \neq j \quad \text{(orthogonality)}, \\ 1 & \text{if } i = j \quad \text{(normality)}. \end{cases}$$

A set $\{\mathbf{x}_1, \mathbf{x}_2, \ldots, \mathbf{x}_n\}$ of vectors is called an **orthonormal basis** for V if it is a basis and orthonormal.

It will be shown later that any inner product space has an orthonormal basis, just like the standard basis for the Euclidean n-space $\mathbb{R}^n$.

Problem 5.6 Determine whether each of the following sets of vectors in $\mathbb{R}^2$ is orthogonal, orthonormal, or neither with respect to the Euclidean inner product.

$$(1)\ \left\{\begin{bmatrix}1\\0\end{bmatrix},\begin{bmatrix}0\\3\end{bmatrix}\right\}\quad(2)\ \left\{\begin{bmatrix}1\\2\end{bmatrix},\begin{bmatrix}0\\3\end{bmatrix}\right\}$$

$$(3)\ \left\{\begin{bmatrix}1\\1\end{bmatrix},\begin{bmatrix}-1\\2\end{bmatrix}\right\}\quad(4)\ \left\{\begin{bmatrix}1/\sqrt{2}\\1/\sqrt{2}\end{bmatrix},\begin{bmatrix}-1/\sqrt{2}\\1/\sqrt{2}\end{bmatrix}\right\}$$

The next theorem shows a simple expression of a vector in terms of an orthonormal basis.

Theorem 5.6 *If* $\{\mathbf{v}_1, \mathbf{v}_2, \ldots, \mathbf{v}_n\}$ *is an orthonormal basis for an inner product space* V *and* $\mathbf{x}$ *is any vector in* V, *then*

$$\mathbf{x} = \langle \mathbf{x}, \mathbf{v}_1\rangle\mathbf{v}_1 + \langle \mathbf{x}, \mathbf{v}_2\rangle\mathbf{v}_2 + \cdots + \langle \mathbf{x}, \mathbf{v}_n\rangle\mathbf{v}_n.$$

Proof: For any vector $\mathbf{x} \in V$, we can write $\mathbf{x} = x_1\mathbf{v}_1 + x_2\mathbf{v}_2 + \cdots + x_n\mathbf{v}_n$, as a linear combination of basis vectors. However, for each $i = 1, \ldots, n$,

$$\begin{aligned}\langle \mathbf{x}, \mathbf{v}_i\rangle &= \langle x_1\mathbf{v}_1 + \cdots + x_n\mathbf{v}_n,\ \mathbf{v}_i\rangle\\ &= x_1\langle \mathbf{v}_1, \mathbf{v}_i\rangle + \cdots + x_i\langle \mathbf{v}_i, \mathbf{v}_i\rangle + \cdots + x_n\langle \mathbf{v}_n, \mathbf{v}_i\rangle\\ &= x_i,\end{aligned}$$

because $\{\mathbf{v}_1, \mathbf{v}_2, \ldots, \mathbf{v}_n\}$ is orthonormal. □

In an inner product space, the coordinate expression of a vector depends on the choice of an ordered basis, and the inner product is just a matrix product of the coordinate vectors with respect to an ordered basis involving some symmetric matrix between them, as we have seen already.

Actually, we will show in Theorem 5.12 in the following section that every inner product space V has an orthonormal basis, say $\alpha = \{\mathbf{v}_1, \mathbf{v}_2, \ldots, \mathbf{v}_n\}$. Then the matrix representation $A = [a_{ij}]$ of the inner product with respect to the orthonormal basis α is the identity matrix, since $a_{ij} = \langle \mathbf{v}_i, \mathbf{v}_j\rangle = \delta_{ij}$. Thus for any vector $\mathbf{x} = \sum x_i\mathbf{v}_i$ and $\mathbf{y} = \sum y_i\mathbf{v}_i$ in V,

$$\langle \mathbf{x}, \mathbf{y}\rangle = \sum_{i=1}^{n}\sum_{j=1}^{n}\delta_{ij}x_iy_j = \sum_{i=1}^{n}x_iy_i.$$

This expression looks like the dot product in the Euclidean space $\mathbb{R}^n$. Thus *any inner product on* V *can be written just like the dot product in* $\mathbb{R}^n$, *if* V *is equipped with an orthonormal basis.*

5.4 Orthogonal projections

Let U be a subspace of a vector space V. Then by Corollary 3.13 there is another subspace W of V such that $V = U \oplus W$, so that any $\mathbf{x} \in V$ has a unique expression as $\mathbf{x} = \mathbf{u} + \mathbf{w}$ for $\mathbf{u} \in U$ and $\mathbf{w} \in W$. As an easy exercise, one can show that a function $T : V \to V$ defined by $T(\mathbf{x}) = T(\mathbf{u}+\mathbf{w}) = \mathbf{u}$ is a linear transformation, whose image $\mathrm{Im}(T) = T(V)$ is the subspace U and kernel $\mathrm{Ker}(T)$ is the subspace W.

Definition 5.6 Let U and W be subspaces of a vector space V. A linear transformation $T : V \to V$ is called the **projection** of V onto the subspace U along W if $V = U \oplus W$ and $T(\mathbf{x}) = \mathbf{u}$ for $\mathbf{x} = \mathbf{u} + \mathbf{w} \in U \oplus W$.

Note that for a given subspace U of V, there exist many projections T depending on the choice of a complementary subspace W of U. However, if we fix a complementary subspace W of U, then a projection T onto U is uniquely determined and by definition $T(\mathbf{u}) = \mathbf{u}$ for any $\mathbf{u} \in U$ and for any choice of W. In other words, $T \circ T = T$ for any projection T of V.

Example 5.7 Let U, V and W be the 1-dimensional subspaces of the Euclidean 2-space $\mathbb{R}^2$ spanned by the vectors $\mathbf{u} = \mathbf{e}_1$, $\mathbf{w} = \mathbf{e}_2$, and $\mathbf{v} = (1, 1)$, respectively.

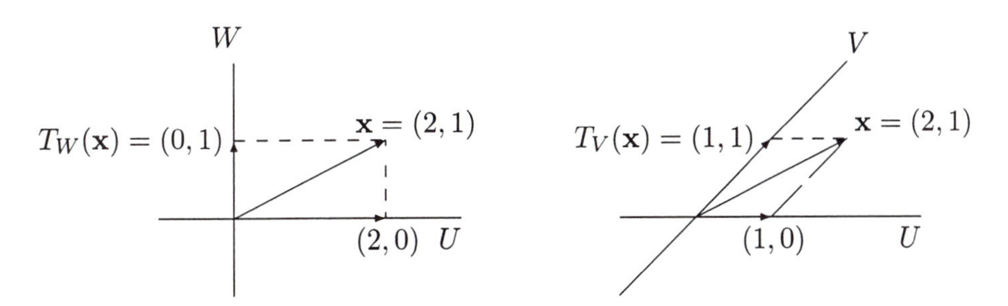

Since the pairs $\{\mathbf{u}, \mathbf{w}\}$ and $\{\mathbf{u}, \mathbf{v}\}$ are linearly independent, the space $\mathbb{R}^2$ can be expressed as the direct sum in two ways: $\mathbb{R}^2 = U \oplus W = U \oplus V$. Thus a vector $\mathbf{x} = (2, 1) \in \mathbb{R}^2$ may be written in two ways:

$$\mathbf{x} = (2,1) = \begin{cases} 2(1,0) + (0,1) & \in \quad U \oplus W \; = \; \mathbb{R}^2, \quad \text{or} \\ (1,0) + (1,1) & \in \quad U \oplus V \; = \; \mathbb{R}^2. \end{cases}$$

Let T_W and T_V denote the projections of $\mathbb{R}^2$ onto W and V along U, respectively. Then

$$T_W(\mathbf{x}) = (0,1) \in V, \quad \text{and} \quad T_V(\mathbf{x}) = (1,1) \in W.$$

It also shows that a projection of $\mathbb{R}^2$ onto the subspace U (= the x-axis) depends on a choice of complementary subspace of U. □

The following shows an algebraic characterization of a projection.

Theorem 5.7 *A linear transformation $T : V \to V$ is a projection onto a subspace U if and only if $T = T^2$* ($= T \circ T$, by definition).

Proof: The necessity is clear, because $T \circ T = T$ for any projection T.

For e sufficiency, let $T^2 = T$. We want to show $V = \text{Im}(T) \oplus \text{Ker}(T)$ and $T(\mathbf{u} + \mathbf{w}) = \mathbf{u}$ for $\mathbf{u} + \mathbf{w} \in \text{Im}(T) \oplus \text{Ker}(T)$. For the first one, we need to prove $\text{Im}(T) \cap \text{Ker}(T) = \{\mathbf{0}\}$ and $V = \text{Im}(T) + \text{Ker}(T)$. Indeed, if $\mathbf{u} \in \text{Im}(T) \cap \text{Ker}(T)$, then there is $\mathbf{x} \in V$ such that $T(\mathbf{x}) = \mathbf{u}$ and $T(\mathbf{u}) = \mathbf{0}$. But

$$\mathbf{u} = T(\mathbf{x}) = T^2(\mathbf{x}) = T(T(\mathbf{x})) = T(\mathbf{u}) = \mathbf{0}$$

proves $\text{Im}(T) \cap \text{Ker}(T) = \{\mathbf{0}\}$. Note that this also shows $T(\mathbf{u}) = \mathbf{u}$ for $\mathbf{u} \in \text{Im}(T)$. Then, $\dim V = \dim(\text{Im}(T)) + \dim(\text{Ker}(T))$ (see Remark (2) in page 138) implies $V = \text{Im}(T) + \text{Ker}(T)$. Now, note that $T(\mathbf{u}+\mathbf{w}) = T(\mathbf{u}) = \mathbf{u}$ for any $\mathbf{u} + \mathbf{w} \in \text{Im}(T) \oplus \text{Ker}(T)$. □

Let $T : V \to V$ be a projection of V, so that $V = \text{Im}(T) \oplus \text{Ker}(T)$. It is not difficult to show that $\text{Im}(Id_V - T) = \text{Ker}(T)$ and $\text{Ker}(Id_V - T) = \text{Im}(T)$.

Corollary 5.8 *A linear transformation $T : V \to V$ is a projection if and only if $Id_V - T$ is a projection. Moreover, if T is the projection of V onto a subspace U along W, then $Id_V - T$ is the projection of V onto W along U.*

Proof: It is enough to show that $(Id_V - T) \circ (Id_V - T) = Id_V - T$. But

$$(Id_V - T) \circ (Id_V - T) = (Id_V - T) - (T - T^2) = Id_V - T.$$

□

Problem 5.7 Let $V = U \oplus W$. Let T_U denote the projection of V onto U along W, and T_W denote the projection of V onto W along U. Prove the following.

(1) For any $\mathbf{x} \in V$, $\mathbf{x} = T_U(\mathbf{x}) + T_W(\mathbf{x})$.
(2) $T_U \circ (Id_V - T_U) = \mathbf{0}$.
(3) $T_U \circ T_W = T_W \circ T_U = \mathbf{0}$.
(4) For any projection $T : V \to V$, $\mathrm{Im}(Id_V - T) = \mathrm{Ker}(T)$ and $\mathrm{Ker}(Id_V - T) = \mathrm{Im}(T)$.

Now, let V be an inner product space and let U be a subspace of V. Recall that there exist many kinds of projections of V onto U depending on the choice of complementary subspace W of U. However, in an inner product space V, there is a particular choice of W, called the *orthogonal complement* of U, along which the projection onto U is called the *orthogonal projection.* Almost all projections used in linear algebra are orthogonal projections.

In an inner product space V, the orthogonality of two vectors can be extended to subspaces of V.

Definition 5.7 Let U and W be subspaces of an inner product space V.

(1) Two subspaces U and W are said to be **orthogonal**, written by $U \perp W$, if $\langle \mathbf{u}, \mathbf{w} \rangle = 0$ for each $\mathbf{u} \in U$ and $\mathbf{w} \in W$.

(2) The set of all vectors in V that are orthogonal to every vector in U is called the **orthogonal complement** of U, denoted by $U^\perp$, *i.e.*,

$$U^\perp = \{\mathbf{v} \in V : \langle \mathbf{v}, \mathbf{u} \rangle = 0 \text{ for all } \mathbf{u} \in U\}.$$

One can easily show that $U^\perp$ is a subspace of V, and $\mathbf{v} \in U^\perp$ if and only if $\langle \mathbf{v}, \mathbf{u} \rangle = 0$ for every $\mathbf{u} \in \beta$, where β is a basis for U. Therefore, clearly $W \perp U$ if and only if $W \subseteq U^\perp$.

Problem 5.8 Show: (1) If $U \perp W$, $U \cap W = \{\mathbf{0}\}$. (2) $U \subseteq W$ if and only if $W^\perp \subseteq U^\perp$.

Theorem 5.9 *Let U be a subspace of an inner product space V. Then*

(1) $\dim U + \dim U^\perp = \dim V$.

(2) $(U^\perp)^\perp = U$.

(3) $V = U \oplus U^\perp$: *that is, for each $\mathbf{x} \in V$, there are unique vectors $\mathbf{x}_U \in U$ and $\mathbf{x}_{U^\perp} \in U^\perp$ such that $\mathbf{x} = \mathbf{x}_U + \mathbf{x}_{U^\perp}$. This is called the* **orthogonal decomposition** *of V (or of $\mathbf{x}$) by U.*

Proof: **(1)** Suppose that $\dim U = k$. Choose a basis $\{\mathbf{v}_1, \ldots, \mathbf{v}_k\}$ for U, and then extend it to a basis $\{\mathbf{v}_1, \ldots, \mathbf{v}_k, \mathbf{v}_{k+1}, \ldots, \mathbf{v}_n\}$ for V, where $n = \dim V$. Then $\mathbf{x} = \sum_{j=1}^n x_j \mathbf{v}_j \in U^\perp$ if and only if $0 = \langle \mathbf{x}, \mathbf{v}_i \rangle = \sum_{j=1}^n a_{ij} x_j$ for $1 \le i \le k$, where $a_{ij} = \langle \mathbf{v}_i, \mathbf{v}_j \rangle$. The latter equations form a homogeneous system of k linear equations in n unknowns, that is, $U^\perp$ is precisely the null space of the $k \times n$ coefficient matrix $B = [a_{ij}]$, which is a submatrix of the matrix representation A of the inner product. Thus, by Theorem 5.5 the rows of B are linearly independent, so B is of rank k. Therefore, the null space has dimension $n - k$, or $\dim U^\perp = n - k = n - \dim U$.

(2) By definition, every vector in U is orthogonal to $U^\perp$, *i.e.*, $U \subseteq (U^\perp)^\perp$. On the other hand, by (1), $\dim (U^\perp)^\perp = n - \dim U^\perp = \dim U$. This proves that $(U^\perp)^\perp = U$.

(3) For a basis $\{\mathbf{v}_1, \ldots, \mathbf{v}_k\}$ for U, take any basis $\{\mathbf{v}_{k+1}, \ldots, \mathbf{v}_n\}$ for $U^\perp$. Since $U \cap U^\perp = \{\mathbf{0}\}$, the set $\{\mathbf{v}_1, \ldots, \mathbf{v}_k, \mathbf{v}_{k+1}, \ldots, \mathbf{v}_n\}$ is linearly independent, so it is a basis for V. Therefore, every vector $\mathbf{x} \in V$ has a unique expression

$$\mathbf{x} = \sum_{i=1}^{k} a_i \mathbf{v}_i + \sum_{j=k+1}^{n} b_j \mathbf{v}_j.$$

Now take $\mathbf{x}_U = \sum_1^k a_i \mathbf{v}_i \in U$ and $\mathbf{x}_{U^\perp} = \sum_{k+1}^n b_j \mathbf{v}_j \in U^\perp$. To show uniqueness, let $\mathbf{x} = \mathbf{u} + \mathbf{w}$ be another expression with $\mathbf{u} \in U$ and $\mathbf{w} \in U^\perp$. Then $\mathbf{x}_U - \mathbf{u} = \mathbf{w} - \mathbf{x}_{U^\perp} \in U \cap U^\perp = \{\mathbf{0}\}$. So, $\mathbf{x}_U = \mathbf{u}$ and $\mathbf{x}_{U^\perp} = \mathbf{w}$. □

Definition 5.8 Let V be an inner product space, and let U be a subspace of V so that $V = U \oplus U^\perp$. Then the projection of V onto U along $U^\perp$ is called the **orthogonal projection** of V onto U, denoted Proj_U. For $\mathbf{x} \in V$, the component vector $\text{Proj}_U(\mathbf{x}) \in U$ is called the **orthogonal projection** of $\mathbf{x}$ into U.

Example 5.8 As in Example 5.7, let U, V and W be subspaces of the Euclidean 2-space $\mathbb{R}^2$ generated by the vectors $\mathbf{u} = \mathbf{e}_1, \mathbf{v} = (1, 1)$, and $\mathbf{w} = \mathbf{e}_2$, respectively. Then clearly $W = U^\perp$ and $V \neq U^\perp$. Hence, for the projections T_V and T_W of $\mathbb{R}^2$ given in Example 5.7, the projection T_W is the orthogonal projection, but the projection T_V is not, so that $T_W = \text{Proj}_W$ and $T_V \neq \text{Proj}_V$. □

Theorem 5.10 *Let U be a subspace of an inner product space V, and let $\mathbf{x} \in V$. Then, the orthogonal projection* $\mathrm{Proj}_U(\mathbf{x})$ *of* $\mathbf{x}$ *satisfies*

$$\|\mathbf{x} - \mathrm{Proj}_U(\mathbf{x})\| \leq \|\mathbf{x} - \mathbf{y}\|$$

for all $\mathbf{y} \in U$. The equality holds if and only if $\mathbf{y} = \mathrm{Proj}_U(\mathbf{x})$.

Proof: Since $\mathbf{x} = \mathrm{Proj}_U(\mathbf{x}) + \mathrm{Proj}_{U^\perp}(\mathbf{x})$ for any vector $\mathbf{x} \in V$, $\mathbf{x} - \mathrm{Proj}_U(\mathbf{x}) = \mathrm{Proj}_{U^\perp}(\mathbf{x}) \in U^\perp$. Thus, for all $\mathbf{y} \in U$,

$$\begin{aligned} \|\mathbf{x} - \mathbf{y}\|^2 &= \|(\mathbf{x} - \mathrm{Proj}_U(\mathbf{x})) + (\mathrm{Proj}_U(\mathbf{x}) - \mathbf{y})\|^2 \\ &= \|\mathbf{x} - \mathrm{Proj}_U(\mathbf{x})\|^2 + \|\mathrm{Proj}_U(\mathbf{x}) - \mathbf{y}\|^2 \\ &\geq \|\mathbf{x} - \mathrm{Proj}_U(\mathbf{x})\|^2, \end{aligned}$$

where the second equality comes from the Pythagorean theorem for $\mathbf{x} - \mathrm{Proj}_U(\mathbf{x}) \perp \mathrm{Proj}_U(\mathbf{x}) - \mathbf{y}$. □

The theorem means that *the orthogonal projection* $\mathrm{Proj}_U(\mathbf{x})$ *of* $\mathbf{x}$ *is the unique vector in U that is closest to* $\mathbf{x}$ in the sense that it minimizes the distance to $\mathbf{x}$ from the vectors in U. Geometrically, the following picture depicts the vector $\mathrm{Proj}_U(\mathbf{x})$:

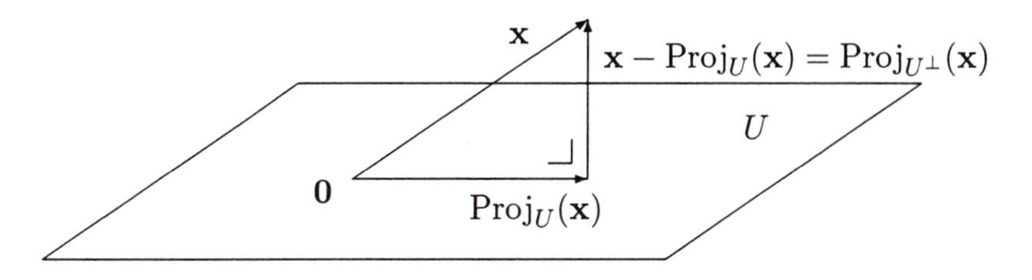

Problem 5.9 Let U and W be subspaces of an inner product space V. Show that
(1) $(U + W)^\perp = U^\perp \cap W^\perp$. (2) $(U \cap W)^\perp = U^\perp + W^\perp$.

Problem 5.10 Let $U \subset \mathbb{R}^4$ with the Euclidean inner product be the subspace spanned by $(1,\ 1,\ 0,\ 0)$ and $(1,\ 0,\ 1,\ 0)$, and $W \subset \mathbb{R}^4$ the subspace spanned by $(0,\ 1,\ 0,\ 1)$ and $(0,\ 0,\ 1,\ 1)$. Find a basis for and the dimension of each of the following subspaces:
(1) $U + W$, (2) $U^\perp$, (3) $U^\perp + W^\perp$, (4) $U \cap W$.

Lemma 5.11 *Let U be a subspace of an inner product space V, and let $\{\mathbf{u}_1, \mathbf{u}_2, \ldots, \mathbf{u}_m\}$ be an orthonormal basis for U. Then, for any $\mathbf{x} \in V$, the orthogonal projection $\mathrm{Proj}_U(\mathbf{x})$ of $\mathbf{x}$ into U is*

$$\mathrm{Proj}_U(\mathbf{x}) = \langle \mathbf{x}, \mathbf{u}_1 \rangle \mathbf{u}_1 + \langle \mathbf{x}, \mathbf{u}_2 \rangle \mathbf{u}_2 + \cdots + \langle \mathbf{x}, \mathbf{u}_m \rangle \mathbf{u}_m.$$

Proof: Let $\mathbf{z} = \langle \mathbf{x}, \mathbf{u}_1 \rangle \mathbf{u}_1 + \langle \mathbf{x}, \mathbf{u}_2 \rangle \mathbf{u}_2 + \cdots + \langle \mathbf{x}, \mathbf{u}_m \rangle \mathbf{u}_m$. It is enough to show that $\mathbf{y} = \mathbf{x} - \mathbf{z}$ is orthogonal to U, because if $\mathbf{y} = \mathbf{x} - \mathbf{z} \in U^\perp$, then $\mathbf{x} = \mathbf{z} + \mathbf{y} \in U \oplus U^\perp$, so the uniqueness of this orthogonal decomposition gives $\mathbf{z} = \mathrm{Proj}_U(\mathbf{x})$. However, for each $j = 1, \ldots, m$,

$$\langle \mathbf{x} - \mathbf{z}, \mathbf{u}_j \rangle = \langle \mathbf{x}, \mathbf{u}_j \rangle - \langle \mathbf{z}, \mathbf{u}_j \rangle = \langle \mathbf{x}, \mathbf{u}_j \rangle - \langle \mathbf{x}, \mathbf{u}_j \rangle = 0,$$

since $\{\mathbf{u}_1, \mathbf{u}_2, \ldots, \mathbf{u}_m\}$ is an orthonormal basis for U. That is, the vector $\mathbf{x} - \mathbf{z} = \mathbf{x} - \sum_1^m \langle \mathbf{x}, \mathbf{u}_i \rangle \mathbf{u}_i$ is orthogonal to U. □

In particular, if $U = V$ in Lemma 5.11, then $\mathrm{Proj}_U(\mathbf{x}) = \mathbf{x}$, and we get Theorem 5.6.

A unit vector $\mathbf{u}$ in an inner product space V determines a 1-dimensional subspace $U = \{r\mathbf{u} : r \in \mathbb{R}\}$. Then, for a vector $\mathbf{x}$ in V, the orthogonal projection of $\mathbf{x}$ into U is simply

$$\mathrm{Proj}_U(\mathbf{x}) = \langle \mathbf{x}, \mathbf{u} \rangle \mathbf{u},$$

where $\langle \mathbf{x}, \mathbf{u} \rangle = \|\mathbf{x}\| \cos\theta$. On the other hand, it is quite clear that $\mathbf{y} = \mathbf{x} - \langle \mathbf{x}, \mathbf{u} \rangle \mathbf{u}$ is a vector orthogonal to $\mathbf{u}$. Thus

$$\mathbf{x} = \langle \mathbf{x}, \mathbf{u} \rangle \mathbf{u} + \mathbf{y} \in U \oplus U^\perp$$

so that $\|\mathbf{x}\|^2 = \|\mathbf{y}\|^2 + |\langle \mathbf{x}, \mathbf{u} \rangle|^2$, which is just the Pythagorean theorem. In particular, if $V = \mathbb{R}^n$ the Euclidean space with the dot product, then

$$\mathrm{Proj}_U(\mathbf{x}) = (\mathbf{u} \cdot \mathbf{x})\mathbf{u} = (\mathbf{u}^T \mathbf{x})\mathbf{u} = \mathbf{u}(\mathbf{u}^T \mathbf{x}) = (\mathbf{u}\mathbf{u}^T)\mathbf{x}.$$

(Here the third equality comes from the matrix products). This equation shows that the matrix $\mathbf{u}\mathbf{u}^T$ is the matrix representation of the orthogonal projection Proj_U with respect to the standard basis for $\mathbb{R}^n$. Further discussions about matrix representations of the orthogonal projections will be given in Section 5.10.

Example 5.9 Let $P(x_0, y_0)$ be a point and $ax + by + c = 0$ a line in the $\mathbb{R}^2$ plane. One might know already from calculus that the nonzero vector $\mathbf{n} = (a, b)$ is perpendicular to the line $ax + by + c = 0$. In fact, for any two points $Q(x_1, y_1)$ and $R(x_2, y_2)$ on the line, the dot product $\overrightarrow{QR} \cdot \mathbf{n} = a(x_2 - x_1) + b(y_2 - y_1) = 0$, that is, $\overrightarrow{QR} \perp \mathbf{n}$.

For any point $P(x_0, y_0)$ in the plane $\mathbb{R}^2$, the distance d between the point $P(x_0, y_0)$ and the line $ax + by + c = 0$ is simply the length of the orthogonal projection of $\overrightarrow{QP}$ into $\mathbf{n}$, for any point $Q(x_1, y_1)$ in the line. Thus,

$$\begin{aligned} d &= \|\mathrm{Proj}_{\mathbf{n}}(\overrightarrow{QP})\| \\ &= \frac{|\overrightarrow{QP} \cdot \mathbf{n}|}{\|\mathbf{n}\|} \\ &= \frac{|a(x_0 - x_1) + b(y_0 - y_1)|}{\sqrt{a^2 + b^2}} \\ &= \frac{|ax_0 + by_0 + c|}{\sqrt{a^2 + b^2}}. \end{aligned}$$

Note that the last equality is due to the fact that the point Q is on the line (*i.e.*, $ax_1 + by_1 + c = 0$). □

Problem 5.11 Let $V = P_3(\mathbb{R})$, the vector space of polynomials of degree ≤ 3 equipped with the inner product

$$\langle f, g \rangle = \int_0^1 f(x)g(x)\,dx \quad \text{for any } f \text{ and } g \text{ in } V.$$

Let W be the subspace of V spanned by $\{1,\ x\}$, and define $f(x) = x^2$. Find the orthogonal projection $\mathrm{Proj}_W(f)$ of f on W.

5.5 The Gram-Schmidt orthogonalization

The construction of the orthogonal projection onto a subspace described in Section 5.4 can be used to find an orthonormal basis from any given basis, as the following example shows.

Example 5.10 Let

$$A = \begin{bmatrix} 1 & 1 & 2 \\ 1 & 2 & 2 \\ 1 & 0 & 4 \\ 1 & 1 & 0 \end{bmatrix}.$$

Find an orthonormal basis for the column space $\mathcal{C}(A)$ of A.

Solution: Let $\mathbf{c}_1$, $\mathbf{c}_2$ and $\mathbf{c}_3$ be the column vectors of A in order from left to right. It is easily verified that they are linearly independent, so they form a basis for the 3-dimensional subspace $\mathcal{C}(A)$ of the Euclidean space $\mathbb{R}^4$, *i.e.*, the column space, but this basis is not orthonormal. To make an orthonormal basis, set

$$\mathbf{v}_1 = \frac{\mathbf{c}_1}{\|\mathbf{c}_1\|} = \frac{\mathbf{c}_1}{2} = \left(\frac{1}{2}, \frac{1}{2}, \frac{1}{2}, \frac{1}{2}\right),$$

which is a unit vector. Clearly, $\mathbf{v}_1$, $\mathbf{c}_2$ and $\mathbf{c}_3$ span the column space $\mathcal{C}(A)$. Let W_1 denote the subspace spanned by $\mathbf{v}_1$. Then

$$\mathrm{Proj}_{W_1}(\mathbf{c}_2) = \langle \mathbf{c}_2, \mathbf{v}_1 \rangle \mathbf{v}_1 = 2\mathbf{v}_1,$$

and $\mathbf{c}_2 - \mathrm{Proj}_{W_1}(\mathbf{c}_2) = \mathbf{c}_2 - 2\mathbf{v}_1 = (0,\ 1,\ -1,\ 0)$ is a nonzero vector orthogonal to $\mathbf{v}_1$. To convert it to a unit vector, we set

$$\mathbf{v}_2 = \frac{\mathbf{c}_2 - 2\mathbf{v}_1}{\|\mathbf{c}_2 - 2\mathbf{v}_1\|} = \frac{1}{\sqrt{2}}(0,\ 1,\ -1,\ 0) = \left(0,\ \frac{1}{\sqrt{2}},\ -\frac{1}{\sqrt{2}},\ 0\right).$$

Since $\mathbf{c}_2 = 2\mathbf{v}_1 + \sqrt{2}\,\mathbf{v}_2$, we still have a spanning set $\{\mathbf{v}_1, \mathbf{v}_2, \mathbf{c}_3\}$ of the column space $\mathcal{C}(A)$ and thus a basis. Let W_2 denote the subspace spanned by $\mathbf{v}_1$ and $\mathbf{v}_2$. Then

$$\mathrm{Proj}_{W_2}(\mathbf{c}_3) = \langle \mathbf{c}_3, \mathbf{v}_1 \rangle \mathbf{v}_1 + \langle \mathbf{c}_3, \mathbf{v}_2 \rangle \mathbf{v}_2 = 4\mathbf{v}_1 - \sqrt{2}\mathbf{v}_2,$$

so $\mathbf{c}_3 - \mathrm{Proj}_{W_2}(\mathbf{c}_3) = \mathbf{c}_3 - 4\mathbf{v}_1 + \sqrt{2}\mathbf{v}_2 = (0,\ 1,\ 1,\ -2)$ is a nonzero vector orthogonal to both $\mathbf{v}_1$ and $\mathbf{v}_2$. In fact,

$$\begin{aligned} \langle \mathbf{c}_3 - 4\mathbf{v}_1 + \sqrt{2}\,\mathbf{v}_2,\ \mathbf{v}_1 \rangle &= \langle \mathbf{c}_3, \mathbf{v}_1 \rangle - 4\langle \mathbf{v}_1, \mathbf{v}_1 \rangle + \sqrt{2}\,\langle \mathbf{v}_2, \mathbf{v}_1 \rangle = 0\,, \\ \langle \mathbf{c}_3 - 4\mathbf{v}_1 + \sqrt{2}\,\mathbf{v}_2,\ \mathbf{v}_2 \rangle &= \langle \mathbf{c}_3, \mathbf{v}_2 \rangle - 4\langle \mathbf{v}_1, \mathbf{v}_2 \rangle + \sqrt{2}\,\langle \mathbf{v}_2, \mathbf{v}_2 \rangle = 0\,, \end{aligned}$$

since $\mathbf{v}_1$ and $\mathbf{v}_2$ are orthogonal. Thus we can normalize the vector $\mathbf{c}_3 - \mathrm{Proj}_{W_2}(\mathbf{c}_3)$ and set

$$\mathbf{v}_3 = \frac{\mathbf{c}_3 - \langle \mathbf{c}_3, \mathbf{v}_1 \rangle \mathbf{v}_1 - \langle \mathbf{c}_3, \mathbf{v}_2 \rangle \mathbf{v}_2}{\|\mathbf{c}_3 - \langle \mathbf{c}_3, \mathbf{v}_1 \rangle \mathbf{v}_1 - \langle \mathbf{c}_3, \mathbf{v}_2 \rangle \mathbf{v}_2\|} = \frac{1}{\sqrt{6}}(0,\ 1,\ 1,\ -2).$$

Then one can easily show that the set $\{\mathbf{v}_1,\ \mathbf{v}_2,\ \mathbf{v}_3\}$ still spans $\mathcal{C}(A)$ and forms an orthonormal basis for it. □

In fact, the orthonormalization process in Example 5.10 indicates how to prove the following general version, called the **Gram-Schmidt orthogonalization.**

Theorem 5.12 *Every inner product space has an orthonormal basis.*

Proof: [Gram-Schmidt orthogonalization process] Let $\{\mathbf{x}_1, \mathbf{x}_2, \ldots, \mathbf{x}_n\}$ be a basis for an n-dimensional inner product space V. Let

$$\mathbf{v}_1 = \frac{\mathbf{x}_1}{\|\mathbf{x}_1\|}, \qquad \mathbf{v}_2 = \frac{\mathbf{x}_2 - \langle \mathbf{x}_2, \mathbf{v}_1 \rangle \mathbf{v}_1}{\|\mathbf{x}_2 - \langle \mathbf{x}_2, \mathbf{v}_1 \rangle \mathbf{v}_1\|}.$$

Of course, $\mathbf{x}_2 - \langle \mathbf{x}_2, \mathbf{v}_1 \rangle \mathbf{v}_1 \neq \mathbf{0}$, because $\{\mathbf{x}_1, \mathbf{x}_2\}$ is linearly independent. Generally, we define by induction on $k = 1, 2, \ldots, n$

$$\mathbf{v}_k = \frac{\mathbf{x}_k - \langle \mathbf{x}_k, \mathbf{v}_1 \rangle \mathbf{v}_1 - \langle \mathbf{x}_k, \mathbf{v}_2 \rangle \mathbf{v}_2 - \cdots - \langle \mathbf{x}_k, \mathbf{v}_{k-1} \rangle \mathbf{v}_{k-1}}{\|\mathbf{x}_k - \langle \mathbf{x}_k, \mathbf{v}_1 \rangle \mathbf{v}_1 - \langle \mathbf{x}_k, \mathbf{v}_2 \rangle \mathbf{v}_2 - \cdots - \langle \mathbf{x}_k, \mathbf{v}_{k-1} \rangle \mathbf{v}_{k-1}\|}.$$

Thus, $\mathbf{v}_k$ is the normalized vector of $\mathbf{x}_k - \mathrm{Proj}_{W_{k-1}}(\mathbf{x}_k)$, where W_{k-1} is the subspace of V spanned by $\{\mathbf{x}_1, \mathbf{x}_2, \ldots, \mathbf{x}_{k-1}\}$ (or equivalently, by $\{\mathbf{v}_1, \mathbf{v}_2, \ldots, \mathbf{v}_{k-1}\}$). Then, the vectors $\mathbf{v}_1, \mathbf{v}_2, \ldots, \mathbf{v}_n$ are orthonormal in the n-dimensional vector space V. Since every orthonormal set is linearly independent, it is an orthonormal basis for V. □

Here is a simpler proof of Theorem 5.9. Suppose that U is a subspace of an inner product space V. Then clearly we have $U \perp U^{\perp}$, by definition. To show $(U^{\perp})^{\perp} = U$, take an orthonormal basis, say $\alpha = \{\mathbf{v}_1, \mathbf{v}_2, \ldots, \mathbf{v}_k\}$, for U by the Gram-Schmidt orthonormalization, and then extend it to an orthonormal basis for V, say $\beta = \{\mathbf{v}_1, \mathbf{v}_2, \ldots, \mathbf{v}_k, \mathbf{v}_{k+1}, \ldots, \mathbf{v}_n\}$, which is always possible. Then clearly $\gamma = \{\mathbf{v}_{k+1}, \ldots, \mathbf{v}_n\}$ forms an (orthonormal) basis for $U^{\perp}$, which means that $(U^{\perp})^{\perp} = U$ and $V = U \oplus U^{\perp}$.

Problem 5.12 Find an orthonormal basis for the subspace of the Euclidean space $\mathbb{R}^3$ given by $x + 2y - 3z = 0$, which is the orthogonal complement of the vector $(1, 2, -3)$ in $\mathbb{R}^3$.

Problem 5.13 Let $V = C[0, 1]$ with the inner product

$$\langle f, g \rangle = \int_0^1 f(x)g(x)dx \quad \text{for any } f \text{ and } g \text{ in } V.$$

Find an orthonormal basis for the subspace spanned by 1, x and x^2.

We can now identify an n-dimensional inner product space V with the Euclidean n-space $\mathbb{R}^n$ via the Gram-Schmidt orthogonalization. In fact, if $(V, \langle\ ,\ \rangle)$ is an inner product space, then by the Gram-Schmidt orthogonalization we can choose an orthonormal basis $\alpha = \{\mathbf{v}_1, \mathbf{v}_2, \ldots, \mathbf{v}_n\}$ for V. With this orthonormal basis α, the natural isomorphism $\Phi : V \to \mathbb{R}^n$ given by $\Phi(\mathbf{v}_i) = [\mathbf{v}_i]_\alpha = \mathbf{e}_i$, $i = 1, \ldots, n$ (see the last remark of Section 4.4) preserves the inner product of vectors: Every vector $\mathbf{x} \in V$ has a unique expression $\mathbf{x} = \sum_{i=1}^n x_i \mathbf{v}_i$ with $x_i = \langle \mathbf{x}, \mathbf{v}_i \rangle$. Thus the coordinate vector of $\mathbf{x}$ with respect to α is a column matrix

$$[\mathbf{x}]_\alpha = \begin{bmatrix} x_1 \\ x_2 \\ \vdots \\ x_n \end{bmatrix},$$

which is a vector in $\mathbb{R}^n$. Moreover, if $\mathbf{y} = \sum_{i=1}^n y_i \mathbf{v}_i$ is another vector in V, then

$$\langle \mathbf{x}, \mathbf{y} \rangle = \left\langle \sum_{i=1}^n x_i \mathbf{v}_i, \sum_{j=1}^n y_j \mathbf{v}_j \right\rangle = \sum_{i=1}^n x_i y_i = [\mathbf{x}]_\alpha^T [\mathbf{y}]_\alpha.$$

The right side of this equation is just the dot product of vectors in the Euclidean space $\mathbb{R}^n$. That is,

$$\langle \mathbf{x}, \mathbf{y} \rangle = [\mathbf{x}]_\alpha^T [\mathbf{y}]_\alpha = \Phi(\mathbf{x}) \cdot \Phi(\mathbf{y})$$

for any $\mathbf{x}$, $\mathbf{y} \in V$. Hence, *the natural isomorphism Φ preserves the inner product,* and *identifies the inner product on V with the dot product on $\mathbb{R}^n$.* In this sense, we may restrict our study of an inner product space to the case of the Euclidean n-space $\mathbb{R}^n$ with the dot product.

A special kind of linear transformation that preserves the inner product such as the natural isomorphism from V to $\mathbb{R}^n$ plays an important role in linear algebra, and we will study this kind of transformation in Section 5.6.

Problem 5.14 Use the Gram-Schmidt orthogonalization on the Euclidean space $\mathbb{R}^4$ to transform the basis

$$\{(0,\ 1,\ 1,\ 0),\ (-1,\ 1,\ 0,\ 0),\ (1,\ 2,\ 0,\ -1),\ (-1,\ 0,\ 0,\ -1)\}$$

into an orthonormal basis.

Problem 5.15 Find the point on the plane $x - y - z = 0$ that is closest to $\mathbf{p} = (1,\ 2,\ 0)$.

5.6 Orthogonal matrices and transformations

In Chapter 4, we saw that a linear transformation can be associated with a matrix, and vice versa. In this section, we are mainly interested in those linear transformations (or matrices) that preserve the lengths of vectors in an inner product space.

Let $A = [\mathbf{c}_1 \ \cdots \ \mathbf{c}_n]$ be an $n \times n$ square matrix, where $\mathbf{c}_1, \ \ldots, \ \mathbf{c}_n \in \mathbb{R}^n$ are the column vectors of A. Then a simple computation shows that

$$A^T A = \begin{bmatrix} -- & \mathbf{c}_1^T & -- \\ & \vdots & \\ -- & \mathbf{c}_n^T & -- \end{bmatrix} \begin{bmatrix} | & & | \\ \mathbf{c}_1 & \cdots & \mathbf{c}_n \\ | & & | \end{bmatrix} = [\mathbf{c}_i^T \mathbf{c}_j] = [\mathbf{c}_i \cdot \mathbf{c}_j].$$

Hence, if the column vectors are orthonormal, $\mathbf{c}_i^T \mathbf{c}_j = \delta_{ij}$, then $A^T A = I_n$, that is, A^T is a left inverse of A, and vice versa. Since A is a square matrix, this left inverse must be the right inverse of A, *i.e.*, $AA^T = I_n$. Equivalently, the row vectors of A are also orthonormal. This argument can now be summarized as follows.

Lemma 5.13 *Let A be an $n \times n$ matrix. The following are equivalent.*

(1) *The column vectors of A are orthonormal.*

(2) $A^T A = I_n$.

(3) $A^T = A^{-1}$.

(4) $AA^T = I_n$.

(5) *The row vectors of A are orthonormal.*

Definition 5.9 A square matrix A is called an **orthogonal** matrix if A satisfies one (and hence all) of the statements in Lemma 5.13.

Therefore, A is orthogonal if and only if A^T is orthogonal.

Example 5.11 It is easy to see that the matrices

$$A = \begin{bmatrix} \cos\theta & -\sin\theta \\ \sin\theta & \cos\theta \end{bmatrix}, \quad B = \begin{bmatrix} \cos\theta & \sin\theta \\ \sin\theta & -\cos\theta \end{bmatrix}$$

are orthogonal, and satisfy

$$A^{-1} = A^T = \begin{bmatrix} \cos\theta & \sin\theta \\ -\sin\theta & \cos\theta \end{bmatrix}, \quad B^{-1} = B^T = \begin{bmatrix} \cos\theta & \sin\theta \\ \sin\theta & -\cos\theta \end{bmatrix}.$$

Note that the linear transformation $T : \mathbb{R}^2 \to \mathbb{R}^2$ defined by $T(\mathbf{x}) = A\mathbf{x}$ is a rotation through the angle θ, while $S : \mathbb{R}^2 \to \mathbb{R}^2$ defined by $S(\mathbf{x}) = B\mathbf{x}$ is the reflection about the line passing through the origin that forms an angle $\theta/2$ with the positive x-axis. □

Example 5.12 Show that every 2×2 orthogonal matrix must be one of the following forms

$$\begin{bmatrix} \cos\theta & -\sin\theta \\ \sin\theta & \cos\theta \end{bmatrix} \quad \text{or} \quad \begin{bmatrix} \cos\theta & \sin\theta \\ \sin\theta & -\cos\theta \end{bmatrix}.$$

Solution: Suppose that $A = \begin{bmatrix} a & b \\ c & d \end{bmatrix}$ is an orthogonal matrix, so that $AA^T = I_2 = A^T A$. From the first equality, we get $a^2 + b^2 = 1$, $ac + bd = 0$, and $c^2 + d^2 = 1$. From the second equality, we get $a^2 + c^2 = 1$, $ab + cd = 0$, and $b^2 + d^2 = 1$. Thus, $b = \pm c$. If $b = -c$, then we get $a = d$. If $b = c$, then we get $a = -d$. Now, choose θ so that $a = \cos\theta$ and $b = \sin\theta$. □

Problem 5.16 Find the inverse of each of the following matrices.

$$(1) \begin{bmatrix} 1 & 0 & 0 \\ 0 & \cos\theta & \sin\theta \\ 0 & -\sin\theta & \cos\theta \end{bmatrix}, \quad (2) \begin{bmatrix} 1/\sqrt{2} & -1/\sqrt{2} & 0 \\ -1/\sqrt{2} & -1/\sqrt{2} & 0 \\ 0 & 0 & 1 \end{bmatrix}.$$

What are they as linear transformations on $\mathbb{R}^3$: rotations, reflections, or other?

Intuitively, any rotation or reflection on the Euclidean space $\mathbb{R}^n$ preserves both the lengths of vectors and the angle of two vectors. In general, any orthogonal matrix A preserves the lengths of vectors:

$$\|A\mathbf{x}\|^2 = A\mathbf{x} \cdot A\mathbf{x} = (A\mathbf{x})^T(A\mathbf{x}) = \mathbf{x}^T A^T A\mathbf{x} = \mathbf{x}^T\mathbf{x} = \|\mathbf{x}\|^2.$$

Definition 5.10 Let V and W be two inner product spaces. A linear transformation $T : V \to W$ is called an **isometry**, or an **orthogonal transformation**, if it preserves the lengths of vectors, that is, for every vector $\mathbf{x} \in V$

$$\|T(\mathbf{x})\| = \|\mathbf{x}\|.$$

Clearly, any orthogonal matrix is an isometry as a linear transformation. If $T : V \to W$ is an isometry, then T is a one-to-one, since the kernel of T is trivial: $T(\mathbf{x}) = \mathbf{0}$ implies $\|\mathbf{x}\| = \|T(\mathbf{x})\| = 0$. Thus, if $\dim V = \dim W$, then an isometry is also an isomorphism.

The following is an interesting characterization of an isometry.

Theorem 5.14 *Let $T : V \to W$ be a linear transformation on an inner product space V to W. Then T is an isometry if and only if T preserves inner products, that is,*

$$\langle T(\mathbf{x}), T(\mathbf{y})\rangle = \langle \mathbf{x}, \mathbf{y}\rangle$$

for any vectors $\mathbf{x}$, $\mathbf{y}$ in V.

Proof: Let T be an isometry. Then $\|T(\mathbf{x})\|^2 = \|\mathbf{x}\|^2$ for any $\mathbf{x} \in V$. Hence,

$$\langle T(\mathbf{x}+\mathbf{y}), T(\mathbf{x}+\mathbf{y})\rangle = \|T(\mathbf{x}+\mathbf{y})\|^2 = \|\mathbf{x}+\mathbf{y}\|^2 = \langle \mathbf{x}+\mathbf{y}, \mathbf{x}+\mathbf{y}\rangle$$

for any $\mathbf{x}, \mathbf{y} \in V$. On the other hand,

$$\begin{aligned} \langle T(\mathbf{x}+\mathbf{y}), T(\mathbf{x}+\mathbf{y})\rangle &= \langle T(\mathbf{x}), T(\mathbf{x})\rangle + 2\langle T(\mathbf{x}), T(\mathbf{y})\rangle + \langle T(\mathbf{y}), T(\mathbf{y})\rangle, \\ \langle \mathbf{x}+\mathbf{y}, \mathbf{x}+\mathbf{y}\rangle &= \langle \mathbf{x}, \mathbf{x}\rangle + 2\langle \mathbf{x}, \mathbf{y}\rangle + \langle \mathbf{y}, \mathbf{y}\rangle, \end{aligned}$$

from which we get $\langle T(\mathbf{x}), T(\mathbf{y})\rangle = \langle \mathbf{x}, \mathbf{y}\rangle$.

The converse is quite clear by choosing $\mathbf{y} = \mathbf{x}$. □

Corollary 5.15 *Let A be an $n \times n$ matrix. Then, A is an orthogonal matrix if and only if $A : \mathbb{R}^n \to \mathbb{R}^n$, as a linear transformation, preserves the dot product. That is, for any vectors $\mathbf{x}$, $\mathbf{y} \in \mathbb{R}^n$,*

$$A\mathbf{x} \cdot A\mathbf{y} = \mathbf{x} \cdot \mathbf{y}.$$

Proof: One way is clear. Suppose that A preserves the dot product. Then for any vectors $\mathbf{x}$, $\mathbf{y} \in \mathbb{R}^n$,

$$A\mathbf{x} \cdot A\mathbf{y} = \mathbf{x}^T A^T A\mathbf{y} = \mathbf{x}^T\mathbf{y} = \mathbf{x} \cdot \mathbf{y}.$$

Take $\mathbf{x} = \mathbf{e}_i$ and $\mathbf{y} = \mathbf{e}_j$. Then this equation is just $[A^T A]_{ij} = \delta_{ij}$. □

Since $d(\mathbf{x}, \mathbf{y}) = \|\mathbf{x} - \mathbf{y}\|$ for any $\mathbf{x}$ and $\mathbf{y}$ in V, one can easily derive the following corollary.

Corollary 5.16 *A linear transformation $T : V \to W$ is an isometry if and only if*

$$d(T(\mathbf{x}), T(\mathbf{y})) = d(\mathbf{x}, \mathbf{y})$$

for any $\mathbf{x}$ and $\mathbf{y}$ in V.

Recall that if θ is the angle between two nonzero vectors $\mathbf{x}$ and $\mathbf{y}$ in an inner product space V, then for any isometry $T : V \to V$,

$$\cos\theta = \frac{\langle \mathbf{x}, \mathbf{y}\rangle}{\|\mathbf{x}\|\|\mathbf{y}\|} = \frac{\langle T\mathbf{x}, T\mathbf{y}\rangle}{\|T\mathbf{x}\|\|T\mathbf{y}\|}.$$

Hence, we have

Corollary 5.17 *An isometry preserves the angle.*

The following problem shows that the converse of Corollary 5.17 is not true in general.

Problem 5.17 Find an example of a linear transformation on the Euclidean space $\mathbb{R}^n$ that preserves the angles but not the lengths of vectors (*i.e.*, not an isometry). Such a linear transformation is called a **dilation**.

We have seen that any orthogonal matrix is an isometry as the linear transformation $T(\mathbf{x}) = A\mathbf{x}$. The following theorem says that the converse is also true, that is, the matrix representation of an isometry with respect to an orthonormal basis is an orthogonal matrix.

Theorem 5.18 *Let $T : V \to W$ be an isometry of an inner product space V to W of the same dimension. Let $\alpha = \{\mathbf{v}_1, \ldots, \mathbf{v}_n\}$ and $\beta = \{\mathbf{w}_1, \ldots, \mathbf{w}_n\}$ be orthonormal bases for V and W, respectively. Then the matrix $[T]_\alpha^\beta$ for T with respect to the basis α and β is an orthogonal matrix.*

Proof: Note that the k-th column vector of the matrix $[T]_\alpha^\beta$ is just $[T(\mathbf{v}_k)]_\beta$. Since T preserves inner products and α is orthonormal, we get

$$[T(\mathbf{v}_k)]_\beta^T[T(\mathbf{v}_\ell)]_\beta = \langle T(\mathbf{v}_k), T(\mathbf{v}_\ell)\rangle = \langle \mathbf{v}_k, \mathbf{v}_\ell\rangle = \delta_{k\ell},$$

which shows that the column vectors of $[T]_\alpha^\beta$ are orthonormal. □

Therefore, a linear transformation $T : V \to V$ is an isometry if and only if $[T]_\alpha$ is an orthogonal matrix for an orthonormal basis α. Moreover, a square matrix A preserves the dot product if and only if it preserves the lengths of vectors.

Problem 5.18 Find values $r > 0$, $s > 0$, a, b and c such that matrix Q is orthogonal.

$$(1)\ Q = \begin{bmatrix} r & s & a \\ 0 & 2s & b \\ r & -s & c \end{bmatrix}, \quad (2)\ Q = \begin{bmatrix} r & -s & a \\ r & 3s & b \\ r & -2s & c \end{bmatrix}.$$

Problem 5.19 (Bessel's Inequality) Let V be an inner product space, and let $\{\mathbf{v}_1, \ldots, \mathbf{v}_m\}$ be a set of orthonormal vectors in V (not necessarily a basis for V). Prove that for any $\mathbf{x}$ in V, $\|\mathbf{x}\|^2 \geq \sum_{i=1}^m |\langle \mathbf{x}, \mathbf{v}_i \rangle|^2$.

Problem 5.20 Determine whether the following linear transformations on Euclidean space $\mathbb{R}^3$ are orthogonal.

(1) $T(x, y, z) = (z, \frac{\sqrt{3}}{2}x + \frac{1}{2}y, \frac{x}{2} - \frac{\sqrt{3}}{2}y)$.
(2) $T(x, y, z) = (\frac{5}{13}x + \frac{11}{13}z, \frac{12}{13}y - \frac{5}{13}z, x)$.

5.7 Relations of fundamental subspaces

We now go back to the study of the system $A\mathbf{x} = \mathbf{b}$ of linear equations. One of the most important applications of the orthogonal projection of vectors onto a subspace is to study the relations or structures of the four fundamental subspaces $\mathcal{N}(A)$, $\mathcal{R}(A)$, $\mathcal{C}(A)$, and $\mathcal{N}(A^T)$ of an $m \times n$ matrix A.

Lemma 5.19 *For any $m \times n$ matrix A, the null space $\mathcal{N}(A)$ and the row space $\mathcal{R}(A)$ are orthogonal in $\mathbb{R}^n$. Similarly, the null space $\mathcal{N}(A^T)$ of A^T and the column space $\mathcal{C}(A) = \mathcal{R}(A^T)$ are orthogonal in $\mathbb{R}^m$.*

Proof: Note that $\mathbf{w} \in \mathcal{N}(A)$ if and only if $A\mathbf{w} = \mathbf{0}$, *i.e.*, for every row vector $\mathbf{r}$ in A, $\mathbf{r} \cdot \mathbf{w} = 0$. For the second statement, do the same with A^T. □

This theorem shows that $\mathcal{N}(A) \perp \mathcal{R}(A)$ and $\mathcal{C}(A) \perp \mathcal{N}(A^T)$, hence $\mathcal{N}(A) \subseteq \mathcal{R}(A)^\perp$ (or $\mathcal{R}(A) \subseteq \mathcal{N}(A)^\perp$) and $\mathcal{N}(A^T) \subseteq \mathcal{C}(A)^\perp$ (or $\mathcal{C}(A) \subseteq \mathcal{N}(A^T)^\perp$), but the equalities between them do not follow immediately. The next theorem shows that we have equalities in both inclusions, that is, the row space $\mathcal{R}(A)$ and the null space $\mathcal{N}(A)$ are orthogonal complements of each other, and the column space $\mathcal{C}(A)$ and the null space $\mathcal{N}(A^T)$ of A^T are orthogonal complements of each other. Note that the above theorem also shows that $\mathcal{N}(A) \cap \mathcal{R}(A) = \{\mathbf{0}\}$ and $\mathcal{C}(A) \cap \mathcal{N}(A^T) = \{\mathbf{0}\}$.

Theorem 5.20 (The second fundamental theorem) *For any $m \times n$ matrix A,*

(1) $\mathcal{N}(A) \oplus \mathcal{R}(A) = \mathbb{R}^n$,
(2) $\mathcal{N}(A^T) \oplus \mathcal{C}(A) = \mathbb{R}^m$.

Proof: **(1)** Since both the row space $\mathcal{R}(A)$ and the null space $\mathcal{N}(A)$ of A are subspaces of $\mathbb{R}^n$, we have $\mathcal{N}(A)+\mathcal{R}(A) \subseteq \mathbb{R}^n$ in general. However,

$$\begin{aligned}\dim(\mathcal{N}(A)+\mathcal{R}(A)) &= \dim \mathcal{N}(A)+\dim \mathcal{R}(A)-\dim(\mathcal{N}(A)\cap\mathcal{R}(A)) \\ &= \dim \mathcal{N}(A)+\dim \mathcal{R}(A) \\ &= \dim \mathcal{N}(A)+\operatorname{rank} A \\ &= n \;=\; \dim \mathbb{R}^n,\end{aligned}$$

since dim(row space) + dim(null space) $= n =$ number of columns. This means that $\mathcal{N}(A)+\mathcal{R}(A)=\mathbb{R}^n$. Actually we have $\mathcal{N}(A)\oplus\mathcal{R}(A)=\mathbb{R}^n$ since $\mathcal{N}(A)\cap\mathcal{R}(A)=\{\mathbf{0}\}$. A similar argument applies to A^T to get (2). □

Corollary 5.21 **(1)** *$\mathcal{N}(A)=\mathcal{R}(A)^{\perp}$, and hence $\mathcal{R}(A)=\mathcal{N}(A)^{\perp}$.*
(2) *$\mathcal{N}(A^T)=\mathcal{C}(A)^{\perp}$, and hence $\mathcal{C}(A)=\mathcal{N}(A^T)^{\perp}$.*

For an $m\times n$ matrix A considered as a linear transformation $A:\mathbb{R}^n\to\mathbb{R}^m$, the decompositions $\mathbb{R}^n=\mathcal{R}(A)\oplus\mathcal{N}(A)$ and $\mathbb{R}^m=\mathcal{C}(A)\oplus\mathcal{N}(A^T)$ given in Theorem 5.20 depict the following figure with $r=\operatorname{rank} A$.

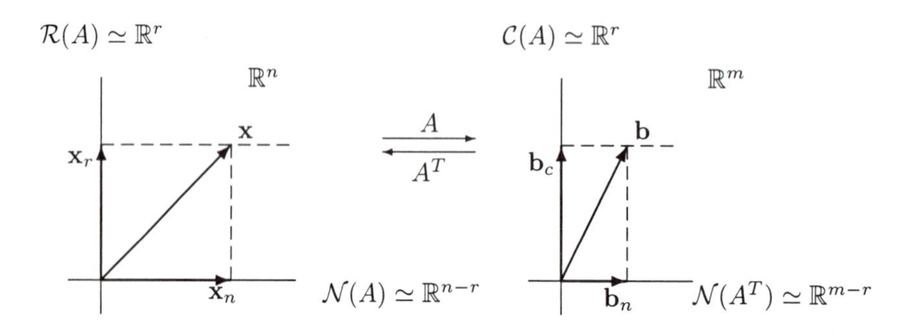

Note that if rank $A=r$, then $\dim\mathcal{R}(A)=r=\dim\mathcal{C}(A)$, $\dim\mathcal{N}(A)=n-r$ and $\dim\mathcal{N}(A^T)=m-r$. The figure shows that for any $\mathbf{b}_c$ in the column space $\mathcal{C}(A)$, which is the range of A, there is an $\mathbf{x}\in\mathbb{R}^n$ such that $A\mathbf{x}=\mathbf{b}_c$. Now there exist unique $\mathbf{x}_r\in\mathcal{R}(A)$ and $\mathbf{x}_n\in\mathcal{N}(A)$ such that $\mathbf{x}=\mathbf{x}_r+\mathbf{x}_n$. Thus $\mathbf{b}_c=A\mathbf{x}=A(\mathbf{x}_r+\mathbf{x}_n)=A\mathbf{x}_r$. Moreover, for any $\mathbf{x}'\in\mathcal{N}(A)$, $A(\mathbf{x}_r+\mathbf{x}')=A\mathbf{x}_r=\mathbf{b}_c$, since $A\mathbf{x}'=\mathbf{0}$. Therefore, the set of all solutions to $A\mathbf{x}=\mathbf{b}_c$ is precisely $\mathbf{x}_r+\mathcal{N}(A)$, which is the $n-r$ dimensional plane parallel to the null space $\mathcal{N}(A)$ and passing through $\mathbf{x}_r$.

In particular, if rank $A = m$, then $\mathcal{N}(A^T) = \{\mathbf{0}\}$ and hence $\mathcal{C}(A) = \mathbb{R}^m$. Thus for any $\mathbf{b} \in \mathbb{R}^m$, the system $A\mathbf{x} = \mathbf{b}$ has solutions of the form $\mathbf{x}_r + \mathbf{x}_n$, where $\mathbf{x}_n \in \mathcal{N}(A)$ is arbitrary and $\mathbf{x}_r \in \mathcal{R}(A)$ is unique (this is the case in the existence Theorem 3.23).

On the other hand, if rank $A = n \le m$, then $\mathcal{N}(A) = \{\mathbf{0}\}$ and hence $\mathcal{R}(A) = \mathbb{R}^n$. Therefore, the system $A\mathbf{x} = \mathbf{b}$ has at most one solution, that is, it has a unique solution $\mathbf{x}_r$ in the row space if $\mathbf{b} \in \mathcal{C}(A)$, and has no solution (that is, the system is inconsistent) if $\mathbf{b} \notin \mathcal{C}(A)$ (this is the case in the uniqueness Theorem 3.24). The latter case occurs when $m > r = \text{rank } A$: that is, $\mathcal{N}(A^T)$ is a nontrivial subspace of $\mathbb{R}^m$.

Problem 5.21 Show that
(1) if $A\mathbf{x} = \mathbf{b}$ and $A^T\mathbf{y} = \mathbf{0}$, then $\mathbf{y}^T\mathbf{b} = 0$, and
(2) if $A\mathbf{x} = \mathbf{0}$ and $A^T\mathbf{y} = \mathbf{c}$, then $\mathbf{x}^T\mathbf{c} = 0$.

Problem 5.22 Given two vectors (1, 2, 1, 2) and (0, -1, -1, 1), find all vectors in $\mathbb{R}^4$ that are perpendicular to them.

Problem 5.23 Find a basis for the orthogonal complement of the row space of A:

$$(1)\ A = \begin{bmatrix} 1 & 2 & 8 \\ 2 & -1 & 1 \\ 3 & 0 & 6 \end{bmatrix}, \qquad (2)\ A = \begin{bmatrix} 0 & 0 & 1 \\ 0 & 0 & 1 \\ 1 & 1 & 1 \end{bmatrix}.$$

5.8 Least square solutions

We consider again a system $A\mathbf{x} = \mathbf{b}$ of linear equations. Recall that the system $A\mathbf{x} = \mathbf{b}$ has at least one solution if and only if $\mathbf{b}$ belongs to the column space $\mathcal{C}(A)$ of A. In this case, such a solution is unique if and only if the null space $\mathcal{N}(A)$ of A is trivial.

Now the problem is *"what happens if* $\mathbf{b} \notin \mathcal{C}(A) \subseteq \mathbb{R}^m$ *so that* $A\mathbf{x} = \mathbf{b}$ *is inconsistent?"* Note that for any $\mathbf{x} \in \mathbb{R}^n$, $A\mathbf{x} \in \mathcal{C}(A)$. Thus the best we can do is to find a vector $\mathbf{x}_0 \in \mathbb{R}^n$ such that $A\mathbf{x}_0$ is *closest* to the given vector $\mathbf{b}$ in $\mathbb{R}^m$, *i.e.*, $\|A\mathbf{x}_0 - \mathbf{b}\|$ is as small as possible. Such a solution vector $\mathbf{x}_0$ gives the best approximation $A\mathbf{x}_0$ to $\mathbf{b}$, and is called a **least square solution** of $A\mathbf{x} = \mathbf{b}$. However, since we have the orthogonal decomposition $\mathbb{R}^m = \mathcal{C}(A) \oplus \mathcal{N}(A^T)$, we know that for any $\mathbf{b} \in \mathbb{R}^m$, $\text{Proj}_{\mathcal{C}(A)}(\mathbf{b}) = \mathbf{b}_c \in \mathcal{C}(A)$

is the closest vector to $\mathbf{b}$ among the vectors in $\mathcal{C}(A)$. Therefore, a least square solution $\mathbf{x}_0 \in \mathbb{R}^n$ satisfies the following:

$$\begin{aligned} A\mathbf{x}_0 &= \mathbf{b}_c &= \mathrm{Proj}_{\mathcal{C}(A)}(\mathbf{b}), \\ \|A\mathbf{x}_0 - \mathbf{b}\| &\le \|A\mathbf{x} - \mathbf{b}\| \end{aligned}$$

for any vector $\mathbf{x}$ in $\mathbb{R}^n$. Since $\mathbf{b}_c \in \mathcal{C}(A)$, there always exists a least square solution $\mathbf{x}_0 \in \mathbb{R}^n$ such that $A\mathbf{x}_0 = \mathbf{b}_c$. It is quite easy to show that all other least square solutions are the vectors in $\mathbf{x}_0 + \mathcal{N}(A)$.

In summary, a least square solution of $A\mathbf{x} = \mathbf{b}$, when $\mathbf{b} \notin \mathcal{C}(A)$, is simply a solution of $A\mathbf{x} = \mathbf{b}_c$, where $\mathbf{b}_c = \mathrm{Proj}_{\mathcal{C}(A)}(\mathbf{b}) \in \mathcal{C}(A)$ in the unique orthogonal decomposition of

$$\mathbf{b} = \mathbf{b}_c + \mathbf{b}_n \in \mathcal{C}(A) \oplus \mathcal{N}(A^T) = \mathbb{R}^m,$$

with $\mathbf{b}_n = \mathbf{b} - \mathbf{b}_c \in \mathcal{N}(A^T)$. That is, to find such a least square solution, we first have to find $\mathbf{b}_c$ and then solve $A\mathbf{x}_0 = \mathbf{b}_c$.

Practically, the computation of $\mathbf{b}_c$ from $\mathbf{b}$ could be quite complicated, since we first have to find an orthonormal basis for $\mathcal{C}(A)$ by using the Gram-Schmidt orthogonalization (whose computation is cumbersome) and then express $\mathbf{b}_c$ with respect to this orthonormal basis for a given $\mathbf{b}$.

To find an easier method, let us examine a least square solution in a little more detail. If $\mathbf{x}_0 \in \mathbb{R}^n$ is a least square solution of $A\mathbf{x} = \mathbf{b}$, *i.e.*, a solution of $A\mathbf{x}_0 = \mathbf{b}_c$, then $A\mathbf{x}_0 - \mathbf{b} = A\mathbf{x}_0 - (\mathbf{b}_c + \mathbf{b}_n) = -\mathbf{b}_n \in \mathcal{N}(A^T)$ holds. Thus, by applying A^T to the equation, we get $A^TA\mathbf{x}_0 = A^T\mathbf{b}$, *i.e.*, $\mathbf{x}_0$ is a solution of the equation

$$A^TA\mathbf{x} = A^T\mathbf{b}.$$

This equation is very interesting because it also gives a sufficient condition of a least square solution as the following theorem shows, and so is defined to be the **normal equation** of $A\mathbf{x} = \mathbf{b}$.

Theorem 5.22 *Let A be an $m \times n$ matrix, and let $\mathbf{b} \in \mathbb{R}^m$ be any vector. Then a vector $\mathbf{x}_0 \in \mathbb{R}^n$ is a least square solution of $A\mathbf{x} = \mathbf{b}$ if and only if $\mathbf{x}_0$ is a solution of the normal equation $A^TA\mathbf{x} = A^T\mathbf{b}$.*

Proof: We only need to show the sufficiency of the normal equation: If $\mathbf{x}_0$ is a solution of the equation $A^TA\mathbf{x} = A^T\mathbf{b}$, then, $A^T(A\mathbf{x}_0 - \mathbf{b}) = \mathbf{0}$, so $A\mathbf{x}_0 - \mathbf{b} = A\mathbf{x}_0 - (\mathbf{b}_c + \mathbf{b}_n) \in \mathcal{N}(A^T)$. This means that, as a vector in $\mathcal{C}(A)$, $A\mathbf{x}_0 - \mathbf{b}_c \in \mathcal{N}(A^T) \cap \mathcal{C}(A) = \{\mathbf{0}\}$. Therefore $A\mathbf{x}_0 = \mathbf{b}_c = \mathrm{Proj}_{\mathcal{C}(A)}(\mathbf{b})$, *i.e.*,

$\mathbf{x}_0$ is a least square solution of $A\mathbf{x} = \mathbf{b}$. □

Note that if the rows of A are linearly independent, then $\operatorname{rank} A = m$ and $\mathcal{C}(A) = \mathbb{R}^m$ (or $\mathcal{N}(A^T) = \mathbf{0}$). Thus, a least square solution of $A\mathbf{x} = \mathbf{b}$ is simply a usual solution.

Example 5.13 Find all the least square solutions to $A\mathbf{x} = \mathbf{b}$, and then determine the orthogonal projection $\mathbf{b}_c$ of $\mathbf{b}$ into the column space $\mathcal{C}(A)$ of A, where

$$A = \begin{bmatrix} 1 & -2 & 1 \\ 2 & -3 & -1 \\ -1 & 1 & 2 \\ 3 & -5 & 0 \end{bmatrix}, \quad \mathbf{b} = \begin{bmatrix} 1 \\ 0 \\ 1 \\ 0 \end{bmatrix}.$$

Solution:

$$A^T A = \begin{bmatrix} 1 & 2 & -1 & 3 \\ -2 & -3 & 1 & -5 \\ 1 & -1 & 2 & 0 \end{bmatrix} \begin{bmatrix} 1 & -2 & 1 \\ 2 & -3 & -1 \\ -1 & 1 & 2 \\ 3 & -5 & 0 \end{bmatrix} = \begin{bmatrix} 15 & -24 & -3 \\ -24 & 39 & 3 \\ -3 & 3 & 6 \end{bmatrix},$$

and

$$A^T\mathbf{b} = \begin{bmatrix} 1 & 2 & -1 & 3 \\ -2 & -3 & 1 & -5 \\ 1 & -1 & 2 & 0 \end{bmatrix} \begin{bmatrix} 1 \\ 0 \\ 1 \\ 0 \end{bmatrix} = \begin{bmatrix} 0 \\ -1 \\ 3 \end{bmatrix}.$$

From the normal equation, a least square solution of $A\mathbf{x} = \mathbf{b}$ is a solution of $A^T A\mathbf{x} = A^T\mathbf{b}$, *i.e.*,

$$\begin{bmatrix} 15 & -24 & -3 \\ -24 & 39 & 3 \\ -3 & 3 & 6 \end{bmatrix} \begin{bmatrix} x_1 \\ x_2 \\ x_3 \end{bmatrix} = \begin{bmatrix} 0 \\ -1 \\ 3 \end{bmatrix}.$$

By solving this system of equations (left for an exercise), we obtain all the least square solutions desired:

$$\mathbf{x} = \begin{bmatrix} x_1 \\ x_2 \\ x_3 \end{bmatrix} = \frac{1}{3}\begin{bmatrix} -8 \\ -5 \\ 0 \end{bmatrix} + t\begin{bmatrix} 5 \\ 3 \\ 1 \end{bmatrix}$$

for any number t. Now

$$\mathbf{b}_c = A\mathbf{x} = \frac{1}{3}\begin{bmatrix} 2 \\ -1 \\ 3 \\ 1 \end{bmatrix} \in \mathcal{C}(A).$$

Note that the vector $\mathbf{x} = \frac{1}{3}\begin{bmatrix} -8 \\ -5 \\ 0 \end{bmatrix}$ is not in $\mathcal{R}(A)$. One needs to do a little more computation to find a least square solution $\mathbf{x} \in \mathcal{R}(A)$. □

Problem 5.24 Find all least square solutions $\mathbf{x}$ in $\mathbb{R}^3$ of $A\mathbf{x} = \mathbf{b}$, where

$$A = \begin{bmatrix} 1 & 0 & 2 \\ 0 & 2 & 2 \\ -1 & 1 & -1 \\ -1 & 2 & 0 \end{bmatrix}, \quad \mathbf{b} = \begin{bmatrix} 3 \\ -3 \\ 0 \\ -3 \end{bmatrix}.$$

Practically, finding the solutions of the normal equation depends very much on A^TA. In the most fortunate case, if the square matrix A^TA is the identity matrix, then the normal equation $A^TA\mathbf{x} = A^T\mathbf{b}$ of the system $A\mathbf{x} = \mathbf{b}$ reduces to $\mathbf{x} = A^T\mathbf{b}$, which is simply a least square solution. Even if A^TA is not the identity matrix, we may still have several simple cases.

Remark: Let us now discuss the solvability of this normal equation. Observe that $A^T : \mathbb{R}^m \to \mathbb{R}^n$ and the row space $\mathcal{R}(A)$ of A and the column space $\mathcal{C}(A^T)$ of A^T are the same. Thus, for any $\mathbf{x}_r \in \mathcal{C}(A^T) = \mathcal{R}(A)$ there exists a vector $\mathbf{b} \in \mathbb{R}^m$ such that $A^T\mathbf{b} = \mathbf{x}_r$. If we write $\mathbf{b} = \mathbf{b}_c + \mathbf{b}_n$ for unique $\mathbf{b}_c \in \mathcal{C}(A)$ and $\mathbf{b}_n \in \mathcal{N}(A^T)$, then $\mathbf{x}_r = A^T\mathbf{b} = A^T\mathbf{b}_c$. Therefore, the restrictions

$$\begin{aligned} \bar{A} &= A|_{\mathcal{R}(A)} : \mathcal{R}(A) \subseteq \mathbb{R}^n \to \mathcal{C}(A) \subseteq \mathbb{R}^m \quad \text{and} \\ \bar{A^T} &= A^T|_{\mathcal{C}(A)} : \mathcal{C}(A) \subseteq \mathbb{R}^m \to \mathcal{R}(A) \subseteq \mathbb{R}^n \end{aligned}$$

are one-to-one and onto transformations, that is, they are invertible. However, even in this case we do not have $\bar{A}\bar{A^T} = I_r$ nor $\bar{A^T}\bar{A} = I_r$ in general.

The transpose A^T of a matrix A satisfies the following equation: For $\mathbf{x} \in \mathbb{R}^n$ and $\mathbf{y} \in \mathbb{R}^m$, $A\mathbf{x} \in \mathbb{R}^m$, so

$$A\mathbf{x} \cdot \mathbf{y} = (A\mathbf{x})^T\mathbf{y} = \mathbf{x}^TA^T\mathbf{y} = \mathbf{x} \cdot A^T\mathbf{y}.$$

The following theorem gives a condition for A^TA to be invertible.

Theorem 5.23 *For any $m \times n$ matrix A, $A^T A$ is a symmetric $n \times n$ square matrix and* $\text{rank}(A^T A) = \text{rank } A$.

Proof: Clearly, $A^T A$ is square and symmetric: $(A^T A)^T = A^T (A^T)^T = A^T A$. Since the number of columns of A and $A^T A$ are both n, we have

$$\text{rank } A + \dim \mathcal{N}(A) = n = \text{rank } (A^T A) + \dim \mathcal{N}(A^T A).$$

Hence, it suffices to show that A and $A^T A$ have exactly the same null space so that $\dim \mathcal{N}(A) = \dim \mathcal{N}(A^T A)$. If $\mathbf{x} \in \mathcal{N}(A)$, then $A\mathbf{x} = \mathbf{0}$ and also $A^T A\mathbf{x} = A^T \mathbf{0} = \mathbf{0}$, so that $\mathbf{x} \in \mathcal{N}(A^T A)$. Conversely, suppose that $A^T A\mathbf{x} = \mathbf{0}$. Then

$$A\mathbf{x} \cdot A\mathbf{x} = (A\mathbf{x})^T (A\mathbf{x}) = \mathbf{x}^T (A^T A\mathbf{x}) = \mathbf{x} \cdot A^T A\mathbf{x} = \mathbf{x} \cdot \mathbf{0} = 0.$$

Hence $A\mathbf{x} = \mathbf{0}$, *i.e.*, $\mathbf{x} \in \mathcal{N}(A)$. □

In the following discussion, we assume that the columns of A are linearly independent, *i.e.*, rank $A = n$, so that $\mathcal{N}(A) = \{\mathbf{0}\}$, or A is one-to-one. Hence the system $A\mathbf{x} = \mathbf{b}_c$ has a unique solution $\mathbf{x}$ in $\mathcal{R}(A) = \mathbb{R}^n$. Moreover, by Theorem 5.23, the square matrix $A^T A$ is also of rank n and it is invertible. In this case, from the normal equation, a least square solution is

$$\mathbf{x} = (A^T A)^{-1} A^T \mathbf{b}.$$

Corollary 5.24 *If the columns of A are linearly independent, then*

(1) *$A^T A$ is invertible so that $(A^T A)^{-1} A^T$ is a left inverse of A,*

(2) *the vector $\mathbf{x} = (A^T A)^{-1} A^T \mathbf{b}$ is the unique least square solution of a system $A\mathbf{x} = \mathbf{b}$, and*

(3) *$A\mathbf{x} = A(A^T A)^{-1} A^T \mathbf{b}$ is the projection $\mathbf{b}_c$ of $\mathbf{b}$ into the column space $\mathcal{C}(A)$.*

By applying Corollary 5.24 to A^T, we can say that, if rank $A = m$ for an $m \times n$ matrix, then AA^T is invertible and $A^T (AA^T)^{-1}$ is a right inverse of A (cf. Remark after Theorem 3.24). Moreover, by using Theorem 5.23, we can show that for a matrix A, $A^T A$ is invertible if and only if the columns of A are linearly independent, and AA^T is invertible if and only if the rows of A are linearly independent.

Example 5.14 Consider the following system of linear equations:

$$A\mathbf{x} = \begin{bmatrix} 1 & 2 \\ 1 & 5 \\ 0 & 0 \end{bmatrix} \begin{bmatrix} x \\ y \end{bmatrix} = \begin{bmatrix} 4 \\ 3 \\ 9 \end{bmatrix} = \mathbf{b}.$$

Clearly, the two columns of A are linearly independent and $\mathcal{C}(A)$ is the xy-plane. Thus $\mathbf{b} \notin \mathcal{C}(A)$. Note that

$$A^TA = \begin{bmatrix} 1 & 1 & 0 \\ 2 & 5 & 0 \end{bmatrix} \begin{bmatrix} 1 & 2 \\ 1 & 5 \\ 0 & 0 \end{bmatrix} = \begin{bmatrix} 2 & 7 \\ 7 & 29 \end{bmatrix},$$

which is invertible. By a simple computation one can obtain

$$(A^TA)^{-1} = \frac{1}{9} \begin{bmatrix} 29 & -7 \\ -7 & 2 \end{bmatrix}.$$

Hence,

$$\mathbf{x} = (A^TA)^{-1}A^T\mathbf{b} = \frac{1}{9} \begin{bmatrix} 29 & -7 \\ -7 & 2 \end{bmatrix} \begin{bmatrix} 7 \\ 23 \end{bmatrix} = \frac{1}{9} \begin{bmatrix} 42 \\ -3 \end{bmatrix} = \begin{bmatrix} 14/3 \\ -1/3 \end{bmatrix}$$

is a least square solution, and the orthogonal projection of $\mathbf{b}$ in $\mathcal{C}(A)$ is

$$\mathbf{b}_c = A\mathbf{x} = \begin{bmatrix} 1 & 2 \\ 1 & 5 \\ 0 & 0 \end{bmatrix} \begin{bmatrix} 14/3 \\ -1/3 \end{bmatrix} = \begin{bmatrix} 4 \\ 3 \\ 0 \end{bmatrix}.$$

□

Problem 5.25 Let W be the subspace of the Euclidean space $\mathbb{R}^3$ spanned by the vectors $\mathbf{v}_1 = (1,\ 1,\ 2)$ and $\mathbf{v}_2 = (1,\ 1,\ -1)$. Find $\mathrm{Proj}_W(\mathbf{b})$ for $\mathbf{b} = (1,\ 3,\ -2)$.

5.9 Application: Polynomial approximations

In this section, one can find a reason for the name of the "*least square*" solutions, and the following example illustrates an application of the least square solution to the determination of the spring constants in physics.

Example 5.15 *Hooke's law* for springs in physics says that for a uniform spring, *the length stretched or compressed is a linear function of the force applied*, that is, the force F applied to the spring is related to the length x stretched or compressed by the equation

$$F = a + kx,$$

where a and k are some constants determined by the spring.

Suppose now that, given a spring of length 6.1 inches, we want to determine the constants a and k under the experimental data: The lengths are found to be 7.6, 8.7 and 10.4 inches when forces of 2, 4 and 6 kilograms, respectively, are applied to the spring. However, by plotting these data

$$(x, F) = (6.1,\ 0),\ (7.6,\ 2),\ (8.7,\ 4),\ (10.4,\ 6),$$

in the xF-plane, one can easily recognize that they are not on a straight line of the form $F = a + kx$ in the xF-plane, which may be caused by experimental errors. This means that the system of linear equations:

$$\begin{cases} F_1 &= a + 6.1k &= 0, \\ F_2 &= a + 7.6k &= 2, \\ F_3 &= a + 8.7k &= 4, \\ F_4 &= a + 10.4k &= 6 \end{cases}$$

is inconsistent (*i.e.*, has no solutions so the second equality in each equation may not be a true equality). Thus, the best thing one can do is to determine the straight line that "fits" the data, that is, the line that minimizes the sum of the squares of the vertical distances from the line to the data: *i.e.*, one needs to minimize

$$(0 - F_1)^2 + (2 - F_2)^2 + (4 - F_3)^2 + (6 - F_4)^2.$$

This quantity is simply the square of the distance between the vector $\mathbf{b} = (0, 2, 4, 6)$ in $\mathbb{R}^4$ and the vectors (F_1, F_2, F_3, F_4) in the column space $\mathcal{C}(A)$ of the 4×2 matrix

$$A = \begin{bmatrix} 1 & 6.1 \\ 1 & 7.6 \\ 1 & 8.7 \\ 1 & 10.4 \end{bmatrix},$$

since the matrix form of the system of linear equations is

$$\begin{bmatrix} F_1 \\ F_2 \\ F_3 \\ F_4 \end{bmatrix} = \begin{bmatrix} 1 & 6.1 \\ 1 & 7.6 \\ 1 & 8.7 \\ 1 & 10.4 \end{bmatrix} \begin{bmatrix} a \\ k \end{bmatrix} \in \mathcal{C}(A).$$

The minimum of the sum of squares is obtained when (F_1, F_2, F_3, F_4) is the projection of the vector $\mathbf{b} = (0, 2, 4, 6)$ into the column space $\mathcal{C}(A)$, that is, what we are looking for is the least square solution of the system, which is now easily computed as

$$\begin{bmatrix} a \\ k \end{bmatrix} = \mathbf{x} = (A^T A)^{-1} A^T \mathbf{b} = \begin{bmatrix} -8.6 \\ 1.4 \end{bmatrix}.$$

It gives $F = -8.6 + 1.4x$. □

In general, a common problem in experimental work is to obtain a polynomial $y = f(x)$ in two variables x and y that "fits" the data of various values of y determined experimentally for inputs x, say

$$(x_1,\ y_1),\ (x_2,\ y_2),\ \ldots,\ (x_n,\ y_n),$$

plotted in the xy-plane. Some possibilities are (1) by a straight line: $y = a+bx$, (2) by a quadratic polynomial: $y = a+bx+cx^2$, or (3) by a polynomial of degree k: $y = a_0 + a_1 x + \cdots + a_k x^k$, etc.

As a general case, suppose that we are looking for a polynomial $y = f(x) = a_0 + a_1 x + a_2 x^2 + \cdots + a_k x^k$ of degree k that passes through the given data, then we obtain a system of linear equations,

$$\begin{cases} f(x_1) = a_0 + a_1 x_1 + a_2 x_1^2 + \cdots + a_k x_1^k & = & y_1 \\ f(x_2) = a_0 + a_1 x_2 + a_2 x_2^2 + \cdots + a_k x_2^k & = & y_2 \\ \qquad\vdots & & \\ f(x_n) = a_0 + a_1 x_n + a_2 x_n^2 + \cdots + a_k x_n^k & = & y_n, \end{cases}$$

or, in matrix form, the system may be written as $A\mathbf{x} = \mathbf{b}$:

$$\begin{bmatrix} 1 & x_1 & x_1^2 & \cdots & x_1^k \\ 1 & x_2 & x_2^2 & \cdots & x_2^k \\ \vdots & & & \ddots & \vdots \\ 1 & x_n & x_n^2 & \cdots & x_n^k \end{bmatrix} \begin{bmatrix} a_0 \\ a_1 \\ \vdots \\ a_k \end{bmatrix} = \begin{bmatrix} y_1 \\ y_2 \\ \vdots \\ y_n \end{bmatrix}.$$

The left side $A\mathbf{x}$ represents the values of the polynomial at x_i's and the right side represents the data obtained from the inputs x_i's in the experiment.

If $n \leq k+1$, then the cases have already been discussed in Section 3.8. In general, this kind of system may be inconsistent (*i.e.*, it may have no solution) if $n > k+1$. This means that there may be no polynomial of degree $k < n-1$ whose graph passes through the n data (x_i, y_i) in the xy-plane. Practically, it is due to the fact that the experimental data usually have some errors.

Thus, the best thing we can do is to find the polynomial $f(x)$ that minimizes the sum of the squares of the vertical distances between the graph of the polynomial and the data. In matrix and vector space language, an inconsistency of the system means that the vector $\mathbf{b} \in \mathbb{R}^n$ representing the data is not in the column space $\mathcal{C}(A)$ of the coefficient matrix A. And minimizing the sum of the squares of the vertical distances between the graph of the polynomial and the data means looking for the least square solution of the system, because for any $\mathbf{c} \in \mathcal{C}(A)$ of the form

$$\begin{bmatrix} 1 & x_1 & x_1^2 & \cdots & x_1^k \\ 1 & x_2 & x_2^2 & \cdots & x_2^k \\ \vdots & & & \ddots & \vdots \\ 1 & x_n & x_n^2 & \cdots & x_n^k \end{bmatrix} \begin{bmatrix} a_0 \\ a_1 \\ \vdots \\ a_k \end{bmatrix} = \begin{bmatrix} a_0 + a_1 x_1 + \cdots + a_k x_1^k \\ a_0 + a_1 x_2 + \cdots + a_k x_2^k \\ \vdots \\ a_0 + a_1 x_n + \cdots + a_k x_n^k \end{bmatrix} = \mathbf{c},$$

we have

$$\begin{aligned} \|\mathbf{b} - \mathbf{c}\|^2 = \ & (y_1 - a_0 - a_1 x_1 - \cdots - a_k x_1^k)^2 + \cdots \\ & + (y_n - a_0 - a_1 x_n - \cdots - a_k x_n^k)^2. \end{aligned}$$

The previous theory says that the orthogonal projection $\mathbf{b}_c$ of $\mathbf{b}$ into the column space of A minimizes this quantity and shows how to find $\mathbf{b}_c$ and a least square solution $\mathbf{x}_0$.

Example 5.16 Find a straight line $y = a + bx$ that fits the given experimental data, (1, 0), (2, 3), (3, 4) and (4, 4), that is, a line $y = a + bx$ that minimizes the sum of squares of the vertical distances $|y_i - a - bx_i|$'s from the line $y = a + bx$ to the data $(x_i,\ y_i)$. By adapting matrix notation

$$A = \begin{bmatrix} 1 & x_1 \\ 1 & x_2 \\ 1 & x_3 \\ 1 & x_4 \end{bmatrix} = \begin{bmatrix} 1 & 1 \\ 1 & 2 \\ 1 & 3 \\ 1 & 4 \end{bmatrix}, \quad \mathbf{x} = \begin{bmatrix} a \\ b \end{bmatrix} \quad \text{and} \quad \mathbf{b} = \begin{bmatrix} 0 \\ 3 \\ 4 \\ 4 \end{bmatrix},$$

we have $A\mathbf{x} = \mathbf{b}$ and want to find a least square solution of $A\mathbf{x} = \mathbf{b}$. But the columns of A are linearly independent, and the least square solution is $\mathbf{x} = (A^TA)^{-1}A^T\mathbf{b}$. Now,

$$A^TA = \begin{bmatrix} 4 & 10 \\ 10 & 30 \end{bmatrix}, \ (A^TA)^{-1} = \begin{bmatrix} \frac{3}{2} & -\frac{1}{2} \\ -\frac{1}{2} & \frac{1}{5} \end{bmatrix}, \ A^T\mathbf{b} = \begin{bmatrix} 11 \\ 34 \end{bmatrix}.$$

Hence, we have

$$\mathbf{x} = (A^TA)^{-1}A^T\mathbf{b} = \begin{bmatrix} -\frac{1}{2} \\ \frac{13}{10} \end{bmatrix},$$

and $y = -\frac{1}{2} + \frac{13}{10}x$ is the desired line. □

Problem 5.26 From Newton's second law of motion, a body near the surface of the earth falls vertically downward according to the equation

$$s(t) = s_0 + \mathbf{v}_0 t + \frac{1}{2}\mathbf{g}t^2,$$

where $s(t)$ is the distance that the body traveled in time t, and s_0, $\mathbf{v}_0$ are the initial displacement and velocity, respectively, of the body, and $\mathbf{g}$ is the gravitational acceleration at the earth's surface. Suppose a weight is released, and the distances that the body has fallen from some reference point were measured to be $s = -0.18,\ 0.31,\ 1.03,\ 2.48,\ 3.73$ feet at times $t = 0.1,\ 0.2,\ 0.3,\ 0.4,\ 0.5$ seconds, respectively. Determine approximate values of s_0, $\mathbf{v}_0$, $\mathbf{g}$ using these data.

5.10 Orthogonal projection matrices

In Section 5.8, we have seen that the orthogonal projection $\mathrm{Proj}_{\mathcal{C}(A)}$ of $\mathbb{R}^m$ on the column space $\mathcal{C}(A)$ of an $m \times n$ matrix A plays an important role in finding a least square solution of $A\mathbf{x} = \mathbf{b}$. Note that any subspace W of $\mathbb{R}^m$ is the column space of such a matrix A, whose columns are the vectors in a basis for W. Therefore, in this section, we only consider the orthogonal projection Proj_W of an inner product space V onto a subspace W, and aim to find its associated matrix, called an **orthogonal projection matrix**, for the projection Proj_W. This will give us a practical way of computing a given orthogonal projection.

First of all, if a subspace W of the Euclidean space $\mathbb{R}^m$ has an orthonormal basis $\beta = \{\mathbf{u}_1, \mathbf{u}_2, \ldots, \mathbf{u}_n\}$, then for any $\mathbf{x} \in \mathbb{R}^m$,

$$\begin{aligned}
\mathrm{Proj}_W(\mathbf{x}) &= (\mathbf{u}_1 \cdot \mathbf{x})\mathbf{u}_1 + (\mathbf{u}_2 \cdot \mathbf{x})\mathbf{u}_2 + \cdots + (\mathbf{u}_n \cdot \mathbf{x})\mathbf{u}_n \\
&= \mathbf{u}_1(\mathbf{u}_1^T\mathbf{x}) + \mathbf{u}_2(\mathbf{u}_2^T\mathbf{x}) + \cdots + \mathbf{u}_n(\mathbf{u}_n^T\mathbf{x}) \\
&= (\mathbf{u}_1\mathbf{u}_1^T + \mathbf{u}_2\mathbf{u}_2^T + \cdots + \mathbf{u}_n\mathbf{u}_n^T)\mathbf{x},
\end{aligned}$$

by Lemma 5.11. Note that in this equation, Proj_W is a linear transformation, but the right side is the usual matrix product of vectors. It implies that if an orthonormal basis for a subspace W is given, the matrix representation (projection matrix) of the orthogonal projection Proj_W with respect to the standard basis α for $\mathbb{R}^m$ is given as

$$[\mathrm{Proj}_W]_\alpha = \mathbf{u}_1\mathbf{u}_1^T + \mathbf{u}_2\mathbf{u}_2^T + \cdots + \mathbf{u}_n\mathbf{u}_n^T.$$

Note that if we denote by $\mathrm{Proj}_{\mathbf{u}_i}$ the orthogonal projection of $\mathbb{R}^m$ on the subspace spanned by the basis vector $\mathbf{u}_i$ for each i, then matrix representation is $\mathbf{u}_i\mathbf{u}_i^T$ (see page 176). Moreover, by using the matrix representations, it can be shown that

$$\mathrm{Proj}_W = \mathrm{Proj}_{\mathbf{u}_1} + \mathrm{Proj}_{\mathbf{u}_2} + \cdots + \mathrm{Proj}_{\mathbf{u}_n}$$

and

$$\mathrm{Proj}_{\mathbf{u}_j} \circ \mathrm{Proj}_{\mathbf{u}_i} = \begin{cases} \mathbf{0} & \text{if } i \neq j, \\ \mathrm{Proj}_{\mathbf{u}_j} & \text{if } i = j. \end{cases}$$

Problem 5.27 Let $\mathbf{u} = (\frac{1}{\sqrt{2}}, \frac{1}{\sqrt{2}})$ be a vector in $\mathbb{R}^2$ which determines 1-dimensional subspace $U = \{a\mathbf{u} = (\frac{a}{\sqrt{2}}, \frac{a}{\sqrt{2}}) : a \in \mathbb{R}\}$. Show that the matrix

$$A = \mathbf{u}\mathbf{u}^T = \frac{1}{2}\begin{bmatrix} 1 \\ 1 \end{bmatrix}[1\,1] = \frac{1}{2}\begin{bmatrix} 1 & 1 \\ 1 & 1 \end{bmatrix},$$

considered as a linear transformation on $\mathbb{R}^2$, is an orthogonal projection onto the subspace U.

Problem 5.28 Show that if $\{\mathbf{v}_1, \mathbf{v}_2, \ldots, \mathbf{v}_m\}$ is an orthonormal basis for $\mathbb{R}^m$, then $\mathbf{v}_1\mathbf{v}_1^T + \mathbf{v}_2\mathbf{v}_2^T + \cdots + \mathbf{v}_m\mathbf{v}_m^T = I_m$.

Definition 5.11 Let W be a subspace of the Euclidean m-space $\mathbb{R}^m$. An $m \times m$ matrix P is called the **(orthogonal) projection matrix** on a subspace W if $\mathrm{Proj}_W(\mathbf{x}) = P\mathbf{x}$ for any vector $\mathbf{x}$ in $\mathbb{R}^m$. Equivalently, P is the matrix representation of the orthogonal projection Proj_W of $\mathbb{R}^m$ onto W with respect to the standard basis for $\mathbb{R}^m$.

It has already been shown that $\mathbf{u}_1\mathbf{u}_1^T + \mathbf{u}_2\mathbf{u}_2^T + \cdots + \mathbf{u}_k\mathbf{u}_k^T$ is a projection matrix for any orthonormal set $\{\mathbf{u}_1, \mathbf{u}_2, \ldots, \mathbf{u}_k\}$ in $\mathbb{R}^m$. Such an expression of the projection matrix on a subspace W can be obtained only when an orthonormal basis for W is known.

Now, let W be an n-dimensional subspace of the Euclidean space $\mathbb{R}^m$, and let $\{\mathbf{v}_1, \mathbf{v}_2, \ldots, \mathbf{v}_n\}$ be a (not necessarily orthonormal) basis for W. If we find an orthonormal basis for W by the Gram-Schmidt orthogonalization, then we can get the projection matrix of the previous form. But the Gram-Schmidt orthogonalization process could be cumbersome and tedious. Sometimes, one can avoid this cumbersome process. Let $A = [\mathbf{v}_1\ \mathbf{v}_2\ \cdots\ \mathbf{v}_n]$ be the $m \times n$ matrix having the basis vectors $\mathbf{v}_i$'s as columns. Clearly, we have $W = \mathcal{C}(A)$. For any vector $\mathbf{b} \in \mathbb{R}^m$, the projection vector $\mathrm{Proj}_W(\mathbf{b})$ is simply the vector $A\mathbf{x}_0$ for a least square solution $\mathbf{x}_0$ of $A\mathbf{x} = \mathbf{b}$ that is a solution of the normal equation $A^TA\mathbf{x} = A^T\mathbf{b}$.

On the other hand, since the columns of A are linearly independent, A^TA is invertible, so $\mathbf{x}_0 = (A^TA)^{-1}A^T\mathbf{b}$, and, furthermore,

$$\mathrm{Proj}_W(\mathbf{b}) = A\mathbf{x}_0 = A(A^TA)^{-1}A^T\mathbf{b},$$

by Corollary 5.24. This means that $A(A^TA)^{-1}A^T$ is the projection matrix on the subspace $W = \mathcal{C}(A)$. Note that this projection matrix is independent of the choice of basis for W due to the uniqueness of the matrix representation of a linear transformation with respect to a fixed basis. Some possible simple computations for the matrix $A(A^TA)^{-1}A^T$ will follow later. This argument proves the following theorem.

Theorem 5.25 *For any subspace W of $\mathbb{R}^m$, the projection matrix P on W can be written as*

$$P = [\mathrm{Proj}_W]_\alpha = A(A^TA)^{-1}A^T$$

for a matrix A whose columns form a basis for W.

Example 5.17 Find the projection matrix P on the plane $2x - y - 3z = 0$ in $\mathbb{R}^3$ and calculate $P\mathbf{b}$ for $\mathbf{b} = (1,\ 0,\ 1)$.

Solution: Choose any basis for the plane $2x - y - 3z = 0$, say,

$$\mathbf{v}_1 = (0,\ 3,\ -1) \quad \text{and} \quad \mathbf{v}_2 = (1,\ 2,\ 0).$$

Let $A = \begin{bmatrix} 0 & 1 \\ 3 & 2 \\ -1 & 0 \end{bmatrix}$ be the matrix with $\mathbf{v}_1$ and $\mathbf{v}_2$ as columns. Then

$$(A^T A)^{-1} = \begin{bmatrix} 10 & 6 \\ 6 & 5 \end{bmatrix}^{-1} = \frac{1}{14}\begin{bmatrix} 5 & -6 \\ -6 & 10 \end{bmatrix}.$$

The projection matrix is

$$\begin{aligned} P &= A(A^T A)^{-1} A^T \\ &= \frac{1}{14}\begin{bmatrix} 0 & 1 \\ 3 & 2 \\ -1 & 0 \end{bmatrix}\begin{bmatrix} 5 & -6 \\ -6 & 10 \end{bmatrix}\begin{bmatrix} 0 & 3 & -1 \\ 1 & 2 & 0 \end{bmatrix} \\ &= \frac{1}{14}\begin{bmatrix} 10 & 2 & 6 \\ 2 & 13 & -3 \\ 6 & -3 & 5 \end{bmatrix}, \end{aligned}$$

and

$$P\mathbf{b} = \frac{1}{14}\begin{bmatrix} 10 & 2 & 6 \\ 2 & 13 & -3 \\ 6 & -3 & 5 \end{bmatrix}\begin{bmatrix} 1 \\ 0 \\ 1 \end{bmatrix} = \frac{1}{14}\begin{bmatrix} 16 \\ -1 \\ 11 \end{bmatrix}.$$

□

Remark: In particular, if the columns of A consist of an orthonormal basis $\alpha = \{\mathbf{u}_1, \ldots, \mathbf{u}_n\}$ for W, then

$$A^T A = \begin{bmatrix} -- & \mathbf{u}_1^T & -- \\ & \vdots & \\ -- & \mathbf{u}_n^T & -- \end{bmatrix}\begin{bmatrix} | & & | \\ \mathbf{u}_1 & \cdots & \mathbf{u}_n \\ | & & | \end{bmatrix} = \begin{bmatrix} 1 & & \\ & \ddots & \\ & & 1 \end{bmatrix} = I_n.$$

since $\mathbf{u}_i^T \mathbf{u}_j = \delta_{ij}$. Hence, the normal equation $A^T A\mathbf{x} = A^T\mathbf{b}$ becomes

$$\mathbf{x} = A^T\mathbf{b} = \begin{bmatrix} -- & \mathbf{u}_1^T & -- \\ & \vdots & \\ -- & \mathbf{u}_n^T & -- \end{bmatrix}\begin{bmatrix} b_1 \\ \vdots \\ b_n \end{bmatrix} = \begin{bmatrix} \langle \mathbf{u}_1, \mathbf{b}\rangle \\ \vdots \\ \langle \mathbf{u}_n, \mathbf{b}\rangle \end{bmatrix},$$

which is just the expression of $\mathrm{Proj}_W(\mathbf{b})$ with respect to the orthonormal basis α for W that are the columns of A.

Corollary 5.26 *Suppose that the column vectors $\{\mathbf{u}_1, \ldots, \mathbf{u}_n\}$ of A form an orthonormal basis for W in $\mathbb{R}^m$. Then we get*

$$P = A(A^TA)^{-1}A^T = AA^T = \mathbf{u}_1\mathbf{u}_1^T + \mathbf{u}_2\mathbf{u}_2^T + \cdots + \mathbf{u}_n\mathbf{u}_n^T.$$

In particular, if A is an $m \times m$ orthogonal matrix, then, for all $\mathbf{b} \in \mathbb{R}^m$, $A\mathbf{x} = \mathbf{b}$ has the unique solution $\mathbf{x} = A^{-1}\mathbf{b} = A^T\mathbf{b}$.

Proof: For any $\mathbf{x} \in \mathbb{R}^m$,

$$\begin{aligned}
P\mathbf{x} = AA^T\mathbf{x} &= \begin{bmatrix} \mathbf{u}_1 & \cdots & \mathbf{u}_n \end{bmatrix} \begin{bmatrix} -- & \mathbf{u}_1^T & -- \\ & \vdots & \\ -- & \mathbf{u}_n^T & -- \end{bmatrix} \mathbf{x} \\
&= \begin{bmatrix} \mathbf{u}_1 & \cdots & \mathbf{u}_n \end{bmatrix} \begin{bmatrix} \mathbf{u}_1^T\mathbf{x} \\ \vdots \\ \mathbf{u}_n^T\mathbf{x} \end{bmatrix} \\
&= (\mathbf{u}_1\mathbf{u}_1^T + \mathbf{u}_2\mathbf{u}_2^T + \cdots + \mathbf{u}_n\mathbf{u}_n^T)\mathbf{x},
\end{aligned}$$

where each $\mathbf{u}_i^T\mathbf{x}$ is a scalar as the inner product of $\mathbf{u}_i$ and $\mathbf{x}$. □

Example 5.18 If $A = [\mathbf{c}_1 \ \mathbf{c}_2]$, where $\mathbf{c}_1 = (1, 0, 0)$, $\mathbf{c}_2 = (0, 1, 0)$, then the column vectors of A are orthonormal, $\mathcal{C}(A)$ is the xy-plane, and the projection of $\mathbf{b} = (x, y, z) \in \mathbb{R}^3$ onto $\mathcal{C}(A)$ is $\mathbf{b}_c = (x, y, 0)$. In fact,

$$P = AA^T = \begin{bmatrix} 1 & 0 \\ 0 & 1 \\ 0 & 0 \end{bmatrix} \begin{bmatrix} 1 & 0 & 0 \\ 0 & 1 & 0 \end{bmatrix} = \begin{bmatrix} 1 & & \\ & 1 & \\ & & 0 \end{bmatrix}.$$ □

Before discussing the computation of $P = [\mathrm{Proj}_W]_\alpha$ with a general basis for W, we exhibit a criterion for a square matrix to be a projection matrix.

Theorem 5.27 *A square matrix P is a projection matrix if and only if it is symmetric and idempotent, i.e., $P^T = P$ and $P^2 = P$.*

Proof: Let P be a projection matrix. Then, by Theorem 5.25, the matrix P can be written as $P = A(A^TA)^{-1}A^T$ for some matrix A whose columns are linearly independent. A simple expansion of $P = A(A^TA)^{-1}A^T$ gives

$$\begin{aligned}
P^T &= \left(A(A^TA)^{-1}A^T\right)^T = A(A^TA)^{-1^T}A^T = A(A^TA)^{-1}A^T = P, \\
P^2 &= PP = \left(A(A^TA)^{-1}A^T\right)\left(A(A^TA)^{-1}A^T\right) = A(A^TA)^{-1}A^T = P.
\end{aligned}$$

We have already shown the second equation in Theorem 5.7.

For the converse, we have the orthogonal decomposition $\mathbb{R}^m = \mathcal{C}(P) \oplus \mathcal{N}(P^T)$ by Theorem 5.20. But $\mathcal{N}(P^T) = \mathcal{N}(P)$ since $P^T = P$. Note that $P^2 = P$ implies $P\mathbf{u} = \mathbf{u}$ for $\mathbf{u} \in \mathcal{C}(P)$ (see Theorem 5.7). $\square$

From Corollary 5.8, if P is a projection matrix on $\mathcal{C}(P)$, then $I - P$ is also a projection matrix on the null space $\mathcal{N}(P)$ $(= \mathcal{C}(I - P))$, which is orthogonal to $\mathcal{C}(P)$ $(= \mathcal{N}(I - P))$.

Example 5.19 Let $P_i : \mathbb{R}^m \to \mathbb{R}^m$ be defined by

$$P_i(x_1, \ldots, x_m) = (0, \ldots, 0, x_i, 0, \ldots, 0),$$

for $i = 1, \ldots, m$. Then each P_i is the projection of $\mathbb{R}^m$ onto the i-th axis, whose matrix form looks like

$$P_i = \begin{bmatrix} \ddots & & & & 0 \\ & 0 & & & \\ & & 1 & & \\ & & & 0 & \\ 0 & & & & \ddots \end{bmatrix}, \qquad I - P_i = \begin{bmatrix} \ddots & & & & 0 \\ & 1 & & & \\ & & 0 & & \\ & & & 1 & \\ 0 & & & & \ddots \end{bmatrix}.$$

When we restrict the image to $\mathbb{R}$, P_i is an element in the dual space $\mathbb{R}^{n*}$, and usually denoted by x_i as the i-th coordinate function (see Example 4.25).

Problem 5.29 Show that any square matrix P that satisfies $P^TP = P$ is a projection matrix.

In general, if $A = [\mathbf{c}_1 \ \cdots \ \mathbf{c}_n]$ is an $m \times n$ matrix with linearly independent column vectors $\mathbf{c}_1, \ldots, \mathbf{c}_n$, then rank $A = n \leq m$ and $\{\mathbf{c}_1, \ldots, \mathbf{c}_n\}$ form a basis for the column space $W = \mathcal{C}(A)$ of dim n in $\mathbb{R}^m$. By using the Gram-Schmidt orthogonalization, one can obtain an orthonormal basis $\{\mathbf{u}_1, \ldots, \mathbf{u}_n\}$ for $\mathcal{C}(A)$ from this basis, so that the matrix $Q = [\mathbf{u}_1 \ \cdots \ \mathbf{u}_n]$ and A have the same column space W. Then, by the Remark on page 199, $\mathrm{Proj}_W = QQ^T$. The computation of the Gram-Schmidt orthogonalization might be messy, but these cases occur frequently in applied science and engineering problems, so we show the process in detail in the following.

From the Gram-Schmidt orthogonalization,

$$\begin{aligned}
\mathbf{q}_1 &= \mathbf{c}_1 \\
\mathbf{q}_2 &= \mathbf{c}_2 - \frac{\langle \mathbf{c}_2, \mathbf{q}_1 \rangle}{\langle \mathbf{q}_1, \mathbf{q}_1 \rangle} \mathbf{q}_1 \\
&\vdots \\
\mathbf{q}_n &= \mathbf{c}_n - \frac{\langle \mathbf{c}_n, \mathbf{q}_{n-1} \rangle}{\langle \mathbf{q}_{n-1}, \mathbf{q}_{n-1} \rangle} \mathbf{q}_{n-1} - \cdots - \frac{\langle \mathbf{c}_n, \mathbf{q}_1 \rangle}{\langle \mathbf{q}_1, \mathbf{q}_1 \rangle} \mathbf{q}_1,
\end{aligned}$$

gives an orthogonal basis $\{\mathbf{q}_1, \ldots, \mathbf{q}_n\}$ for $\mathcal{C}(A)$. By taking normalization of these vectors, we obtain an orthonormal basis $\{\mathbf{u}_1, \ldots, \mathbf{u}_n\}$ for $\mathcal{C}(A)$, where $\mathbf{u}_i = \mathbf{q}_i/\|\mathbf{q}_i\|$. Rewriting these equations gives us

$$\begin{aligned}
\mathbf{c}_1 &= \mathbf{q}_1 &&= \|\mathbf{q}_1\|\mathbf{u}_1 \\
\mathbf{c}_2 &= a_{21}\mathbf{q}_1 + \mathbf{q}_2 &&= b_{21}\mathbf{u}_1 + b_{22}\mathbf{u}_2 \\
&\vdots \\
\mathbf{c}_n &= a_{n1}\mathbf{q}_1 + \cdots + a_{nn-1}\mathbf{q}_{n-1} + \mathbf{q}_n &&= b_{n1}\mathbf{u}_1 + \cdots + b_{nn}\mathbf{u}_n,
\end{aligned}$$

where $a_{ij} = \frac{\langle \mathbf{c}_i, \mathbf{q}_j \rangle}{\langle \mathbf{q}_j, \mathbf{q}_j \rangle}$ for $i > j$, $a_{ii} = 1$, and $b_{ij} = a_{ij}\|\mathbf{q}_j\|$ for $i \geq j$. Hence,

$$A = [\mathbf{c}_1 \ \cdots \ \mathbf{c}_n] = [\mathbf{u}_1 \ \cdots \ \mathbf{u}_n] \begin{bmatrix} b_{11} & b_{21} & \cdots & b_{n1} \\ 0 & b_{22} & \cdots & b_{n2} \\ & & \ddots & \\ 0 & & 0 & b_{nn} \end{bmatrix} = QR.$$

The matrix $Q = [\mathbf{u}_1 \ \cdots \ \mathbf{u}_n]$ is an $m \times n$ matrix with orthonormal columns, called the **orthogonal part** of A, and

$$R = \begin{bmatrix} b_{11} & b_{21} & \cdots & b_{n1} \\ 0 & b_{22} & \cdots & b_{n2} \\ & & \ddots & \\ 0 & & 0 & b_{nn} \end{bmatrix}$$

is an invertible upper triangular matrix, called the **upper triangular part** of A (note that all the diagonal $b_{ii} \neq 0$). Such an $A = QR$ is called the QR **factorization** of an $m \times n$ matrix A, when rank $A = n$. With this decomposition of A, the projection matrix can now be calculated easily as

$$P = A(A^TA)^{-1}A^T = QR(R^TQ^TQR)^{-1}R^TQ^T = QQ^T,$$

and $\mathbf{x} = (A^TA)^{-1}A^T\mathbf{b} = R^{-1}Q^T\mathbf{b}$.

Example 5.20 Let us find the projection matrix for

$$A = [\mathbf{c}_1\ \mathbf{c}_2\ \mathbf{c}_3] = \begin{bmatrix} 1 & 1 & 0 \\ 1 & 0 & 1 \\ 0 & 1 & 1 \\ 0 & 0 & 1 \end{bmatrix}.$$

Solution: We first find the decomposition of A into Q and R, the orthogonal part and the upper triangular part:

$$\begin{aligned} \mathbf{q}_1 &= \mathbf{c}_1 = (1,\ 1,\ 0,\ 0) \\ \mathbf{q}_2 &= \mathbf{c}_2 - \frac{\langle \mathbf{c}_2, \mathbf{q}_1\rangle}{\langle \mathbf{q}_1, \mathbf{q}_1\rangle}\,\mathbf{q}_1 = \left(\frac{1}{2},\ -\frac{1}{2},\ 1,\ 0\right) \\ \mathbf{q}_3 &= \mathbf{c}_3 - \frac{\langle \mathbf{c}_3, \mathbf{q}_2\rangle}{\langle \mathbf{q}_2, \mathbf{q}_2\rangle}\,\mathbf{q}_2 - \frac{\langle \mathbf{c}_3, \mathbf{q}_1\rangle}{\langle \mathbf{q}_1, \mathbf{q}_1\rangle}\,\mathbf{q}_1 = \left(-\frac{2}{3},\ \frac{2}{3},\ \frac{2}{3},\ 1\right), \end{aligned}$$

and $\|\mathbf{q}_1\| = \sqrt{2}$, $\|\mathbf{q}_2\| = \sqrt{3/2}$, $\|\mathbf{q}_3\| = \sqrt{7/3}$. Hence,

$$\begin{aligned} \mathbf{u}_1 &= \frac{\mathbf{q}_1}{\|\mathbf{q}_1\|} = \left(\frac{1}{\sqrt{2}},\ \frac{1}{\sqrt{2}},\ 0,\ 0\right) \\ \mathbf{u}_2 &= \frac{\mathbf{q}_2}{\|\mathbf{q}_2\|} = \left(\frac{1}{\sqrt{6}},\ -\frac{1}{\sqrt{6}},\ \frac{\sqrt{2}}{\sqrt{3}},\ 0\right) \\ \mathbf{u}_3 &= \frac{\mathbf{q}_3}{\|\mathbf{q}_3\|} = \left(-\frac{2}{\sqrt{21}},\ \frac{2}{\sqrt{21}},\ \frac{2}{\sqrt{21}},\ \frac{\sqrt{3}}{\sqrt{7}}\right). \end{aligned}$$

Then $\mathbf{c}_1 = \sqrt{2}\mathbf{u}_1$, $\mathbf{c}_2 = \frac{1}{\sqrt{2}}\mathbf{u}_1 + \sqrt{\frac{3}{2}}\mathbf{u}_2$, $\mathbf{c}_3 = \frac{1}{\sqrt{2}}\mathbf{u}_1 + \frac{1}{\sqrt{6}}\mathbf{u}_2 + \sqrt{\frac{7}{3}}\mathbf{u}_3$. Therefore,

$$\begin{aligned} A &= \begin{bmatrix} 1 & 1 & 0 \\ 1 & 0 & 1 \\ 0 & 1 & 1 \\ 0 & 0 & 1 \end{bmatrix} = [\mathbf{u}_1\ \mathbf{u}_2\ \mathbf{u}_3]\begin{bmatrix} \sqrt{2} & 1/\sqrt{2} & 1/\sqrt{2} \\ 0 & \sqrt{3}/\sqrt{2} & 1/\sqrt{6} \\ 0 & 0 & \sqrt{7}/\sqrt{3} \end{bmatrix} \\ &= \begin{bmatrix} 1/\sqrt{2} & 1/\sqrt{6} & -2/\sqrt{21} \\ 1/\sqrt{2} & -1/\sqrt{6} & 2/\sqrt{21} \\ 0 & \sqrt{2}/\sqrt{3} & 2/\sqrt{21} \\ 0 & 0 & \sqrt{3}/\sqrt{7} \end{bmatrix}\begin{bmatrix} \sqrt{2} & 1/\sqrt{2} & 1/\sqrt{2} \\ 0 & \sqrt{3}/\sqrt{2} & 1/\sqrt{6} \\ 0 & 0 & \sqrt{7}/\sqrt{3} \end{bmatrix} = QR, \end{aligned}$$

and

$$P = QQ^T = \begin{bmatrix} 6/7 & 1/7 & 1/7 & -2/7 \\ 1/7 & 6/7 & -1/7 & 2/7 \\ -1/7 & -1/7 & 6/7 & 2/7 \\ -2/7 & 2/7 & 2/7 & 3/7 \end{bmatrix}.$$

□

Problem 5.30 Find the 2×2 matrix P that projects the xy-plane onto the line $y = x$.

Problem 5.31 Find the projection matrix P of $\mathbb{R}^3$ onto the column space $\mathcal{C}(A)$ for $A = \begin{bmatrix} 1 & 1 \\ 1 & 0 \\ 0 & 1 \end{bmatrix}$.

Problem 5.32 Find the matrix for orthogonal projection from $\mathbb{R}^3$ to the plane spanned by the vectors $(1,\ 1,\ 1)$ and $(1,\ 0,\ 2)$.

Problem 5.33 Find the projection matrix P on the $x_1,\ x_2,\ x_4$ coordinate subspace of $\mathbb{R}^4$.

Problem 5.34 Find the QR factorization of the matrix $\begin{bmatrix} \sin\theta & \cos\theta \\ \cos\theta & 0 \end{bmatrix}$.

5.11 Exercises

5.1. Decide which of the following functions on $\mathbb{R}^2$ are inner products and which are not. For $\mathbf{x} = (x_1,\ x_2)$, $\mathbf{y} = (y_1,\ y_2)$,
(1) $\langle \mathbf{x}, \mathbf{y} \rangle = x_1 y_1 x_2 y_2$,
(2) $\langle \mathbf{x}, \mathbf{y} \rangle = 4x_1 y_1 + 4x_2 y_2 - x_1 y_2 - x_2 y_1$,
(3) $\langle \mathbf{x}, \mathbf{y} \rangle = x_1 y_2 - x_2 y_1$,
(4) $\langle \mathbf{x}, \mathbf{y} \rangle = x_1 y_1 + 3x_2 y_2$,
(5) $\langle \mathbf{x}, \mathbf{y} \rangle = x_1 y_1 - x_1 y_2 - x_2 y_1 + 3x_2 y_2$.

5.2. Show that the function $\langle A, B \rangle = \operatorname{tr}(A^T B)$ for $A, B \in M_{n\times n}(\mathbb{R})$ defines an inner product on $M_{n\times n}(\mathbb{R})$.

5.3. Find the angle between the vectors $(4,\ 7,\ 9,\ 1,\ 3)$ and $(2,\ 1,\ 1,\ 6,\ 8)$ in $\mathbb{R}^5$.

5.4. Determine the values of k so that the given vectors are orthogonal with respect to the Euclidean inner product in $\mathbb{R}^4$.

$$(1) \left\{ \begin{bmatrix} 2 \\ 3 \\ k \\ 4 \end{bmatrix}, \begin{bmatrix} 1 \\ k \\ 3 \\ -5 \end{bmatrix} \right\}, \quad (2) \left\{ \begin{bmatrix} 2 \\ 8 \\ 4 \\ k \end{bmatrix}, \begin{bmatrix} 2 \\ -6 \\ 2 \\ k \end{bmatrix} \right\}.$$

5.5. Consider the space $C[0,\ 1]$ with the inner product defined by

$$\langle f, g \rangle = \int_0^1 f(x)g(x)dx.$$

Compute the length of each vector and the cosine of the angle between each pair of vectors in each of the following:

(1) $f(x) = 1,\ g(x) = x$;
(2) $f(x) = x^m,\ g(x) = x^n$, where $m,\ n$ are nonnegative integers;
(3) $f(x) = \sin \pi m x,\ g(x) = \sin \pi n x$, where $m,\ n$ are integers.

5.6. Prove that

$$(a_1 + \cdots + a_n)^2 \le n(a_1^2 + \cdots + a_n^2)$$

for any real numbers $a_1,\ a_2,\ \ldots,\ a_n$. When does equality hold?

5.7. Let $V = P_2([0,\ 1])$, the space of polynomials of degree ≤ 2 on $[0,\ 1]$. Equip V with the inner product

$$\langle f, g\rangle = \int_0^1 f(t)g(t)dt.$$

(1) Compute $\langle f, g\rangle$ and $\|f\|$ for $f(x) = x + 2$ and $g(x) = x^2 - 2x - 3$.
(2) Find the orthogonal complement of the subspace of scalar polynomials.

5.8. Find an orthonormal basis for $\mathbb{R}^3$ with the Euclidean inner product by applying the Gram-Schmidt orthogonalization to the vectors $\mathbf{x} = (1,\ 0,\ 1)$, $\mathbf{x}_2 = (1,\ 0,\ -1)$, $\mathbf{x}_3 = (0,\ 3,\ 4)$.

5.9. Show that if $\mathbf{u}$ is orthogonal to $\mathbf{v}$, then every scalar multiple of $\mathbf{u}$ is also orthogonal to $\mathbf{v}$. Find a unit vector orthogonal to $\mathbf{v}_1 = (1,\ 1,\ 2)$ and $\mathbf{v}_2 = (0,\ 1,\ 3)$ in $\mathbb{R}^3$.

5.10. Determine the orthogonal projection of $\mathbf{v}_1$ onto $\mathbf{v}_2$ for the following vectors in the n-space $\mathbb{R}^n$ with the Euclidean inner product.

(1) $\mathbf{v}_1 = (1,\ 2,\ 3)$, $\mathbf{v}_2 = (1,\ 1,\ 2)$,
(2) $\mathbf{v}_1 = (1,\ 2,\ 1)$, $\mathbf{v}_2 = (2,\ 1,\ -1)$,
(3) $\mathbf{v}_1 = (1,\ 0,\ 1,\ 0)$, $\mathbf{v}_2 = (0,\ 2,\ 2,\ 0)$.

5.11. Let $S = \{\mathbf{v}_i\}$, where $\mathbf{v}_i$'s are given below. For each S, find a basis for $S^\perp$ with respect to the Euclidean inner product on $\mathbb{R}^n$.

(1) $\mathbf{v}_1 = (0,\ 1,\ 0)$, $\mathbf{v}_2 = (0,\ 0,\ 1)$,
(2) $\mathbf{v}_1 = (1,\ 1,\ 0)$, $\mathbf{v}_2 = (1,\ 1,\ 1)$,
(3) $\mathbf{v}_1 = (1,\ 0,\ 1,\ 2)$, $\mathbf{v}_2 = (1,\ 1,\ 1,\ 1)$, $\mathbf{v}_3 = (2,\ 2,\ 0,\ 1)$.

5.12. Which of the following matrices are orthogonal?

$$(1)\ \begin{bmatrix} 1/2 & -1/3 \\ -1/2 & 1/3 \end{bmatrix}, \qquad (2)\ \begin{bmatrix} 4/5 & -3/5 \\ -3/5 & 4/5 \end{bmatrix},$$

$$(3)\ \begin{bmatrix} 1/\sqrt{2} & 0 & -1/\sqrt{2} \\ 0 & -1/\sqrt{2} & 1/\sqrt{2} \\ -1/\sqrt{2} & 1/\sqrt{2} & 0 \end{bmatrix}, \quad (4)\ \begin{bmatrix} 1/\sqrt{2} & 1/\sqrt{3} & -1/\sqrt{6} \\ 1/\sqrt{2} & -1/\sqrt{3} & 1/\sqrt{6} \\ 0 & 1/\sqrt{3} & 2/\sqrt{6} \end{bmatrix}.$$

5.13. Consider $\mathbb{R}^4$ with the Euclidean inner product. Let W be the subspace of $\mathbb{R}^4$ consisting of all vectors that are orthogonal to both $\mathbf{x} = (1,\ 0,\ -1,\ 1)$ and $\mathbf{y} = (2,\ 3,\ -1,\ 2)$. Find a basis for W.

5.14. Let V be an inner product space. For vectors $\mathbf{x}$ and $\mathbf{y}$ in V, establish the following identities:

(1) $\langle \mathbf{x}, \mathbf{y}\rangle = \frac{1}{4}\|\mathbf{x}+\mathbf{y}\|^2 - \frac{1}{4}\|\mathbf{x}-\mathbf{y}\|^2$ (polarization identity),

(2) $\langle \mathbf{x}, \mathbf{y}\rangle = \frac{1}{2}\left(\|\mathbf{x}+\mathbf{y}\|^2 - \|\mathbf{x}\|^2 - \|\mathbf{y}\|^2\right)$ (Polarization identity),

(3) $\|\mathbf{x}+\mathbf{y}\|^2 + \|\mathbf{x}-\mathbf{y}\|^2 = 2(\|\mathbf{x}\|^2 + \|\mathbf{y}\|^2)$ (parallelogram equality).

5.15. Show that $\mathbf{x}+\mathbf{y}$ is perpendicular to $\mathbf{x}-\mathbf{y}$ if and only if $\|\mathbf{x}\| = \|\mathbf{y}\|$.

5.16. Let A be the $m \times n$ matrix whose columns are $\mathbf{c}_1, \ldots, \mathbf{c}_n$ in $\mathbb{R}^m$. Prove that the volume of the n-dimensional parallelepiped $\mathcal{P}(A)$ determined by those vectors $\mathbf{c}_j$'s in $\mathbb{R}^m$ is given by

$$\mathrm{vol}(A) = \sqrt{\det(A^T A)}\,.$$

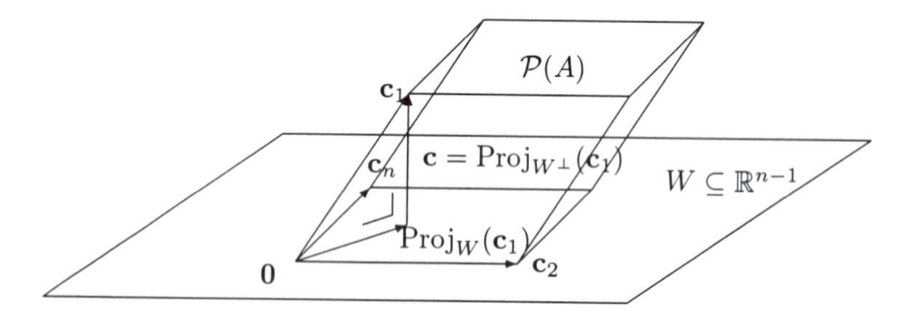

(Note that the volume of the n-dimensional parallelepiped determined by $\mathbf{c}_1$, ..., $\mathbf{c}_n$ in $\mathbb{R}^m$ is by definition the multiplication of the volume of the $(n-1)$-dimensional parallelepiped (base) determined by $\mathbf{c}_2, \ldots, \mathbf{c}_n$ and the height of $\mathbf{c}_1$ from the plane W which is spanned by $\mathbf{c}_2, \ldots, \mathbf{c}_n$. Here, the height is the length of the vector $\mathbf{c} = \mathbf{c}_1 - \mathrm{Proj}_W(\mathbf{c}_1)$, which is orthogonal to W. If the vectors are linearly dependent, then the parallelepiped is degenerate, *i.e.*, it is contained in a subspace of dimension less than n.)

5.17. Find the volume of the three-dimensional tetrahedron in $\mathbb{R}^4$ whose vertices are at $(0,0,0,0)$, $(1,0,0,0)$, $(0,1,2,2)$ and $(0,0,1,2)$.

5.18. For an orthogonal matrix A, show that $\det A = \pm 1$. Give an example of an orthogonal matrix A for which $\det A = -1$.

5.19. Find orthonormal bases for the row space and the null space of each of the following matrices.

$$(1) \begin{bmatrix} 2 & 4 & 3 \\ 1 & 1 & 1 \\ 2 & 0 & 1 \end{bmatrix}, \quad (2) \begin{bmatrix} 1 & 4 & 0 \\ -2 & -3 & 1 \\ 0 & 0 & 2 \end{bmatrix}, \quad (3) \begin{bmatrix} 1 & 1 & 0 & 0 \\ 1 & 0 & 1 & 0 \\ 1 & 0 & 0 & 1 \\ 1 & 0 & 0 & 0 \end{bmatrix}.$$

5.20. Let A be an $m \times n$ matrix of rank r. Find a relation of m, n and r so that $A\mathbf{x} = \mathbf{b}$ has infinitely many solutions for every $\mathbf{b} \in \mathbb{R}^m$.

5.21. Find the equation of the straight line that best fits the data of the four points $(0,\ 1)$, $(1,\ 3)$, $(2,\ 4)$, and $(3,\ 4)$.

5.22. Find the cubic polynomial that best fits the data of the five points $(-1,\ -14)$, $(0,\ -5)$, $(1,\ -4)$, $(2,\ 1)$, and $(3,\ 22)$.

5.23. Let W be the subspace of $\mathbb{R}^4$ spanned by the vectors $\mathbf{x}_i$'s given in each of the following problems. Find the projection matrix P for the subspace W and the null space $\mathcal{N}(P)$ of P. Compute $P\mathbf{b}$ for $\mathbf{b}$ given in each problem.

(1) $\mathbf{x}_1 = (1,\ 1,\ 1,\ 1)$, $\mathbf{x}_2 = (1,\ -1,\ 1,\ -1)$, $\mathbf{x}_3 = (-1,\ 1,\ 1,\ 0)$, and $\mathbf{b} = (1,\ 2,\ 1,\ 1)$.
(2) $\mathbf{x}_1 = (0,\ -2,\ 2,\ 1)$, $\mathbf{x}_2 = (2,\ 0,\ -1,\ 2)$, and $\mathbf{b} = (1,\ 1,\ 1,\ 1)$.
(3) $\mathbf{x}_1 = (2,\ 0,\ 3,\ -6)$, $\mathbf{x}_2 = (-3,\ 6,\ 8,\ 0)$, and $\mathbf{b} = (-1,\ 2,\ -1,\ 1)$.

5.24. Find the projection matrix for the row space and the null space of each of the following matrices:

$$(1) \begin{bmatrix} \frac{2}{\sqrt{5}} & -\frac{1}{\sqrt{5}} \\ \frac{1}{\sqrt{5}} & \frac{2}{\sqrt{5}} \end{bmatrix}, \quad (2) \begin{bmatrix} 2 & 4 & 1 \\ 1 & 1 & 1 \end{bmatrix}, \quad (3) \begin{bmatrix} 1 & 4 & 0 \\ 0 & 0 & 2 \\ 2 & 3 & -1 \end{bmatrix}.$$

5.25. Consider the space $C[-1,\ 1]$ with the inner product defined by

$$\langle f, g \rangle = \int_{-1}^{1} f(x)g(x)dx.$$

A function $f \in C[-1,\ 1]$ is *even* if $f(-x) = f(x)$, or *odd* if $f(-x) = -f(x)$. Let U and V be the sets of all even functions and odd functions in $C[-1,\ 1]$, respectively.

(1) Prove that U and V are subspaces and $C[-1,\ 1] = U + V$.
(2) Prove that $U \perp V$.
(3) Prove that for any $f \in C[-1,\ 1]$, $\|f\|^2 = \|h\|^2 + \|g\|^2$ where $f = h + g \in U \oplus V$.

5.26. Determine whether the following statements are true or false, in general, and justify your answers.

(1) Two vectors $\mathbf{x}$ and $\mathbf{y}$ in an inner product space are linearly independent if and only if the angle between $\mathbf{x}$ and $\mathbf{y}$ is not zero.
(2) If V is perpendicular to W, then $V^{\perp}$ is perpendicular to $W^{\perp}$.
(3) Every permutation matrix is an orthogonal matrix.
(4) The projection of $\mathbb{R}^m$ on a subspace W is a linear transformation of $\mathbb{R}^m$ into itself.
(5) Two different subspaces of $\mathbb{R}^m$ may have the same projection matrix.
(6) An $n \times n$ symmetric matrix A is a projection matrix if and only if $A^2 = I$.
(7) For any $m \times n$ matrix A and $\mathbf{b} \in \mathbb{R}^m$, $A^T A\mathbf{x} = A^T\mathbf{b}$ always has a solution.
(8) An inner product can be defined on every vector space.
(9) Let V be an inner product space. Then $\|\mathbf{x} - \mathbf{y}\| \geq \|\mathbf{x}\| - \|\mathbf{y}\|$ for any vectors $\mathbf{x}$ and $\mathbf{y}$ in V.
(10) The least square solution of $A\mathbf{x} = \mathbf{b}$ is unique for any symmetric matrix A.
(11) Every system of linear equations has a least square solution.
(12) The least square solution of $A\mathbf{x} = \mathbf{b}$ is the orthogonal projection of $\mathbf{b}$ on the column space A.

Chapter 6

Eigenvectors and Eigenvalues

6.1 Introduction

Gaussian elimination plays a fundamental role in solving a system $A\mathbf{x} = \mathbf{b}$ of linear equations. In order to solve a system of linear equations, Gaussian elimination reduces the augmented matrix to a (reduced) row-echelon form by using elementary row operations that preserve row and null spaces.

In this chapter, as another method of simplifying a square matrix, we examine which matrices can be similar to diagonal matrices, and what the transition matrices are in this case. The tools are *eigenvalues* and *eigenvectors.* In fact, they play important roles in their own right in mathematics and have far-reaching applications not only in mathematics, but also other fields of science and engineering. Some specific applications with a square matrix A are **(1)** solving systems $A\mathbf{x} = \mathbf{b}$ of linear equations, **(2)** checking the invertibility of A or estimation of $\det A$, **(3)** calculating a power A^n or the limit of a matrix series $\sum_{n=1}^{\infty} A^n$, **(4)** solving systems of linear differential equations or difference equations, **(5)** finding a simple form of the matrix representation of a linear transformation, etc. One might notice that some of the problems listed above are easy if A is diagonal.

We begin by introducing eigenvalues and eigenvectors of a square matrix A. For an $n \times n$ square matrix A, there may exist a nonzero vector that is transformed by A into a scalar multiple of itself.

Definition 6.1 Let A be an $n \times n$ square matrix. A nonzero vector $\mathbf{x}$ in the n-space $\mathbb{R}^n$ is called an **eigenvector** (or **characteristic vector**) of A

if there is a scalar λ in $\mathbb{R}$ such that

$$A\mathbf{x} = \lambda\mathbf{x}.$$

The scalar λ is called an **eigenvalue** (or **characteristic value**) of A, and we say $\mathbf{x}$ **belongs** to λ.

Geometrically, an eigenvector of a matrix A is a nonzero vector $\mathbf{x}$ in the n-space $\mathbb{R}^n$ to which $A\mathbf{x}$ is parallel. Algebraically, an eigenvector $\mathbf{x}$ is a nontrivial solution of the homogeneous system $(\lambda I - A)\mathbf{x} = \mathbf{0}$ of linear equations, that is, an eigenvector $\mathbf{x}$ is a nonzero vector in the null space $\mathcal{N}(\lambda I - A)$. There are two unknowns in this equation: an eigenvalue λ and an eigenvector $\mathbf{x}$. To find those unknowns, first we should find an eigenvalue λ by using the fact that the equation $(\lambda I - A)\mathbf{x} = \mathbf{0}$ has a nontrivial solution $\mathbf{x}$ if and only if λ satisfies the equation

$$\det(\lambda I - A) = 0.$$

Note that the left side is a polynomial of degree n in λ, called the **characteristic polynomial** of A. Thus the eigenvalues are simply the roots of the equation $\det(\lambda I - A) = 0$.

Thus, to find eigenvectors of A, first find the roots (or eigenvalues of A) of the equation $\det(\lambda I - A) = 0$, and then solve the homogeneous system $(\lambda I - A)\mathbf{x} = \mathbf{0}$ for each eigenvalue λ. In summary, by referring to Theorem 3.25 we have the following theorem.

Theorem 6.1 *If A is an $n \times n$ matrix, then the following are equivalent:*

(1) *λ is an eigenvalue of A;*

(2) $\det(\lambda I - A) = 0$ *(or* $\det(A - \lambda I) = 0$*);*

(3) *$\lambda I - A$ is singular;*

(4) *the homogeneous system $(\lambda I - A)\mathbf{x} = \mathbf{0}$ has a nontrivial solution.*

Hence, the eigenvectors of A belonging to an eigenvalue λ are just the nonzero vectors $\mathbf{x}$ in the null space $\mathcal{N}(\lambda I - A)$. We call this null space the **eigenspace** of A belonging to λ, and denote it $E(\lambda)$.

Example 6.1 Find the eigenvalues and eigenvectors of

$$A = \begin{bmatrix} 2 & \sqrt{2} \\ \sqrt{2} & 1 \end{bmatrix}.$$

Solution: The characteristic polynomial is

$$\det(\lambda I - A) = \det\begin{bmatrix} \lambda - 2 & -\sqrt{2} \\ -\sqrt{2} & \lambda - 1 \end{bmatrix} = \lambda^2 - 3\lambda = \lambda(\lambda - 3),$$

whence the eigenvalues are $\lambda_1 = 0$ and $\lambda_2 = 3$. To determine the eigenvectors belonging to λ_i's, we should solve the homogeneous system of equations $(\lambda_i I - A)\mathbf{x} = \mathbf{0}$. Let us take $\lambda_1 = 0$ first; then the system of equations $(\lambda_1 I - A)\mathbf{x} = \mathbf{0}$ becomes

$$\begin{cases} -2\,x_1 & - & \sqrt{2}\,x_2 & = & 0, \\ -\sqrt{2}\,x_1 & - & x_2 & = & 0, \end{cases} \qquad \text{or} \quad x_1 = -\frac{1}{\sqrt{2}}\,x_2.$$

Hence, $\mathbf{x}_1 = (x_1,\ x_2) = (-1,\ \sqrt{2})$ is an eigenvector belonging to $\lambda_1 = 0$, and the eigenvectors of A belonging to $\lambda_1 = 0$ are nonzero vectors of the form $t\mathbf{x}_1$, $t \in \mathbb{R}$. (Here, we can take any nonzero solution $(x_1,\ x_2)$ as an eigenvector $\mathbf{x}_1$ belonging to $\lambda_1 = 0$.)

For $\lambda_2 = 3$, the system of equations $(\lambda_2 I - A)\mathbf{x} = \mathbf{0}$ becomes

$$\begin{cases} x_1 & - & \sqrt{2}\,x_2 & = & 0, \\ -\sqrt{2}\,x_1 & + & 2\,x_2 & = & 0, \end{cases} \qquad \text{or} \quad x_1 = \sqrt{2}\,x_2.$$

Thus, by a similar calculation, $\mathbf{x}_2 = (\sqrt{2},\ 1)$ is an eigenvector belonging to $\lambda_2 = 3$ and the eigenvectors of A belonging to $\lambda_2 = 3$ are the nonzero vectors of the form $t\mathbf{x}_2$, $t \in \mathbb{R}$. Note that the eigenvectors $\mathbf{x}_1$ and $\mathbf{x}_2$ belonging to the eigenvalues λ_1 and λ_2 are linearly independent. □

Example 6.2 Find a basis for the eigenspaces of

$$A = \begin{bmatrix} 3 & -2 & 0 \\ -2 & 3 & 0 \\ 0 & 0 & 5 \end{bmatrix}.$$

Solution: The characteristic polynomial of A is $(\lambda - 1)(\lambda - 5)^2$, so that the eigenvalues of A are $\lambda_1 = 1$ and $\lambda_2 = 5$ with multiplicity 2. Thus, there are two eigenspaces of A. By definition, $\mathbf{x} = (x_1,\ x_2,\ x_3)$ is an eigenvector of A belonging to λ if and only if $\mathbf{x}$ is a nontrivial solution of the homogeneous system $(\lambda I - A)\mathbf{x} = \mathbf{0}$:

$$\begin{bmatrix} \lambda - 3 & 2 & 0 \\ 2 & \lambda - 3 & 0 \\ 0 & 0 & \lambda - 5 \end{bmatrix}\begin{bmatrix} x_1 \\ x_2 \\ x_3 \end{bmatrix} = \begin{bmatrix} 0 \\ 0 \\ 0 \end{bmatrix}.$$

If $\lambda_1 = 1$, then the system becomes

$$\begin{bmatrix} -2 & 2 & 0 \\ 2 & -2 & 0 \\ 0 & 0 & -4 \end{bmatrix} \begin{bmatrix} x_1 \\ x_2 \\ x_3 \end{bmatrix} = \begin{bmatrix} 0 \\ 0 \\ 0 \end{bmatrix}.$$

Solving this system yields $x_1 = t$, $x_2 = t$, $x_3 = 0$ for $t \in \mathbb{R}$. Thus, the eigenvectors belonging to $\lambda_1 = 1$ are nonzero vectors of the form

$$\mathbf{x} = \begin{bmatrix} t \\ t \\ 0 \end{bmatrix} = t \begin{bmatrix} 1 \\ 1 \\ 0 \end{bmatrix}, \quad t \in \mathbb{R},$$

so that $(1,\ 1,\ 0)$ is a basis for the eigenspace $E(\lambda_1)$ belonging to $\lambda_1 = 1$.

If $\lambda_2 = 5$, then the system becomes

$$\begin{bmatrix} 2 & 2 & 0 \\ 2 & 2 & 0 \\ 0 & 0 & 0 \end{bmatrix} \begin{bmatrix} x_1 \\ x_2 \\ x_3 \end{bmatrix} = \begin{bmatrix} 0 \\ 0 \\ 0 \end{bmatrix}.$$

Solving this system yields $x_1 = -s$, $x_2 = s$, $x_3 = t$ for $s,\ t \in \mathbb{R}$. Thus, the eigenvectors of A belonging to $\lambda_2 = 5$ are nonzero vectors of the form

$$\mathbf{x} = \begin{bmatrix} -s \\ s \\ t \end{bmatrix} = \begin{bmatrix} -s \\ s \\ 0 \end{bmatrix} + \begin{bmatrix} 0 \\ 0 \\ t \end{bmatrix} = s \begin{bmatrix} -1 \\ 1 \\ 0 \end{bmatrix} + t \begin{bmatrix} 0 \\ 0 \\ 1 \end{bmatrix}$$

for $s,\ t \in \mathbb{R}$. Since $(-1,\ 1,\ 0)$ and $(0,\ 0,\ 1)$ are linearly independent, they form a basis for the eigenspace $E(\lambda_2)$ belonging to $\lambda_2 = 5$. □

Example 6.3 Consider the matrix $A = \begin{bmatrix} 1 & 2 \\ 0 & 1 \end{bmatrix}$. Like the above example, a simple computation shows that the characteristic polynomial is $(\lambda - 1)^2$ so that the eigenvalue $\lambda = 1$ is of multiplicity 2. However, there is only one linearly independent eigenvector $\mathbf{x} = (1,\ 0)$ belonging to $\lambda = 1$. This kind of matrix will be discussed in Chapter 7. □

Note that the equation $\det(\lambda I - A) = 0$ may have complex roots, which are called **complex eigenvalues**. However, the complex numbers are not scalars of the real vector space. In many cases, it is necessary to deal with those complex numbers, that is, we need to expand the set of scalars to include complex numbers. This expansion of the set of scalars to the set of complex numbers leads us to work with complex vector spaces, which will be treated in the next chapter. In this chapter, we restrict the discussion to the case of real eigenvalues, even though the entire discussion in this chapter applies in the same way to general complex vector spaces.

Example 6.4 The characteristic polynomial of the matrix

$$A = \begin{bmatrix} \cos\theta & -\sin\theta \\ \sin\theta & \cos\theta \end{bmatrix}$$

is $\lambda^2 - 2\cos\theta\lambda + (\cos^2\theta + \sin^2\theta)$. Thus, the eigenvalues are $\lambda_i = \cos\theta \pm i\sin\theta$, which are complex numbers, so this matrix as a rotation of $\mathbb{R}^2$ has no real eigenvalues unless $\theta = n\pi$, $n = 0, \pm 1, \pm 2, \ldots$. □

Problem 6.1 Let A be a 2×2 matrix whose characteristic polynomial is $\det(\lambda I - A) = \lambda^2 + b\lambda + c$. Show that $b = -\text{tr}\, A$ and $c = \det A$.

Problem 6.2 Let λ be an eigenvalue of A and $\mathbf{x}$ an eigenvector belonging to λ. Use mathematical induction to show that λ^m is an eigenvalue of A^m and $\mathbf{x}$ is an eigenvector of A^m belonging to λ^m for each $m = 1, 2, \ldots$.

Remark: (1) If A is an upper triangular matrix, then the diagonal entries are exactly the eigenvalues of A. In fact, the characteristic polynomial satisfies

$$\begin{aligned} \det(\lambda I - A) &= \det \begin{bmatrix} \lambda - a_{11} & * & \cdots & * \\ & \ddots & \ddots & \vdots \\ & & \ddots & * \\ 0 & & & \lambda - a_{nn} \end{bmatrix} \\ &= (\lambda - a_{11}) \cdots (\lambda - a_{nn}) = 0. \end{aligned}$$

(2) Let A and B be square matrices similar to each other (*i.e.*, there exists a nonsingular matrix Q such that $B = Q^{-1}AQ$). Then

$$\begin{aligned} \det(\lambda I - B) &= \det\left(Q^{-1}(\lambda I)Q - Q^{-1}AQ\right) \\ &= \det\left(Q^{-1}(\lambda I - A)Q\right) \\ &= \det Q^{-1} \det(\lambda I - A) \det Q \\ &= \det(\lambda I - A). \end{aligned}$$

Therefore, similar matrices have the same characteristic polynomial with the same roots. In particular, the eigenvalues are invariant under the similarity. However, their eigenvectors might be different: When $B = Q^{-1}AQ$, $\mathbf{x}$ is an eigenvector of B belonging to λ if and only if $Q\mathbf{x}$ is an eigenvector of A belonging to λ, since $AQ = QB$, and $QB\mathbf{x} = \lambda Q\mathbf{x}$.

(3) If an $n \times n$ matrix A has n eigenvalues $\lambda_1, \ldots, \lambda_n$ counting multiplicities, then the characteristic polynomial of A can be factorized as

$$\det(\lambda I - A) = (\lambda - \lambda_1) \cdots (\lambda - \lambda_n).$$

If we take $\lambda = 0$, then we get $\det(-A) = (-1)^n \det A = (-1)^n \lambda_1 \cdots \lambda_n$. Therefore, $\det A = \lambda_1 \cdots \lambda_n$, the product of the n eigenvalues.

(4) On the other hand,

$$\begin{aligned}(\lambda - \lambda_1) \cdots (\lambda - \lambda_n) &= \det(\lambda I - A) \\ &= \det \begin{bmatrix} \lambda - a_{11} & -a_{12} & \cdots & -a_{1n} \\ -a_{21} & \lambda - a_{22} & \cdots & -a_{2n} \\ \vdots & \vdots & & \vdots \\ -a_{n1} & -a_{n2} & \cdots & \lambda - a_{nn} \end{bmatrix},\end{aligned}$$

which is a polynomial of the form $p(\lambda) = \lambda^n + c_{n-1}\lambda^{n-1} + \cdots + c_1\lambda + c_0$ in λ. We can compute the coefficient c_{n-1} of λ^{n-1} in two ways by expending both sides, and get

$$\lambda_1 + \cdots + \lambda_n = a_{11} + \cdots + a_{nn} = \text{tr}A.$$

This shows that the trace of A is the sum of the n eigenvalues. Thus we have the following theorem:

Theorem 6.2 *Let A be an $n \times n$ square matrix. Then*

(1) *the eigenvalues are invariant under the similarity.*

Moreover, if A has n eigenvalues counting multiplicities, then

(2) *the determinant of A is the product of the n eigenvalues, and*

(3) *the trace of A is the sum of the n eigenvalues.*

In Theorem 6.2 (2)-(3), we assume that the matrix A has n (real) eigenvalues counting multiplicities. But, by allowing the scalars to be complex numbers, which will be done in the next chapter, every $n \times n$ matrix has n eigenvalues counting multiplicities, so that (2) and (3) in Theorem 6.2 are true for any square matrix.

Corollary 6.3 *The determinant and the trace of A are invariant under similarity.*

Recall that any square matrix A is singular if and only if $\det A = 0$. However, $\det A$ is the product of its n eigenvalues. Thus a square matrix A is singular if and only if zero is an eigenvalue of A, or A is invertible if and only if zero is not an eigenvalue of A. The following corollaries are easy consequences of this fact.

Corollary 6.4 *For any $n \times n$ matrices A and B, the following are equivalent.*

(1) *Zero is an eigenvalue of AB.*

(2) *A or B is singular.*

(3) *Zero is an eigenvalue of BA.*

Corollary 6.5 *For any $n \times n$ matrices A and B, AB and BA have the same eigenvalues.*

Proof: By Corollary 6.4, zero is an eigenvalue of AB if and only if it is an eigenvalue of BA. Let λ be a nonzero eigenvalue of AB with $(AB)\mathbf{x} = \lambda\mathbf{x}$ for a nonzero vector $\mathbf{x}$. Then the vector $B\mathbf{x}$ is not zero, since $\lambda \neq 0$, but

$$(BA)(B\mathbf{x}) = B(\lambda\mathbf{x}) = \lambda(B\mathbf{x}).$$

This means that $B\mathbf{x}$ is an eigenvector of BA belonging to the eigenvalue λ, and λ is an eigenvalue of BA. Similarly, any nonzero eigenvalue of BA is also an eigenvalue of AB. □

Problem 6.3 Find the matrices A and B such that $\det A = \det B$, $\mathrm{tr}A = \mathrm{tr}B$, but A is not similar to B.

Problem 6.4 Let $\lambda_1, \lambda_2, \ldots, \lambda_n$ be the eigenvalues of an $n \times n$ matrix A. Then

(1) A is invertible if and only if $\lambda_i \neq 0$, for all $i = 1, 2, \ldots, n$.

(2) If A is invertible, then the inverse A^{-1} has eigenvalues $\frac{1}{\lambda_1}, \frac{1}{\lambda_2}, \ldots, \frac{1}{\lambda_n}$.

Problem 6.5 Show that A and A^T have the same eigenvalues. Do they necessarily have the same eigenvectors?

Problem 6.6 For any $n \times n$ matrices A and B, show that AB and BA are similar if A or B is nonsingular.

6.2 Diagonalization of matrices

In this section, we are going to show what kind of square matrices are similar to diagonal matrices. That is, given a square matrix A, we want to know whether there exists an invertible matrix Q such that $Q^{-1}AQ$ is a diagonal matrix, and if so, how one can find such a matrix Q. It is closely related to the eigenvalues and eigenvectors of A.

Recall that an $n \times n$ matrix A is a linear transformation on the n-space $\mathbb{R}^n$, whose matrix representation with respect to the standard basis is the matrix A itself. If we take another basis for $\mathbb{R}^n$, then we get another matrix representation D, which is similar to A by a transition matrix Q as shown in Section 4.6. In some cases, one can find a "good" basis for $\mathbb{R}^n$ so that the matrix D is a diagonal matrix. In this case, the similarity $D = Q^{-1}AQ$ gives an easy way to solve some problems related to the matrix A. For instance, let $A\mathbf{x} = \mathbf{b}$ be a system of linear equations with a square matrix A, and suppose that there is an invertible matrix Q such that $Q^{-1}AQ$ is a diagonal matrix D. Then the system $A\mathbf{x} = \mathbf{b}$ can be written as $QDQ^{-1}\mathbf{x} = \mathbf{b}$, or equivalently $DQ^{-1}\mathbf{x} = Q^{-1}\mathbf{b}$. Hence, for $\mathbf{c} = Q^{-1}\mathbf{b}$ the solution $\mathbf{y}$ of $D\mathbf{y} = \mathbf{c}$ yields the solution $\mathbf{x} = Q\mathbf{y}$ of the original problem. Note that $D\mathbf{y} = \mathbf{c}$ can be solved easily.

In this section, we shall discuss what we mean by a "good" basis and how to find it.

Definition 6.2 A square matrix A is said to be **diagonalizable** if there exists an invertible matrix Q such that $Q^{-1}AQ$ is a diagonal matrix (*i.e.*, A is similar to a diagonal matrix).

The next theorem characterizes a diagonalizable matrix, and the proof shows a practical way of diagonalizing a matrix.

Theorem 6.6 *Let A be an $n \times n$ matrix. Then A is diagonalizable if and only if A has n linearly independent eigenvectors.*

Proof: ($\Rightarrow$) Suppose A is diagonalizable. Then there is an invertible matrix Q such that $Q^{-1}AQ$ is a diagonal matrix D, say

$$Q^{-1}AQ = D = \begin{bmatrix} \lambda_1 & 0 & \cdots & 0 \\ 0 & \lambda_2 & \cdots & 0 \\ \vdots & \vdots & \ddots & \vdots \\ 0 & 0 & \cdots & \lambda_n \end{bmatrix},$$

or, equivalently, $AQ = QD$. Let $\mathbf{x}_1, \ldots, \mathbf{x}_n$ denote the column vectors of Q. Since

$$\begin{aligned} AQ &= [A\mathbf{x}_1 \; A\mathbf{x}_2 \; \cdots \; A\mathbf{x}_n] \\ QD &= [\lambda_1\mathbf{x}_1 \; \lambda_2\mathbf{x}_2 \; \cdots \; \lambda_n\mathbf{x}_n], \end{aligned}$$

the matrix equation $AQ = QD$ implies $A\mathbf{x}_i = \lambda_i\mathbf{x}_i$ for $i = 1, \ldots, n$. Moreover, since Q is invertible, their column vectors are nonzero and are linearly independent, that is, the $\mathbf{x}_i$'s are n linearly independent eigenvectors of A.

($\Leftarrow$) Assume that A has n linearly independent eigenvectors $\mathbf{x}_1, \ldots, \mathbf{x}_n$ belonging to the eigenvalues $\lambda_1, \ldots, \lambda_n$, respectively, so that $A\mathbf{x}_i = \lambda_i\mathbf{x}_i$ for $i = 1, \ldots, n$. If we define a matrix Q as

$$Q = [\mathbf{x}_1 \; \mathbf{x}_2 \; \cdots \; \mathbf{x}_n]$$

with $\mathbf{x}_j$ as the j-th column vector, then the same equation shows $AQ = QD$, where D is the diagonal matrix having the eigenvalues $\lambda_1, \ldots, \lambda_n$ on the diagonal. Since the column vectors of Q are assumed to be linearly independent, Q is invertible, so $Q^{-1}AQ = D$. □

Remark: (1) Theorem 6.6 says that to diagonalize a matrix A what we need is a basis consisting of eigenvectors of A, by which we meant a "good" basis. Moreover, the proof of the theorem reveals a method for diagonalizing an $n \times n$ matrix A.

Step 1. Find n linearly independent eigenvectors $\mathbf{x}_1, \mathbf{x}_2, \ldots, \mathbf{x}_n$ of A.

Step 2. Form the matrix Q having $\mathbf{x}_1, \mathbf{x}_2, \ldots, \mathbf{x}_n$ as its column vectors.

Step 3. The matrix $Q^{-1}AQ$ will be a diagonal form with $\lambda_1, \ldots, \lambda_n$ as its successive diagonal entries, where λ_j is the eigenvalue associated with the eigenvector $\mathbf{x}_j$, $j = 1, 2, \ldots, n$.

(2) Let α denote the standard basis for $\mathbb{R}^n$ and let $\beta = \{\mathbf{x}_1, \mathbf{x}_2, \ldots, \mathbf{x}_n\}$ be the basis for $\mathbb{R}^n$ consisting of n linearly independent eigenvectors of A. Then the matrix

$$Q = [\mathbf{x}_1 \; \mathbf{x}_2 \; \cdots \; \mathbf{x}_n] = [[\mathbf{x}_1]_\alpha \; [\mathbf{x}_2]_\alpha \; \cdots \; [\mathbf{x}_n]_\alpha] = [Id]_\beta^\alpha$$

is the transition matrix from β to α, and the matrix representation of A, as a linear transformation, with respect to β, is

$$[A]_\beta = [Id]_\alpha^\beta [A]_\alpha [Id]_\beta^\alpha = Q^{-1}AQ = \begin{bmatrix} \lambda_1 & & 0 \\ & \ddots & \\ 0 & & \lambda_n \end{bmatrix}.$$

Note that the diagonal entries λ_i's are the eigenvalues of A.

(3) Not all matrices are diagonalizable. A standard example is $A = \begin{bmatrix} 0 & 1 \\ 0 & 0 \end{bmatrix}$. Its eigenvalues are $\lambda_1 = \lambda_2 = 0$. Hence, if A is diagonalizable, then

$$Q^{-1}AQ = \begin{bmatrix} \lambda_1 & 0 \\ 0 & \lambda_2 \end{bmatrix} = \mathbf{0}$$

for some invertible matrix Q, and then A must be the zero matrix. Since A is not the zero matrix, no invertible matrix Q can be achieved so that $Q^{-1}AQ$ is diagonal.

Example 6.5 Diagonalize the following matrix:

$$A = \begin{bmatrix} 1 & -3 & 3 \\ 0 & -5 & 6 \\ 0 & -3 & 4 \end{bmatrix}.$$

Solution: A direct calculation gives that the eigenvalues of A are $\lambda_1 = 1$, $\lambda_2 = 1$ and $\lambda_3 = -2$, and their associated eigenvectors are

$$\mathbf{x}_1 = (1,\ 0,\ 0),\ \mathbf{x}_2 = (0,\ 1,\ 1) \quad \text{and} \quad \mathbf{x}_3 = (1,\ 2,\ 1),$$

respectively. They are linearly independent, and the first two vectors $\mathbf{x}_1$, $\mathbf{x}_2$ form a basis for the eigenspace $E(1)$ belonging to $\lambda_1 = \lambda_2 = 1$, and $\mathbf{x}_3$ forms a basis for the eigenspace $E(-2)$ belonging to $\lambda_3 = -2$. Thus, the matrix

$$P = \begin{bmatrix} 1 & 0 & 1 \\ 0 & 1 & 2 \\ 0 & 1 & 1 \end{bmatrix}$$

diagonalizes A. In fact, one can verify that

$$\begin{aligned} P^{-1}AP &= \begin{bmatrix} 1 & -1 & 1 \\ 0 & -1 & 2 \\ 0 & 1 & -1 \end{bmatrix} \begin{bmatrix} 1 & -3 & 3 \\ 0 & -5 & 6 \\ 0 & -3 & 4 \end{bmatrix} \begin{bmatrix} 1 & 0 & 1 \\ 0 & 1 & 2 \\ 0 & 1 & 1 \end{bmatrix} \\ &= \begin{bmatrix} 1 & 0 & 0 \\ 0 & 1 & 0 \\ 0 & 0 & -2 \end{bmatrix}. \end{aligned}$$

What would happen if one chose different eigenvectors belonging to the eigenvalues 1 and -2? According to the proof of Theorem 6.6, nothing would happen: Any matrix whose columns are linearly independent eigenvectors will diagonalize A. For example, $\{(-1, 0, 0), (0, -1, -1)\}$ is another basis for $E(1)$, and $\{(2, 4, 2)\}$ is also a basis for $E(-2)$. The matrix

$$Q = \begin{bmatrix} -1 & 0 & 2 \\ 0 & -1 & 4 \\ 0 & -1 & 2 \end{bmatrix} \quad \text{also diagonalizes } A \text{ as } Q^{-1}AQ = \begin{bmatrix} 1 & 0 & 0 \\ 0 & 1 & 0 \\ 0 & 0 & -2 \end{bmatrix}.$$

A change in the order of the eigenvectors in constructing a transition matrix Q does not change the diagonalizability of A, but the eigenvalues appearing on the main diagonal of the resulting diagonal matrix would appear in accordance with the order of the eigenvectors in the transition matrix. For example, let

$$S = \begin{bmatrix} 1 & 1 & 0 \\ 0 & 2 & 1 \\ 0 & 1 & 1 \end{bmatrix}.$$

Then S will diagonalize A, because it has linearly independent eigenvectors as columns, but we find that

$$S^{-1} = \begin{bmatrix} 1 & -1 & 1 \\ 0 & 1 & -1 \\ 0 & -1 & 2 \end{bmatrix} \quad \text{and} \quad S^{-1}AS = \begin{bmatrix} 1 & 0 & 0 \\ 0 & -2 & 0 \\ 0 & 0 & 1 \end{bmatrix}.$$

□

Problem 6.7 Show that the following matrices are not diagonalizable.

$$(1)\, A = \begin{bmatrix} \lambda & 1 & 0 \\ 0 & \lambda & 1 \\ 0 & 0 & \lambda \end{bmatrix}, \quad (2)\, B = \begin{bmatrix} \lambda & 0 & 0 \\ 1 & \lambda & 0 \\ 0 & 1 & \lambda \end{bmatrix}, \quad \lambda \text{ is any scalar.}$$

Problem 6.8 Find a 2×2 matrix A whose eigenvalues are 2 and 3, and whose eigenvectors are $(2, 1)$ and $(3, 2)$, respectively.

Theorem 6.6 shows how to diagonalize a matrix and what the diagonal matrix is when the matrix has a full set of linearly independent eigenvectors. We next consider when a square matrix A can have a full set of linearly independent eigenvectors. This problem is closely related to the eigenvalues of the matrix, because eigenvectors can be found practically after the eigenvalues have been computed.

The following theorem shows that if an $n \times n$ matrix has n distinct (real) eigenvalues, then it has n linearly independent eigenvectors so that it is always diagonalizable and the diagonal matrix has the eigenvalues on the main diagonal.

Theorem 6.7 *Let $\lambda_1, \lambda_2, \ldots, \lambda_k$ be distinct eigenvalues of a matrix A and $\mathbf{x}_1, \mathbf{x}_2, \ldots, \mathbf{x}_k$ eigenvectors belonging to them, respectively. Then $\{\mathbf{x}_1, \mathbf{x}_2, \ldots, \mathbf{x}_k\}$ is linearly independent.*

Proof: Let r be the largest integer such that $\{\mathbf{x}_1, \ldots, \mathbf{x}_r\}$ is linearly independent. If $r = k$, then there is nothing to prove. Suppose not, *i.e.*, $1 \le r < k$. Then $\{\mathbf{x}_1, \ldots, \mathbf{x}_{r+1}\}$ is linearly dependent. Thus, there are scalars $c_1, c_2, \ldots, c_{r+1}$ with $c_{r+1} \neq 0$ such that

$$c_1\mathbf{x}_1 + c_2\mathbf{x}_2 + \cdots + c_{r+1}\mathbf{x}_{r+1} = \mathbf{0}. \qquad (1)$$

Multiplying both sides by A, and, using

$$A\mathbf{x}_1 = \lambda_1\mathbf{x}_1, \ A\mathbf{x}_2 = \lambda_2\mathbf{x}_2, \ \ldots, \ A\mathbf{x}_{r+1} = \lambda_{r+1}\mathbf{x}_{r+1},$$

we get

$$c_1\lambda_1\mathbf{x}_1 + c_2\lambda_2\mathbf{x}_2 + \cdots + c_{r+1}\lambda_{r+1}\mathbf{x}_{r+1} = \mathbf{0}. \qquad (2)$$

Multiplying both sides of (1) by λ_{r+1} and subtracting the resulting equation from (2) yields

$$c_1(\lambda_1 - \lambda_{r+1})\mathbf{x}_1 + c_2(\lambda_2 - \lambda_{r+1})\mathbf{x}_2 + \cdots + c_r(\lambda_r - \lambda_{r+1})\mathbf{x}_r = \mathbf{0}.$$

Since $\{\mathbf{x}_1, \mathbf{x}_2, \ldots, \mathbf{x}_r\}$ is linearly independent and $\lambda_1, \lambda_2, \ldots, \lambda_{r+1}$ are all distinct, it follows that $c_1 = c_2 = \cdots = c_r = 0$. Substituting these values in (1) yields $c_{r+1} = 0$, which is a contradiction to the assumption. $\square$

As a consequence of Theorems 6.6 and 6.7, we obtain the following.

Theorem 6.8 *If an $n \times n$ matrix A has n distinct eigenvalues, then A is diagonalizable.*

It follows from Theorem 6.8 that if $\mathbf{x}_1, \mathbf{x}_2, \ldots, \mathbf{x}_n$ are eigenvectors of an $n \times n$ matrix A belonging to n distinct eigenvalues $\lambda_1, \lambda_2, \ldots, \lambda_n$, respectively, then they form a basis for $\mathbb{R}^n$ and the matrix representation of A with respect to this basis should be a diagonal matrix as shown in Remark (2) on page 217.

Of course, some matrices can have eigenvalues with multiplicities > 1 so that the number of distinct eigenvalues is strictly less than n. In this case, if such a matrix still has n linearly independent eigenvectors, then it is also diagonalizable, because for a diagonalization all we need is n linearly independent eigenvectors. In some cases, such a matrix does not have n linearly independent eigenvectors (see Example 6.3), so a diagonalization is impossible. This case will be discussed in Section 9.1.

Example 6.6 Compute A^{100} for $A = \begin{bmatrix} 1 & 4 \\ 3 & 2 \end{bmatrix}$.

Solution: Its eigenvalues are 5 and -2 with associated eigenvectors $(1,1)$ and $(-4,3)$, respectively. Hence $Q = \begin{bmatrix} 1 & -4 \\ 1 & 3 \end{bmatrix}$ diagonalizes A, *i.e.*,

$$Q^{-1}AQ = \begin{bmatrix} 5 & 0 \\ 0 & -2 \end{bmatrix} \quad \text{or} \quad A = Q\begin{bmatrix} 5 & 0 \\ 0 & -2 \end{bmatrix}Q^{-1}.$$

Therefore,

$$\begin{aligned} A^{100} &= Q\begin{bmatrix} 5^{100} & 0 \\ 0 & (-2)^{100} \end{bmatrix}Q^{-1} \\ &= \frac{1}{7}\begin{bmatrix} 3\cdot 5^{100} + 4\cdot 2^{100} & 4\cdot 5^{100} - 4\cdot 2^{100} \\ 3\cdot 5^{100} - 3\cdot 2^{100} & 4\cdot 5^{100} + 3\cdot 2^{100} \end{bmatrix}. \end{aligned}$$

□

Problem 6.9 For the matrix $A = \begin{bmatrix} 5 & -4 & 4 \\ 12 & -11 & 12 \\ 4 & -4 & 5 \end{bmatrix}$,

(1) diagonalize the matrix A,
(2) find the eigenvalues of $A^{10} + A^7 + 5A$.

6.3 Application: Difference equations

The discrete analogs of differential equations are called *difference equations.* They represent a mathematical model of dynamic processes that change over time and are widely used in such areas as economics, electrical engineering, and ecology.

Let us begin with a classical example. Early in the thirteenth century, Fibonacci posed the following problem: "Suppose that a newly born pair of rabbits produces no offspring during the first month of their lives, but each pair gives birth to a new pair once a month from the second month onward. Starting with one $(= x_1)$ newly born pair in the first month, how many pairs of rabbits can be bred in a given time, assuming no rabbit dies?"

Initially, there is one pair. After one month there is still one pair, but a month later it gives a birth, so there are two pairs. If at the end of n months there are x_n pairs, then after $n+1$ months the number will be the

x_n pairs plus the number of offspring of the x_{n-1} pairs who were alive at $n-1$ months. Therefore, we have for $n \geq 2$,

$$x_{n+1} = x_n + x_{n-1}.$$

It is convenient to set $x_0 = 0$. Hence the first several terms of the sequence become

$$0,\ 1,\ 1,\ 2,\ 3,\ 5,\ 8,\ 13,\ \cdots,$$

which is called the **Fibonacci sequence**; each term is called a **Fibonacci number**.

In general, the **linear difference equation** (or **recurrence relation**) **of order** k is written as

$$x_n = a_1 x_{n-1} + a_2 x_{n-2} + \cdots + a_k x_{n-k},$$

where a_i, $i = 1, \ldots, k$ are constants with a_1 and a_k nonzero.

The equation for the Fibonacci sequence above is an example of the linear difference equation. A solution to the difference equation is any sequence $\{x_n : n \geq 0\}$ of numbers that satisfies the equation. Of course, a solution can be found by simply writing out enough terms of the sequence, but that can be an awful task. Linear algebra gives us an easier approach to this problem.

Instead of discussing the general method of finding a solution to a difference equation, we restrict our discussion to the case of the Fibonacci sequence.

Example 6.7 Find the 1997th Fibonacci number.

Solution: A standard trick is to consider a trivial extra equation $x_n = x_n$ together with the given equation:

$$\begin{cases} x_{n+1} &= x_n + x_{n-1} \\ x_n &= x_n. \end{cases}$$

Equivalently in matrix notation,

$$\begin{bmatrix} x_{n+1} \\ x_n \end{bmatrix} = \begin{bmatrix} 1 & 1 \\ 1 & 0 \end{bmatrix} \begin{bmatrix} x_n \\ x_{n-1} \end{bmatrix},$$

which is of the form

$$\mathbf{x}_n = A\mathbf{x}_{n-1} = A^n \mathbf{x}_0, \quad n = 1,\ 2,\ \ldots,$$

where $\mathbf{x}_n = \begin{bmatrix} x_{n+1} \\ x_n \end{bmatrix}$ and $A = \begin{bmatrix} 1 & 1 \\ 1 & 0 \end{bmatrix}$. Thus, the problem is reduced to computing A^n. However, by a simple computation, we obtain the eigenvalues $\lambda_1 = \frac{1}{2}(1+\sqrt{5})$, $\lambda_2 = \frac{1}{2}(1-\sqrt{5})$ of A and their associated eigenvectors $\mathbf{v}_1 = (\lambda_1,\ 1)$, $\mathbf{v}_2 = (\lambda_2,\ 1)$, respectively. Moreover, the transition matrix and its inverse are found to be

$$Q = [\mathbf{v}_1\ \mathbf{v}_2] = \begin{bmatrix} \frac{1+\sqrt{5}}{2} & \frac{1-\sqrt{5}}{2} \\ 1 & 1 \end{bmatrix}, \quad Q^{-1} = \frac{1}{\sqrt{5}} \begin{bmatrix} 1 & -\frac{1-\sqrt{5}}{2} \\ -1 & \frac{1+\sqrt{5}}{2} \end{bmatrix}.$$

With $D = \begin{bmatrix} \frac{1+\sqrt{5}}{2} & 0 \\ 0 & \frac{1-\sqrt{5}}{2} \end{bmatrix}$,

$$A^n = QD^nQ^{-1} = Q \begin{bmatrix} \left(\frac{1+\sqrt{5}}{2}\right)^n & 0 \\ 0 & \left(\frac{1-\sqrt{5}}{2}\right)^n \end{bmatrix} Q^{-1}.$$

For instance, if $n = 1997$, then

$$\begin{aligned} \begin{bmatrix} x_{1998} \\ x_{1997} \end{bmatrix} = \mathbf{x}_{1997} &= A^{1997} \begin{bmatrix} 1 \\ 0 \end{bmatrix} = QD^{1997}Q^{-1}\mathbf{x}_0 \\ &= \frac{1}{\sqrt{5}} \begin{bmatrix} \lambda_1^{1998} - \lambda_2^{1998} \\ \lambda_1^{1997} - \lambda_2^{1997} \end{bmatrix}. \end{aligned}$$

Therefore, we get

$$x_{1997} = \frac{1}{\sqrt{5}} \left(\left(\frac{1+\sqrt{5}}{2}\right)^{1997} - \left(\frac{1-\sqrt{5}}{2}\right)^{1997} \right).$$

Note that since x_{1997} must be an integer, we look for the nearest integer to this huge number. Since $\left(\frac{1-\sqrt{5}}{2}\right)^k < \frac{1}{2}$, actually very small, for large k, it must be the integer nearest to $\frac{1}{\sqrt{5}}\left(\frac{1+\sqrt{5}}{2}\right)^{1997}$. Historically, the number $\frac{1+\sqrt{5}}{2}$, which is very close to the ratio $\frac{x_{1998}}{x_{1997}}$, is called the **golden mean**. □

A more general form of the linear difference equation can be written in a matrix equation as

$$\mathbf{x}_k = A\mathbf{x}_{k-1} = A^k\mathbf{x}_0, \ k = 1, 2, \ldots,$$

where A is an $n \times n$ matrix and $\mathbf{x}_0 \in \mathbb{R}^n$, which is also called a **linear difference equation**. In fact, this equation defines a sequence of vectors $\{\mathbf{x}_k\}_{k=0}^{\infty}$, and a solution of this equation is reduced to a simple computation of A^k.

When A is a diagonalizable matrix, it is easy to compute the solution. Let Q be an invertible matrix so that

$$Q^{-1}AQ = \begin{bmatrix} \lambda_1 & & 0 \\ & \ddots & \\ 0 & & \lambda_n \end{bmatrix} = D.$$

Then $\mathbf{x}_k = A^k\mathbf{x}_0 = QD^kQ^{-1}\mathbf{x}_0$ for $k = 1, 2, \ldots$. Note that the columns $\mathbf{v}_i$ of Q are the eigenvectors of A, so by setting $Q^{-1}\mathbf{x}_0 = \mathbf{a} = [a_1 \; a_2 \; \cdots \; a_n]^T$, we get

$$\begin{aligned} \mathbf{x}_k = QD^kQ^{-1}\mathbf{x}_0 &= \begin{bmatrix} | & | & & | \\ \mathbf{v}_1 & \mathbf{v}_2 & \cdots & \mathbf{v}_n \\ | & | & & | \end{bmatrix} \begin{bmatrix} \lambda_1^k & & & 0 \\ & \lambda_2^k & & \\ & & \ddots & \\ 0 & & & \lambda_n^k \end{bmatrix} \begin{bmatrix} a_1 \\ a_2 \\ \vdots \\ a_n \end{bmatrix} \\ &= a_1\lambda_1^k\mathbf{v}_1 + a_2\lambda_2^k\mathbf{v}_2 + \cdots + a_n\lambda_n^k\mathbf{v}_n. \end{aligned}$$

Hence, if $|\lambda_i| < 1$ for all i, then the vector $\mathbf{x}_k$ must approach the zero vector as k increases. On the other hand, if there exists an eigenvalue λ_i with $|\lambda_i| > 1$, this vector $\mathbf{x}_k$ may grow exponentially in magnitude.

Therefore, we have three possible cases for the process given by $\mathbf{x}_k = A\mathbf{x}_{k-1}$, $k = 1, 2, \cdots$. The process is said to be

(1) unstable if A has an eigenvalue λ with $|\lambda| > 1$,

(2) stable if $|\lambda| < 1$ for all eigenvalues of A,

(3) neutrally stable if the maximum value of the eigenvalues of A is 1.

Problem 6.10 Let $\{a_n\}$ be a sequence with $a_0 = 1$, $a_1 = 2$, $a_2 = 0$, and the recurrence relation $a_n = 2a_{n-1} + a_{n-2} - 2a_{n-3}$ for $n \geq 3$. Find the n-th term a_n.

The following example shows a special type of difference equation called a **Markov process.**

Example 6.8 (Markov process) We start with x_0 people outside a big city and y_0 people inside the city. Suppose that each year 20% of the people outside the city move in, and 10% of the people inside move out. Then we are concerned about the "eventual" distribution of the population.

At the end of the first year, the distribution of the population will be

$$\begin{cases} x_1 &= 0.8x_0 + 0.1y_0 \\ y_1 &= 0.2x_0 + 0.9y_0. \end{cases}$$

Or, in matrix form,

$$\mathbf{x}_1 = \begin{bmatrix} x_1 \\ y_1 \end{bmatrix} = \begin{bmatrix} 0.8 & 0.1 \\ 0.2 & 0.9 \end{bmatrix} \begin{bmatrix} x_0 \\ y_0 \end{bmatrix} = A\mathbf{x}_0.$$

Thus if $\mathbf{x}_n = (x_n, y_n)$ denotes the distribution of the population after n years in the country in where the city is, we get $\mathbf{x}_n = A^n\mathbf{x}_0$.

In this formulation, the problem can be summarized as follows:

(1) The total number of people stays fixed.

(2) The numbers x_n, y_n can never become negative.

A process having these two properties is called a **Markov process.** In general, a Markov process is a repeated application of a matrix A to an initial state $\mathbf{x}_0$, where the matrix A satisfies the following two conditions:

(1′) entries of each column of A add up to 1;

(2′) entries of A are all nonnegative so that the powers A^n are all nonnegative.

Such a matrix A is called a **stochastic matrix.**

Now, to solve the problem, we first find the eigenvalues and eigenvectors of A. They are $\lambda_1 = 1$, $\lambda_2 = 0.7$ and $\mathbf{v}_1 = (1, 2)$, $\mathbf{v}_2 = (-1, 1)$, respectively, so that

$$Q = \begin{bmatrix} 1 & -1 \\ 2 & 1 \end{bmatrix}, \quad \text{and} \quad Q^{-1} = \frac{1}{3}\begin{bmatrix} 1 & 1 \\ -2 & 1 \end{bmatrix}.$$

Hence, $Q^{-1}AQ = \begin{bmatrix} 1 & 0 \\ 0 & 0.7 \end{bmatrix} = D$ and

$$\begin{aligned} \mathbf{x}_n &= A^n\mathbf{x}_0 = QD^nQ^{-1}\mathbf{x}_0 \\ &= (x_0 + y_0)\begin{bmatrix} 1/3 \\ 2/3 \end{bmatrix} + (-2x_0 + y_0)(0.7)^n \begin{bmatrix} -1/3 \\ 1/3 \end{bmatrix} \\ &\to (x_0 + y_0)\begin{bmatrix} 1/3 \\ 2/3 \end{bmatrix}, \quad \text{as } n \to \infty. \end{aligned}$$

□

Let A be a stochastic matrix for a Markov process. Then the entries of each column of A add up to 1. This means that the sum of each column of $A - I$ is 0, or equivalently $\mathbf{r}_1 + \mathbf{r}_2 + \cdots + \mathbf{r}_n = \mathbf{0}$ for the row vectors $\mathbf{r}_i$ of $A - I$. This is a nontrivial linear combination of the row vectors, and these row vectors are linearly dependent, so that $\det(A - I) = 0$. Consequently, $\lambda = 1$ is an eigenvalue of A. If $\mathbf{x}$ is an eigenvector of A belonging to $\lambda = 1$, then $A\mathbf{x} = \mathbf{x}$. Hence, it is called the **steady state**.

Problem 6.11 Suppose that land use in a city in 1990 is

$$\begin{array}{ll} \text{Residential} & x_0 = 30\% \\ \text{Commercial} & y_0 = 20\% \\ \text{Industrial} & z_0 = 50\%. \end{array}$$

Denote by x_k, y_k, z_k the percentage of residential, commercial, and industrial, respectively, after k years, and assume that the stochastic matrix is given as follows:

$$\begin{bmatrix} x_{k+1} \\ y_{k+1} \\ z_{k+1} \end{bmatrix} = \begin{bmatrix} 0.8 & 0.1 & 0 \\ 0.1 & 0.7 & 0.1 \\ 0.1 & 0.2 & 0.9 \end{bmatrix} \begin{bmatrix} x_k \\ y_k \\ z_k \end{bmatrix}.$$

Find the land use in the city after 50 years. This problem has two essential properties of Markov process: The total area of the city stays fixed, and each portion of area can never become negative.

Problem 6.12 A car rental company has three branch offices in different cities. When a car is rented at one of the offices, it may be returned to any of three offices. This company started business with 900 cars, and initially an equal number of cars was distributed to each office. When the week-by-week distribution of cars is governed by a stochastic matrix

$$A = \begin{bmatrix} 0.6 & 0.1 & 0.2 \\ 0.2 & 0.2 & 0.2 \\ 0.2 & 0.7 & 0.6 \end{bmatrix},$$

determine the number of cars at each office in the k-th week. Also, find $\lim_{k\to\infty} A^k$.

6.4 Application: Differential equations I

The diagonalization of matrices may be used to solve systems of linear differential equations. For those who are not familiar with differential equations, we begin this section with some basic preliminaries about them.

Let $y = f(t)$ be a real-valued differentiable function on an interval $I = [a, b]$ containing 0. From elementary calculus, it is easy to see that the differential equation $y'(t) = \frac{df(t)}{dt} = 5y(t)$ has the **general solution** $y = ce^{5t}$, where c is an arbitrary constant. If we are given an additional condition $y(0) = y_0 = 3$ for $0 \in I$, called an **initial condition**, then the solution of $y' = 5y$ is $y = 3e^{5t}$, called a **particular solution**.

This computation can be extended to a system of n linear differential equations (called a homogeneous linear system of order n) with constant coefficients, which is by definition of the form

$$\begin{cases} y_1' &= a_{11}y_1 + a_{12}y_2 + \cdots + a_{1n}y_n \\ y_2' &= a_{21}y_1 + a_{22}y_2 + \cdots + a_{2n}y_n \\ &\vdots \\ y_n' &= a_{n1}y_1 + a_{n2}y_2 + \cdots + a_{nn}y_n, \end{cases}$$

where $y_i = f_i(t)$, for $i = 1, 2, \ldots, n$, are real-valued differentiable functions on an interval $I = [a, b]$ and $y_i' = \frac{df_i(t)}{dt}$ is its derivative. In most cases, we may assume that the interval I contains 0, and some initial conditions are given as $f_i(0) = d_i$ at $0 \in I$.

Let $\mathbf{y}(t)$ be the vector whose entries are the functions $f_i(t)$'s. Then its derivative is defined by

$$\mathbf{y}'(t) = \begin{bmatrix} y_1'(t) \\ y_2'(t) \\ \vdots \\ y_n'(t) \end{bmatrix}.$$

If A is the matrix of the coefficient in the system, then the matrix form of the system is written as

$$\mathbf{y}'(t) = A\mathbf{y}(t),$$

with an initial condition $\mathbf{y}_0 = \mathbf{y}(0) = (d_1, \ldots, d_n) \in \mathbb{R}^n$. It is well-known that this system has a unique solution $\mathbf{y}(t)$ depending on an initial condition $\mathbf{y}_0$ by the fundamental theorem of ordinary differential equations.

In a more general type of system of linear differential equations, the entries of the coefficient matrix could be functions. However, for our purpose we restrict our attention to systems with constant coefficients.

Example 6.9 Consider the following three systems:

$$\begin{cases} y_1' &= 2y_1 - 3y_2 \\ y_2' &= 2y_1 + y_2 \end{cases}, \quad \begin{cases} y_1' &= ty_1 + t^2y_2 \\ y_2' &= 2y_1 + t^3y_2 \end{cases}, \quad \begin{cases} y_1' &= 2y_1 - 3y_2^2 \\ y_2' &= y_1^2 + 5y_2 \end{cases}.$$

The first two systems are linear, but the coefficients of the second are functions. The third is not linear. □

Some basic facts on a system $\mathbf{y}'(t) = A\mathbf{y}(t)$ of linear differential equations are listed below.

(1) Write k solutions $\{\mathbf{y}_1(t), \cdots, \mathbf{y}_k(t)\}$ of the system as

$$\mathbf{y}_1(t) = \begin{bmatrix} y_{11}(t) \\ y_{21}(t) \\ \vdots \\ y_{n1}(t) \end{bmatrix}, \cdots, \mathbf{y}_k(t) = \begin{bmatrix} y_{1k}(t) \\ y_{2k}(t) \\ \vdots \\ y_{nk}(t) \end{bmatrix}.$$

Then each solution $\mathbf{y}_i(t)$, for $i = 1, \ldots, k$, draws a curve in $\mathbb{R}^n$ passing through the initial vector $\mathbf{y}_{i0} = \mathbf{y}_i(0) = (d_{i1}, \ldots, d_{in})$ as t varies in the interval I, and it is uniquely determined by the given initial condition $\mathbf{y}_{i0}$. By definition, this set of k solutions is said to be linearly independent on the interval I if, at each $t \in I$, $\{\mathbf{y}_1(t), \cdots, \mathbf{y}_k(t)\}$ is linearly independent in $\mathbb{R}^n$.

Example 6.10 For the system

$$\mathbf{y}'(t) = \begin{bmatrix} y_1'(t) \\ y_2'(t) \end{bmatrix} = \begin{bmatrix} 0 & 1 \\ 1 & 0 \end{bmatrix} \begin{bmatrix} y_1(t) \\ y_2(t) \end{bmatrix} = \begin{bmatrix} y_2(t) \\ y_1(t) \end{bmatrix} = A\mathbf{y}(t),$$

one can easily verify that the vector functions

$$\mathbf{y}_1(t) = \begin{bmatrix} y_{11}(t) \\ y_{21}(t) \end{bmatrix} = \begin{bmatrix} e^t \\ e^t \end{bmatrix}, \text{ and } \mathbf{y}_2(t) = \begin{bmatrix} y_{12}(t) \\ y_{22}(t) \end{bmatrix} = \begin{bmatrix} e^t \\ -e^t \end{bmatrix}$$

are solutions of the system. Suppose $c_1\mathbf{y}_1(t) + c_2\mathbf{y}_2(t) = \mathbf{0}$ for all $t \in I$. Then at $t = 0$, in particular, it becomes $c_1 \begin{bmatrix} 1 \\ 1 \end{bmatrix} + c_2 \begin{bmatrix} 1 \\ -1 \end{bmatrix} = \mathbf{0}$. Then clearly we have $c_1 = c_2 = 0$, that is, they are linearly independent. □

(2) If $\mathbf{y}_1, \cdots, \mathbf{y}_k$ are solutions of the system, then

$$\begin{aligned} (c_1\mathbf{y}_1 + \cdots + c_k\mathbf{y}_k)' &= c_1\mathbf{y}_1' + \cdots + c_k\mathbf{y}_k' \\ &= c_1A\mathbf{y}_1 + \cdots + c_kA\mathbf{y}_k \\ &= A(c_1\mathbf{y}_1 + \cdots + c_k\mathbf{y}_k) \end{aligned}$$

implies that their linear combination $c_1\mathbf{y}_1 + \cdots + c_k\mathbf{y}_k$ is also a solution for any constants c_i's. Moreover, if $\mathbf{y}_1(t), \cdots, \mathbf{y}_n(t)$ are n linearly independent solutions of the system, then their initial vectors $\mathbf{y}_{i0}$'s are linearly independent, so they form a basis for $\mathbb{R}^n$. Any initial condition $\mathbf{y}_0$ as a vector in $\mathbb{R}^n$

is a linear combination of $\mathbf{y}_{i0}$'s, $i = 1, \ldots, n$: $\mathbf{y}_0 = c_1\mathbf{y}_{10} + \cdots + c_n\mathbf{y}_{n0}$. Then one can easily see that the vector function

$$\mathbf{y}(t) = c_1\mathbf{y}_1(t) + \cdots + c_n\mathbf{y}_n(t)$$

is the solution of the system satisfying the initial condition $\mathbf{y}_0$. In fact, any solution can be obtained in this form, and this form of the solution is called the **general solution** of the system. A set of linearly independent n solutions is called a **fundamental set** of solutions, and they may be determined by a set of n linearly independent initial vectors $\mathbf{y}_{i0}$'s.

Given a set of n solutions $\mathbf{y}_1(t), \cdots, \mathbf{y}_n(t)$ of the system $\mathbf{y}'(t) = A\mathbf{y}(t)$, the linear independence of the set can be determined as follows: Let

$$Y(t) = [\mathbf{y}_1(t) \cdots \mathbf{y}_n(t)] = \begin{bmatrix} y_{11}(t) & y_{12}(t) & \cdots & y_{1n}(t) \\ y_{21}(t) & y_{22}(t) & \cdots & y_{2n}(t) \\ \vdots & \vdots & & \vdots \\ y_{n1}(t) & y_{n2}(t) & \cdots & y_{nn}(t) \end{bmatrix}.$$

Clearly, the n solutions are linearly independent on I if and only if $\det Y(t) \neq 0$ for all $t \in I$. However, the following lemma says that $\det Y(t) \neq 0$ for all $t \in I$ if and only if $\det Y(t) \neq 0$ at one point $t \in I$. The determinant of $Y(t)$ is called the **Wronskian** of the solutions, denoted by $W(t) = \det Y(t)$ for $t \in I$.

Lemma 6.9 $W'(t) = \operatorname{tr}(A)W(t)$.

Proof:

$$\begin{aligned}
W'(t) &= (\det Y(t))' = \sum_{\sigma \in S_n} \operatorname{sgn}(\sigma)(y_{1\sigma(1)} \cdots y_{n\sigma(n)})' \\
&= \sum_{\sigma} \operatorname{sgn}(\sigma) y'_{1\sigma(1)} \cdots y_{n\sigma(n)} + \cdots + \sum_{\sigma} \operatorname{sgn}(\sigma) y_{1\sigma(1)} \cdots y'_{n\sigma(n)} \\
&= (y'_{11}Y_{11} + \cdots + y'_{1n}Y_{1n}) + \cdots + (y'_{n1}Y_{n1} + \cdots + y'_{nn}Y_{nn}) \\
&= \sum_i^n \sum_j^n y'_{ij}Y_{ij} = \sum_i^n \left(\sum_j^n y'_{ij}[\operatorname{adj} Y]_{ji} \right) = \sum_i^n [Y' \cdot \operatorname{adj} Y]_{ii} \\
&= \operatorname{tr}(Y' \cdot \operatorname{adj} Y) = \operatorname{tr}(A \cdot (Y \cdot \operatorname{adj} Y)) \\
&= \operatorname{tr}(\det Y(t) A) = \operatorname{tr}(A)W(t),
\end{aligned}$$

where $Y_{ij}(t)$ is the cofactor of y_{ij}, and the equalities in the last two lines are due to the fact that

$$\begin{aligned} Y'(t) &= [\mathbf{y}_1'(t) \cdots \mathbf{y}_n'(t)] = A[\mathbf{y}_1(t) \cdots \mathbf{y}_n(t)] = AY(t), \\ Y(t)\,\mathrm{adj}Y(t) &= \det Y(t) I_n = W(t) I_n. \end{aligned}$$

□

This lemma shows that the Wronskian $W(t)$ satisfies a differential equation of the form $y' = ay$ discussed at the beginning of this section. Thus it must be an exponential function of the form $W(t) = W_0 e^{(\mathrm{tr}A)t}$ with an initial condition $W(0) = W_0$. This means that the value of $W(t)$ is zero for all t or never zero on I depending on whether or not $W_0 = 0$. Thus, the n solutions are linearly independent for all $t \in I$ if and only if the n solutions are linearly independent at some $t \in I$, *i.e.*, the linear dependence of the n solutions may be checked at any convenient point. This again justifies that it is good enough to begin with a set of n linearly independent initial vectors to find the general solution of $\mathbf{y}'(t) = A\mathbf{y}(t)$. That is, if the n initial vectors (conditions) $\mathbf{y}_{10}(0) = \mathbf{y}_{10}, \cdots, \mathbf{y}_{n0}(0) = \mathbf{y}_{n0}$ are linearly independent, then the solutions determined by them form a fundamental set.

6.5 Application: Differential equations II

We now get to the problem of solving a system of homogeneous linear differential equations $\mathbf{y}'(t) = A\mathbf{y}(t)$ with the given initial condition $\mathbf{y}_0 = (d_1, \ldots, d_n)$. This problem may be considered in three steps: (1) A is a diagonal matrix D, (2) A is diagonalizable, and (3) A is any square matrix.

(1) First suppose that A is a diagonal matrix D. Then

$$\begin{bmatrix} y_1'(t) \\ \vdots \\ y_n'(t) \end{bmatrix} = \begin{bmatrix} \lambda_1 & & 0 \\ & \ddots & \\ 0 & & \lambda_n \end{bmatrix} \begin{bmatrix} y_1(t) \\ \vdots \\ y_n(t) \end{bmatrix} = \begin{bmatrix} \lambda_1 y_1(t) \\ \vdots \\ \lambda_n y_n(t) \end{bmatrix}$$

means that we are given n simple first order linear equations:

$$y_i'(t) = \lambda_i y_i(t), \quad i = 1, 2, \ldots, n,$$

whose solutions are known as $y_i(t) = d_i e^{\lambda_i t}$ with $y_i(0) = d_i$, for $i = 1, \ldots, n$. In matrix notation,

$$\mathbf{y}(t) = \begin{bmatrix} y_1(t) \\ \vdots \\ y_n(t) \end{bmatrix} = \begin{bmatrix} e^{\lambda_1 t} & & 0 \\ & \ddots & \\ 0 & & e^{\lambda_n t} \end{bmatrix} \begin{bmatrix} d_1 \\ \vdots \\ d_n \end{bmatrix} = e^{tD}\mathbf{y}_0,$$

where e^{tD} is by definition,

$$e^{tD} = \begin{bmatrix} e^{\lambda_1 t} & & 0 \\ & \ddots & \\ 0 & & e^{\lambda_n t} \end{bmatrix}.$$

Remark: Actually, the above solution is the general solution of the system. Indeed, if we take n initial conditions to be the standard basis $\{\mathbf{e}_1, \ldots, \mathbf{e}_n\}$ for $\mathbb{R}^n$, then for each initial vector $\mathbf{e}_i = (0, \ldots, 1, \ldots, 0)$ for $i = 1, \ldots, n$, we get the solution $\mathbf{y}_i(t) = e^{\lambda_i t}\mathbf{e}_i$. Since the initial conditions $\mathbf{y}_i(0) = \mathbf{e}_i$ are linearly independent eigenvectors of the diagonal matrix D, the set $\{\mathbf{y}_i(t) = e^{\lambda_i t}\mathbf{e}_i \; : \; i = 1,\; 2,\; \ldots,\; n\}$ is a fundamental set. Any initial condition can be written as

$$\mathbf{y}_0 = d_1\mathbf{e}_1 + \cdots + d_n\mathbf{e}_n,$$

so the general solution is of the form

$$\begin{aligned} \mathbf{y}(t) &= d_1\mathbf{y}_1(t) + \cdots + d_n\mathbf{y}_n(t) = d_1 e^{\lambda_1 t}\mathbf{e}_1 + \cdots + d_n e^{\lambda_n t}\mathbf{e}_n \\ &= \begin{bmatrix} e^{\lambda_1 t} & & 0 \\ & \ddots & \\ 0 & & e^{\lambda_n t} \end{bmatrix} \begin{bmatrix} d_1 \\ \vdots \\ d_n \end{bmatrix} = e^{tD}\mathbf{y}_0. \end{aligned}$$

Example 6.11 One of the fundamental problems of mathematical ecology is the *predator-prey* problem. Let $x(t)$ and $y(t)$ denote the populations at time t of two species in a specified region, one of which x preys upon the other y. For example, $x(t)$ and $y(t)$ may be the number of sharks and small fish, respectively, in a restricted region of the ocean. Without the fish (preys) the population of the sharks (predators) will decrease, and without the sharks, the population of the fish will increase. A mathematical model showing their interactions and whether an ecological balance exists can be written as the following system of differential equations:

$$\begin{cases} x'(t) &= \;\; ax(t) - bx(t)y(t), \\ y'(t) &= -cy(t) + dx(t)y(t). \end{cases}$$

In this equation, the coefficients a and c are the birth rate of x and the death rate of y, respectively. The nonlinear $x(t)y(t)$ terms in the two equations mean the interaction of the two species, so the coefficients b and d are the measures of the effect of the interaction between them. A study of this general system of differential equations leads to a very interesting development in the theory of dynamical systems and can be found in any book on ordinary differential equations. Here, we restrict our study to the case of x and y very small, *i.e.*, near the origin in the plane. In this case, one can neglect the nonlinear terms in the equations, so the system is assumed to be given as follows:

$$\begin{bmatrix} x'(t) \\ y'(t) \end{bmatrix} = \begin{bmatrix} a & 0 \\ 0 & -c \end{bmatrix} \begin{bmatrix} x(t) \\ y(t) \end{bmatrix}.$$

Thus the eigenvalues are $\lambda_1 = a$, $\lambda_2 = -c$, and their eigenvectors $\mathbf{e}_1$, $\mathbf{e}_2$, respectively. Thus, its general solution is

$$\begin{bmatrix} x(t) \\ y(t) \end{bmatrix} = \begin{bmatrix} d_1 e^{at} \\ d_2 e^{-ct} \end{bmatrix} = \begin{bmatrix} e^{at} & 0 \\ 0 & e^{-ct} \end{bmatrix} \begin{bmatrix} d_1 \\ d_2 \end{bmatrix} = d_1 e^{at}\mathbf{e}_1 + d_2 e^{-ct}\mathbf{e}_2. \quad \square$$

(2) We next assume that a matrix A in the system $\mathbf{y}'(t) = A\mathbf{y}(t)$ is diagonalizable, that is, it has n linearly independent eigenvectors $\mathbf{v}_1, \ldots, \mathbf{v}_n$ belonging to the eigenvalues $\lambda_1, \ldots, \lambda_n$, respectively. Then the transition matrix $Q = [\mathbf{v}_1 \ \cdots \ \mathbf{v}_n]$ diagonalizes A and

$$A = QDQ^{-1} = Q \begin{bmatrix} \lambda_1 & & 0 \\ & \ddots & \\ 0 & & \lambda_n \end{bmatrix} Q^{-1}.$$

Thus the system becomes $Q^{-1}\mathbf{y}' = DQ^{-1}\mathbf{y}$. If we take a change of variables by the new vector $\mathbf{x} = Q^{-1}\mathbf{y}$ (or $\mathbf{y} = Q\mathbf{x}$), then we obtain a new system

$$\mathbf{x}' = D\mathbf{x},$$

with an initial condition $\mathbf{x}_0 = Q^{-1}\mathbf{y}_0 = (c_1, \ldots, c_n)$. Since D is diagonal, its general solution is

$$\mathbf{x} = e^{tD}\mathbf{x}_0 = c_1 e^{\lambda_1 t}\mathbf{e}_1 + \cdots + c_n e^{\lambda_n t}\mathbf{e}_n.$$

The general solution of the original system $\mathbf{y}' = A\mathbf{y}$ is

$$\begin{aligned} \mathbf{y} = Q\mathbf{x} &= Qe^{tD}Q^{-1}\mathbf{y}_0 \\ &= [\mathbf{v}_1 \cdots \mathbf{v}_n] \begin{bmatrix} e^{\lambda_1 t} & & 0 \\ & \ddots & \\ 0 & & e^{\lambda_n t} \end{bmatrix} \begin{bmatrix} c_1 \\ \vdots \\ c_n \end{bmatrix} \\ &= c_1 e^{\lambda_1 t}\mathbf{v}_1 + c_2 e^{\lambda_2 t}\mathbf{v}_2 + \cdots + c_n e^{\lambda_n t}\mathbf{v}_n \\ &= c_1\mathbf{y}_1(t) + c_2\mathbf{y}_2(t) + \cdots + c_n\mathbf{y}_n(t). \end{aligned}$$

Remark: Note that each vector function $\mathbf{y}_i(t) = e^{\lambda_i t}\mathbf{v}_i$ is the solution of the system determined by the initial condition $\mathbf{y}_i(0) = \mathbf{v}_i$, for $i = 1, \ldots, n$, which are linearly independent eigenvectors of A. Hence, they form a fundamental set of solutions.

Thus, we have obtained the following theorem:

Theorem 6.10 *If an $n \times n$ matrix A has n linearly independent eigenvectors, then for any $\mathbf{y}_0 \in \mathbb{R}^n$, the system of linear differential equations*

$$\mathbf{y}'(t) = A\mathbf{y}(t) \textit{ with initial condition } \mathbf{y}(0) = \mathbf{y}_0$$

has a unique solution of the form $\mathbf{y} = Qe^{tD}Q^{-1}\mathbf{y}_0$, where $A = QDQ^{-1}$.

Remark: When A is diagonalizable, the procedure for solving a system $\mathbf{y}'(t) = A\mathbf{y}(t)$ with an initial condition $\mathbf{y}_0$ at $t = 0$ may be summarized as follows:

Step 1. Find the eigenvalues and n linearly independent eigenvectors of A.

Step 2. Construct the transition matrix Q with the eigenvectors so that $A = QDQ^{-1}$.

Step 3. Take a change of variables by $\mathbf{y} = Q\mathbf{x}$ to get a new system $\mathbf{x}' = D\mathbf{x}$, whose solution is $\mathbf{x} = e^{tD}\mathbf{x}_0$.

Step 4. Use the substitution $\mathbf{y} = Q\mathbf{x}$ to get the solution $\mathbf{y} = Q\mathbf{x} = Qe^{tD}Q^{-1}\mathbf{y}_0$.

Consequently, when A is diagonalizable, the general solution of $\mathbf{y}'(t) = A\mathbf{y}(t)$ may be obtained directly from a basis of eigenvectors and eigenvalues of A without looking for an individual fundamental set.

Example 6.12 Solve the system of linear differential equations

$$\left\{\begin{array}{rcrcrcr} y_1' & = & 5y_1 & - & 4y_2 & + & 4y_3 \\ y_2' & = & 12y_1 & - & 11y_2 & + & 12y_3 \\ y_3' & = & 4y_1 & - & 4y_2 & + & 5y_3, \end{array}\right.$$

and also find its particular solution satisfying the initial conditions $y_1(0) = 0$, $y_2(0) = 3$ and $y_3(0) = 2$.

Solution: In matrix form, the system may be written as $\mathbf{y}' = A\mathbf{y}$ with

$$A = \begin{bmatrix} 5 & -4 & 4 \\ 12 & -11 & 12 \\ 4 & -4 & 5 \end{bmatrix}.$$

(1) The eigenvalues of A are $\lambda_1 = \lambda_2 = 1$, and $\lambda_3 = -3$, and their eigenvectors are $\mathbf{v}_1 = (1, 1, 0)$, $\mathbf{v}_2 = (-1, 0, 1)$ and $\mathbf{v}_3 = (1, 3, 1)$, respectively, which are clearly linearly independent (see Problem 6.9).

(2) The matrix $Q = [\mathbf{v}_1\ \mathbf{v}_2\ \mathbf{v}_3] = \begin{bmatrix} 1 & -1 & 1 \\ 1 & 0 & 3 \\ 0 & 1 & 1 \end{bmatrix}$ diagonalizes A:

$$Q^{-1}AQ = D = \begin{bmatrix} 1 & 0 & 0 \\ 0 & 1 & 0 \\ 0 & 0 & -3 \end{bmatrix}.$$

(3) Then $\mathbf{y}' = A\mathbf{y}$ is transformed to $\mathbf{x}' = D\mathbf{x}$ by a change of variables $\mathbf{y} = Q\mathbf{x}$. The system $\mathbf{x}' = D\mathbf{x}$ consists of three equations: $x_1' = x_1$, $x_2' = x_2$, $x_3' = -3x_3$. These are easily solved to give $x_1 = ae^t$, $x_2 = be^t$, $x_3 = ce^{-3t}$, where a, b, and c are arbitrary constants. In matrix form, the solution is written as $\mathbf{x} = e^{Dt}\mathbf{x}_0$ with $\mathbf{x}_0 = (a, b, c)$, *i.e.*,

$$\begin{bmatrix} x_1 \\ x_2 \\ x_3 \end{bmatrix} = \begin{bmatrix} e^t & 0 & 0 \\ 0 & e^t & 0 \\ 0 & 0 & e^{-3t} \end{bmatrix} \begin{bmatrix} a \\ b \\ c \end{bmatrix} = \begin{bmatrix} ae^t \\ be^t \\ ce^{-3t} \end{bmatrix}.$$

(4) Since $\mathbf{y} = Q\mathbf{x}$, we get

$$\begin{aligned} \begin{bmatrix} y_1 \\ y_2 \\ y_3 \end{bmatrix} &= \begin{bmatrix} 1 & -1 & 1 \\ 1 & 0 & 3 \\ 0 & 1 & 1 \end{bmatrix} \begin{bmatrix} ae^t \\ be^t \\ ce^{-3t} \end{bmatrix} = \begin{bmatrix} ae^t - be^t + ce^{-3t} \\ ae^t + 3ce^{-3t} \\ be^t + ce^{-3t} \end{bmatrix} \\ &= ae^t \begin{bmatrix} 1 \\ 1 \\ 0 \end{bmatrix} + be^t \begin{bmatrix} -1 \\ 0 \\ 1 \end{bmatrix} + ce^{-3t} \begin{bmatrix} 1 \\ 3 \\ 1 \end{bmatrix} \\ &= ae^t\mathbf{v}_1 + be^t\mathbf{v}_2 + ce^{-3t}\mathbf{v}_3. \end{aligned}$$

For the initial conditions $\mathbf{y}_0 = [0\ 3\ 2]^T$,

$$\begin{bmatrix} a \\ b \\ c \end{bmatrix} = \mathbf{x}_0 = Q^{-1}\mathbf{y}_0 = \begin{bmatrix} 3 & -2 & 3 \\ 1 & -1 & 2 \\ -1 & 1 & -1 \end{bmatrix} \begin{bmatrix} 0 \\ 3 \\ 2 \end{bmatrix} = \begin{bmatrix} -1 \\ 0 \\ 1 \end{bmatrix}.$$

Thus the particular solution is

$$\begin{bmatrix} y_1 \\ y_2 \\ y_3 \end{bmatrix} = -e^t \begin{bmatrix} 1 \\ 1 \\ 0 \end{bmatrix} + e^{-3t} \begin{bmatrix} 1 \\ 3 \\ 1 \end{bmatrix} = \begin{bmatrix} -e^t + e^{-3t} \\ 3e^{-3t} \\ e^t + e^{-3t} \end{bmatrix}.$$

□

(3) A system $\mathbf{y}' = A\mathbf{y}$ of linear differential equations with a general matrix A will be discussed in Section 6.7.

Problem 6.13 Solve the system $\begin{cases} y_1' &= y_1 + y_2 \\ y_2' &= 4y_1 - 2y_2 . \end{cases}$

Problem 6.14 Solve the system $\begin{cases} y_1' &= 4y_1 + y_3 \\ y_2' &= -2y_1 + y_2 \\ y_3' &= -2y_1 + y_3 , \end{cases}$
and find the particular solution of the system satisfying the initial conditions $y_1(0) = -1,\ y_2(0) = 1,\ y_3(0) = 0$.

Problem 6.15 Solve the system $\begin{cases} y_1' &= y_1 - y_2 + 2y_3 \\ y_2' &= 3y_1 + 4y_3 \\ y_3' &= 2y_1 + y_2 \end{cases}$
with initial conditions $y_1(0) = 0,\ y_2(0) = 2,\ y_3(0) = 1$.

6.6 Exponential matrices

Consider again a system of linear differential equations $\mathbf{y}' = A\mathbf{y}$ with initial condition $\mathbf{y}(0) = \mathbf{y}_0$ for a square matrix A. If A is diagonalizable, say $Q^{-1}AQ = D$, the system is solvable and has a unique solution of the form $\mathbf{y} = Qe^{tD}Q^{-1}\mathbf{y}_0$, by Theorem 6.10. Now let A be any square matrix, not necessarily diagonalizable. It will be shown later that the solution of the system $\mathbf{y}' = A\mathbf{y}$ with $\mathbf{y}(0) = \mathbf{y}_0$ is still of the form

$$\mathbf{y} = e^{tA}\mathbf{y}_0,$$

provided that e^{tA} is made meaningful. In particular, if A is diagonalizable so that $A = QDQ^{-1}$, then this form should be

$$e^{tA} = e^{tQDQ^{-1}} = Qe^{tD}Q^{-1}.$$

Motivated from the Maclaurin series of the exponential function e^x, we define the exponential matrix of a matrix.

Definition 6.3 For any square matrix A, the **exponential matrix** of A is defined as the series

$$e^A = \sum_{k=0}^{\infty} \frac{A^k}{k!} = I + A + \frac{A^2}{2!} + \frac{A^3}{3!} + \cdots .$$

Example 6.13 If $A = D$ is a diagonal matrix, then for any integer k

$$D = \begin{bmatrix} \lambda_1 & & 0 \\ & \ddots & \\ 0 & & \lambda_n \end{bmatrix}, \quad \text{and} \quad D^k = \begin{bmatrix} \lambda_1^k & & 0 \\ & \ddots & \\ 0 & & \lambda_n^k \end{bmatrix}.$$

Thus, the exponential matrix e^D is

$$\begin{aligned} e^D &= I + D + \frac{1}{2!}D^2 + \frac{1}{3!}D^3 + \cdots \\ &= \begin{bmatrix} \sum_{k=0}^{\infty} \frac{\lambda_1^k}{k!} & & 0 \\ & \ddots & \\ 0 & & \sum_{k=0}^{\infty} \frac{\lambda_n^k}{k!} \end{bmatrix} = \begin{bmatrix} e^{\lambda_1} & & 0 \\ & \ddots & \\ 0 & & e^{\lambda_n} \end{bmatrix}, \end{aligned}$$

which coincides with the previous definition. □

Note that the computation of the powers A^k is not simple in general. However, if A is diagonalizable, then it is easy to see that $A^k = QD^kQ^{-1}$, which is relatively simple to compute.

In any case, it should be verified that e^A exists for any square matrix A, that is, each entry of the series form of e^A is convergent. For this we first discuss the convergence of sequences of matrices in general.

Definition 6.4 A sequence of matrices $A_1, A_2, A_3, \ldots$ of the same size is said to **converge** to a matrix L if each sequence of the (i, j)-entries of $A_1, A_2, A_3, \ldots$ converges to the (i, j)-entry of L for all i, j. In this case, we write

$$L = \lim_{k \to \infty} A_k.$$

Such a matrix L is called a **limit** of the sequence $A_1, A_2, A_3, \ldots$.

Theorem 6.11 *Let $A_1, A_2, A_3, \ldots$ be a sequence of $m \times n$ matrices such that $\lim_{k\to\infty} A_k = L$. Then*

$$\lim_{k \to \infty} BA_k = BL \quad \text{and} \quad \lim_{k \to \infty} A_kC = LC$$

for any matrices B and C for which the products are defined.

Proof: By comparing the (i, j)-entries of both sides

$$\begin{aligned}\lim_{k\to\infty}[BA_k]_{ij} &= \lim_{k\to\infty}\left(\sum_{\ell=1}^{m}[B]_{i\ell}[A_k]_{\ell j}\right) = \sum_{\ell=1}^{m}[B]_{i\ell}\lim_{k\to\infty}[A_k]_{\ell j} \\ &= \sum_{\ell=1}^{m}[B]_{i\ell}[L]_{\ell j} = [BL]_{ij},\end{aligned}$$

we get $\lim_{k\to\infty} BA_k = BL$. Similarly $\lim_{k\to\infty} A_kC = LC$. □

For example, if A is a diagonalizable matrix such that $A = QDQ^{-1}$ for some invertible matrix Q, then, for each integer $k \geq 0$, $A^k = QD^kQ^{-1}$ implies that

$$\lim_{k\to\infty} A^k = Q(\lim_{k\to\infty} D^k)Q^{-1} = Q\begin{bmatrix}\lim_{k\to\infty}\lambda_1^k & & 0 \\ & \ddots & \\ 0 & & \lim_{k\to\infty}\lambda_n^k\end{bmatrix}Q^{-1}.$$

Thus, $\lim_{k\to\infty} A^k$ exists if and only if $\lim_{k\to\infty}\lambda_i^k$ exists for $i = 1, \ldots, n$.

Problem 6.16 Let $A = \begin{bmatrix} 1 & 1 \\ 0 & 1 \end{bmatrix}$. Find $\lim_{k\to\infty} A^k$ if it exists. (Note that the matrix A is not diagonalizable.)

Definition 6.5 A series of matrices $A_0 + A_1 + A_2 + \cdots$ is said to **converge** to a matrix L if L is the limit of the sequence $\{S_m = \sum_{k=0}^{m} A_k \mid m = 0, 1, 2, \ldots\}$ of the partial sums, that is, $\lim_{m\to\infty} S_m = L$. In this case, we write

$$A_0 + A_1 + A_2 + \cdots = \sum_{k=0}^{\infty} A_k = L.$$

Example 6.14 The sequence of the partial sums of a geometric series

$$A^0 + A^1 + A^2 + \cdots$$

of a square matrix A is $\{S_m = \sum_{k=0}^{m} A^k\}$, and so if $A = QDQ^{-1}$ is diagonalizable, then

$$S_m = \sum_{k=0}^{m} A^k = \sum_{k=0}^{m} QD^kQ^{-1} = Q(\sum_{k=0}^{m} D^k)Q^{-1}$$

$$= Q \begin{bmatrix} \frac{1-\lambda_1^{m+1}}{1-\lambda_1} & & 0 \\ & \ddots & \\ 0 & & \frac{1-\lambda_n^{m+1}}{1-\lambda_n} \end{bmatrix} Q^{-1}.$$

Thus, the sequence converges if and only if $|\lambda_i| < 1$ for all i. □

In particular, the exponential matrix e^A is defined to be the limit of the sequence

$$\left\{ \sum_{k=0}^{m} \frac{A^k}{k!} \ : \ m = 0, 1, 2, \cdots \right\}.$$

The existence of e^A for any square matrix A can easily be shown as follows: Let M be a number such that $|a_{ij}| \le M$ for all (i, j)-entries a_{ij} of A (such a number exists since A has only finite number of entries). Then all (i, j)-entries of A^k are bounded by $n^{k-1}M^k$ for all k, and hence each entry of e^A is bounded by

$$\sum_{k=0}^{\infty} \frac{1}{k!} n^{k-1} M^k = \frac{1}{n} e^{nM},$$

so by the comparison test, each entry of $e^A = \sum_{n=0}^{\infty} \frac{A^k}{k!}$ is absolutely convergent for any square matrix A.

Again, if A is diagonalizable with $Q^{-1}AQ = D$, then by Theorem 6.11,

$$\begin{aligned} e^A = e^{QDQ^{-1}} &= I + QDQ^{-1} + \frac{(QDQ^{-1})^2}{2!} + \frac{(QDQ^{-1})^3}{3!} + \cdots \\ &= Q(I + D + \frac{D^2}{2!} + \frac{D^3}{3!} + \cdots)Q^{-1} \\ &= Qe^D Q^{-1}. \end{aligned}$$

In general, the computation of e^A is not easy at all if A is not diagonalizable. When A is a triangular 2×2 matrix, it is relatively easy.

Example 6.15 If $A = \begin{bmatrix} 1 & 1 \\ 0 & 3 \end{bmatrix}$, then

$$\begin{aligned} e^A &= I + A + \frac{1}{2}A^2 + \cdots \\ &= \begin{bmatrix} 1 & 0 \\ 0 & 1 \end{bmatrix} + \begin{bmatrix} 1 & 1 \\ 0 & 3 \end{bmatrix} + \frac{1}{2} \begin{bmatrix} 1 & 1+3 \\ 0 & 3^2 \end{bmatrix} + \cdots = \begin{bmatrix} e^1 & * \\ 0 & e^3 \end{bmatrix}. \end{aligned}$$

It is a good exercise to calculate the missing entry $*$ directly from the definition. □

The following properties of the exponential matrices are easy to prove, so are left for exercises.

Theorem 6.12 **(1)** *$e^{A+B} = e^A e^B$ provided that $AB = BA$.*

(2) *$e^{Q^{-1}AQ} = Q^{-1}e^A Q$ for any invertible matrix Q.*

(3) *If $\lambda_1, \lambda_2, \ldots, \lambda_n$ are the eigenvalues of a matrix A with their associated eigenvectors $\mathbf{v}_1, \mathbf{v}_2, \ldots, \mathbf{v}_n$, then e^{λ_i}'s are the eigenvalues of e^A with the same associated eigenvectors $\mathbf{v}_i$'s for $i = 1, 2, \ldots, n$. Moreover, $\det e^A = e^{\lambda_1} \cdots e^{\lambda_n} = e^{tr(A)} \neq 0$ for any square matrix A.*

(4) *In particular, the matrix e^A is never singular for any square matrix A, and $(e^A)^{-1} = e^{-A}$.*

Problem 6.17 Prove the above four properties of e^A.

Problem 6.18 Prove that if A is skew-symmetric, then e^A is orthogonal.

Example 6.16 For $A = \begin{bmatrix} 2 & 3 \\ 0 & 2 \end{bmatrix}$, one can compute e^A as a simple application of property (1). We first write it as $A = 2I + N$, where $N = \begin{bmatrix} 0 & 3 \\ 0 & 0 \end{bmatrix}$. Since $(2I)N = N(2I)$, by (1)

$$e^A = e^{2I} e^N.$$

From the direct computation of the series expansion, we get $e^{2I} = e^2 I$. Moreover, since $N^k = \mathbf{0}$ for $k \geq 2$, $e^N = I + N + \frac{N^2}{2!} + \cdots = I + N = \begin{bmatrix} 1 & 3 \\ 0 & 1 \end{bmatrix}$. Thus,

$$e^A = e^2 I(I + N) = e^2 \begin{bmatrix} 1 & 3 \\ 0 & 1 \end{bmatrix} = \begin{bmatrix} e^2 & 3e^2 \\ 0 & e^2 \end{bmatrix}.$$ □

Problem 6.19 Compute e^A for $A = \begin{bmatrix} 2 & 3 & 0 \\ 0 & 2 & 3 \\ 0 & 0 & 2 \end{bmatrix}$.

6.7 Application: Differential equations III

One of the most prominent applications of exponential matrices is to the theory of linear differential equations.

Lemma 6.13 **(1)** *Let $A(t)$ and $B(t)$ be matrices, whose entries are all differentiable functions in t, such that their product is defined. Then $A(t)B(t)$ is differentiable, and its derivative is*

$$\frac{d}{dt}(A(t)B(t)) = \frac{dA(t)}{dt}B(t) + A(t)\frac{dB(t)}{dt}.$$

(2) *For any $t \in \mathbb{R}$ and any square matrix A, the exponential matrix*

$$e^{tA} = I + tA + \frac{t^2}{2!}A^2 + \frac{t^3}{3!}A^3 + \cdots$$

is a differentiable function of t, and $\frac{d}{dt}e^{tA} = Ae^{tA}$.

Proof: (1) is a usual computation. (2) By definition,

$$\frac{d}{dt}e^{tA} = \lim_{h\to 0}\frac{e^{(t+h)A} - e^{tA}}{h} = \lim_{h\to 0}\left(\frac{e^{hA} - I}{h}\right)e^{tA},$$

since $(tA)(hA) = (hA)(tA)$. One can now easily show $\lim_{h\to 0}\frac{e^{hA}-I}{h} = A$. □

Lemma 6.13 implies that $\mathbf{y} = e^{tA}\mathbf{y}_0$ is the solution of the system $\mathbf{y}' = A\mathbf{y}$. In particular, if A is diagonalizable, say $Q^{-1}AQ = D$, then from (2) in Theorem 6.12 section we get

$$\mathbf{y} = e^{tA}\mathbf{y}_0 = e^{tQDQ^{-1}}\mathbf{y}_0 = Qe^{tD}Q^{-1}\mathbf{y}_0,$$

as in Theorem 6.10. The following theorem is a direct consequence of Lemma 6.13.

Theorem 6.14 *For any $n \times n$ matrix A, the linear differential equation $\mathbf{y}' = A\mathbf{y}$ with initial condition $\mathbf{y}(0) = \mathbf{y}_0$ has the solution*

$$\mathbf{y}(t) = e^{tA}\mathbf{y}_0.$$

If A is not diagonalizable, then it is not easy to compute the matrix e^{tA}. For this case, we introduce the Jordan canonical form of A in Chapter 9, with which the computation of e^{tA} is made relatively easy. The matrices in the following examples will be treated in general in Chapter 9.

Example 6.17 Solve the system $\mathbf{y}' = A\mathbf{y}$ of linear differential equations with initial condition $\mathbf{y}(0) = \mathbf{y}_0$, where

$$A = \begin{bmatrix} \lambda & 1 \\ 0 & \lambda \end{bmatrix}, \qquad \mathbf{y}_0 = \begin{bmatrix} a \\ b \end{bmatrix}.$$

Solution: First note that A has an eigenvalue λ of multiplicity 2 and is not diagonalizable. Now we write it as

$$A = \begin{bmatrix} \lambda & 0 \\ 0 & \lambda \end{bmatrix} + \begin{bmatrix} 0 & 1 \\ 0 & 0 \end{bmatrix} = \lambda I + N.$$

Then, by the same argument as in Example 6.16,

$$e^{tA} = e^{t(\lambda I + N)} = e^{\lambda t} e^{tN} = e^{\lambda t} \begin{bmatrix} 1 & t \\ 0 & 1 \end{bmatrix}.$$

Therefore, the solution is

$$\mathbf{y} = e^{tA}\mathbf{y}_0 = e^{\lambda t} \begin{bmatrix} 1 & t \\ 0 & 1 \end{bmatrix} \begin{bmatrix} a \\ b \end{bmatrix} = \begin{bmatrix} (a+bt)e^{\lambda t} \\ be^{\lambda t} \end{bmatrix}.$$

In terms of components, $y_1 = (a+bt)e^{\lambda t}$, $y_2 = be^{\lambda t}$. □

Example 6.18 Find the general solution of the system $\mathbf{y}' = A\mathbf{y}$, where

$$A = \begin{bmatrix} a & -b \\ b & a \end{bmatrix}.$$

Solution: Note that the eigenvalues of A are $a \pm ib$, which are not real unless $b = 0$, in which case the matrix is already in diagonal form. If $b \neq 0$, the diagonalization discussed in this section does not apply since we have complex eigenvalues which are going to be discussed in Chapter 9. However, there is another method of solving the system without using diagonalization. We first write A as

$$A = \begin{bmatrix} a & -b \\ b & a \end{bmatrix} = a \begin{bmatrix} 1 & 0 \\ 0 & 1 \end{bmatrix} + b \begin{bmatrix} 0 & -1 \\ 1 & 0 \end{bmatrix} = aI + bJ,$$

then clearly $IJ = JI$, so $e^{tA} = e^{atI+btJ} = e^{at}e^{btJ}$. Since

$$J^2 = \begin{bmatrix} -1 & 0 \\ 0 & -1 \end{bmatrix} = -I, \quad J^3 = \begin{bmatrix} 0 & 1 \\ -1 & 0 \end{bmatrix} = -J, \quad J^4 = \begin{bmatrix} 1 & 0 \\ 0 & 1 \end{bmatrix} = I,$$

one can deduce $J^k = J^{k+4}$ for all $k = 1, 2, \cdots$, and, moreover,

$$\begin{aligned} e^{btJ} &= I + \frac{btJ}{1!} + \frac{(bt)^2J^2}{2!} + \frac{(bt)^3J^3}{3!} + \frac{(bt)^4J^4}{4!} + \cdots \\ &= \begin{bmatrix} 1 - \frac{(bt)^2}{2!} + \frac{(bt)^4}{4!} - \cdots & -(bt) + \frac{(bt)^3}{3!} - \frac{(bt)^5}{5!} + \cdots \\ (bt) - \frac{(bt)^3}{3!} + \frac{(bt)^5}{5!} - \cdots & 1 - \frac{(bt)^2}{2!} + \frac{(bt)^4}{4!} - \cdots \end{bmatrix} \\ &= \begin{bmatrix} \cos bt & -\sin bt \\ \sin bt & \cos bt \end{bmatrix} \end{aligned}$$

for any constant b and t. Thus, the general solution of $\mathbf{y}' = A\mathbf{y}$ is

$$\mathbf{y} = e^{tA}\mathbf{c} = e^{at}e^{btJ}\mathbf{c} = e^{at} \begin{bmatrix} \cos bt & -\sin bt \\ \sin bt & \cos bt \end{bmatrix} \begin{bmatrix} c_1 \\ c_2 \end{bmatrix}.$$

In terms of components,

$$\begin{cases} y_1 = e^{at}(c_1 \cos bt - c_2 \sin bt) \\ y_2 = e^{at}(c_1 \sin bt + c_2 \cos bt). \end{cases}$$

□

Problem 6.20 Solve the system $\mathbf{y}' = A\mathbf{y}$ with initial condition $\mathbf{y}(0) = \mathbf{y}_0$ by computing $e^{tA}\mathbf{y}_0$ for

(1) $A = \begin{bmatrix} 1 & -2 \\ 0 & -1 \end{bmatrix}$, $\mathbf{y}_0 = \begin{bmatrix} 1 \\ 1 \end{bmatrix}$. (2) $A = \begin{bmatrix} 1 & 1 & 1 \\ 0 & 0 & 1 \\ 0 & 0 & -1 \end{bmatrix}$, $\mathbf{y}_0 = \begin{bmatrix} 1 \\ 1 \\ 1 \end{bmatrix}$.

6.8 Diagonalization of linear transformations

Recall that two matrices are similar if and only if they can be the matrix representations of the same linear transformation, and similar matrices have the same eigenvalues. In this section, we aim to find a basis α so that the matrix representation of a linear transformation with respect to α is a diagonal matrix. First, we start with the eigenvalues and the eigenvectors of a linear transformation.

Definition 6.6 Let V be an n-dimensional vector space, and let $T : V \to V$ be a linear transformation on V. Then the **eigenvalues** and **eigenvectors** of T can be defined by the same equation, $T\mathbf{x} = \lambda\mathbf{x}$, with a nonzero vector $\mathbf{x} \in V$.

Practically, the eigenvalues of T are computed as those of the matrix representation $[T]_\alpha$ of T with respect to a basis α for V. In fact, this is well defined, since $[T]_\alpha$ is similar to $[T]_\beta$ for any other basis β for V and their eigenvalues are the same by Theorem 6.2.

The eigenvectors of T can also be found from the eigenvectors of its matrix representation. Let $\alpha = \{\mathbf{v}_1, \mathbf{v}_2, \ldots, \mathbf{v}_n\}$ be a basis for V. Then the natural isomorphism $\Phi : V \to \mathbb{R}^n$ identifies the associated matrix $A = [T]_\alpha : \mathbb{R}^n \to \mathbb{R}^n$ with the linear transformation $T : V \to V$ via the following commutative diagram.

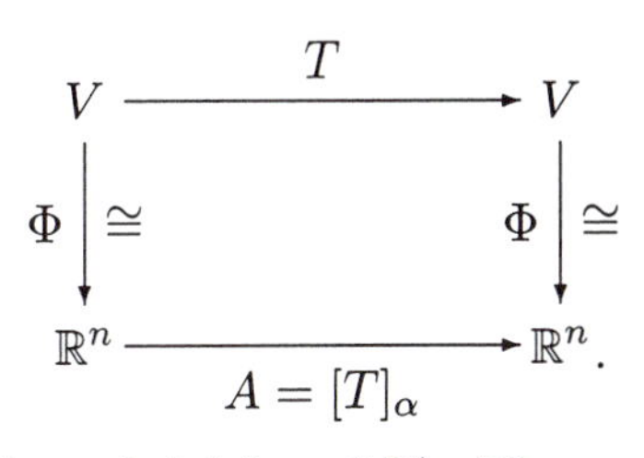

Let λ be an eigenvalue of A (also of T). Then, $\mathbf{x} = (x_1, x_2, \ldots, x_n) \in \mathbb{R}^n$ is an eigenvector of the matrix A belonging to λ ($A\mathbf{x} = \lambda\mathbf{x}$) if and only if $\Phi^{-1}(\mathbf{x}) = \mathbf{v} = x_1\mathbf{v}_1 + x_2\mathbf{v}_2 + \cdots + x_n\mathbf{v}_n \in V$ is an eigenvector of T ($T(\mathbf{v}) = \lambda\mathbf{v}$), because the commutativity of the diagram shows

$$[T(\mathbf{v})]_\alpha = [T]_\alpha[\mathbf{v}]_\alpha = A\mathbf{x} = \lambda\mathbf{x} = [\lambda\mathbf{v}]_\alpha.$$

Therefore, if $\mathbf{x}_1, \mathbf{x}_2, \ldots, \mathbf{x}_k$ are linearly independent eigenvectors of $A = [T]_\alpha$, then $\Phi^{-1}(\mathbf{x}_1), \Phi^{-1}(\mathbf{x}_2), \ldots, \Phi^{-1}(\mathbf{x}_k)$ are linearly independent eigenvectors of T. Hence, the linear transformation T has a diagonal matrix representation if and only if it has n linear independent eigenvectors, by Theorem 6.6.

The following example illustrates how to find a diagonal matrix representation of a linear transformation on a vector space.

Example 6.19 Let $T : P_2(\mathbb{R}) \to P_2(\mathbb{R})$ be the linear transformation defined by

$$(Tf)(x) = f(x) + xf'(x) + f'(x).$$

Find a basis for $P_2(\mathbb{R})$ with respect to which the matrix of T is diagonal.

Solution: First of all, we find the eigenvalues and the eigenvectors of T. Take a basis for the vector space $P_2(\mathbb{R})$, say $\alpha = \{1,\ x,\ x^2\}$. Then the matrix of T with respect to α is

$$[T]_\alpha = \begin{bmatrix} 1 & 1 & 0 \\ 0 & 2 & 2 \\ 0 & 0 & 3 \end{bmatrix},$$

which is upper triangular. Hence, the eigenvalues of T are $\lambda_1 = 1$, $\lambda_2 = 2$ and $\lambda_3 = 3$. By a simple computation, one can verify that the vectors $\mathbf{x}_1 = (1,\ 0,\ 0)$, $\mathbf{x}_2 = (1,\ 1,\ 0)$, and $\mathbf{x}_3 = (1,\ 2,\ 1)$ are eigenvectors of $[T]_\alpha$ in $\mathbb{R}^3$ belonging to eigenvalues $\lambda_1,\ \lambda_2,\ \lambda_3$, respectively. Their corresponding eigenvectors of T in $P_2(\mathbb{R})$ are $f_1(x) = 1$, $f_2(x) = 1 + x$, $f_3(x) = 1 + 2x + x^2$, respectively. Since the eigenvalues $\lambda_1,\ \lambda_2,\ \lambda_3$ are all distinct, the eigenvectors $\{\mathbf{x}_1,\ \mathbf{x}_2,\ \mathbf{x}_3\}$ of $[T]_\alpha$ are linearly independent and so are $\beta = \{f_1,\ f_2,\ f_3\}$ in $P_2(\mathbb{R})$. Thus, each f_i is a basis for the eigenspace $E(\lambda_i)$ of T belonging to λ_i for $i = 1,\ 2,\ 3$. Thus the transition matrix is

$$Q = [Id]^\alpha_\beta = [\mathbf{x}_1\ \mathbf{x}_2\ \mathbf{x}_3] = [\,[f_1]_\alpha\ [f_2]_\alpha\ [f_3]_\alpha\,] = \begin{bmatrix} 1 & 1 & 1 \\ 0 & 1 & 2 \\ 0 & 0 & 1 \end{bmatrix}.$$

Hence, by changing the basis α to β, the matrix representation of T is a diagonal matrix:

$$[T]_\beta = [Id]^\beta_\alpha [T]_\alpha [Id]^\alpha_\beta = Q^{-1}[T]_\alpha Q = \begin{bmatrix} 1 & 0 & 0 \\ 0 & 2 & 0 \\ 0 & 0 & 3 \end{bmatrix} = D.$$

basis $\mathbf{e}_i$'s □

Note that, if $T = A$ is an $n \times n$ square matrix written in column vectors, $A = [\mathbf{c}_1 \ \cdots \ \mathbf{c}_n]$, then the linear transformation $A : \mathbb{R}^n \to \mathbb{R}^n$ is given by $A(\mathbf{e}_i) = \mathbf{c}_i$, $i = 1, \ldots, n$, so that A itself is just the matrix representation with respect to the standard basis $\alpha = \{\mathbf{e}_1, \ldots, \mathbf{e}_n\}$ for $\mathbb{R}^n$, say $A = [A]_\alpha$. Now if there is a basis $\beta = \{\mathbf{x}_1, \ldots, \mathbf{x}_n\}$ of n linearly independent eigenvectors of A, then the natural isomorphism $\Phi : \mathbb{R}^n \to \mathbb{R}^n$ defined by $\Phi(\mathbf{x}_i) = \mathbf{e}_i$ is simply a change of basis by the transition matrix $Q = [Id]_\beta^\alpha$ and the matrix representation of A with respect to β is a diagonal matrix:

$$D = [A]_\beta = Q^{-1}[A]_\alpha Q = Q^{-1}AQ.$$

Problem 6.21 Let T be the linear transformation on $\mathbb{R}^3$ defined by

$$T(x, y, z) = (4x + z, \ 2x + 3y + 2z, \ x + 4z).$$

Find all the eigenvalues and their eigenvectors of T and diagonalize T.

Problem 6.22 Let $M_{2\times 2}(\mathbb{R})$ be the vector space of all real 2×2 matrices and let T be the linear transformation on $M_{2\times 2}(\mathbb{R})$ defined by

$$T \begin{bmatrix} a & b \\ c & d \end{bmatrix} = \begin{bmatrix} a+b+d & a+b+c \\ b+c+d & a+c+d \end{bmatrix}.$$

Find the eigenvalues and basis for each of the eigenspaces of T, and diagonalize T.

Problem 6.23 Let $T : P_2(\mathbb{R}) \to P_2(\mathbb{R})$ be the linear transformation defined by $T(f(x)) = f(x) + xf'(x)$. Find all the eigenvalues of T and find a basis α for $P_2(\mathbb{R})$ so that $[T]_\alpha$ is a diagonal matrix.

6.9 Exercises

6.1. Find the eigenvalues and eigenvectors for the given matrix, if they exist.

$$(1) \begin{bmatrix} 6 & 0 \\ -2 & 2 \end{bmatrix}, \qquad (2) \begin{bmatrix} 1 & -4 & -1 \\ 3 & 2 & 3 \\ 1 & 1 & 3 \end{bmatrix},$$

$$(3) \begin{bmatrix} 0 & 1 & 0 & 1 \\ 1 & 0 & 1 & 0 \\ 0 & 1 & 0 & 1 \\ 1 & 0 & 1 & 0 \end{bmatrix}, \qquad (4) \begin{bmatrix} 1 & 1 & 1 & 1 \\ 1 & 1 & 1 & 1 \\ 1 & 1 & 1 & 1 \\ 1 & 1 & 1 & 1 \end{bmatrix},$$

$$(5) \begin{bmatrix} 2 & -1 & 0 & -1 \\ -1 & 2 & -1 & 0 \\ 0 & -1 & 2 & -1 \\ -1 & 0 & -1 & 2 \end{bmatrix}, \quad (6) \begin{bmatrix} 1 & -5 & 0 & 0 \\ 5 & 1 & 0 & 0 \\ 0 & 0 & 1 & -2 \\ 0 & 0 & 2 & 1 \end{bmatrix}.$$

6.2. Find the characteristic polynomial, eigenvalues and eigenvectors of the matrix

$$A = \begin{bmatrix} -2 & 0 & 0 \\ 3 & 2 & 3 \\ 4 & -1 & 6 \end{bmatrix}.$$

6.3. Show that a 2×2 matrix $A = \begin{bmatrix} a & b \\ c & d \end{bmatrix}$ has

(1) two distinct real eigenvalues if $(a-d)^2 + 4bc > 0$,
(2) one eigenvalue if $(a-d)^2 + 4bc = 0$,
(3) no real eigenvalues if $(a-d)^2 + 4bc < 0$,
(4) only real eigenvalues if it is symmetric (*i.e.*, $b = c$).

6.4. Suppose that a 3×3 matrix A has eigenvalues -1, 0, 1 with eigenvectors $\mathbf{u}$, $\mathbf{v}$, $\mathbf{w}$, respectively. Describe the null space $\mathcal{N}(A)$, and the column space $\mathcal{C}(A)$.

6.5. If a 3×3 matrix A has eigenvalues 1, 2, 3, what are the eigenvectors of $B = (A - I)(A - 2I)(A - 3I)$?

6.6. Show that any 2×2 skew-symmetric nonzero matrix has no real eigenvalue.

6.7. Find a 3×3 matrix that has the eigenvalues $\lambda_1 = 1$, $\lambda_2 = 2$, $\lambda_3 = 3$ with the associated eigenvectors $\mathbf{x}_1 = (2, -1, 0)$, $\mathbf{x}_2 = (-1, 2, -1)$, $\mathbf{x}_3 = (0, -1, 2)$.

6.8. Let P be the projection matrix that projects $\mathbb{R}^n$ onto a subspace W. Find the eigenvalues and the eigenspaces for P.

6.9. Let $\mathbf{u}$, $\mathbf{v}$ be $n \times 1$ column vectors, and let $A = \mathbf{u}\mathbf{v}^T$. Show that $\mathbf{u}$ is an eigenvector of A, and find the eigenvalues and the eigenvectors of A.

6.10. Show that if λ is an eigenvalue of an idempotent $n \times n$ matrix A (*i.e.*, $A^2 = A$), then λ must be either 0 or 1.

6.11. Prove that if A is an idempotent matrix, then tr A = rank A.

6.12. Let $A = [a_{ij}]$ be an $n \times n$ matrix with eigenvalues $\lambda_1, \ldots, \lambda_n$. Show that

$$\lambda_j = a_{jj} + \sum_{i \neq j} (a_{ii} - \lambda_i) \qquad \text{for } j = 1, \ldots, n.$$

6.13. Prove that if two diagonalizable matrices A and B have the same eigenvectors (*i.e.*, there exists an invertible matrix Q such that both $Q^{-1}AQ$ and $Q^{-1}BQ$ are diagonal; such matrices A and B are said to be **simultaneously diagonalizable**), then $AB = BA$. In fact, the converse is also true. (See Exercise **7.17**.) Prove the converse with an assumption that the eigenvalues of A are all distinct.

6.14. Show that the matrix $A = \begin{bmatrix} 0 & 0 & 0 & \cdots & 0 & -c_n \\ 1 & 0 & 0 & \cdots & 0 & -c_{n-1} \\ 0 & 1 & 0 & \cdots & 0 & -c_{n-2} \\ \vdots & \vdots & \vdots & & \vdots & \vdots \\ 0 & 0 & 0 & \cdots & 1 & -c_1 \end{bmatrix}$ has the characteristic polynomial $p(\lambda) = \lambda^n + c_1\lambda^{n-1} + \cdots + c_{n-1}\lambda + c_n$.
(This shows that every *monic* polynomial is the characteristic polynomial of some matrix. The matrix A is called the *companion matrix* of $p(\lambda)$.)

6.15. Let $D : P_3(\mathbb{R}) \to P_3(\mathbb{R})$ be the differentiation defined by $Df(x) = f'(x)$ for $f \in P_3(\mathbb{R})$. Find all eigenvalues and eigenvectors of D and of D^2.

6.16. Let $T : P_2(\mathbb{R}) \to P_2(\mathbb{R})$ be the linear transformation defined by

$$T(a_2x^2 + a_1x + a_0) = (a_0 + a_1)x^2 + (a_1 + a_2)x + (a_0 + a_2).$$

Find a basis for $P_2(\mathbb{R})$ with respect to which the matrix representation for T is diagonal.

6.17. Determine whether or not each of the following matrices is diagonalizable.

$$(1) \begin{bmatrix} 2 & 1 & -1 \\ 1 & 0 & 2 \\ -1 & 2 & 3 \end{bmatrix} \quad (2) \begin{bmatrix} 2 & 0 & 0 \\ 1 & 2 & 0 \\ 0 & 1 & 2 \end{bmatrix} \quad (3) \begin{bmatrix} 3 & 0 & 2 \\ 0 & 2 & 0 \\ -2 & 0 & -1 \end{bmatrix}$$

6.18. Find an orthogonal matrix Q and a diagonal matrix D such that $Q^TAQ = D$ for

$$(1)\ A = \begin{bmatrix} -3 & 2 & 4 \\ 2 & -6 & 2 \\ 4 & 2 & -3 \end{bmatrix}, \quad (2)\ A = \begin{bmatrix} 1 & 2 & 0 \\ 2 & 2 & 2 \\ 0 & 2 & 3 \end{bmatrix}, \quad (3)\ A = \begin{bmatrix} 1 & 0 & 0 \\ 0 & 1 & 1 \\ 0 & 1 & 1 \end{bmatrix}.$$

6.19. Calculate $A^{10}\mathbf{x}$ for $A = \begin{bmatrix} 1 & 2 & -1 \\ 0 & 5 & -2 \\ 0 & 6 & -2 \end{bmatrix}, \quad \mathbf{x} = \begin{bmatrix} 2 \\ 4 \\ 7 \end{bmatrix}$.

6.20. For $n \geq 1$, let a_n denote the number of subsets of $\{1,\ 2,\ \ldots,\ n\}$ that contain no consecutive integers. Find the number a_n for all $n \geq 1$.

6.21. Evaluate $\det A_n$, where

$$A_n = \begin{bmatrix} 1 & 1 & 0 & 0 & \cdots & 0 & 0 & 0 & 0 \\ 1 & 1 & 1 & 0 & \cdots & 0 & 0 & 0 & 0 \\ 0 & 1 & 1 & 1 & \cdots & 0 & 0 & 0 & 0 \\ \vdots & \vdots & \vdots & \vdots & \ddots & \vdots & \vdots & \vdots & \vdots \\ 0 & 0 & 0 & 0 & \cdots & 1 & 1 & 1 & 0 \\ 0 & 0 & 0 & 0 & \cdots & 0 & 1 & 1 & 1 \\ 0 & 0 & 0 & 0 & \cdots & 0 & 0 & 1 & 1 \end{bmatrix}$$

is the $n \times n$ $\{0, 1\}$-matrix with 1's on the main diagonal and its two parallel side diagonals.

6.22. Let $A = \begin{bmatrix} 0 & 0.3 \\ 0.6 & x \end{bmatrix}$. Find a value x so that A has an eigenvalue $\lambda = 1$. For $\mathbf{x}_0 = (1,\ 1)$, calculate $\lim_{k\to\infty} \mathbf{x}_k$, where $\mathbf{x}_k = A\mathbf{x}_{k-1}$, $k = 1, 2, \cdots$.

6.23. Compute e^A for

$$(1)\ A = \begin{bmatrix} 1 & 1 \\ 0 & 0 \end{bmatrix}, \quad (2)\ A = \begin{bmatrix} 3 & 1 \\ 1 & 3 \end{bmatrix}.$$

6.24. In 1985, the initial status of the car owners in a city was reported as follows: 40% of the car owners drove large cars, 20% drove medium-sized cars, and 40% drove small cars. In 1995, 70% of the large-car owners in 1985 still owned large cars, but 30% had changed to a medium-sized car. Of those who owned medium-sized cars in 1985, 10% had changed to large cars, 70% continued to drive medium-sized cars, and 20% had changed to small cars. Finally, of those who owned the small cars in 1985, 10% had changed to medium-sized cars and 90% still owned small cars in 1995. Assuming that these trends continue, and that no car owners are born, die or otherwise add realism to the problem, determine the percentage of car owners who will own cars of each size in 2025.

6.25. Let $A = \begin{bmatrix} 1 & 1 \\ 0 & 2 \end{bmatrix}$.

(1) Compute e^A directly from the expansion.

(2) Compute e^A by diagonalizing A.

6.26. Let $A(t)$ be a matrix whose entries are all differentiable functions in t and invertible for all t. Compute the following:

$$(1)\ \frac{d}{dt}(A(t)^3), \qquad (2)\ \frac{d}{dt}(A(t)^{-1}).$$

6.27. Solve $\mathbf{y}' = A\mathbf{y}$, where

$$(1)\ A = \begin{bmatrix} -6 & 24 & 8 \\ -1 & 8 & 4 \\ 2 & -12 & -6 \end{bmatrix} \quad \text{and} \quad \mathbf{y}(1) = \begin{bmatrix} 2 \\ 1 \\ 0 \end{bmatrix}.$$

$$(2)\ A = \begin{bmatrix} 1 & -1 \\ 2 & 3 \end{bmatrix} \quad \text{and} \quad \mathbf{y}(0) = \begin{bmatrix} 1 \\ 0 \end{bmatrix}.$$

6.28. Let $f(\lambda) = \det(\lambda I - A)$ be the characteristic polynomial of A. Evaluate $f(A)$ for

$$(1)\ A = \begin{bmatrix} 3 & 1 & 1 \\ 2 & 4 & 2 \\ 1 & 1 & 3 \end{bmatrix}, \qquad (2)\ A = \begin{bmatrix} 1 & 2 & 2 \\ 1 & 2 & -1 \\ -1 & 1 & 4 \end{bmatrix}.$$

In fact, $f(A) = \mathbf{0}$ for any square matrix A and its characteristic polynomial $f(\lambda)$ (this is the Cayley-Hamilton theorem).

6.29. Determine whether the following statements are true or false, in general, and justify your answers.

(1) If B is obtained from A by interchanging two rows, then B is similar to A.
(2) If A and B are diagonalizable, so is AB.
(3) Every invertible matrix is diagonalizable.
(4) Every diagonalizable matrix is invertible.
(5) Every permutation matrix is orthogonal.
(6) Interchanging the rows of a 2×2 matrix reverses the signs of its eigenvalues.
(7) A matrix A cannot be similar to $A + I$.
(8) The eigenvalues of $A + B$ equal the sum of the eigenvalues of A and B.
(9) The sum of the eigenvalues of $A+B$ equals the sum of all the individual eigenvalues of A and B.

Chapter 7

Complex Vector Spaces

7.1 Introduction

A real matrix has real coefficients in its characteristic polynomial, but the eigenvalues may fail to be real. For instance, the matrix $A = \begin{bmatrix} 1 & -1 \\ 1 & 1 \end{bmatrix}$ has no real eigenvalues, but it has the complex eigenvalues $\lambda = 1 \pm i$. Thus, it is indispensable to work with complex numbers to find the full set of eigenvalues and eigenvectors. Therefore, it is natural to extend the concept of real vector spaces to that of complex vector spaces, and then develop the basic properties of complex vector spaces. With this extension, all the square matrices of order n will have n eigenvalues.

The Eucidean n-dimensional complex vector space is the set $\mathbb{C}^n$ of vectors with n complex components:

$$\mathbb{C}^n = \{(z_1, z_2, \ldots, z_n) : z_i \in \mathbb{C},\ i = 1, 2, \ldots, n\}.$$

In this complex vector space $\mathbb{C}^n$, the addition and the scalar multiplication are given as follows:

$$\begin{aligned} (z_1, z_2, \ldots, z_n) + (z_1', z_2', \ldots, z_n') &= (z_1 + z_1', z_2 + z_2', \ldots, z_n + z_n') \\ k(z_1, z_2, \ldots, z_n) &= (kz_1, kz_2, \ldots, kz_n), \quad \text{for} \quad k \in \mathbb{C}. \end{aligned}$$

The standard basis for $\mathbb{C}^n$ is again $\{\mathbf{e}_1, \ldots, \mathbf{e}_n\}$ as the real case, but the scalars are now complex numbers so that any vector $\mathbf{z}$ in $\mathbb{C}^n$ is of the form $\mathbf{z} = \sum_{k=1}^n z_k \mathbf{e}_k$ with $z_k = x_k + iy_k \in \mathbb{C}$, *i.e.*, $\mathbf{z} = \mathbf{x} + i\mathbf{y}$ with $\mathbf{x},\ \mathbf{y} \in \mathbb{R}^n$.

In a complex vector space, linear combinations are defined exactly the same as real space except the scalars are replaced by complex numbers.

Thus the same is true for the linear independence, spanning spaces, basis, dimension, and subspace. For complex matrices, whose entries are complex numbers, the matrix sum and product follow the same rules as real matrices. The same is true for the concept of a linear transformation $T : V \to W$ from a complex vector space V to a complex vector space W. The definitions of the kernel and the image of a linear transformation remain the same as those in the real case, as well as the facts about null spaces, column spaces, matrix representations of linear transformations, and similarity.

However, if we are concerned about the inner product, there should be a modification from the real case. Note that the *absolute value* (or *modulus*) of a complex number $z = x + iy$ is defined as the nonnegative real number $|z| = (\overline{z}z)^{\frac{1}{2}} = \sqrt{x^2 + y^2}$, where $\overline{z}$ is the complex conjugate of z. Accordingly, the length of a vector $\mathbf{z} = (z_1, \ldots, z_n)$ in the n-space $\mathbb{C}^n$ with $z_k = x_k + iy_k \in \mathbb{C}$ has to be modified: if we would take an inner product in $\mathbb{C}^n$ as $\|\mathbf{z}\|^2 = z_1^2 + \cdots + z_n^2$, then a nonzero vector $(1, i)$ in $\mathbb{C}^2$ would have zero length: $1^2 + i^2 = 0$. In any case, a modified definition should coincide with the old definition, when the vectors and matrices were real. The following is the usual definition of an inner product in $\mathbb{C}^n$.

Definition 7.1 If $\mathbf{u} = [u_1\ u_2\ \cdots\ u_n]^T$ and $\mathbf{v} = [v_1\ v_2\ \cdots\ v_n]^T$ are vectors in the n-space $\mathbb{C}^n$ with $u_k, v_k \in \mathbb{C}$, then their **inner product** $\mathbf{u} \cdot \mathbf{v}$ is defined by

$$\mathbf{u} \cdot \mathbf{v} = \overline{u_1}v_1 + \overline{u_2}v_2 + \cdots + \overline{u_n}v_n = \overline{\mathbf{u}}^T\mathbf{v},$$

where $\overline{\mathbf{u}} = [\overline{u}_1\ \overline{u}_2\ \cdots\ \overline{u}_n]^T$. The **length** (or **magnitude**) of a vector $\mathbf{u}$ in $\mathbb{C}^n$ is defined by

$$\|\mathbf{u}\| = (\mathbf{u} \cdot \mathbf{u})^{\frac{1}{2}} = \sqrt{|u_1|^2 + |u_2|^2 + \cdots + |u_n|^2},$$

where $|u_k|^2 = \overline{u}_k u_k$, and the **distance** between two vectors $\mathbf{u}$ and $\mathbf{v}$ in $\mathbb{C}^n$ is defined by

$$d(\mathbf{u}, \mathbf{v}) = \|\mathbf{u} - \mathbf{v}\|.$$

In an (abstract) complex vector space, we can also define an inner product by adopting the basic properties of the Euclidean inner product on $\mathbb{C}^n$ as axioms.

Definition 7.2 A (complex) **inner product** (or **Hermitian inner product**) on a complex vector space V is a function that associates a complex number $\langle \mathbf{u}, \mathbf{v} \rangle$ with each pair of vectors $\mathbf{u}$ and $\mathbf{v}$ in V in such a way that the following conditions are satisfied: For any vectors $\mathbf{u}$, $\mathbf{v}$ and $\mathbf{w}$ in V and any scalar k in $\mathbb{C}$,

(1) $\langle \mathbf{u}, \mathbf{v} \rangle = \overline{\langle \mathbf{v}, \mathbf{u} \rangle}$,
(2) $\langle \mathbf{u} + \mathbf{v}, \mathbf{w} \rangle = \langle \mathbf{u}, \mathbf{w} \rangle + \langle \mathbf{v}, \mathbf{w} \rangle$,
(3) $\langle k\mathbf{u}, \mathbf{v} \rangle = \bar{k}\langle \mathbf{u}, \mathbf{v} \rangle$,
(4) $\langle \mathbf{v}, \mathbf{v} \rangle \geq 0$, and $\langle \mathbf{v}, \mathbf{v} \rangle = 0$ if and only if $\mathbf{v} = \mathbf{0}$.

A complex vector space together with an inner product is called a **complex inner product space** or a **unitary space**.

The following additional properties follow immediately from the definition of an inner product space:

(5) $\langle \mathbf{0}, \mathbf{v} \rangle = \langle \mathbf{v}, \mathbf{0} \rangle = 0$,
(6) $\langle \mathbf{u}, \mathbf{v} + \mathbf{w} \rangle = \langle \mathbf{u}, \mathbf{v} \rangle + \langle \mathbf{u}, \mathbf{w} \rangle$,
(7) $\langle \mathbf{u}, k\mathbf{v} \rangle = k\langle \mathbf{u}, \mathbf{v} \rangle$.

Remark: There is another way to define an inner product on a complex vector space. If we redefine the Euclidean inner product $\mathbf{u} \cdot \mathbf{v}$ on $\mathbb{C}^n$ by

$$\mathbf{u} \cdot \mathbf{v} = u_1\overline{v_1} + u_2\overline{v_2} + \cdots + u_n\overline{v_n},$$

then the third equation in Definition 7.2 should be modified to be

(3′) $\langle \mathbf{u}, k\mathbf{v} \rangle = \bar{k}\langle \mathbf{u}, \mathbf{v} \rangle$, so that $\langle k\mathbf{u}, \mathbf{v} \rangle = k\langle \mathbf{u}, \mathbf{v} \rangle$.

But these two different definitions do not induce any essential difference in a complex vector space.

In a complex inner product space, as in a real inner product space, the **length** (or **magnitude**) of a vector $\mathbf{u}$ and the **distance** between two vectors $\mathbf{u}$ and $\mathbf{v}$ are defined by

$$\|\mathbf{u}\| = \langle \mathbf{u}, \mathbf{u} \rangle^{\frac{1}{2}}, \quad d(\mathbf{u}, \mathbf{v}) = \|\mathbf{u} - \mathbf{v}\|,$$

respectively.

Example 7.1 Let $C_{\mathbb{C}}[a,\ b]$ denote the set of all complex-valued continuous functions defined on $[a,\ b]$. Thus an element in $C_{\mathbb{C}}[a,\ b]$ is of the form $\mathbf{f}(x) = f_1(x) + if_2(x)$, where $f_1(x)$ and $f_2(x)$ are real-valued continuous on $[a, b]$. Note that $\mathbf{f}$ is continuous if and only if each component function f_i is continuous. From the theory of continuous functions, it is quite easy to see that $C_{\mathbb{C}}[a,\ b]$ is a complex vector space under the sum and scalar multiplication of functions. For a vector $\mathbf{f}(x) = f_1(x) + if_2(x)$ in $C_{\mathbb{C}}[a, b]$, we define

$$\int_a^b \mathbf{f}(x)dx = \int_a^b [f_1(x) + if_2(x)]\,dx = \int_a^b f_1(x)dx + i\int_a^b f_2(x)dx.$$

We leave it as an exercise to show that, for vectors $\mathbf{f}(x) = f_1(x) + if_2(x)$ and $\mathbf{g}(x) = g_1(x) + ig_2(x)$ in the complex vector space $C_{\mathbb{C}}[a, b]$, the following formula defines an inner product on $C_{\mathbb{C}}[a, b]$:

$$\begin{aligned}\langle \mathbf{f}, \mathbf{g}\rangle &= \int_a^b \overline{\mathbf{f}(x)}\mathbf{g}(x)dx \\ &= \int_a^b [f_1(x) - if_2(x)]\,[g_1(x) + ig_2(x)]\,dx \\ &= \int_a^b [f_1(x)g_1(x) + f_2(x)g_2(x)]\,dx \\ &\quad + i\int_a^b [f_1(x)g_2(x) - f_2(x)g_1(x)]\,dx.\end{aligned}$$

□

Problem 7.1 Show that the Euclidean inner product on $\mathbb{C}^n$ satisfies all the inner product axioms.

The definitions of such terms as orthogonal sets, orthogonal complements, orthonormal sets, and orthonormal basis remain the same in complex inner product spaces as in real inner product spaces. Moreover, the Gram-Schmidt orthogonalization is still valid in complex inner product spaces, and can be used to convert an arbitrary basis into an orthonormal basis. If V is an n-dimensional complex vector space, then by taking an orthonormal basis for V, there is a natural isometry from V to $\mathbb{C}^n$ that preserves the inner product as in the real case. Hence, without lose of generality, we may only work on $\mathbb{C}^n$ with the Euclidean inner product, and we use $\cdot$ and $\langle\ ,\ \rangle$ interchangeably.

On the other hand, we may consider the set $\mathbb{C}^n$ as a real vector space by defining addition and scalar multiplication as

$$\begin{aligned}(z_1, z_2, \ldots, z_n) + (z_1', z_2', \ldots, z_n') &= (z_1 + z_1', z_2 + z_2', \ldots, z_n + z_n') \\ r(z_1, z_2, \ldots, z_n) &= (rz_1, rz_2, \ldots, rz_n), \quad \text{for } r \in \mathbb{R}.\end{aligned}$$

Two vectors $\mathbf{e}_1 = (1, 0, \ldots, 0)$ and $i\mathbf{e}_1 = (i, 0, \ldots, 0)$ are linearly dependent when we consider $\mathbb{C}^n$ as a complex vector space. However, they are linearly independent if we consider $\mathbb{C}^n$ as a real vector space. In general,

$$\{\mathbf{e}_1,\ \ldots,\ \mathbf{e}_n,\ i\mathbf{e}_1,\ \ldots,\ i\mathbf{e}_n\}$$

forms a basis for $\mathbb{C}^n$ considered as a real vector space. In this way, $\mathbb{C}^n$ is naturally identified with the $2n$-dimensional real vector space $\mathbb{R}^{2n}$. That is, $\dim \mathbb{C}^n = n$ when $\mathbb{C}^n$ is considered as a complex vector space, but $\dim \mathbb{C}^n = 2n$ when $\mathbb{C}^n$ is considered as a real vector space.

Note that when $\mathbb{C}^n$ is considered as a $2n$-dimensional real vector space, the space $\mathbb{R}^n = \{(x_1, x_2, \ldots, x_n) : x_i \in \mathbb{R}\}$ is a subspace of $\mathbb{C}^n$, but not when $\mathbb{C}^n$ is considered as an n-dimensional complex vector space.

Example 7.2 Consider the complex vector space $\mathbb{C}^3$ with the Euclidean inner product. Apply the Gram-Schmidt orthogonalization to convert the basis $\mathbf{u}_1 = (i,\ i,\ i)$, $\mathbf{u}_2 = (0,\ i,\ i)$, $\mathbf{u}_3 = (0,\ 0,\ i)$ into an orthonormal basis.

Solution: Step 1. Set

$$\mathbf{v}_1 = \frac{\mathbf{u}_1}{\|\mathbf{u}_1\|} = \frac{(i,\ i,\ i)}{\sqrt{3}} = \left(\frac{i}{\sqrt{3}},\ \frac{i}{\sqrt{3}},\ \frac{i}{\sqrt{3}}\right).$$

Step 2. Let W_1 denote the subspace spanned by $\mathbf{v}_1$. Then

$$\begin{aligned} \mathbf{u}_2 - \mathrm{Proj}_{W_1}\mathbf{u}_2 &= \mathbf{u}_2 - \langle \mathbf{u}_2, \mathbf{v}_1 \rangle \mathbf{v}_1 \\ &= (0,\ i,\ i) - \frac{2}{\sqrt{3}}\left(\frac{i}{\sqrt{3}},\ \frac{i}{\sqrt{3}},\ \frac{i}{\sqrt{3}}\right) \\ &= \left(-\frac{2i}{3},\ \frac{i}{3},\ \frac{i}{3}\right). \end{aligned}$$

Therefore,

$$\mathbf{v}_2 = \frac{\mathbf{u}_2 - \mathrm{Proj}_{W_1}\mathbf{u}_2}{\|\mathbf{u}_2 - \mathrm{Proj}_{W_1}\mathbf{u}_2\|} = \frac{3}{\sqrt{6}}\left(-\frac{2i}{3},\ \frac{i}{3},\ \frac{i}{3}\right) = \left(-\frac{2i}{\sqrt{6}},\ \frac{i}{\sqrt{6}},\ \frac{i}{\sqrt{6}}\right).$$

Step 3. Let W_2 denote the subspace spanned by $\{\mathbf{v}_1,\ \mathbf{v}_2\}$. Then

$$\begin{aligned} &\mathbf{u}_3 - \mathrm{Proj}_{W_2}\mathbf{u}_3 \\ &= \mathbf{u}_3 - \langle \mathbf{u}_3, \mathbf{v}_1 \rangle \mathbf{v}_1 - \langle \mathbf{u}_3, \mathbf{v}_2 \rangle \mathbf{v}_2 \\ &= (0,\ 0,\ i) - \frac{1}{\sqrt{3}}\left(\frac{i}{\sqrt{3}},\ \frac{i}{\sqrt{3}},\ \frac{i}{\sqrt{3}}\right) - \frac{1}{\sqrt{6}}\left(-\frac{2i}{\sqrt{6}},\ \frac{i}{\sqrt{6}},\ \frac{i}{\sqrt{6}}\right) \\ &= \left(0,\ -\frac{i}{2},\ \frac{i}{2}\right). \end{aligned}$$

Therefore,

$$\mathbf{v}_3 = \frac{\mathbf{u}_3 - \mathrm{Proj}_{W_2}\mathbf{u}_3}{\|\mathbf{u}_3 - \mathrm{Proj}_{W_2}\mathbf{u}_3\|} = \sqrt{2}\left(0,\ -\frac{i}{2},\ \frac{i}{2}\right) = \left(0,\ -\frac{i}{\sqrt{2}},\ \frac{i}{\sqrt{2}}\right).$$

Thus,

$$\mathbf{v}_1 = \left(\frac{i}{\sqrt{3}}, \frac{i}{\sqrt{3}}, \frac{i}{\sqrt{3}}\right), \ \mathbf{v}_2 = \left(-\frac{2i}{\sqrt{6}}, \frac{i}{\sqrt{6}}, \frac{i}{\sqrt{6}}\right), \ \mathbf{v}_3 = \left(0, -\frac{i}{\sqrt{2}}, \frac{i}{\sqrt{2}}\right)$$

form an orthonormal basis for $\mathbb{C}^3$. □

Example 7.3 Let $C_{\mathbb{C}}[0,\ 2\pi]$ be the complex vector space with the inner product given in Example 7.1, and let W be the set of vectors in $C_{\mathbb{C}}[0,\ 2\pi]$ of the form

$$e^{ikx} = \cos kx + i \sin kx,$$

where k is an integer. The set W is orthogonal. In fact, if

$$\mathbf{g}_k(x) = e^{ikx} \quad \text{and} \quad \mathbf{g}_\ell(x) = e^{i\ell x}$$

are vectors in W, then

$$\begin{aligned}
\langle \mathbf{g}_k, \mathbf{g}_\ell \rangle &= \int_0^{2\pi} \overline{e^{ikx}} e^{i\ell x} dx = \int_0^{2\pi} e^{-ikx} e^{i\ell x} dx = \int_0^{2\pi} e^{i(\ell-k)x} dx \\
&= \int_0^{2\pi} \cos(\ell-k)x dx + i \int_0^{2\pi} \sin(\ell-k)x dx \\
&= \begin{cases} \left[\frac{1}{\ell-k}\sin(\ell-k)x\right]_0^{2\pi} + i\left[\frac{-1}{\ell-k}\cos(\ell-k)x\right]_0^{2\pi}, & \text{if } k \neq \ell, \\ \int_0^{2\pi} dx, & \text{if } k = \ell. \end{cases} \\
&= \begin{cases} 0, & \text{if } k \neq \ell, \\ 2\pi, & \text{if } k = \ell. \end{cases}
\end{aligned}$$

Thus, the vectors in W are orthogonal and each vector has length $\sqrt{2\pi}$. By normalizing each vector in the orthogonal set W, we obtain an orthonormal set. Therefore, the vectors

$$\mathbf{f}_k(x) = \frac{1}{\sqrt{2\pi}} e^{ikx}, \quad k = 0,\ \pm 1,\ \pm 2,\ \ldots,$$

form an orthonormal set in the complex vector space $C_{\mathbb{C}}[0,\ 2\pi]$. □

Problem 7.2 Prove that in a complex inner product space V,

(1) (Cauchy-Schwarz inequality) $|\langle \mathbf{x}, \mathbf{y} \rangle|^2 \le \langle \mathbf{x}, \mathbf{x} \rangle \langle \mathbf{y}, \mathbf{y} \rangle$,
(2) (Triangular inequality) $\|\mathbf{x}+\mathbf{y}\| \le \|\mathbf{x}\| + \|\mathbf{y}\|$,
(3) (Pythagorean theorem) $\|\mathbf{x}+\mathbf{y}\|^2 = \|\mathbf{x}\|^2 + \|\mathbf{y}\|^2$ if $\mathbf{x}$ and $\mathbf{y}$ are orthogonal.

The definitions of eigenvalues and eigenvectors in a complex vector space are the same as in the real case, but the eigenvalues can now be complex numbers. Hence for any $n \times n$ (real or complex) matrix A the characteristic polynomial $\det(\lambda I - A)$ has always exactly n complex roots (*i.e.*, eigenvalues) counting multiplicities.

For example, consider a rotation matrix

$$A = \begin{bmatrix} \cos\theta & -\sin\theta \\ \sin\theta & \cos\theta \end{bmatrix}$$

with real entries. This matrix has two complex eigenvalues for any $\theta \in \mathbb{R}$, but no real eigenvalues unless $\theta = k\pi$, for an integer k.

Therefore, the theorems in Chapter 6 regarding eigenvalues and eigenvectors remain true without requiring the existence of n eigenvalues explicitly, and exactly the same proofs as in the real case are valid since the argument in the proofs is not concerned with what the scalars are. For example, one can have a theorem like *"for an $n \times n$ matrix A, the eigenvectors belonging to distinct eigenvalues are linearly independent"*, and *"if the n eigenvalues of A are distinct, then the eigenvectors belonging to them form a basis for $\mathbb{C}^n$ so that A is diagonalizable"*.

An $n \times n$ real matrix A can be considered as a linear transformation on both $\mathbb{R}^n$ and $\mathbb{C}^n$:

$$\begin{array}{lll} T : \mathbb{R}^n \to \mathbb{R}^n & \text{defined by} & T(\mathbf{x}) = A\mathbf{x}, \\ S : \mathbb{C}^n \to \mathbb{C}^n & \text{defined by} & S(\mathbf{x}) = A\mathbf{x}. \end{array}$$

Since the entries are all real, the coefficients of the characteristic polynomial of A, $p(\lambda) = \det(\lambda I - A)$, are all real. Thus, if λ is a root of $p(\lambda) = 0$, then its conjugate $\bar{\lambda}$ is also a root because $p(\bar{\lambda}) = \overline{p(\lambda)} = 0$. In particular, any $n \times n$ real matrix A has at least one real eigenvalue if n is odd.

Moreover, if $\mathbf{x}$ is an eigenvector belonging to the complex eigenvalue λ, then the complex conjugate $\bar{\mathbf{x}}$ is an eigenvector belonging to $\bar{\lambda}$. In fact, if $A\mathbf{x} = \lambda\mathbf{x}$ with $\mathbf{x} \neq \mathbf{0}$, then

$$A\bar{\mathbf{x}} = \overline{A\mathbf{x}} = \overline{\lambda\mathbf{x}} = \bar{\lambda}\bar{\mathbf{x}},$$

where $\bar{\mathbf{x}}$ denotes the vector whose entries are the complex conjugates of the corresponding entries in $\mathbf{x}$.

Using this fact, the following example shows that any 2×2 matrix with no real eigenvalues can be written as a scalar multiple of a rotation.

Example 7.4 Let A be a 2×2 real matrix with no real eigenvalues. Then it has two complex eigenvalues $\lambda = a + ib$ and $\bar{\lambda} = a - ib$ with $a, b \in \mathbb{R}$ and $b \neq 0$. Denote their associated eigenvectors in $\mathbb{C}^2$ by $\mathbf{x} = \mathbf{u} + i\mathbf{v}$ and $\bar{\mathbf{x}} = \mathbf{u} - i\mathbf{v}$ with $\mathbf{u}, \mathbf{v} \in \mathbb{R}^2$, respectively. Clearly, we have

$$\begin{aligned} \mathbf{u} &= \tfrac{1}{2}(\mathbf{x} + \bar{\mathbf{x}}), & \mathbf{v} &= -\tfrac{i}{2}(\mathbf{x} - \bar{\mathbf{x}}), \\ a &= \frac{1}{2}(\lambda + \bar{\lambda}), & b &= -\frac{i}{2}(\lambda - \bar{\lambda}). \end{aligned}$$

Since $\lambda \neq \bar{\lambda}$, by the same argument as in Theorem 6.7, $\mathbf{x}$ and $\bar{\mathbf{x}}$ are linearly independent in the complex vector space $\mathbb{C}^2$, so they are when $\mathbb{C}^2$ is considered as a real vector space. This implies that $\mathbf{u}$ and $\mathbf{v}$ are linearly independent real vectors in $\mathbb{R}^2$ (see Problem 7.3 below), which is regarded as a subspace of the real vector space $\mathbb{C}^2$. Thus $\alpha = \{\mathbf{u}, \mathbf{v}\}$ is a basis for the real vector space $\mathbb{R}^2$, and we have

$$\begin{aligned} A\mathbf{u} &= \frac{1}{2}(A\mathbf{x} + A\bar{\mathbf{x}}) = \frac{1}{2}(\lambda\mathbf{x} + \bar{\lambda}\bar{\mathbf{x}}) \\ &= \lambda\left(\frac{\mathbf{u} + i\mathbf{v}}{2}\right) + \bar{\lambda}\left(\frac{\mathbf{u} - i\mathbf{v}}{2}\right) \\ &= a\mathbf{u} - b\mathbf{v}, \end{aligned}$$

and similarly $A\mathbf{v} = b\mathbf{u} + a\mathbf{v}$. It implies that the matrix representation of the linear transformation $A : \mathbb{R}^2 \to \mathbb{R}^2$ with respect to the basis α is

$$[A]_\alpha = \begin{bmatrix} a & b \\ -b & a \end{bmatrix}.$$

That is, any 2×2 matrices which have no real eigenvalues is similar to a matrix of the above form. By setting $r = \sqrt{a^2 + b^2} > 0$, we get $a = r\cos\theta$ and $b = r\sin\theta$ for some $\theta \in \mathbb{R}$, so

$$\begin{bmatrix} r\cos\theta & r\sin\theta \\ -r\sin\theta & r\cos\theta \end{bmatrix}.$$

□

Problem 7.3 Let $\mathbf{x}$ and $\mathbf{y}$ be two vectors in a vector space V. Show that $\mathbf{x}$ and $\mathbf{y}$ are linearly independent if and only if $\mathbf{x} + \mathbf{y}$ and $\mathbf{x} - \mathbf{y}$ are linearly independent.

Problem 7.4 Find the eigenvalues and the eigenvectors of

$$(1) \begin{bmatrix} i & 0 & 1 \\ 0 & 2 & 0 \\ 1 & 0 & -i \end{bmatrix}, \qquad (2) \begin{bmatrix} 1 & i & 1+i \\ -i & 2 & 0 \\ 1-i & 0 & 1 \end{bmatrix}.$$

Problem 7.5 Prove that an $n \times n$ complex matrix A is diagonalizable if and only if A has n linearly independent eigenvectors in the complex vector space $\mathbb{C}^n$.

7.2 Hermitian and unitary matrices

Recall that the dot product of real vectors $\mathbf{x}, \mathbf{y} \in \mathbb{R}^n$ is given by $\mathbf{x} \cdot \mathbf{y} = \mathbf{x}^T\mathbf{y}$ in matrix notation. For complex vectors $\mathbf{u}, \mathbf{v} \in \mathbb{C}^n$, the inner product is defined by $\mathbf{u} \cdot \mathbf{v} = \bar{u}_1 v_1 + \cdots + \bar{u}_n v_n = \bar{\mathbf{u}}^T\mathbf{v}$. That is, we need the conjugate transpose, not just the transpose.

Definition 7.3 Let A be a complex matrix. Then the matrix

$$A^H = \overline{A}^T,$$

the complex conjugate transpose of A, is called the **adjoint** of A.

Note that $\overline{A}$ is the matrix whose entries are the complex conjugates of the corresponding entries in A. Thus, $[a_{ij}]^H = [\overline{a_{ji}}]$. With this notation, the Euclidean inner product on $\mathbb{C}^n$ can be written as

$$\mathbf{u} \cdot \mathbf{v} = \overline{\mathbf{u}}^T\mathbf{v} = \mathbf{u}^H\mathbf{v}.$$

Problem 7.6 For any matrices A and B such that AB is defined, show that $(AB)^H = B^H A^H$.

Problem 7.7 Prove that if A is invertible, then so is A^H, and $(A^H)^{-1} = (A^{-1})^H$.

For complex matrices, the notion of symmetry, skew-symmetry, and orthogonal real matrices are replaced by *Hermitian*, *skew-Hermitian* and *unitary* matrices, respectively.

Definition 7.4 A complex square matrix A is said to be **Hermitian** (or **self-adjoint**) if $A^H = A$, or **skew-Hermitian** if $A^H = -A$.

Examples of Hermitian and skew-Hermitian matrices are

$$A = \begin{bmatrix} 2 & 4+i \\ 4-i & 3 \end{bmatrix}, \quad B = \begin{bmatrix} i & 1+i \\ -1+i & -i \end{bmatrix}.$$

those of a skew-Hermitian A Hermitian matrix with real entries is just a real symmetric matrix, and conversely, any real symmetric matrix is Hermitian.

Like real matrices, any $m \times n$ (complex) matrix A can be considered as a linear transformation from $\mathbb{C}^n$ to $\mathbb{C}^m$, and we note that

$$(A\mathbf{x}) \cdot \mathbf{y} = (A\mathbf{x})^H\mathbf{y} = \mathbf{x}^H A^H \mathbf{y} = \mathbf{x} \cdot (A^H\mathbf{y})$$

for any $\mathbf{x} \in \mathbb{C}^n$ and $\mathbf{y} \in \mathbb{C}^m$. The following theorem lists some important properties of Hermitian matrices.

Theorem 7.1 *Let A be a Hermitian matrix.*

(1) *For any (complex) vector $\mathbf{x} \in \mathbb{C}^n$, $\mathbf{x}^H A\mathbf{x}$ is real.*

(2) *Every (complex) eigenvalue of A is real. In particular, an $n \times n$ real symmetric matrix has precisely n real eigenvalues.*

(3) *The eigenvectors of A belonging to different eigenvalues are mutually orthogonal.*

Proof: (1) Since $\mathbf{x}^H A\mathbf{x}$ is 1×1 matrix, $\overline{(\mathbf{x}^H A\mathbf{x})} = (\mathbf{x}^H A\mathbf{x})^H = \mathbf{x}^H A\mathbf{x}$.

(2) If $A\mathbf{x} = \lambda\mathbf{x}$, then $\mathbf{x}^H A\mathbf{x} = \mathbf{x}^H \lambda\mathbf{x} = \lambda\mathbf{x}^H\mathbf{x} = \lambda\|\mathbf{x}\|^2$. The left side is real and $\|\mathbf{x}\|^2$ is real and positive, because $\mathbf{x} \neq \mathbf{0}$. Therefore, λ must be real.

(3) Let $\mathbf{x}$ and $\mathbf{y}$ be eigenvectors of A belonging to eigenvalues λ and μ, respectively. Let $\lambda \neq \mu$. Because $A = A^H$ and λ is real, we get

$$\lambda(\mathbf{x}\cdot\mathbf{y}) = (\lambda\mathbf{x})\cdot\mathbf{y} = A\mathbf{x}\cdot\mathbf{y} = \mathbf{x}\cdot A\mathbf{y} = \mu(\mathbf{x}\cdot\mathbf{y}).$$

Since $\lambda \neq \mu$, it gives that $\mathbf{x}\cdot\mathbf{y} = \mathbf{x}^H\mathbf{y} = 0$, *i.e.*, $\mathbf{x}$ is orthogonal to $\mathbf{y}$. □

In particular, eigenvectors belonging to different eigenvalues of a real symmetric matrix are orthogonal.

Remark: Condition (1) in Theorem 7.1 (*i.e.*, $\mathbf{x}^H A\mathbf{x}$ is real for any complex vector $\mathbf{x} \in \mathbb{C}^n$) is equivalent to saying that the diagonals of A are real:

$$\begin{aligned} \mathbf{x}^H A\mathbf{x} &= [\bar{x}_1 \ \cdots \ \bar{x}_n] \begin{bmatrix} a_{11} & \cdots & a_{1n} \\ & \ddots & \\ a_{n1} & \cdots & a_{nn} \end{bmatrix} \begin{bmatrix} x_1 \\ \vdots \\ x_n \end{bmatrix} \\ &= \sum_{i,j} a_{ij}\bar{x}_i x_j \\ &= \sum_i a_{ii}|x_i|^2 + C + \bar{C}, \end{aligned}$$

where $C = \sum_{i<j} a_{ij}\bar{x}_i x_j$. Since $C+\bar{C} \in \mathbb{R}$, all $a_{ii} \in \mathbb{R}$ if and only if $\mathbf{x}^H A\mathbf{x} \in \mathbb{R}$ for any $\mathbf{x} \in \mathbb{C}^n$.

Problem 7.8 Prove that the determinant of any Hermitian matrix is real.

Problem 7.9 Let $\mathbf{x}$ be a nonzero vector in the complex vector space $\mathbb{C}^n$, and $A = \mathbf{x}\mathbf{x}^H$. Show that A is Hermitian, and find all the eigenvalues and their eigenspaces for A.

Note that if A is Hermitian, then the matrix iA is skew-Hermitian; similarly, if A is skew-Hermitian, then iA is Hermitian. Therefore, the following theorem is a direct consequence of these facts and Theorem 7.1. The proof is left for an exercise.

Theorem 7.2 *Let A be a skew-Hermitian matrix.*

(1) *For any complex vector $\mathbf{x} \neq \mathbf{0}$, $\mathbf{x}^H A\mathbf{x}$ is purely imaginary, and the diagonal entries of A are purely imaginary.*

(2) *Every eigenvalue of A is purely imaginary. In particular, a real skew-symmetric matrix has purely imaginary n eigenvalues.*

(3) *The eigenvectors of A belonging to different eigenvalues are mutually orthogonal.*

Problem 7.10 Prove Theorem 7.2 by using Theorem 7.1, and prove (3) directly.

Problem 7.11 Show that $A = B + iC$ (B and C real matrices) is skew-Hermitian if and only if B is skew-symmetric and C is symmetric.

Problem 7.12 Let A and B be either both Hermitian or both skew-Hermitian.
(1) AB is Hermitian if and only if $AB = BA$.
(2) AB is skew-Hermitian if and only if $AB = -BA$.

Recall that a real matrix Q is said to be orthogonal if the column vectors of Q are orthonormal (*i.e.*, $Q^T Q = I$). The same is true for complex matrices (see Lemma 5.13).

Lemma 7.3 *For a complex square matrix U, the following are equivalent:*

(1) *the column vectors of U are orthonormal;*

(2) $U^H U = I$;

(3) $U^{-1} = U^H$;

(4) $UU^H = I$;

(5) *the row vectors of U are orthonormal.*

The complex analogue to an orthogonal matrix is a *unitary* matrix.

Definition 7.5 A complex square matrix U is said to be **unitary** if it satisfies any one (and hence, all) of the conditions in Lemma 7.3.

Like a real orthogonal matrix, any unitary matrix preserves length.

Theorem 7.4 *Let U be an $n \times n$ unitary matrix.*

(1) *U preserves the inner products (and hence the length) on $\mathbb{C}^n$ for all vectors $\mathbf{x}$ and $\mathbf{y}$, $\langle U\mathbf{x}, U\mathbf{y}\rangle = \langle \mathbf{x}, \mathbf{y}\rangle$ (or equivalently, $\|U\mathbf{x}\| = \|\mathbf{x}\|$).*

(2) *If λ is an eigenvalue of U, then $|\lambda| = 1$.*

(3) *The eigenvectors of U belonging to different eigenvalues are mutually orthogonal.*

Proof: **(1)** $\langle U\mathbf{x}, U\mathbf{y}\rangle = \langle \mathbf{x}, U^H U\mathbf{y}\rangle = \langle \mathbf{x}, \mathbf{y}\rangle$, and setting $\mathbf{x} = \mathbf{y}$ gives us $\|U\mathbf{x}\|^2 = \|\mathbf{x}\|^2$.

(2) For $U\mathbf{x} = \lambda\mathbf{x}$, $\langle \mathbf{x}, \mathbf{x}\rangle = \langle U\mathbf{x}, U\mathbf{x}\rangle = \bar{\lambda}\lambda\langle \mathbf{x}, \mathbf{x}\rangle = |\lambda|^2\langle \mathbf{x}, \mathbf{x}\rangle$.

(3) Let $U\mathbf{x} = \lambda\mathbf{x}$, $U\mathbf{y} = \mu\mathbf{y}$, and $\lambda \neq \mu$. Since U is unitary, we have $\lambda\bar{\lambda} = 1 = \mu\bar{\mu}$, and $U^{-1}\mathbf{y} = \mu^{-1}\mathbf{y} = \bar{\mu}\mathbf{y}$. Therefore,

$$\bar{\lambda}\langle \mathbf{x}, \mathbf{y}\rangle = \langle U\mathbf{x}, \mathbf{y}\rangle = (U\mathbf{x})^H\mathbf{y} = \mathbf{x}^H U^{-1}\mathbf{y} = \mathbf{x}^H(\bar{\mu}\mathbf{y}) = \bar{\mu}\langle \mathbf{x}, \mathbf{y}\rangle$$

holds, and $\lambda \neq \mu$ implies $\langle \mathbf{x}, \mathbf{y}\rangle = 0$. □

Theorem 7.5 *A transition matrix from one orthonormal basis to another in a complex vector space is unitary.*

Proof: Let $\alpha = \{\mathbf{v}_1, \ldots, \mathbf{v}_n\}$ and $\beta = \{\mathbf{w}_1, \ldots, \mathbf{w}_n\}$ be two orthonormal bases, and let $Q = [q_{ij}]$ be the transition matrix from the basis β to the basis α. By definition,

$$\mathbf{w}_j = \sum_{\ell=1}^{n} q_{\ell j}\mathbf{v}_\ell.$$

Thus,

$$\begin{aligned}
\delta_{ij} = \langle \mathbf{w}_i, \mathbf{w}_j\rangle &= \left\langle \sum_{k=1}^{n} q_{ki}\mathbf{v}_k, \sum_{\ell=1}^{n} q_{\ell j}\mathbf{v}_\ell \right\rangle \\
&= \sum_{k=1}^{n} \overline{q_{ki}} \sum_{\ell=1}^{n} q_{\ell j}\langle \mathbf{v}_k, \mathbf{v}_\ell\rangle \\
&= \sum_{k=1}^{n} \overline{q_{ki}} q_{kj} = \sum_{k=1}^{n} [Q^H]_{ik}^H [Q]_{kj}.
\end{aligned}$$

This means that the columns of Q are orthonormal and Q is unitary. □

As in the real case, it is true that two matrices representing the same linear transformation on a complex vector space with respect to different bases are similar. If the two bases are both orthonormal, then the transition matrix is unitary (or orthogonal).

Problem 7.13 Show that $|\det U| = 1$ for any unitary matrix U.

Problem 7.14 Show that

$$A = \begin{bmatrix} \frac{1+i}{4} & \frac{1+i}{4} \\ \frac{1-i}{4} & \frac{-1+i}{4} \end{bmatrix}$$

is unitary but neither Hermitian nor skew-Hermitian.

Problem 7.15 Show that if U is a unitary matrix, then so is U^H.

Problem 7.16 Show that if A and B are unitary, so is AB.

Problem 7.17 Describe all 3×3 matrices that are simultaneously Hermitian, unitary, and diagonal. How many are there?

7.3 Unitarily diagonalizable matrices

In the previous section, we saw that if an $n \times n$ square matrix A is Hermitian, skew-Hermitian or unitary, then the eigenvectors belonging to distinct eigenvalues are mutually orthogonal. Hence, if such a matrix A has n distinct eigenvalues, then there exists an orthonormal basis α for $\mathbb{C}^n$ consisting of eigenvectors of A so that the matrix representation $[A]_\alpha$ is diagonal, *i.e.*, A is diagonalizable by a unitary matrix. In this section, it will be shown that any Hermitian, skew-Hermitian or unitary matrix has n orthonormal eigenvectors even if the eigenvalues are not all distinct. In particular, it is always diagonalizable by a unitary matrix.

Definition 7.6 (1) Two real matrices A and B are **orthogonally similar** if there exists an orthogonal matrix P such that $P^{-1}AP = B$. A matrix is **orthogonally diagonalizable** if it is orthogonally similar to a diagonal matrix.

(2) Two complex matrices A and B are **unitarily similar** if there exists a unitary matrix U such that $U^{-1}AU = B$. A matrix is **unitarily diagonalizable** if it is unitarily similar to a diagonal matrix.

We begin with a classical theorem due to Schur (1909) concerning orthogonal and unitary similarity.

Lemma 7.6 (Schur's lemma) **(1)** *If an $n \times n$ real matrix A has only real eigenvalues, then A is orthogonally similar to an upper triangular matrix.*

(2) *Every $n \times n$ complex matrix is unitarily similar to an upper triangular matrix.*

Proof: We will prove the second assertion only. The proof of the first is similar. We will prove it by mathematical induction on n. Clearly, it is true for $n = 1$. Assume now that the theorem holds for $n = r - 1$. Let A be an $r \times r$ matrix and let λ_1 be an eigenvalue of A with a normalized eigenvector $\mathbf{x}$. Extend it to an orthonormal basis by the Gram-Schmidt orthogonalization, say $\{\mathbf{x},\ \mathbf{u}_2,\ \ldots,\ \mathbf{u}_r\}$. Set the unitary matrix $U_1 = [\mathbf{x}\ \mathbf{u}_2 \cdots\ \mathbf{u}_r]$ with these basis vectors in its columns. A direct computation of the product $U_1^{-1}AU_1$ shows

$$\begin{aligned}
U_1^{-1}AU_1 &= U_1^H AU_1 = U_1^H[A\mathbf{x}\ A\mathbf{u}_2\ \cdots\ A\mathbf{u}_r] \\
&= \begin{bmatrix} -- & \bar{\mathbf{x}}^T & -- \\ -- & \bar{\mathbf{u}}_2^T & -- \\ & \vdots & \\ -- & \bar{\mathbf{u}}_r^T & -- \end{bmatrix} \begin{bmatrix} | & | & & | \\ \lambda_1\mathbf{x} & A\mathbf{u}_2 & \cdots & A\mathbf{u}_r \\ | & | & & | \end{bmatrix} \\
&= \begin{bmatrix} \lambda_1 & | & & * & \\ -- & + & -- & -- & -- \\ 0 & | & & & \\ \vdots & | & & B & \\ 0 & | & & & \end{bmatrix},
\end{aligned}$$

where B is an $(r-1) \times (r-1)$ matrix. By the inductive hypothesis there exists an $(r-1) \times (r-1)$ unitary matrix U_2 such that $U_2^{-1}BU_2$ is an upper triangular matrix with diagonal entries $\lambda_2,\ \lambda_3,\ \cdots,\ \lambda_r$. Define

$$U = U_1 \begin{bmatrix} 1 & 0 \\ 0 & U_2 \end{bmatrix}.$$

Then it is easy to check that U is also a unitary matrix, so

$$U^{-1}AU = U^H AU = \begin{bmatrix} \lambda_1 & * \\ 0 & U_2^H BU_2 \end{bmatrix} = \begin{bmatrix} \lambda_1 & & & \\ & \lambda_2 & & * \\ 0 & & \ddots & \\ & & & \lambda_r \end{bmatrix}. \qquad \square$$

Schur's lemma is a cornerstone in the study of complex matrices.

Theorem 7.7 *If A is either a Hermitian, a skew-Hermitian or a unitary matrix, then it is unitarily diagonalizable.*

Proof: By Schur's lemma, $U^H AU = B$ is an upper triangular matrix for some unitary matrix U. However,

$$B^H = U^H A^H U = \begin{cases} U^H(\pm A)U = \pm B & \text{if } A \text{ is (skew-) Hermitian} \\ U^{-1}A^{-1}U = B^{-1} & \text{if } A \text{ is unitary,} \end{cases}$$

where the right sides of the equalities depend on whether A is either a Hermitian, a skew-Hermitian or a unitary matrix. This means that B is also either a Hermitian, a skew-Hermitian or a unitary matrix depending on whether A is either a Hermitian, a skew-Hermitian or a unitary matrix, respectively.

It is quite easy to show that an upper triangular matrix that is also Hermitian or skew-Hermitian must already be a diagonal matrix. Moreover, if an upper triangular matrix B is unitary, then one can easily show that B is already a diagonal matrix by comparing the diagonal entries of both sides of $B^H B = BB^H$. □

Note that, in the similarity condition $U^{-1}AU(= U^H AU) = D$ of A to a diagonal matrix D through a unitary matrix U, the equation $AU = UD$ shows that the column vectors of U constitute a set of n orthonormal eigenvectors of A while the diagonal entries of D are eigenvalues of A as shown in Theorem 6.6. Therefore, by Theorems 7.1, 7.2 and 7.4, all the diagonal entries of D are real, purely imaginary or of unit length depending on the types (Hermitian, skew-Hermitian or unitary, respectively) of the matrix A.

Example 7.5 Let

$$A = \begin{bmatrix} 2 & 1-i \\ 1+i & 1 \end{bmatrix}.$$

Then A is Hermitian, and the eigenvalues of A are $\lambda_1 = 3$ and $\lambda_2 = 0$ with associated eigenvectors $\mathbf{x}_1 = (1-i,\ 1)$ and $\mathbf{x}_2 = (-1,\ 1+i)$. Let

$$\begin{aligned} \mathbf{u}_1 &= \frac{\mathbf{x}_1}{\|\mathbf{x}_1\|} = \frac{1}{\sqrt{3}}(1-i,\ 1), \\ \mathbf{u}_2 &= \frac{\mathbf{x}_2}{\|\mathbf{x}_2\|} = \frac{1}{\sqrt{3}}(-1,\ 1+i), \end{aligned}$$

and let

$$U = \frac{1}{\sqrt{3}} \begin{bmatrix} 1-i & -1 \\ 1 & 1+i \end{bmatrix}.$$

Then U is a unitary matrix and diagonalizes A:

$$\begin{aligned} U^H AU &= \frac{1}{3} \begin{bmatrix} 1+i & 1 \\ -1 & 1-i \end{bmatrix} \begin{bmatrix} 2 & 1-i \\ 1+1 & 1 \end{bmatrix} \begin{bmatrix} 1-i & -1 \\ 1 & 1+i \end{bmatrix} \\ &= \begin{bmatrix} 3 & 0 \\ 0 & 0 \end{bmatrix}. \end{aligned}$$

□

Since all the real symmetric matrices are Hermitian matrices, they are unitarily diagonalizable by Theorem 7.7. However, the following theorem says more than that.

Theorem 7.8 *Let A be an $n \times n$ real matrix. Then the following are equivalent.*

(1) *A is symmetric.*

(2) *A is orthogonally diagonalizable.*

(3) *A has a full set of n orthonormal eigenvectors.*

Proof: **(1)** $\Leftrightarrow$ **(2)**: If A is real and symmetric, then it is a Hermitian matrix, so it has only real eigenvalues. By Schur's lemma 7.6, A is orthogonally similar to an upper triangular matrix, which must be already diagonal. Hence it is orthogonally diagonalizable. Conversely, if A is orthogonally diagonalizable, *i.e.*, there is an orthogonal matrix Q such that $Q^{-1}AQ = D$, then $A = QDQ^{-1} = QDQ^T$ is clearly symmetric.

(2) $\Leftrightarrow$ **(3)** : If A is diagonalized by an orthogonal matrix Q, then the column vectors of Q are eigenvectors of A. Hence A has a full set of n orthonormal eigenvectors. Conversely, if A has a full set of n orthonormal eigenvectors, then these eigenvectors form an orthogonal transition matrix Q that diagonalizes A. □

Corollary 7.9 *If A is a real symmetric matrix, then the dimension of the eigenspace $E(\lambda) = \mathcal{N}(\lambda I - A)$ belonging to an eigenvalue λ is equal to the multiplicity of λ as a root of the characteristic polynomial.*

Therefore, if a real matrix A is symmetric, then it is always diagonalizable, even more, orthogonally. Moreover, they are all that can be "orthogonally" diagonalized. Even though not all matrices are diagonalizable, certain nonsymmetric matrices may still have a full set of linearly independent eigenvectors so that they are diagonalizable, but in this case the eigenvectors cannot be orthogonal. That is, the transition matrix Q cannot be an orthogonal matrix. For example, one can verify that the matrix $A = \begin{bmatrix} 1 & 2 \\ 1 & 1 \end{bmatrix}$ is nonsymmetric, and has two linearly independent eigenvectors so that it is diagonalizable, but not orthogonally diagonalizable.

Problem 7.18 Show that the nonsymmetric matrix

$$A = \begin{bmatrix} 1 & 0 & -1 \\ 0 & 1 & 0 \\ 0 & 0 & 2 \end{bmatrix}$$

is diagonalizable.

Remark: The procedure for orthogonal diagonalization of a symmetric matrix A can be summarized as follows.

Step 1. Find a basis for each eigenspace of A.

Step 2. Apply the Gram-Schmidt orthogonalization to each of these bases to obtain an orthonormal basis for each eigenspace.

Step 3. Form the matrix Q whose columns are the basis vectors constructed in *Step 2*; this matrix orthogonally diagonalizes A.

The justification of this procedure should be clear, because eigenvectors belonging to *distinct* eigenvalues are orthogonal, while an application of the Gram-Schmidt orthogonalization assures that the eigenvectors obtained within the *same* eigenspace are orthonormal. Thus, the entire set of eigenvectors obtained by this procedure is orthonormal.

Example 7.6 Find an orthogonal matrix Q that diagonalizes

$$A = \begin{bmatrix} 4 & 2 & 2 \\ 2 & 4 & 2 \\ 2 & 2 & 4 \end{bmatrix}.$$

Solution: The characteristic polynomial of A is

$$\det(\lambda I - A) = \det \begin{bmatrix} \lambda - 4 & -2 & -2 \\ -2 & \lambda - 4 & -2 \\ -2 & -2 & \lambda - 4 \end{bmatrix} = (\lambda - 2)^2(\lambda - 8).$$

Thus, the eigenvalues of A are $\lambda = 2$ and $\lambda = 8$. By the method used in Example 6.2, it can be shown that

$$\mathbf{u}_1 = (-1,\ 1,\ 0) \quad \text{and} \quad \mathbf{u}_2 = (-1,\ 0,\ 1)$$

form a basis for the eigenspace belonging to $\lambda = 2$. Applying the Gram-Schmidt orthogonalization to $\{\mathbf{u}_1,\ \mathbf{u}_2\}$ yields the following orthonormal eigenvectors (verify):

$$\mathbf{v}_1 = \frac{1}{\sqrt{2}}(-1,\ 1,\ 0) \quad \text{and} \quad \mathbf{v}_2 = \frac{1}{\sqrt{6}}(-1,\ -1,\ 2).$$

The eigenspace belonging to $\lambda = 8$ has $\mathbf{u}_3 = (1,\ 1,\ 1)$ as a basis. The normalization of $\mathbf{u}_3$ yields $\mathbf{v}_3 = \dfrac{1}{\sqrt{3}}(1,\ 1,\ 1)$. Finally, using $\mathbf{v}_1,\ \mathbf{v}_2$, and $\mathbf{v}_3$ as column vectors, we obtain

$$Q = \begin{bmatrix} -\frac{1}{\sqrt{2}} & -\frac{1}{\sqrt{6}} & \frac{1}{\sqrt{3}} \\ \frac{1}{\sqrt{2}} & -\frac{1}{\sqrt{6}} & \frac{1}{\sqrt{3}} \\ 0 & \frac{2}{\sqrt{6}} & \frac{1}{\sqrt{3}} \end{bmatrix},$$

which orthogonally diagonalizes A. (It is suggested that the readers verify that $Q^T AQ$ is actually a diagonal matrix.) □

7.4 Normal matrices

We have seen that Hermitian, skew-Hermitian and unitary matrices are all unitarily diagonalizable. However, it turns out that they do not constitute the entire class of unitarily diagonalizable matrices, whereas in the class of real matrices the real symmetric matrices are the only matrices with real entries that are orthogonally diagonalizable. That is, there are many unitarily diagonalizable matrices that are neither one of the above-mentioned classes of matrices. Actually, all unitarily diagonalizable matrices belong to the following class of matrices, called *normal* matrices.

Definition 7.7 A complex square matrix A is called **normal** if

$$AA^H = A^H A.$$

Note that all the Hermitian, unitary and skew-Hermitian matrices are normal. There are matrices that are normal but are none of these.

Example 7.7 The 2×2 matrix

$$A = \begin{bmatrix} 1 & i \\ i & 1 \end{bmatrix}$$

is normal, but is neither Hermitian, skew-Hermitian, unitary, nor a diagonal matrix. However, one can easily check that this matrix is unitarily diagonalizable. In fact,

$$U^{-1}AU = \frac{1}{2}\begin{bmatrix} 1 & 1 \\ 1 & -1 \end{bmatrix}\begin{bmatrix} 1 & i \\ i & 1 \end{bmatrix}\begin{bmatrix} 1 & 1 \\ 1 & -1 \end{bmatrix} = \begin{bmatrix} 1+i & 0 \\ 0 & 1-i \end{bmatrix} = D. \quad \square$$

Problem 7.19 Which of following matrices are Hermitian, skew-Hermitian or normal?

$$(1)\ \begin{bmatrix} 1 & 1+i \\ 1-i & 0 \end{bmatrix};\quad (2)\ \begin{bmatrix} 1 & 1 & 1 \\ 1 & 1 & 1 \\ 1 & 1 & 1 \end{bmatrix};\quad (3)\ \begin{bmatrix} i & 1 & -i \\ -1 & 2i & 0 \\ -i & 0 & -i \end{bmatrix};$$

$$(4)\ \begin{bmatrix} -i & 3 \\ 3 & 2 \end{bmatrix};\quad (5)\ \begin{bmatrix} 0 & 0 & i \\ 0 & i & 0 \\ i & 0 & 0 \end{bmatrix};\quad (6)\ \begin{bmatrix} 3 & 1+i & i \\ 1-i & 1 & 3 \\ -i & 3 & 1 \end{bmatrix}.$$

As a matter of fact, the theorem below shows that normal matrices are all classified as unitarily diagonalizable matrices. We begin with a lemma.

Lemma 7.10 *If an upper triangular matrix T is normal, then it must be a diagonal matrix.*

Proof: We first make a direct computation of the equation $TT^H = T^H T$:

$$\begin{bmatrix} t_{11} & & t_{1n} \\ & \ddots & \\ 0 & & t_{nn} \end{bmatrix}\begin{bmatrix} \overline{t_{11}} & & 0 \\ & \ddots & \\ \overline{t_{1n}} & & \overline{t_{nn}} \end{bmatrix} = \begin{bmatrix} \overline{t_{11}} & & 0 \\ & \ddots & \\ \overline{t_{1n}} & & \overline{t_{nn}} \end{bmatrix}\begin{bmatrix} t_{11} & & t_{1n} \\ & \ddots & \\ 0 & & t_{nn} \end{bmatrix},$$

and then compare the corresponding diagonal entries of both sides, *i.e.*,

$$(TT^H)_{11} = |t_{11}|^2 + \cdots + |t_{1n}|^2 = (T^H T)_{11} = |t_{11}|^2,$$

which implies $t_{12} = \cdots = t_{1n} = 0$. Inductively, assume that $t_{i-1i} = \cdots = t_{i-1n} = 0$ has been shown for $i = 1, \ldots, k$. Then

$$(TT^H)_{kk} = |t_{kk}|^2 + \cdots + |t_{kn}|^2$$

and

$$(T^HT)_{kk} = |t_{1k}|^2 + \cdots + |t_{k-1k}|^2 + |t_{kk}|^2 = |t_{kk}|^2,$$

because $t_{1k} = \cdots = t_{k-1k} = 0$ by induction hypothesis. But $TT^H = T^HT$ yields $t_{kk+1} = \cdots = t_{kn} = 0$. Thus, we get $t_{ii+1} = \cdots = t_{in} = 0$ for all $i = 1, \ldots, n$, which shows that all the entries of T off the diagonal are zero, *i.e.*, T is a diagonal matrix. □

Theorem 7.11 *If A is a complex square matrix, then the following are equivalent:*

(1) *A is normal;*

(2) *A is unitarily diagonalizable;*

(3) *A has a full set of n orthonormal eigenvectors.*

Proof: (1) ⇔ (2) Suppose that A is normal. By Schur's lemma again, we can find a unitary matrix U so that $T = U^HAU$ is an upper triangular matrix. Then T is also normal, since

$$\begin{aligned} TT^H &= U^HAUU^HA^HU = U^HAA^HU = U^HA^HAU \\ &= U^HA^HUU^HAU = T^HT. \end{aligned}$$

Thus by Lemma 7.10, T is already diagonal so that A is unitarily diagonalizable. Conversely, if A is unitarily diagonalizable, *i.e.*, $U^HAU = D$ for a unitary matrix U, then

$$\begin{aligned} AA^H &= UDU^HUD^HU^H = UDD^HU^H = UD^HDU^H \\ &= UD^HU^HUDU^H = A^HA. \end{aligned}$$

That is, A is normal.

The proof of **(2)** ⇔ **(3)** is the same as in the proof of Theorem 7.8. □

Note that there are many nonnormal complex matrices that are still diagonalizable, but of course not unitarily. Readers are suggested to find such an example.

Recall that an $n \times n$ real matrix A is the sum $S+T$ of a symmetric matrix $S = \frac{1}{2}(A + A^T)$ and a skew-symmetric matrix $T = \frac{1}{2}(A - A^T)$. The same can be said about a complex matrix. Let A be a complex matrix. Then it is the sum $A = H_1 + iH_2$, where

$$H_1 = \frac{1}{2}(A + A^H), \quad H_2 = -\frac{i}{2}(A - A^H) \text{ or } iH_2 = \frac{1}{2}(A - A^H).$$

Both H_1 and H_2 are Hermitian, so iH_2 is skew-Hermitian.

Problem 7.20 Determine whether or not the matrix

$$A = \begin{bmatrix} 1 & i & i \\ i & 1 & i \\ i & i & 1 \end{bmatrix}$$

is unitarily diagonalizable. If it is, find a unitary matrix U that diagonalizes A.

Problem 7.21 For any real matrices H_1 and H_2 of the same size, show that $A = H_1 + iH_2$ is normal if and only if $H_1H_2 = H_2H_1$.

Problem 7.22 For any unitarily diagonalizable matrix A, prove that

(1) A is Hermitian if and only if A has only real eigenvalues;
(2) A is skew-Hermitian if and only if A has only purely imaginary eigenvalues;
(3) A is unitary if and only if $|\lambda| = 1$ for any eigenvalue λ of A.

7.5 The spectral theorem

As we saw in the previous section, normal matrices are the matrices that are unitarily diagonalizable. That is, A is normal if and only if there exists a basis α for $\mathbb{C}^n$ consisting of orthonormal eigenvectors of A such that the matrix representation $[A]_\alpha$ of A with respect to α is diagonal.

Theorem 7.12 (Spectral theorem) *Let A be a normal matrix, and let $\{\mathbf{u}_1, \mathbf{u}_2, \ldots, \mathbf{u}_n\}$ be a set of orthonormal eigenvectors belonging to the eigenvalues $\lambda_1, \lambda_2, \ldots, \lambda_n$ of A, respectively. Then A can be written in the following form:*

$$A = UDU^H = \lambda_1 P_1 + \lambda_2 P_2 + \cdots + \lambda_n P_n,$$

where $P_i = \mathbf{u}_i\mathbf{u}_i^H$ is the orthogonal projection matrix onto the subspace spanned by the eigenvector $\mathbf{u}_i$ for each $i = 1, \cdots, n$.

The above expression of A is called the **spectral decomposition** of A into projections.

Proof: Note that $U = [\mathbf{u}_1\ \mathbf{u}_2\ \cdots\ \mathbf{u}_n]$ is the unitary matrix that transforms A into a diagonal matrix D, *i.e.*, $U^{-1}AU = U^HAU = D$. Then

$$\begin{aligned} A &= UDU^H = [\mathbf{u}_1\ \mathbf{u}_2\ \cdots\ \mathbf{u}_n] \begin{bmatrix} \lambda_1 \mathbf{u}_1^H \\ \vdots \\ \lambda_n \mathbf{u}_n^H \end{bmatrix} \\ &= \lambda_1 \mathbf{u}_1 \mathbf{u}_1^H + \cdots + \lambda_n \mathbf{u}_n \mathbf{u}_n^H \\ &= \lambda_1 P_1 + \cdots + \lambda_n P_n, \end{aligned}$$

where

$$P_i = \mathbf{u}_i \mathbf{u}_i^H = \begin{bmatrix} u_{1i} \\ \vdots \\ u_{ni} \end{bmatrix} [\bar{u}_{1i}\ \cdots\ \bar{u}_{ni}] = \begin{bmatrix} |u_{1i}|^2 & \cdots & u_{1i}\bar{u}_{ni} \\ \vdots & \ddots & \vdots \\ u_{ni}\bar{u}_{1i} & \cdots & |u_{ni}|^2 \end{bmatrix},$$

which is a Hermitian matrix.

Now, for any $\mathbf{x} \in \mathbb{C}^n$ and $i, j = 1, \ldots, n$,

$$\begin{aligned} P_i\mathbf{x} &= \mathbf{u}_i \mathbf{u}_i^H \mathbf{x} = \langle \mathbf{u}_i, \mathbf{x} \rangle \mathbf{u}_i, \\ P_iP_j &= \mathbf{u}_i \mathbf{u}_i^H \mathbf{u}_j \mathbf{u}_j^H \ = \ \langle \mathbf{u}_i, \mathbf{u}_j \rangle \mathbf{u}_i \mathbf{u}_j^H \\ &= \begin{cases} \mathbf{u}_i \mathbf{u}_i^H = P_i & \text{if } i = j, \\ \mathbf{u}_i \mathbf{u}_j^H = \mathbf{0} & \text{if } i \neq j, \end{cases} \\ (P_1 + \cdots + P_n)\mathbf{x} &= P_1\mathbf{x} + \cdots + P_n\mathbf{x} = \sum_{i=1}^n \langle \mathbf{u}_i, \mathbf{x} \rangle \mathbf{u}_i = \mathbf{x} = Id(\mathbf{x}). \end{aligned}$$

Therefore, each P_i is nothing but the orthogonal projection onto the subspace spanned by $\mathbf{u}_i$ which is the eigenspace $E(\lambda_i) = \mathcal{N}(\lambda_i I - A)$. □

Note that the equation $P_1 + \cdots + P_n = Id$ means that if we restrict the images of the P_i's to be the span of $\mathbf{u}_i$ which is just $\mathbb{R}$, then $(P_1, \ldots, P_n)$ is another orthogonal coordinate expression just like $(z_1, \ldots, z_n)$ of the $\mathbb{C}^n$, but in this case, with respect to the orthonormal basis $\{\mathbf{u}_1,\ \mathbf{u}_2,\ \ldots,\ \mathbf{u}_n\}$ (see Sections 5.4 and 5.10).

Note that any $\mathbf{x} \in \mathbb{C}^n$ has the unique expression $\mathbf{x} = \sum \langle \mathbf{u}_i, \mathbf{x} \rangle \mathbf{u}_i$ as a linear combination of the orthonormal basis vector $\mathbf{u}_i$'s, and

$$A\mathbf{x} = \lambda_1 P_1 \mathbf{x} + \lambda_2 P_2 \mathbf{x} + \cdots + \lambda_n P_n \mathbf{x}$$

$$\begin{aligned} &= \lambda_1 \mathbf{u}_1(\mathbf{u}_1^H \mathbf{x}) + \cdots + \lambda_n \mathbf{u}_n(\mathbf{u}_n^H \mathbf{x}) \\ &= \lambda_1 \langle \mathbf{u}_1, \mathbf{x} \rangle \mathbf{u}_1 + \cdots + \lambda_n \langle \mathbf{u}_n, \mathbf{x} \rangle \mathbf{u}_n. \end{aligned}$$

If an eigenvalue λ has multiplicity ℓ, *i.e.*, $\lambda = \lambda_{i_1} = \cdots = \lambda_{i_\ell}$, with a set of ℓ orthonormal eigenvectors $\mathbf{u}_{i_1}, \ldots, \mathbf{u}_{i_\ell}$, then they form an orthonormal basis for the eigenspace $E(\lambda)$, and

$$P_\lambda = P_{i_1} + \cdots + P_{i_\ell} = \mathbf{u}_{i_1}\mathbf{u}_{i_1}^H + \cdots + \mathbf{u}_{i_\ell}\mathbf{u}_{i_\ell}^H$$

is the orthogonal projection matrix onto $E(\lambda)$. Therefore, counting the multiplicity of each eigenvalue, every normal matrix A has the unique **spectral decomposition** into projections

$$A = \lambda_1 P_{\lambda_1} + \cdots + \lambda_k P_{\lambda_k},$$

for $k \le n$, where λ_i's are all distinct.

Corollary 7.13 *Let A be a normal matrix.*

(1) *The eigenvectors of A belonging to different eigenvalues are mutually orthogonal.*

(2) *If an eigenvalue λ of A has multiplicity k, then the eigenspace $\mathcal{N}(A - \lambda I)$ belonging to λ is of dimension k.*

Corollary 7.14 *Let A be a normal matrix with the spectral decomposition $A = \lambda_1 P_{\lambda_1} + \cdots + \lambda_k P_{\lambda_k}$. Then, for any positive integer m,*

$$A^m = \lambda_1^m P_{\lambda_1} + \cdots + \lambda_k^m P_{\lambda_k}.$$

Moreover, if A is invertible, then for any positive integer ℓ

$$A^{-\ell} = \frac{1}{\lambda_1^\ell} P_{\lambda_1} + \cdots + \frac{1}{\lambda_k^\ell} P_{\lambda_k}.$$

Example 7.8 Find the spectral decomposition of

$$A = \begin{bmatrix} 4 & 2 & 2 \\ 2 & 4 & 2 \\ 2 & 2 & 4 \end{bmatrix}.$$

Solution: From Example 7.6, the spectral decomposition is

$$A = 2(P_1 + P_2) + 8P_3,$$

where the projections are

$$\begin{aligned} P_1 &= \mathbf{v}_1\mathbf{v}_1^H = \frac{1}{2}\begin{bmatrix} -1 \\ 1 \\ 0 \end{bmatrix}[-1\ 1\ 0] = \frac{1}{2}\begin{bmatrix} 1 & -1 & 0 \\ -1 & 1 & 0 \\ 0 & 0 & 0 \end{bmatrix}, \\ P_2 &= \mathbf{v}_2\mathbf{v}_2^H = \frac{1}{6}\begin{bmatrix} -1 \\ -1 \\ 2 \end{bmatrix}[-1\ -1\ 2] = \frac{1}{6}\begin{bmatrix} 1 & 1 & -2 \\ 1 & 1 & -2 \\ -2 & -2 & 4 \end{bmatrix}, \\ P_3 &= \mathbf{v}_3\mathbf{v}_3^H = \frac{1}{3}\begin{bmatrix} 1 \\ 1 \\ 1 \end{bmatrix}[1\ 1\ 1] = \frac{1}{3}\begin{bmatrix} 1 & 1 & 1 \\ 1 & 1 & 1 \\ 1 & 1 & 1 \end{bmatrix}. \end{aligned}$$

Hence,

$$P_1' = P_1 + P_2 = \frac{1}{3}\begin{bmatrix} 2 & -1 & -1 \\ -1 & 2 & -1 \\ -1 & -1 & 2 \end{bmatrix}$$

is the projection onto the eigenspace $E(2)$ belonging to $\lambda = 2$, P_3 is the projection onto the eigenspace $E(8)$ belonging to $\lambda = 8$, and

$$A = \begin{bmatrix} 4 & 2 & 2 \\ 2 & 4 & 2 \\ 2 & 2 & 4 \end{bmatrix} = \frac{2}{3}\begin{bmatrix} 2 & -1 & -1 \\ -1 & 2 & -1 \\ -1 & -1 & 2 \end{bmatrix} + \frac{8}{3}\begin{bmatrix} 1 & 1 & 1 \\ 1 & 1 & 1 \\ 1 & 1 & 1 \end{bmatrix}.$$

□

Problem 7.23 Given $A = \begin{bmatrix} 0 & 2 & -1 \\ 2 & 3 & -2 \\ -1 & -2 & 0 \end{bmatrix}$, find an orthogonal matrix Q that diagonalizes A, and find the spectral decomposition of A.

Example 7.9 Find the spectral decomposition of a normal matrix

$$A = \begin{bmatrix} 0 & 0 & i \\ 0 & i & 0 \\ i & 0 & 0 \end{bmatrix}.$$

Solution: Since A is normal ($AA^H = A^HA$), it is unitarily diagonalizable. The characteristic polynomial of A is

$$\det(\lambda I - A) = \det\begin{bmatrix} \lambda & 0 & -i \\ 0 & \lambda - i & 0 \\ -i & 0 & \lambda \end{bmatrix} = (\lambda - i)^2(\lambda + i).$$

Hence, the eigenvalues are $i = \lambda_1 = \lambda_2$ of multiplicity 2 and $-i = \lambda_3$ of multiplicity 1. By a simple computation using the Gram-Schmidt orthogonalization, one can find that

$$\mathbf{v}_1 = \frac{1}{\sqrt{2}}\begin{bmatrix} 1 \\ 0 \\ 1 \end{bmatrix}, \quad \mathbf{v}_2 = \begin{bmatrix} 0 \\ 1 \\ 0 \end{bmatrix}, \quad \mathbf{v}_3 = \frac{1}{\sqrt{2}}\begin{bmatrix} -1 \\ 0 \\ 1 \end{bmatrix}$$

are orthonormal eigenvectors of A belonging to the eigenvalues λ_1, λ_2 and λ_3, respectively. Now, the spectral decomposition is $A = i(P_1 + P_2) - iP_3$, where the projection matrices are

$$P_1 = \mathbf{v}_1\mathbf{v}_1^H = \frac{1}{2}\begin{bmatrix} 1 \\ 0 \\ 1 \end{bmatrix}[1\ 0\ 1] = \frac{1}{2}\begin{bmatrix} 1 & 0 & 1 \\ 0 & 0 & 0 \\ 1 & 0 & 1 \end{bmatrix},$$

$$P_2 = \mathbf{v}_2\mathbf{v}_2^H = \begin{bmatrix} 0 \\ 1 \\ 0 \end{bmatrix}[0\ 1\ 0] = \begin{bmatrix} 0 & 0 & 0 \\ 0 & 1 & 0 \\ 0 & 0 & 0 \end{bmatrix},$$

$$P_3 = \mathbf{v}_3\mathbf{v}_3^H = \frac{1}{2}\begin{bmatrix} -1 \\ 0 \\ 1 \end{bmatrix}[-1\ 0\ 1] = \frac{1}{2}\begin{bmatrix} 1 & 0 & -1 \\ 0 & 0 & 0 \\ -1 & 0 & 1 \end{bmatrix}.$$

Hence,

$$A = i(P_1 + P_2) - iP_3 = \frac{i}{2}\begin{bmatrix} 1 & 0 & 1 \\ 0 & 2 & 0 \\ 1 & 0 & 1 \end{bmatrix} - \frac{i}{2}\begin{bmatrix} 1 & 0 & -1 \\ 0 & 0 & 0 \\ -1 & 0 & 1 \end{bmatrix}.$$

□

Problem 7.24 Find the spectral decomposition of each of the following matrices:

$$(1)\ A = \begin{bmatrix} 2 & 1 \\ 1 & 2 \end{bmatrix}; \qquad (2)\ B = \begin{bmatrix} 1 & 2+i \\ 2-i & 3 \end{bmatrix};$$

$$(3)\ C = \begin{bmatrix} 1 & 0 & 0 & 0 \\ 0 & 2 & 0 & 0 \\ 0 & 0 & 2 & 0 \\ 0 & 0 & 0 & 4 \end{bmatrix}; \qquad (4)\ D = \begin{bmatrix} 1 & 1 & 1 & 1 \\ 1 & 0 & 0 & 0 \\ 1 & 0 & 0 & 0 \\ 1 & 0 & 0 & 0 \end{bmatrix}.$$

7.6 Exercises

7.1. Calculate $\|\mathbf{x}\|$ for

$$(1)\ \mathbf{x} = \begin{bmatrix} 1+i \\ 2 \end{bmatrix}, \quad (2)\ \mathbf{x} = \begin{bmatrix} 1-2i \\ i \\ 3+i \end{bmatrix}.$$

7.2. Construct an orthonormal basis for $\mathbb{C}^2$ from $\{(i,\ 4+2i),\ (5+6i,\ 1)\}$ by applying the Gram-Schmidt orthogonalization.

7.3. Find the rank of the matrix $A = \begin{bmatrix} i & 1 & 1-i & 1+i \\ 1-i & 1+i & 1 & 2+i \\ 1+3i & 1-i & 2-i & 1+4i \end{bmatrix}$.

7.4. Find the eigenvalues and eigenvectors for each of the following matrices:

$$(1)\ \begin{bmatrix} -2 & -1 \\ 5 & 2 \end{bmatrix}, \qquad (2)\ \begin{bmatrix} 0 & i \\ -i & 0 \end{bmatrix},$$

$$(3)\ \begin{bmatrix} 5 & -5 & -5 \\ -1 & 4 & 2 \\ 3 & -5 & -3 \end{bmatrix}, \quad (4)\ \begin{bmatrix} 0 & -i & 0 \\ i & 1 & i \\ 0 & -i & 0 \end{bmatrix}.$$

7.5. Find the third column vector $\mathbf{v}$ so that $U = \begin{bmatrix} \frac{1}{\sqrt{3}} & \frac{1}{\sqrt{2}} & | \\ \frac{1}{\sqrt{3}} & 0 & \mathbf{v} \\ \frac{1}{\sqrt{3}} & -\frac{1}{\sqrt{2}} & | \end{bmatrix}$ is unitary.

How much freedom is there in this choice?

7.6. Find a real matrix A such that $A + rI$ is invertible for all $r \in \mathbb{R}$. Does there exist a square matrix A such that $A + cI$ is invertible for all $c \in \mathbb{C}$?

7.7. Find a unitary matrix whose first row is

(1) $k(1,\ 1-i)$ where k is a number, (2) $(\frac{1}{2},\ \frac{i}{2},\ \frac{1-i}{2})$.

7.8. Let $V = \mathbb{C}^2$ with the Euclidean inner product. Let T be the linear transformation on V with the matrix representation $A = \begin{bmatrix} 1 & i \\ 1 & 1 \end{bmatrix}$ with respect to the standard basis. Show that T is normal and find a set of orthonormal eigenvectors of T.

7.9. Prove that the following matrices are unitarily similar:

$$\begin{bmatrix} \cos\theta & -\sin\theta \\ \sin\theta & \cos\theta \end{bmatrix}, \quad \begin{bmatrix} e^{i\theta} & 0 \\ 0 & e^{-i\theta} \end{bmatrix},$$ where θ is a real number.

7.10. For each of the following real symmetric matrices A, find a real orthogonal matrix Q such that Q^TAQ is diagonal:

$$(1)\ \begin{bmatrix} 1 & 1 \\ 1 & 1 \end{bmatrix}, \quad (2)\ \begin{bmatrix} 1 & 2 \\ 2 & 1 \end{bmatrix}.$$

7.11. For each of the following Hermitian matrices A, find a unitary matrix U such that $U^H AU$ is diagonal.

$$(1)\ \begin{bmatrix} 1 & i \\ -i & 2 \end{bmatrix}, \quad (2)\ \begin{bmatrix} 1 & 2+3i \\ 2-3i & -1 \end{bmatrix}, \quad (3)\ \begin{bmatrix} 1 & i & 2+i \\ -i & 2 & 1-i \\ 2-i & 1+i & 2 \end{bmatrix}.$$

7.12. Find the diagonal matrices to which the following matrices are unitarily similar. Determine whether each of them is Hermitian, unitary or orthogonal.

$$(1)\ \frac{1}{2}\begin{bmatrix} 1+i & 1-i \\ 1-i & 1+i \end{bmatrix}, \quad (2)\ \begin{bmatrix} 0.6 & -0.8 \\ 0.8 & 0.6 \end{bmatrix}, \quad (3)\ \begin{bmatrix} 1 & i & 0 \\ -i & 1 & i \\ 0 & -i & 1 \end{bmatrix}.$$

7.13. For a skew-Hermitian matrix A, show that

(1) $A - I$ is invertible, (2) e^A is unitary.

7.14. Let U be a unitary matrix. Prove that U and U^T have the same set of eigenvalues.

7.15. Verify that $A = \begin{bmatrix} 2 & i \\ i & 2 \end{bmatrix}$ is normal. Diagonalize A by a unitary matrix U.

7.16. Show that the nonsymmetric real matrix

$$A = \begin{bmatrix} 1 & 0 & 0 \\ 1 & 3 & 3 \\ -2 & -4 & -5 \end{bmatrix} \text{ can be diagonalized.}$$

7.17. Suppose that A, B are diagonalizable $n \times n$ matrices. Prove that $AB = BA$ if and only if A and B can be diagonalized simultaneously by the same matrix Q, *i.e.*, $Q^{-1}AQ$ and $Q^{-1}BQ$ are diagonal matrices.

7.18. Find the spectral decomposition of $A = \begin{bmatrix} 2 & 1 & 1 \\ 1 & 2 & 1 \\ 1 & 1 & 2 \end{bmatrix}$.

7.19. Let A and B be 2×2 symmetric matrices. Prove that A and B are similar if and only if $\det A = \det B$ and $\operatorname{tr} A = \operatorname{tr} B$.

7.20. Let A be a real symmetric $n \times n$ matrix and λ an eigenvalue of A with multiplicity m. Show that $\dim \mathcal{N}(A - \lambda I) = m$.

7.21. Show that a matrix A is nilpotent, *i.e.*, $A^n = \mathbf{0}$ for some integer $n \geq 1$, if and only if its eigenvalues are all zero.

7.22. Determine whether the following statements are true or false, in general, and justify your answers.

(1) A Hermitian matrix is always unitarily similar to a diagonal matrix.

(2) An orthogonal matrix is always unitarily similar to a real diagonal matrix.

(3) For any $m \times n$ matrix A, AA^H and A^HA have the same eigenvalues.

(4) If a triangular matrix is similar to a diagonal matrix, it is already diagonal.
(5) If all the columns of a square matrix A are orthonormal, then A is diagonalizable.
(6) Every permutation matrix is diagonalizable.
(7) Every permutation matrix is Hermitian.
(8) A nonzero nilpotent matrix cannot be symmetric.
(9) Every square matrix is unitarily similar to a triangular matrix.
(10) If A is a Hermitian matrix, then $A + iI$ is invertible.
(11) If A is a real matrix, then $A + iI$ is invertible.
(12) If A is an orthogonal matrix, then $A + \frac{1}{2}I$ is invertible.
(13) Every unitarily diagonalizable matrix is Hermitian.
(14) Every diagonalizable matrix is normal.

Chapter 8

Quadratic Forms

8.1 Introduction

The reader should now be well aware of the important roles of matrices in the study of linear equations, which can be expressed in the form

$$a_1x_1 + a_2x_2 + \cdots + a_nx_n = b.$$

The left side $a_1x_1 + a_2x_2 + \cdots + a_nx_n = \mathbf{a}^T\mathbf{x}$ of the equation is a (homogeneous) polynomial of degree 1 in n variables, called a **linear form**. In this chapter, we study a (homogeneous) polynomial of degree 2 in several variables, called a **quadratic form**, and show that matrices also play an important role in the study of a quadratic form. Quadratic forms arise in a variety of applications, including geometry, vibrations of mechanical systems, statistics, and electrical engineering, *etc.* A more general form of a quadratic form is a *bilinear form* which is described in Section 8.7. The inner product of a vector space and the determinant function on $M_{2\times 2}(\mathbb{R})$ are typical examples of bilinear forms. It turns out that a quadratic form (or bilinear form) may be associated with a real symmetric matrix, and *vice versa.*

A *quadratic equation* in two variables x and y is an equation of the form

$$ax^2 + 2bxy + cy^2 + dx + ey + f = 0,$$

in which the left side consists of a constant term f, a linear form $dx + ey$, and a quadratic form $ax^2 + 2bxy + cy^2$. Note that this quadratic form may be written in matrix notation as

$$ax^2 + 2bxy + cy^2 = [\ x \quad y\ ] \begin{bmatrix} a & b \\ b & c \end{bmatrix} \begin{bmatrix} x \\ y \end{bmatrix} = \mathbf{x}^T A \mathbf{x},$$

where

$$\mathbf{x} = \begin{bmatrix} x \\ y \end{bmatrix} \quad \text{and} \quad A = \begin{bmatrix} a & b \\ b & c \end{bmatrix}.$$

Note also that the matrix A is taken to be a (real) symmetric matrix.

Geometrically, the solution set of a quadratic equation in x and y usually represents a *conic section*, such as an ellipse, a parabola or a hyperbola in the xy-plane.

Definition 8.1 **(1)** An equation of the form

$$f(\mathbf{x}) = \sum_{i=1}^{n} \sum_{j=1}^{n} a_{ij} x_i x_j + \sum_{i=1}^{n} b_i x_i + c = 0,$$

where a_{ij}, b_j and c are real constants, is called a **quadratic equation** in n variables $x_1, x_2, \ldots, x_n$. In matrix form, it can be written as

$$f(\mathbf{x}) = \mathbf{x}^T A \mathbf{x} + \mathbf{b}^T \mathbf{x} + c = 0,$$

where $A = [a_{ij}]$, $\mathbf{x} = [x_1 \ \cdots \ x_n]^T$ and $\mathbf{b} = [b_1 \ \cdots \ b_n]^T$ in $\mathbb{R}^n$.

(2) A **linear form** is a polynomial of degree 1 in n variables $x_1, x_2, \ldots, x_n$ of the form

$$\mathbf{b}^T \mathbf{x} = \sum_{i=1}^{n} b_i x_i,$$

where $\mathbf{x} = [x_1 \ \cdots \ x_n]^T$ and $\mathbf{b} = [b_1 \ \cdots \ b_n]^T$ in $\mathbb{R}^n$.

(3) A **quadratic form** is a (homogeneous) polynomial of degree 2 in n variables $x_1, x_2, \ldots, x_n$ of the form

$$q(\mathbf{x}) = \mathbf{x}^T A \mathbf{x} = [x_1 \ x_2 \ \cdots \ x_n][a_{ij}] \begin{bmatrix} x_1 \\ x_2 \\ \vdots \\ x_n \end{bmatrix} = \sum_{i=1}^{n} \sum_{j=1}^{n} a_{ij} x_i x_j,$$

where $\mathbf{x} = [x_1 \ x_2 \ \cdots \ x_n]^T \in \mathbb{R}^n$ and $A = [a_{ij}]$ is a real $n \times n$ matrix.

Remark: **(1)** A quadratic equation is said to be *consistent* if it has a solution, *i.e.*, there is a vector $\mathbf{x} \in \mathbb{R}^n$ such that $f(\mathbf{x}) = 0$. Otherwise, it is said to be inconsistent. For instance, the equation $2x^2 + 3y^2 = -1$ is inconsistent. In the following discussion we will consider only consistent equations.

(2) A linear form is simply the dot product in $\mathbb{R}^n$ with a fixed vector $\mathbf{b} \in \mathbb{R}^n$. (This has been discussed in the previous chapters.)

(3) The matrix A in the definition of a quadratic form is any square matrix, but it can be restricted to be a symmetric matrix. In fact, any square matrix A is the sum of a symmetric part B and a skew-symmetric part C, say

$$A = B + C, \quad \text{where } B = \frac{1}{2}(A + A^T), \quad C = \frac{1}{2}(A - A^T).$$

For the skew-symmetric matrix C, we have

$$\mathbf{x}^T C\mathbf{x} = \left(\mathbf{x}^T C\mathbf{x}\right)^T = \mathbf{x}^T C^T \mathbf{x} = -\mathbf{x}^T C\mathbf{x}.$$

Hence, as a real number, $\mathbf{x}^T C\mathbf{x} = 0$. Therefore,

$$q(\mathbf{x}) = \mathbf{x}^T A\mathbf{x} = \mathbf{x}^T(B + C)\mathbf{x} = \mathbf{x}^T B\mathbf{x}.$$

This means that, without loss of generality, one may assume that *the matrix A in the definition of a quadratic form is a symmetric matrix.*

(4) From the definition of a quadratic form, one can see that, fixing a basis like the standard basis for $\mathbb{R}^n$, a quadratic form is associated with a unique symmetric matrix, which is called the **matrix representation** of the quadratic form q with respect to the basis chosen (the standard basis for $\mathbb{R}^n$). On the other hand, any (real) symmetric matrix A gives rise to a quadratic form $\mathbf{x}^T A\mathbf{x}$. For example, for a symmetric matrix $\begin{bmatrix} 8 & 2 \\ 2 & -1 \end{bmatrix}$, the equation

$$\begin{bmatrix} x_1 & x_2 \end{bmatrix} \begin{bmatrix} 8 & 2 \\ 2 & -1 \end{bmatrix} \begin{bmatrix} x_1 \\ x_2 \end{bmatrix}$$

defines a quadratic form $8x_1^2 + 4x_1x_2 - x_2^2$.

Problem 8.1 Find the symmetric matrices representing the quadratic forms
(1) $9x_1^2 - x_2^2 + 4x_3^2 + 6x_1x_2 - 8x_1x_3 + 2x_2x_3$,
(2) $x_1x_2 + x_1x_3 + x_2x_3$,
(3) $x_1^2 + x_2^2 - x_3^2 - x_4^2 + 2x_1x_2 - 10x_1x_4 + 4x_3x_4$.

8.2 Diagonalization of a quadratic form

In this section, we discuss how to sketch the solutions of a quadratic equation. To study the solution of a quadratic equation $f(\mathbf{x}) = 0$, we first consider an equation $\mathbf{x}^T A\mathbf{x} = c$ without a linear form.

This quadratic form may be rewritten as the sum of two parts:

$$q(\mathbf{x}) = \mathbf{x}^T A\mathbf{x} = \sum_{i=1}^{n} a_{ii}x_i^2 + 2\sum_{i<j} a_{ij}x_i x_j,$$

in which the first part $\sum_{i=1}^{n} a_{ii}x_i^2$ is called the **(perfect) square terms** and the second part $\sum_{i\neq j} a_{ij}x_i x_j$ is called the **cross-product terms**. Actually, what makes it hard to sketch the quadratic surface is the cross product terms. However, the symmetric matrix A can be orthogonally diagonalized, *i.e.*, there exists an orthogonal matrix P such that

$$P^T AP = P^{-1}AP = D = \begin{bmatrix} \lambda_1 & & & 0 \\ & \lambda_2 & & \\ & & \ddots & \\ 0 & & & \lambda_n \end{bmatrix}.$$

Here, the diagonal entries λ_i's are the eigenvalues of A and the column vectors of P are their associated eigenvectors of A. Then we get, by setting $\mathbf{x} = P\mathbf{y}$,

$$\mathbf{x}^T A\mathbf{x} = \mathbf{y}^T(P^T AP)\mathbf{y} = \mathbf{y}^T D\mathbf{y} = \lambda_1 y_1^2 + \lambda_2 y_2^2 + \cdots + \lambda_n y_n^2,$$

which is a quadratic form without the cross-product terms. Consequently, we have proven the following theorem.

Theorem 8.1 *Let $\mathbf{x}^T A\mathbf{x}$ be a quadratic form in $\mathbf{x} = [x_1\ x_2\ \cdots\ x_n]^T \in \mathbb{R}^n$ for a symmetric matrix A. Then there is a change of coordinates of $\mathbf{x}$ into $\mathbf{y} = P^T\mathbf{x} = [y_1\ y_2\ \cdots\ y_n]^T$ such that*

$$\mathbf{x}^T A\mathbf{x} = \mathbf{y}^T D\mathbf{y} = \lambda_1 y_1^2 + \lambda_2 y_2^2 + \cdots + \lambda_n y_n^2,$$

where P is an orthogonal matrix and $P^T AP = D$. □

Remark: (1) Recall that in Theorem 8.1 the columns of the matrix P are the orthonormal eigenvectors of A and $\mathbf{y}$ is just the coordinate expression of $\mathbf{x}$ with respect to the orthonormal eigenvectors of A. In fact, $P = [Id]_\beta^\alpha$, where β is a basis consisting of orthonormal eigenvectors of A and α is the standard basis.

(2) The solution set of a quadratic equation of the form $\mathbf{x}^T A\mathbf{x} = c$ is a hypersurface in $\mathbb{R}^n$, that is, a curved surface that can be parameterized in $n-1$ variables. These are called $n-1$-dimensional *quadratic surfaces*, with axes in the directions of eigenvectors. In particular, if $n = 2$, the solution set of a quadratic equation is called a quadratic curve, or more commonly a **conic section**. When $n = 3$, the **quadratic surfaces** are ellipsoids or hyperboloids depending on the signs of the eigenvalues of A. Of course, a paraboloid is also a quadratic surface, but it appears when a linear form is present in the quadratic equation. The determination of the quadratic hypersurface depends on the signs of the eigenvalues of A.

Example 8.1 Determine the conic section $3x^2 + 2xy + 3y^2 - 8 = 0$.

Solution: This equation can be written in the form

$$\begin{bmatrix} x & y \end{bmatrix} \begin{bmatrix} 3 & 1 \\ 1 & 3 \end{bmatrix} \begin{bmatrix} x \\ y \end{bmatrix} = 8.$$

The matrix $A = \begin{bmatrix} 3 & 1 \\ 1 & 3 \end{bmatrix}$ has eigenvalues $\lambda_1 = 2$ and $\lambda_2 = 4$ with associated unit eigenvectors

$$\mathbf{v}_1 = \left(\frac{1}{\sqrt{2}}, -\frac{1}{\sqrt{2}}\right) \quad \text{and} \quad \mathbf{v}_2 = \left(\frac{1}{\sqrt{2}}, \frac{1}{\sqrt{2}}\right),$$

respectively, which form an orthonormal basis β. If α denotes the standard basis, then the transition matrix

$$P = [Id]^\alpha_\beta = [\mathbf{v}_1 \quad \mathbf{v}_2] = \frac{1}{\sqrt{2}} \begin{bmatrix} 1 & 1 \\ -1 & 1 \end{bmatrix} = \begin{bmatrix} \cos 45^\circ & \sin 45^\circ \\ -\sin 45^\circ & \cos 45^\circ \end{bmatrix},$$

which is a rotation through 45° in clockwise direction such that $P^T = P^{-1}$. It gives a change of coordinates, $\mathbf{x} = P\mathbf{y}$, *i.e.*,

$$\begin{bmatrix} x \\ y \end{bmatrix} = \frac{1}{\sqrt{2}} \begin{bmatrix} 1 & 1 \\ -1 & 1 \end{bmatrix} \begin{bmatrix} x' \\ y' \end{bmatrix} = \begin{bmatrix} \frac{1}{\sqrt{2}}x' + \frac{1}{\sqrt{2}}y' \\ -\frac{1}{\sqrt{2}}x' + \frac{1}{\sqrt{2}}y' \end{bmatrix}.$$

Thus, we get

$$\begin{aligned} 3x^2 + 2xy + 3y^2 &= \mathbf{x}^T A\mathbf{x} = \mathbf{y}^T P^T AP\mathbf{y} \\ &= \mathbf{y}^T \begin{bmatrix} 2 & 0 \\ 0 & 4 \end{bmatrix} \mathbf{y} = 2(x')^2 + 4(y')^2 = 8, \end{aligned}$$

or

$$\frac{(x')^2}{4} + \frac{(y')^2}{2} = 1.$$

Its solution set is just an ellipse with axes $\mathbf{v}_1 = P^T\mathbf{e}_1$ and $\mathbf{v}_2 = P^T\mathbf{e}_2$. □

Definition 8.2 The **inertia** of a symmetric matrix A is a triple of integers denoted by $\text{In}(A) = (p,\ q,\ k)$, where p, q and k are the numbers of positive, negative and zero eigenvalues of A, respectively.

It turns out that the inertia $\text{In}(A)$ completely determines the geometric type of the quadratic form in the following sense. Since $\text{In}(-A) = (q,\ p,\ k)$ if $\text{In}(A) = (p,\ q,\ k)$ and the equation $\mathbf{x}^T A\mathbf{x} = c$ is inconsistent if $p = 0$ and $c > 0$, it suffices to consider the cases of $c \geq 0$ and $p > 0$. Excluding those inconsistent cases we have the following characterization of the solution sets for $n = 2$ and 3:

For $n = 2$, there are only three possible cases for $\text{In}(A)$:

$\text{In}(A)$	The solution of $\mathbf{x}^T A\mathbf{x} = c$	
$(p,\ q,\ k)$	$c > 0$	$c = 0$
(2, 0, 0)	ellipse	a point
(1, 1, 0)	hyperbola	two lines crossing at $\mathbf{0}$
(1, 0, 1)	two parallel lines	a line

For $n = 3$, there are six possibilities:

$\text{In}(A)$	The solution of $\mathbf{x}^T A\mathbf{x} = c$	
$(p,\ q,\ k)$	$c > 0$	$c = 0$
(3, 0, 0)	ellipsoid	a point
(2, 1, 0)	one-sheeted hyperboloid	elliptic cone
(2, 0, 1)	elliptic cylinder	a line
(1, 2, 0)	two-sheeted hyperboloid	elliptic cone
(1, 1, 1)	hyperbolic cylinder	two planes crossing in a line
(1, 0, 2)	two parallel planes	a plane

In general, $\text{In}(A)$ will have $n(n+1)/2$ possibilities, each characterizing a different geometric type of a quadratic form. For example, if $\text{In}(A) = (n,\ 0,\ 0)$ and $c > 0$, *i.e.*, the eigenvalues of A are all positive, then the quadratic form describes an ellipsoid in $\mathbb{R}^n$, *etc.*

Example 8.2 Determine the quadratic surface for $2xy + 2xz = 1$.

Solution: The matrix for the given quadratic form is

$$A = \begin{bmatrix} 0 & 1 & 1 \\ 1 & 0 & 0 \\ 1 & 0 & 0 \end{bmatrix},$$

so the eigenvalues of A are found to be $\lambda_1 = \sqrt{2}$, $\lambda_2 = -\sqrt{2}$, $\lambda_3 = 0$, and the associated orthonormal eigenvectors are

$$\mathbf{v}_1 = \left(\frac{1}{\sqrt{2}}, \frac{1}{2}, \frac{1}{2}\right), \quad \mathbf{v}_2 = \left(-\frac{1}{\sqrt{2}}, \frac{1}{2}, \frac{1}{2}\right), \quad \text{and} \quad \mathbf{v}_3 = \left(0, -\frac{1}{\sqrt{2}}, \frac{1}{\sqrt{2}}\right),$$

respectively. Hence, an orthogonal matrix P that diagonalizes A is

$$P = \frac{1}{2}\begin{bmatrix} \sqrt{2} & -\sqrt{2} & 0 \\ 1 & 1 & -\sqrt{2} \\ 1 & 1 & \sqrt{2} \end{bmatrix},$$

and with the change of coordinates $\mathbf{x} = P\mathbf{y}$, that is,

$$x = \frac{1}{\sqrt{2}}(x' - y'), \quad y = \frac{1}{2}(x' + y' - \sqrt{2}z'), \quad z = \frac{1}{2}(x' + y' + \sqrt{2}z')$$

the equation is transformed to $\sqrt{2}(x')^2 - \sqrt{2}(y')^2 = 1$, which is a hyperbolic cylinder. Note that $\text{In}(A) = (1,\ 1,\ 1)$. □

Consider a general form of quadratic equation

$$\mathbf{x}^T A\mathbf{x} + \mathbf{b}^T\mathbf{x} = c.$$

(1) If it does not have a linear form, *i.e.*, $\mathbf{b} = \mathbf{0}$, then, as we have seen above, a *parabolic* curve or surface does not appear as a solution of the quadratic form.

(2) Suppose that it has a nonzero linear form, *i.e.*, $\mathbf{b} \neq \mathbf{0}$. If the matrix A is invertible, then, by taking a change of variables as $\mathbf{y} = \mathbf{x} + \frac{1}{2}A^{-1}\mathbf{b}$ (or

$\mathbf{x} = \mathbf{y} - \frac{1}{2}A^{-1}\mathbf{b}$), the given quadratic equation is transformed into a new quadratic equation $\mathbf{y}^T A\mathbf{y} = d$ without a linear form, where $d = c + \frac{1}{4}\mathbf{b}^T A^{-1}\mathbf{b}$. However, if A is singular, the solution of the quadratic equation depends not only on the inertia of A, but also on the type of linear form, and the parabolic curve or surface appear as solutions to the quadratic equation with a nonzero linear form. For example, the equation $x^2 - z = c$ has a singular quadratic form for which $\text{In}(A) = (1,\ 0,\ 2)$ and also has a nonzero linear form that cannot be removed by any change of variables. The solution of this equation is a **parabolic cylinder** when $n = 3$.

Example 8.3 Classify the conic section for $3x^2 - 6xy + 4y^2 + 2x - 2y = 0$.

Solution: The matrix for the quadratic form $3x^2 - 6xy + 4y^2$ is

$$A = \begin{bmatrix} 3 & -3 \\ -3 & 4 \end{bmatrix}.$$

Its inverse is

$$A^{-1} = \frac{1}{3}\begin{bmatrix} 4 & 3 \\ 3 & 3 \end{bmatrix} \quad \text{and} \quad \mathbf{b} = \begin{bmatrix} 2 \\ -2 \end{bmatrix}.$$

With the change of coordinates $\mathbf{y} = \mathbf{x} + \frac{1}{2}A^{-1}\mathbf{b}$, that is

$$x' = x + \frac{1}{3}, \quad y' = y,$$

the equation is transformed to a new equation $3(x')^2 - 6x'y' + 4(y')^2 = \frac{1}{3}$. Clearly, the matrix representation of the new quadratic form is also A, and its eigenvalues are $\frac{1}{2}(7 \pm \sqrt{37})$. Therefore, $\text{In}(A) = (2,\ 0,\ 0)$ and the solution of the equation is an ellipse. □

Example 8.4 In analytic geometry, a general quadratic curve (or conic section) is represented by a quadratic equation in two variables as

$$ax^2 + 2bxy + cy^2 + dx + ey + f = 0$$

with the symmetric matrix $A = \begin{bmatrix} a & b \\ b & c \end{bmatrix}$. We present here the classification of the conic sections according to the coefficients.

(1) If $b = 0$, then A is already a diagonal matrix with the eigenvalues a and c, and the equation becomes

$$ax^2 + cy^2 + dx + ey + f = 0.$$

(i) If $a = 0 = c$, then the conic section is a line in the plane.

(ii) If $a \neq 0 = c$, then it is a parabola when $e \neq 0$, or one or two lines when $e = 0$.

(iii) If $a \neq 0 \neq c$, then the quadratic equation becomes

$$ax^2 + cy^2 + dx + ey + f = a(x-p)^2 + c(y-q)^2 + h = 0$$

for some constants p, q, and h. If $h = 0$, the cases are easily classified (try). Suppose $h \neq 0$. Then, the conic section is a circle if $a = c$, an ellipse if $ac > 0$, or a hyperbola if $ac < 0$.

(2) Suppose that $b \neq 0$. Since A is symmetric, it can be diagonalized by an orthogonal matrix P whose columns are orthonormal eigenvectors, and the diagonal matrix has eigenvalues λ_1 and λ_2 on the diagonal. By a coordinate change by P, the quadratic equation becomes

$$ax^2 + 2bxy + cy^2 + dx + ey + f = \lambda_1 u^2 + \lambda_2 v^2 + d'u + e'v + f = 0.$$

for some constants d' and e'. Hence, the classification of the conic sections is reduced to case (1) according to the various possible cases of the eigenvalues of A. However, the eigenvalues are given as

$$\lambda_1 = \frac{(a+c) + \sqrt{(a-c)^2 + 4b^2}}{2}, \quad \lambda_2 = \frac{(a+c) - \sqrt{(a-c)^2 + 4b^2}}{2},$$

which are determined by the coefficients a, b, and c. Hence one can classify the conic section according to the various possible cases a, b, and c (see Exercise **8.12**).

(3) The axes of the conic section are the directions of the eigenvectors, which are orthogonal to each other. Since we only need to find axis lines, but not the direction vectors, we may choose them to be the rotation of the standard coordinate x- and y-axes which are determined by $\mathbf{e}_1, \mathbf{e}_2$. Now, a pair of orthogonal eigenvectors are found to be

$$\mathbf{v}_i = \begin{bmatrix} v_{i1} \\ v_{i2} \end{bmatrix} = \begin{bmatrix} b \\ -(a - \lambda_i) \end{bmatrix}$$

for $i = 1, 2$. The slope of $\mathbf{v}_1$ from the x-axis is

$$\begin{aligned} \frac{-(a-\lambda_1)}{b} &= -\frac{a-c}{2b} + \sqrt{\left(\frac{a-c}{2b}\right)^2 + 1} \\ &= -\cot 2\theta + \operatorname{cosec} 2\theta \;=\; \tan\theta, \end{aligned}$$

where $\cot 2\theta = \frac{a-c}{2b}$ for some θ. Since $b \neq 0$ and $a - c > 0$, we may assume that $0 < \theta < \pi$. This means that if we set $\tan 2\theta = \frac{2b}{a-c}$ with $-\frac{\pi}{2} < \theta < \frac{\pi}{2}$, then θ is the rotation angle we were looking for. Therefore, the orthonormal eigenvectors $\mathbf{u}_1$ and $\mathbf{u}_2$ of A may be chosen as the rotation of the standard basis through the angle θ. The transition matrix is now

$$P = [\mathbf{u}_1 \quad \mathbf{u}_2] = \begin{bmatrix} \cos\theta & -\sin\theta \\ \sin\theta & \cos\theta \end{bmatrix}, \quad \text{and} \quad P^T AP = P^{-1}AP = \begin{bmatrix} \lambda_1 & 0 \\ 0 & \lambda_2 \end{bmatrix}.$$

By a change of coordinates $\begin{bmatrix} x \\ y \end{bmatrix} = P \begin{bmatrix} x' \\ y' \end{bmatrix}$, the quadratic equation becomes

$$ax^2 + 2bxy + cy^2 + dx + ey + f \;=\; \lambda_1 x'^2 + \lambda_2 y'^2 + d'x' + e'y' + f \;=\; 0,$$

where $d' = d\ \cos\theta + e\ \sin\theta$ and $e' = -d\ \sin\theta + e\ \cos\theta$. $\square$

Problem 8.2 Sketch the graph of each of the following quadratic equations:
(1) $2x^2 + 2y^2 + 6yz + 10z^2 = 9$;
(2) $x^2 - 8xy + 16y^2 - 3z^2 = 8$;
(3) $4x^2 + 12xy + 9y^2 + 3x - 4 = 0$.

8.3 Congruence relation

As we have seen so far, in a quadratic equation $\mathbf{x}^T A\mathbf{x} + \mathbf{b}^T\mathbf{x} + c = 0$, the linear form may be eliminated by a change of variables when A is invertible, and then by another change of variables the equation can be transformed into a simple form $\mathbf{y}^T A\mathbf{y} = c$ having only square terms. Hence, the geometric types of quadratic equations may be easily classified. However, these transformations of equations contain basis changes by some orthogonal matrices.

Let us now consider a change of basis (or variables) and a relation between two different matrix representations of a quadratic form. Usually a quadratic form $\mathbf{x}^T A\mathbf{x}$ with a symmetric matrix A is expressed in the coordinates of $\mathbf{x}$ with respect to the standard basis $\alpha = \{\mathbf{e}_1, \mathbf{e}_2, \ldots, \mathbf{e}_n\}$ for $\mathbb{R}^n$. Let $\beta = \{\mathbf{e}'_1, \mathbf{e}'_2, \ldots, \mathbf{e}'_n\}$ be another basis for $\mathbb{R}^n$. Then any vector $\mathbf{x}$ in $\mathbb{R}^n$ has two coordinate representations $[\mathbf{x}]_\alpha$ and $[\mathbf{x}]_\beta$ through the equations

$$x_1\mathbf{e}_1 + x_2\mathbf{e}_2 + \cdots + x_n\mathbf{e}_n \;=\; \mathbf{x} \;=\; y_1\mathbf{e}'_1 + y_2\mathbf{e}'_2 + \cdots + y_n\mathbf{e}'_n.$$

They are related as $[\mathbf{x}]_\alpha = P[\mathbf{x}]_\beta$, where $P = [Id]_\beta^\alpha$ is the transition matrix from β to α. This is just a change of coordinates (or variables). If we set notations $\mathbf{x} = [\mathbf{x}]_\alpha$ and $\mathbf{y} = [\mathbf{x}]_\beta$, then the quadratic form can be written as

$$q(\mathbf{x}) = \mathbf{x}^T A\mathbf{x} = \mathbf{y}^T(P^T AP)\mathbf{y} = \mathbf{y}^T B\mathbf{y},$$

where $\mathbf{y}^T B\mathbf{y}$ is the expression of $\mathbf{x}^T B\mathbf{x}$ in a new basis (or a coordinate system) β, and $B = P^T AP$.

Definition 8.3 Two $n \times n$ matrices A and B are said to be **congruent** if there exists an invertible matrix P such that $P^T AP = B$.

It is easily seen that the congruence relation is an equivalence relation in the vector space $M_{n\times n}(\mathbb{R})$, and *any two matrix representations of a quadratic form with respect to different bases are congruent.*

Remark: (1) Clearly, two orthogonally similar matrices are congruent, but the converse is not true in general. By Theorem 8.1, a symmetric matrix A is congruent to a diagonal matrix D by an orthogonal matrix P. However, it can be congruent to many diagonal matrices (not necessarily by orthogonal matrices). In fact, if $P^T AP = D$ by an orthogonal matrix P, then the matrix $Q = kP$, $k \neq 0$, also diagonalizes A to a different diagonal matrix via a congruent relation:

$$Q^T AQ = (kP)^T A(kP) = k^2 P^T AP = k^2 D,$$

which is another diagonal matrix with diagonal entries $k^2\lambda_1, k^2\lambda_2, \ldots, k^2\lambda_n$. In this case, if $k \neq \pm 1$, Q is not an orthogonal matrix and the resulting diagonal entries are not the eigenvalues of A anymore.

(2) Sylvester's law of inertia (Theorem 8.13 in Section 8.7) says that even though a symmetric matrix A may be congruent to various diagonal matrices, the numbers of positive, negative and zero diagonal entries, respectively, of the congruent diagonal matrices do not change no matter what the diagonalizing matrices are. That is, *any two congruent matrices have the same inertia.*

There is a practical way of diagonalizing a symmetric matrix through the congruence relation by using the elementary operation of adding a row (or a column) to a constant multiple of another.

Suppose that we have diagonalized a symmetric matrix A by an invertible matrix P through the congruence relation $P^T AP = D$. Since both P and

P^T are invertible matrices, P^T can be written as a product of elementary matrices E_i's, say $P^T = E_k \cdots E_2 E_1$. Then we have

$$D = P^T AP = E_k \cdots E_2 E_1 A E_1^T E_2^T \cdots E_k^T.$$

Note also that multiplying an elementary matrix to A on the left (on the right) is the same as executing an elementary row (column, respectively) operation to A. Clearly, if E is an elementary matrix, so is E^T. Moreover, multiplying an elementary matrix E^T on the right of A is equal to performing the same operation on the i-th column of A as the row operation E operates on the i-th row. Since A is symmetric, this means that the operations EAE^T will have the same effect on the diagonally opposite entries of A simultaneously. For instance, if

$$A = \begin{bmatrix} 1 & 1 & 2 \\ 1 & 0 & 3 \\ 2 & 3 & 6 \end{bmatrix},$$

then by the elementary matrix

$$E = \begin{bmatrix} 1 & 0 & 0 \\ -1 & 1 & 0 \\ 0 & 0 & 1 \end{bmatrix},$$

which adds -1 times the first row to the second row, yielding

$$EAE^T = \begin{bmatrix} 1 & 0 & 0 \\ -1 & 1 & 0 \\ 0 & 0 & 1 \end{bmatrix} \begin{bmatrix} 1 & 1 & 2 \\ 1 & 0 & 3 \\ 2 & 3 & 6 \end{bmatrix} \begin{bmatrix} 1 & -1 & 0 \\ 0 & 1 & 0 \\ 0 & 0 & 1 \end{bmatrix} = \begin{bmatrix} 1 & 0 & 2 \\ 0 & -1 & 1 \\ 2 & 1 & 6 \end{bmatrix}.$$

Thus, the above equation implies that the operations performed on the left side of A (*i.e.*, $E_k \cdots E_2 E_1$) are nothing but a Gaussian elimination on A to get an upper triangular matrix $P^T A$ and those on the right side (*i.e.*, $E_1^T E_2^T \cdots E_k^T$) are the corresponding column operations to yield a diagonal matrix D. In summary, if we take a Gaussian-elimination of A by the elementary matrices $E_1, \ldots, E_k$, then $E_k \cdots E_1 A E_1^T \cdots E_k^T = D$ implies $P = E_1^T \cdots E_k^T$, *i.e.*,

$$\begin{aligned} [A \mid I] &\rightarrow [E_1 A E_1^T \mid I E_1^T] \rightarrow [E_2 E_1 A E_1^T E_2^T \mid I E_1^T E_2^T] \rightarrow \cdots \\ &\rightarrow [E_k \cdots E_1 A E_1^T \cdots E_k^T \mid I E_1^T \cdots E_k^T] = [D \mid P]. \end{aligned}$$

Remark: **(1)** Notice that in this case P need not be an orthogonal matrix, and the diagonal entries of D need not be eigenvalues of A.

(2) Be careful not to apply the same argument for the diagonalization of symmetric matrices through the similarity $P^{-1}AP = D$, because multiplying E^{-1} on the right of A is not the same column operation as E (try it yourself) so that the operations EAE^{-1} do not work for the diagonalization of A.

Example 8.5 Let

$$A = \begin{bmatrix} 1 & 1 & 2 \\ 1 & 0 & 3 \\ 2 & 3 & 6 \end{bmatrix}$$

be a symmetric matrix. The preceding method produces

$$\begin{aligned}
[A \mid I] &= \left[\begin{array}{ccc|ccc} 1 & 1 & 2 & 1 & 0 & 0 \\ 1 & 0 & 3 & 0 & 1 & 0 \\ 2 & 3 & 6 & 0 & 0 & 1 \end{array}\right] \\
\longrightarrow [E_2E_1AE_1^TE_2^T \mid IE_1^TE_2^T] &= \left[\begin{array}{ccc|ccc} 1 & 0 & 0 & 1 & -1 & -2 \\ 0 & -1 & 1 & 0 & 1 & 0 \\ 0 & 1 & 2 & 0 & 0 & 1 \end{array}\right] \\
\longrightarrow [E_3E_2E_1AE_1^TE_2^TE_3^T \mid IE_1^TE_2^TE_3^T] &= \left[\begin{array}{ccc|ccc} 1 & 0 & 0 & 1 & -1 & -3 \\ 0 & -1 & 0 & 0 & 1 & 1 \\ 0 & 0 & 3 & 0 & 0 & 1 \end{array}\right] \\
&= [D \mid P],
\end{aligned}$$

where

$$E_1 = \begin{bmatrix} 1 & 0 & 0 \\ -1 & 1 & 0 \\ 0 & 0 & 1 \end{bmatrix}, \quad E_2 = \begin{bmatrix} 1 & 0 & 0 \\ 0 & 1 & 0 \\ -2 & 0 & 1 \end{bmatrix}, \quad E_3 = \begin{bmatrix} 1 & 0 & 0 \\ 0 & 1 & 0 \\ 0 & 1 & 1 \end{bmatrix}.$$

One can check that $P^TAP = D$ by a direct computation. This example shows how to get $\text{In}(A) = (2,\ 1,\ 0)$ without computing the eigenvalues of A. □

Problem 8.3 Find an invertible matrix P such that P^TAP is diagonal for each of the following matrices:

$$(1)\ A = \begin{bmatrix} 0 & 1 & -1 \\ 1 & 1 & 0 \\ -1 & 0 & 2 \end{bmatrix}; \quad (2)\ A = \begin{bmatrix} 1 & -3 & 1 \\ -3 & 4 & 2 \\ 1 & 2 & 5 \end{bmatrix}; \quad (3)\ A = \begin{bmatrix} 0 & 0 & 1 \\ 0 & 1 & 0 \\ 1 & 0 & 0 \end{bmatrix}.$$

8.4 Extrema of quadratic forms

In a calculus course, one uses the second derivative test to see whether a given function $y = f(x)$ takes a local maximum or a local minimum at a critical point. In this section, we show a similar test for a function of more than one variable and also show how quadratic forms arise and how they can be used in this context.

Let $f(\mathbf{x})$ be a real-valued function on $\mathbb{R}^n$. A point $\mathbf{x}_0$ in $\mathbb{R}^n$ at which either a first partial derivative of f fails to exist or the first partial derivatives of f are all zero is called a **critical point** of f. If $f(\mathbf{x})$ has either a local maximum or a local minimum at a point $\mathbf{x}_0$ and all the first partial derivatives of f exist at $\mathbf{x}_0$, then all of them must be zero, *i.e.*, $f_{x_i}(\mathbf{x}_0) = 0$, for all $i = 1, \ldots, n$. Thus if $f(\mathbf{x})$ has first partial derivatives everywhere, its local maxima and minima will occur at critical points.

Let us first consider a function of two variables: $f(\mathbf{x})$, $\mathbf{x} = (x, y) \in \mathbb{R}^2$, which has a critical point $\mathbf{x}_0 = (x_0, y_0) \in \mathbb{R}^2$. If f has continuous third partial derivatives in a neighborhood of $\mathbf{x}_0$, it can be expanded in a Taylor series about that point: for $\mathbf{x} = (x_0 + h, y_0 + k)$,

$$\begin{aligned} f(\mathbf{x}) &= f(x_0 + h, y_0 + k) = f(\mathbf{x}_0) + (hf_x(\mathbf{x}_0) + kf_y(\mathbf{x}_0)) \\ &\quad + \frac{1}{2}\left(h^2 f_{xx}(\mathbf{x}_0) + 2hk f_{xy}(\mathbf{x}_0) + k^2 f_{yy}(\mathbf{x}_0)\right) + R \\ &= f(\mathbf{x}_0) + \frac{1}{2}(ah^2 + 2bhk + ck^2) + R, \end{aligned}$$

where

$$a = f_{xx}(\mathbf{x}_0), \qquad b = f_{xy}(\mathbf{x}_0), \qquad c = f_{yy}(\mathbf{x}_0),$$

and the remainder R is given by

$$R = \frac{1}{6}\left(h^3 f_{xxx}(\mathbf{z}) + 3h^2 k f_{xxy}(\mathbf{z}) + 3hk^2 f_{xyy}(\mathbf{z}) + k^3 f_{yyy}(\mathbf{z})\right),$$

with $\mathbf{z} = (x_0 + \theta h, y_0 + \theta k)$ for some $0 < \theta < 1$.

If h and k are sufficiently small, $|R|$ will be smaller than the absolute value of $\frac{1}{2}(ah^2 + 2bhk + ck^2)$, and hence $f(\mathbf{x}) - f(\mathbf{x}_0)$ and $ah^2 + 2bhk + ck^2$ will have the same sign. Note that the expression

$$q(h, k) = ah^2 + 2bhk + ck^2 = [h\ k]H \begin{bmatrix} h \\ k \end{bmatrix}$$

is a quadratic form in the variables h and k, where

$$H = \begin{bmatrix} a & b \\ b & c \end{bmatrix} = \begin{bmatrix} f_{xx}(\mathbf{x}_0) & f_{xy}(\mathbf{x}_0) \\ f_{xy}(\mathbf{x}_0) & f_{yy}(\mathbf{x}_0) \end{bmatrix}$$

is a symmetric matrix, called the **Hessian** of f at $\mathbf{x}_0 = (x_0, y_0)$. Hence, $f(x, y)$ has a local minimum (or maximum) at $\mathbf{x}_0$ if and only if the quadratic form $q(h, k)$ is positive (or negative, respectively) for all sufficiently small (h, k). The critical point $\mathbf{x}_0$ is called a *saddle point* if $q(h, k)$ takes both positive and negative values. Thus at this point $f(x, y)$ has neither a local minimum nor a local maximum. (This is *the second derivative test* for a local extrema of $f(x, y)$.)

In particular, a quadratic form

$$q(\mathbf{x}) = \mathbf{x}^T A\mathbf{x} = \begin{bmatrix} x & y \end{bmatrix} \begin{bmatrix} a & b \\ b & c \end{bmatrix} \begin{bmatrix} x \\ y \end{bmatrix} = ax^2 + 2bxy + cy^2,$$

for $\mathbf{x} = [x\ y]^T \in \mathbb{R}^2$ is itself a function of two variables, and its first partial derivatives are

$$\begin{aligned} q_x &= 2ax + 2by \\ q_y &= 2bx + 2cy. \end{aligned}$$

By setting these equal to zero, we see that $\mathbf{0} = (0, 0)$ is a critical point of q. If $ac - b^2 \neq 0$, this will be the only critical point of q. Note the Hessian of q is

$$H = 2\begin{bmatrix} a & b \\ b & c \end{bmatrix} = 2A.$$

Thus H is nonsingular if and only if $ac - b^2 \neq 0$.

Since $q(\mathbf{0}) = 0$, it follows that the quadratic form q takes the *global minimum* at $\mathbf{0}$ if and only if

$$q(\mathbf{x}) = \mathbf{x}^T A\mathbf{x} > 0 \qquad \text{for all} \qquad \mathbf{x} \neq \mathbf{0},$$

and q takes the *global maximum* at $\mathbf{0}$ if and only if

$$q(\mathbf{x}) = \mathbf{x}^T A\mathbf{x} < 0 \qquad \text{for all} \qquad \mathbf{x} \neq \mathbf{0}.$$

If $\mathbf{x}^T A\mathbf{x}$ takes both positive and negative values, then $\mathbf{0}$ is a *saddle point*. Thus, if A is nonsingular, the quadratic form q will have either the global minimum, the global maximum or a saddle point at $\mathbf{0}$.

This argument leads us to the following general definition for a symmetric matrix.

Definition 8.4 Let $A = [a_{ij}] \in M_{n \times n}(\mathbb{R})$ be a symmetric matrix and let $\mathbf{x} = (x_1, x_2, \ldots, x_n) \in \mathbb{R}^n$. Then, A is said to be

(1) positive definite if $\mathbf{x}^T A\mathbf{x} = \sum_{i,j} a_{ij}x_i x_j > 0$ for all nonzero $\mathbf{x}$,

(2) positive semidefinite if $\mathbf{x}^T A\mathbf{x} = \sum_{i,j} a_{ij}x_i x_j \geq 0$ for all $\mathbf{x}$,

(3) negative definite if $\mathbf{x}^T A\mathbf{x} = \sum_{i,j} a_{ij}x_i x_j < 0$ for all nonzero $\mathbf{x}$,

(4) negative semidefinite if $\mathbf{x}^T A\mathbf{x} = \sum_{i,j} a_{ij}x_i x_j \leq 0$ for all $\mathbf{x}$,

(5) indefinite if $\mathbf{x}^T A\mathbf{x}$ takes both positive and negative values.

For example, the real symmetric matrix

$$\begin{bmatrix} 2 & -1 & 0 \\ -1 & 2 & -1 \\ 0 & -1 & 2 \end{bmatrix}$$

is positive definite, because the quadratic form satisfies

$$\begin{aligned} \mathbf{x}^T A\mathbf{x} &= [x_1\ x_2\ x_3] \begin{bmatrix} 2 & -1 & 0 \\ -1 & 2 & -1 \\ 0 & -1 & 2 \end{bmatrix} \begin{bmatrix} x_1 \\ x_2 \\ x_3 \end{bmatrix} \\ &= [x_1\ x_2\ x_3] \begin{bmatrix} 2x_1 - x_2 \\ -x_1 + 2x_2 - x_3 \\ -x_2 + 2x_3 \end{bmatrix} \\ &= x_1(2x_1 - x_2) + x_2(-x_1 + 2x_2 - x_3) + x_3(-x_2 + 2x_3) \\ &= 2x_1^2 - 2x_1x_2 + 2x_2^2 - 2x_2x_3 + 2x_3^2 \\ &= x_1^2 + (x_1 - x_2)^2 + (x_2 - x_3)^2 + x_3^2 > 0 \end{aligned}$$

unless $x_1 = x_2 = x_3 = 0$.

To determine whether or not a matrix A is positive definite, one can diagonalize A so that

$$\mathbf{x}^T A\mathbf{x} = \sum_{i,j=1}^{n} a_{ij}x_i x_j = \mathbf{y}^T D\mathbf{y} = \sum_{i=1}^{n} \lambda_i y_i^2,$$

where $\mathbf{y} = P^T\mathbf{x}$ for an orthogonal matrix P and the λ_i's are eigenvalues of A. Therefore, $\mathbf{x}^T A\mathbf{x} > 0$ for all nonzero $\mathbf{x} \in \mathbb{R}^n$ if and only if all the λ_i's are positive.

Consequently we have the following characterization of positive definite matrices:

Theorem 8.2 *A real symmetric $n \times n$ matrix A is positive definite if and only if all the eigenvalues of A are positive.*

In particular, if A is positive definite, $\det A > 0$. If the eigenvalues of A are all negative, then $-A$ must be positive definite and consequently A must be negative definite. If A has eigenvalues that differ in sign, then A is indefinite. Indeed, if λ_1 is a positive eigenvalue of A and $\mathbf{x}_1$ is an eigenvector belonging to λ_1, then

$$\mathbf{x}_1^T A\mathbf{x}_1 = \lambda_1 \mathbf{x}_1^T \mathbf{x}_1 = \lambda_1 \|\mathbf{x}_1\|^2 > 0,$$

and if λ_2 is a negative eigenvalue with eigenvector $\mathbf{x}_2$, then

$$\mathbf{x}_2^T A\mathbf{x}_2 = \lambda_2 \mathbf{x}_2^T \mathbf{x}_2 = \lambda_2 \|\mathbf{x}_2\|^2 < 0.$$

If A is definite, then $\mathbf{0}$ is the only critical point of a quadratic form $q(\mathbf{x}) = \mathbf{x}^T A\mathbf{x}$, and $q(\mathbf{0}) = 0$ is the *global minimum* if A is positive definite and the *global maximum* if A is negative definite. If A is indefinite, then $\mathbf{0}$ is a *saddle point*. Hence, if a function f of two variables has a nonsingular Hessian H at a critical point $\mathbf{x}_0 = (x_0, y_0)$ which has nonzero eigenvalues λ_1 and λ_2, then we can say the *second derivative test* for $f(\mathbf{x})$ as follows:

(1) f has a minimum at $\mathbf{x}_0$ if $\lambda_1 > 0$ and $\lambda_2 > 0$,

(2) f has a maximum at $\mathbf{x}_0$ if $\lambda_1 < 0$ and $\lambda_2 < 0$,

(3) f has a saddle point at $\mathbf{x}_0$ if λ_1 and λ_2 have different signs.

Example 8.6 For $q(x, y) = 2x^2 - 4xy + 5y^2$, determine the nature of the critical point $(0, 0)$.

Solution: The matrix A of the quadratic form is

$$\begin{bmatrix} 2 & -2 \\ -2 & 5 \end{bmatrix}.$$

Its eigenvalues are $\lambda_1 = 6$ and $\lambda_2 = 1$. Since both eigenvalues are positive, A is positive definite and hence $(0, 0)$ is a global minimum. □

Example 8.7 Find and describe all critical points of the function

$$f(x, y) = \frac{1}{3}x^3 + xy^2 - 4xy + 1.$$

Solution: The first partial derivatives of f are

$$f_x = x^2 + y^2 - 4y, \quad f_y = 2xy - 4x \;=\; 2x(y-2).$$

Setting $f_y = 0$, we get $x = 0$ or $y = 2$. Setting $f_x = 0$, we see that if $x = 0$, then y must be either 0 or 4, and if $y = 2$, then $x = \pm 2$. Thus $(0,\ 0)$, $(0,\ 4)$, $(2,\ 2)$, $(-2,\ 2)$ are the critical points of f. To classify these critical points, we compute the second partial derivatives:

$$f_{xx} = 2x, \quad f_{xy} = 2y - 4, \quad f_{yy} = 2x.$$

For each critical point (x_0, y_0), we determine the eigenvalues λ_1 and λ_2 of the Hessian

$$H = \begin{bmatrix} 2x_0 & 2y_0 - 4 \\ 2y_0 - 4 & 2x_0 \end{bmatrix}.$$

These values are summarized in the following table:

Critical Point $(x_0,\ y_0)$	λ_1	λ_2	Description
$(0,\ 0)$	4	-4	saddle point
$(0,\ 4)$	4	-4	saddle point
$(2,\ 2)$	4	4	local minimum
$(-2,\ 2)$	-4	-4	local maximum

□

In general, the above arguments can be extended to *the second derivative test* for functions of more than two variables: Let $f(\mathbf{x}) = f(x_1,\ \ldots,\ x_n)$ be a real-valued function whose third partial derivatives are all continuous. If $\mathbf{x}_0$ is a critical point of f, the **Hessian** of f at $\mathbf{x}_0$ is the symmetric matrix $H = H(\mathbf{x}_0) = [h_{ij}]$ given by

$$h_{ij} = \frac{\partial^2 f}{\partial x_i \partial x_j}(\mathbf{x}_0).$$

The critical point can be classified as follows:

(1) f has a local minimum at $\mathbf{x}_0$ if $H(\mathbf{x}_0)$ is positive definite,

(2) f has a local maximum at $\mathbf{x}_0$ if $H(\mathbf{x}_0)$ is negative definite,

(3) $\mathbf{x}_0$ is a saddle point of f if $H(\mathbf{x}_0)$ is indefinite.

Example 8.8 Find the local extrema of the function

$$f(x,\ y,\ z) = x^2 + xz - 3\cos y + z^2.$$

Solution: The first partial derivatives of f are

$$f_x = 2x + z, \quad f_y = 3\sin y, \quad f_z = x + 2z.$$

It follows that $(x,\ y,\ z)$ is a critical point of f if and only if $x = z = 0$ and $y = n\pi$, where n is an integer. Let $\mathbf{x}_0 = (0,\ 2k\pi,\ 0)$. The Hessian of f at $\mathbf{x}_0$ is given by

$$H(\mathbf{x}_0) = \begin{bmatrix} 2 & 0 & 1 \\ 0 & 3 & 0 \\ 1 & 0 & 2 \end{bmatrix}.$$

The eigenvalues of $H(\mathbf{x}_0)$ are 3, 3, and 1. Since the eigenvalues are all positive, it follows that $H(\mathbf{x}_0)$ is positive definite and hence f has a local minimum at $\mathbf{x}_0$. On the other hand, at a critical point of the form $\mathbf{x}_1 = (0,\ (2k-1)\pi,\ 0)$, the Hessian will be

$$H(\mathbf{x}_1) = \begin{bmatrix} 2 & 0 & 1 \\ 0 & -3 & 0 \\ 1 & 0 & 2 \end{bmatrix}.$$

The eigenvalues of $H(\mathbf{x}_1)$ are -3, 3, and 1. It follows that $H(\mathbf{x}_1)$ is indefinite and hence $\mathbf{x}_1$ is a saddle point of f. □

Problem 8.4 For each of the following functions, determine whether the given critical point corresponds to a local minimum, local maximum, or saddle point:

(1) $f(x,\ y) = 3x^2 - xy + y^2$ at $(0,\ 0)$;

(2) $f(x,\ y,\ z) = x^3 + xyz + y^2 - 3x$ at $(1,\ 0,\ 0)$.

Problem 8.5 Which of the following matrices are positive definite? negative definite? indefinite?

$$(1)\ \begin{bmatrix} 1 & 2 & 1 \\ 2 & 1 & 1 \\ 1 & 1 & 2 \end{bmatrix}, \qquad (2)\ \begin{bmatrix} 2 & 0 & 0 \\ 0 & 5 & 3 \\ 0 & 3 & 5 \end{bmatrix}, \qquad (3)\ \begin{bmatrix} 3 & -1 & 0 \\ -1 & 2 & 1 \\ 0 & 1 & 3 \end{bmatrix}.$$

8.5 Application: Quadratic optimization

One of the most important problems in applied mathematics is the optimization (minimization or maximization) of a real-valued function f of n variables subject to constraints on the variables. For example, when the function f is a linear form subject to constraints in the form of linear equalities and/or inequalities, the optimization problem is known as linear programming. Those optimization problems are extensively used in the military, industrial, governmental planning fields, among others.

In this section, we consider an optimization problem of a quadratic form in n variables. If there are no constraints on the variables, then such an optimization problem was discussed in Section 8.4.

As a quadratic optimization problem with constraints, we consider a very special one: Finding the maximum and minimum values of a quadratic form $q(\mathbf{x}) = \mathbf{x}^T A\mathbf{x}$ on the unit sphere. Advanced calculus tells us that such extrema of $q(\mathbf{x})$ on the unit sphere always exist.

Theorem 8.3 *Let A be a symmetric $n \times n$ matrix whose eigenvalues are $\lambda_1 \geq \lambda_2 \geq \cdots \geq \lambda_n$ in descending order. If $\mathbf{x}$ is constrained so that $\|\mathbf{x}\| = 1$ relative to the Euclidean inner product on $\mathbb{R}^n$, then*

(1) $\lambda_1 \geq \mathbf{x}^T A\mathbf{x} \geq \lambda_n$,

(2) $\mathbf{x}^T A\mathbf{x} = \lambda$ *if $\mathbf{x}$ is an eigenvector of A belonging to an eigenvalue λ.*

Proof: **(1)** Since A is symmetric, there is an orthonormal basis $\alpha = \{\mathbf{v}_1, \mathbf{v}_2, \ldots, \mathbf{v}_n\}$ for $\mathbb{R}^n$ consisting of eigenvectors of A belonging to the eigenvalues $\lambda_1, \lambda_2, \ldots, \lambda_n$, respectively. If $\langle\ ,\ \rangle$ denotes the Euclidean inner product, then any $\mathbf{x}$ in $\mathbb{R}^n$ may be expressed as

$$\mathbf{x} = \langle \mathbf{x}, \mathbf{v}_1\rangle \mathbf{v}_1 + \langle \mathbf{x}, \mathbf{v}_2\rangle \mathbf{v}_2 + \cdots + \langle \mathbf{x}, \mathbf{v}_n\rangle \mathbf{v}_n.$$

Thus,

$$\begin{aligned} A\mathbf{x} &= \langle \mathbf{x}, \mathbf{v}_1\rangle A\mathbf{v}_1 + \langle \mathbf{x}, \mathbf{v}_2\rangle A\mathbf{v}_2 + \cdots + \langle \mathbf{x}, \mathbf{v}_n\rangle A\mathbf{v}_n \\ &= \lambda_1\langle \mathbf{x}, \mathbf{v}_1\rangle \mathbf{v}_1 + \lambda_2\langle \mathbf{x}, \mathbf{v}_2\rangle \mathbf{v}_2 + \cdots + \lambda_n\langle \mathbf{x}, \mathbf{v}_n\rangle \mathbf{v}_n. \end{aligned}$$

If $\|\mathbf{x}\|^2 = \langle \mathbf{x}, \mathbf{v}_1\rangle^2 + \langle \mathbf{x}, \mathbf{v}_2\rangle^2 + \cdots + \langle \mathbf{x}, \mathbf{v}_n\rangle^2 = 1$, then we obtain

$$\begin{aligned} \mathbf{x}^T A\mathbf{x} = \langle \mathbf{x}, A\mathbf{x}\rangle &= \lambda_1\langle \mathbf{x}, \mathbf{v}_1\rangle^2 + \lambda_2\langle \mathbf{x}, \mathbf{v}_2\rangle^2 + \cdots + \lambda_n\langle \mathbf{x}, \mathbf{v}_n\rangle^2 \\ &\leq \lambda_1\langle \mathbf{x}, \mathbf{v}_1\rangle^2 + \lambda_1\langle \mathbf{x}, \mathbf{v}_2\rangle^2 + \cdots + \lambda_1\langle \mathbf{x}, \mathbf{v}_n\rangle^2 \\ &= \lambda_1\left(\langle \mathbf{x}, \mathbf{v}_1\rangle^2 + \langle \mathbf{x}, \mathbf{v}_2\rangle^2 + \cdots + \langle \mathbf{x}, \mathbf{v}_n\rangle^2\right), \\ &= \lambda_1, \end{aligned}$$

since λ_1 is the largest eigenvalue. Similarly, one can show $\lambda_n \leq \mathbf{x}^T A\mathbf{x}$.

(2) If $\mathbf{x}$ is an eigenvector of A belonging to λ and $\|\mathbf{x}\| = 1$, then

$$\mathbf{x}^T A\mathbf{x} = \langle \mathbf{x}, A\mathbf{x}\rangle = \langle \mathbf{x}, \lambda\mathbf{x}\rangle = \lambda\langle \mathbf{x}, \mathbf{x}\rangle = \lambda\|\mathbf{x}\|^2 = \lambda. \qquad \square$$

It follows from the preceding theorem that, subject to the constraint

$$\|\mathbf{x}\| = (x_1^2 + x_2^2 + \cdots + x_n^2)^{\frac{1}{2}} = 1,$$

the quadratic form $\mathbf{x}^T A\mathbf{x}$ has the maximum value λ_1 (the largest eigenvalue) and the minimum value λ_n (the smallest eigenvalue). This is very important in vibration problems ranging from aerodynamics to particle physics.

Corollary 8.4 *Let A be a symmetric matrix. For any nonzero vector $\mathbf{x} \in \mathbb{R}^n$, we get*

$$\lambda_1\|\mathbf{x}\|^2 \geq \mathbf{x}^T A\mathbf{x} \geq \lambda_n\|\mathbf{x}\|^2,$$

where λ_1 and λ_n are the largest and the smallest eigenvalues of A, respectively.

Example 8.9 Find the maximum and minimum values of the quadratic form

$$x_1^2 + x_2^2 + 4x_1x_2$$

subject to the constraint $x_1^2 + x_2^2 = 1$, and determine values of x_1 and x_2 at which the maximum and minimum occur.

Solution: The quadratic form can be written as

$$x_1^2 + x_2^2 + 4x_1x_2 = \mathbf{x}^T A\mathbf{x} = \begin{bmatrix} x_1 & x_2 \end{bmatrix} \begin{bmatrix} 1 & 2 \\ 2 & 1 \end{bmatrix} \begin{bmatrix} x_1 \\ x_2 \end{bmatrix}.$$

The eigenvalues of A are $\lambda = 3$ and $\lambda = -1$, which are the largest and smallest eigenvalues, respectively. Their corresponding eigenvectors are

$$\pm\left(\frac{1}{\sqrt{2}}, \frac{1}{\sqrt{2}}\right), \qquad \pm\left(\frac{1}{\sqrt{2}}, -\frac{1}{\sqrt{2}}\right),$$

respectively. Note that those extreme values of the quadratic form occur at those unit eigenvectors.

Thus, subject to the constraint $x_1^2 + x_2^2 = 1$, the maximum value of the quadratic form is $\lambda = 3$, which occurs at $\mathbf{x} = \pm(1/\sqrt{2},\ 1/\sqrt{2})$, and the minimum value is $\lambda = -1$, which occurs at $\mathbf{x} = \pm(1/\sqrt{2},\ -1/\sqrt{2})$. $\square$

Problem 8.6 Find the maximum and minimum values of the quadratic form

$$2x_1^2 + 2x_2^2 + 3x_1x_2$$

subject to the constraint $x_1^2 + x_2^2 = 1$, and determine values of x_1 and x_2 at which the maximum and minimum occur.

Problem 8.7 Find the maximum and minimum of the following quadratic forms subject to the constraint $x_1^2 + x_2^2 + x_3^2 = 1$ and determine the values of x_1, x_2, and x_3 at which the maximum and minimum occur:
(1) $x_1^2 + x_2^2 + 2x_3^2 - 2x_1x_2 + 4x_1x_3 + 4x_2x_3$,
(2) $2x_1^2 + x_2^2 + x_3^2 + 2x_1x_3 + 2x_1x_2$.

8.6 Definite forms

So far, we have seen that it is important to determine whether or not a symmetric matrix A is positive definite. In most cases, the definition does not help much. But we have seen that Theorem 8.2 gives us a practical characterization of positive definite matrices: *A is positive definite if and only if all eigenvalues of A are positive.* We will find some other practical criteria in terms of the determinant of the matrix. For this, we again look at the quadratic form in two variables, $q(x, y) = ax^2 + 2bxy + cy^2$, which may be rewritten in a complete square form as

$$q(\mathbf{x}) = ax^2 + 2bxy + cy^2 = a\left(x + \frac{b}{a}y\right)^2 + \left(c - \frac{b^2}{a}\right)y^2.$$

We see that q is positive definite, *i.e.*, $q(\mathbf{x}) = \mathbf{x}^T A\mathbf{x} > 0$ for any nonzero vector $\mathbf{x} = (x,\ y) \in \mathbb{R}^2$, if and only if $a > 0$ and $ac > b^2$, or equivalently, the determinants of

$$[a] \quad \text{and} \quad \begin{bmatrix} a & b \\ b & c \end{bmatrix}$$

are positive.

The natural generalization of the above conditions will involve all n-submatrices of A, called the **principal submatrices** of A, which are defined as the upper left square submatrices

$$A_1 = \begin{bmatrix} a_{11} \end{bmatrix},\ A_2 = \begin{bmatrix} a_{11} & a_{12} \\ a_{21} & a_{22} \end{bmatrix},\ A_3 = \begin{bmatrix} a_{11} & a_{12} & a_{13} \\ a_{21} & a_{22} & a_{23} \\ a_{31} & a_{32} & a_{33} \end{bmatrix},\ \ldots,\ A_n = A.$$

With this construction, we have the following characterization of positive definite matrices.

Theorem 8.5 *The following are equivalent for a real symmetric matrix A:*

(1) *A is positive definite, i.e., $\mathbf{x}^T A\mathbf{x} > 0$ for all nonzero vector $\mathbf{x}$;*

(2) *all the eigenvalues of A are positive;*

(3) *all the principal submatrices A_k's have positive determinants;*

(4) *all the pivots (without row interchanges) are positive;*

(5) *there exists a nonsingular matrix W such that $A = W^T W$.*

Proof: **(1)** $\Leftrightarrow$ **(2)** was shown.

(2) $\Rightarrow$ **(3)** If A has positive eigenvalues $\lambda_1, \lambda_2, \ldots, \lambda_n$, then $\det A = \lambda_1\lambda_2\cdots\lambda_n > 0$. To prove the same result for all the submatrices A_k, we show that if A is positive definite, so is every A_k. For each $k = 1, \ldots, n$, consider all the vectors whose last $n - k$ components are zero, say $\mathbf{x} = [x_1 \ \cdots \ x_k \ 0 \ \cdots \ 0]^T = [\mathbf{x}_k \ \mathbf{0}]^T$, where $\mathbf{x}_k$ is any vector in $\mathbb{R}^k$. Then

$$\mathbf{x}^T A\mathbf{x} = [\mathbf{x}_k \ \mathbf{0}]^T \begin{bmatrix} A_k & * \\ & * \end{bmatrix} \begin{bmatrix} \mathbf{x}_k \\ \mathbf{0} \end{bmatrix} = \mathbf{x}_k^T A_k \mathbf{x}_k.$$

Thus $\mathbf{x}^T A\mathbf{x} > 0$ for all such nonzero $\mathbf{x}$ if and only if $\mathbf{x}_k^T A_k \mathbf{x}_k > 0$ for all nonzero $\mathbf{x}_k \in \mathbb{R}^k$; that is, A_k's are positive definite, all eigenvalues of A_k are positive, and its determinant is positive.

(3) $\Rightarrow$ **(4)** Recall that the symmetric matrix A can be factorized uniquely into the form

$$A = LDL^T,$$

where L is a lower triangular matrix with 1's on its diagonal and D is the diagonal matrix with the pivots d_k of A on the diagonal. But the k-th pivot d_k is exactly the ratio of $\det A_k$ to $\det A_{k-1}$:

$$d_k = \frac{\det A_k}{\det A_{k-1}}.$$

Hence, all d_k's are positive.

(4) $\Rightarrow$ **(5)** Let $A = LDL^T$ as above with

$$D = \begin{bmatrix} d_1 & & & 0 \\ & d_2 & & \\ & & \ddots & \\ 0 & & & d_n \end{bmatrix}, \quad d_i > 0.$$

Define

$$\sqrt{D} = \begin{bmatrix} \sqrt{d_1} & & & 0 \\ & \sqrt{d_2} & & \\ & & \ddots & \\ 0 & & & \sqrt{d_n} \end{bmatrix}.$$

Then, clearly $\det(\sqrt{D}) > 0$, $D = \sqrt{D}\sqrt{D}$ and $(\sqrt{D})^T = \sqrt{D}$. Hence,

$$\begin{aligned} A &= LDL^T = (L\sqrt{D})(\sqrt{D}L^T) = (L\sqrt{D})(L\sqrt{D})^T \\ &= W^T W, \end{aligned}$$

where $W = (L\sqrt{D})^T$, which is nonsingular since L and $\sqrt{D}$ are.

(5) $\Rightarrow$ **(1)** If A is real symmetric and $A = W^T W$, where W is nonsingular, then for $\mathbf{x} \neq \mathbf{0}$ we have

$$\mathbf{x}^T A\mathbf{x} = \mathbf{x}^T (W^T W)\mathbf{x} = (W\mathbf{x})^T (W\mathbf{x}) = \|W\mathbf{x}\|^2 > 0,$$

because $W\mathbf{x} \neq 0$. □

Problem 8.8 State the corresponding conditions to the ones in Theorem 8.5 for the negative definite forms.

Problem 8.9 Determine which one of the following matrices A and B is positive definite. For the positive definite one, find a nonsingular matrix W such that it is $W^T W$.

$$A = \begin{bmatrix} 2 & -1 & -1 \\ -1 & 2 & -1 \\ -1 & -1 & 2 \end{bmatrix}, \quad B = \begin{bmatrix} 2 & -1 & 0 \\ -1 & 2 & 1 \\ 0 & 1 & 2 \end{bmatrix}.$$

Problem 8.10 Let A be a positive definite matrix. Prove that $C^T AC$ is also positive definite for any nonsingular matrix C.

We now consider semidefinite matrices. One can easily establish the following analogous theorem.

Theorem 8.6 *The following are equivalent for a real symmetric matrix A:*

(1) *A is positive semidefinite, i.e., $\mathbf{x}^T A\mathbf{x} \geq 0$ for all vectors $\mathbf{x}$;*

(2) *all the eigenvalues of A are nonnegative;*

(3) *all the principal submatrices A_k's have nonnegative determinants;*

(4) *all the pivots (without row exchanges) are nonnegative;*

(5) *there exists a matrix W, possibly singular, such that $A = W^T W$.*

Problem 8.11 State the corresponding conditions to the ones in Theorem 8.6 for the negative semidefinite forms.

Problem 8.12 Show that the determinant of a negative definite $n \times n$ symmetric matrix is positive if n is even and negative if n is odd.

8.7 Bilinear forms

In the study of a system of linear equations, an essential thing is to know which properties of a matrix remain unchanged (*i.e.*, *invariant*) under the elementary row operations (or Gauss-Jordan eliminations). For example, the row space $\mathcal{R}(A)$, the null space $\mathcal{N}(A)$ and the rank of A are *invariant* under the elementary row operations.

If we understand a matrix A as a linear transformation on a vector space, a change of basis for the vector space to diagonalize A gives rise to a similarity relation $S^{-1}AS = D$ by some nonsingular matrix S, and we know that similar matrices have the same eigenvalues and determinants.

However, as we mentioned in Section 8.1, the diagonalization of a symmetric matrix A, considered as a quadratic form on a vector space, can be obtained by the conjugation $P^T AP = D$ by some invertible matrix P which is also a change of basis (*i.e.*, change of variables) for the vector space. In this case there may be various matrices P that diagonalize A so that the diagonal entries of D need not be eigenvalues of A and may vary depending on P: that is, the eigenvalue is no longer invariant under the congruence relation.

However, Sylvester's law of inertia says that the inertia or the numbers of positive and negative signs of the diagonal entries are unchanged whatever P is, that is, they are invariant under the congruence relation. In this section we will prove this. For this, we extend the quadratic forms to more general forms:

Definition 8.5 A **bilinear form** on a pair of vector spaces V and W is a real-valued bilinear function b on $V \times W$ ($b : V \times W \to \mathbb{R}$) satisfying

$$\begin{aligned} b(k\mathbf{x} + \mathbf{x}', \ \mathbf{y}) &= kb(\mathbf{x}, \ \mathbf{y}) + b(\mathbf{x}', \ \mathbf{y}) \\ b(\mathbf{x}, \ k\mathbf{y} + \mathbf{y}') &= kb(\mathbf{x}, \ \mathbf{y}) + b(\mathbf{x}, \ \mathbf{y}') \end{aligned}$$

for any $\mathbf{x}$, $\mathbf{x}'$ in V, $\mathbf{y}$, $\mathbf{y}'$ in W and any scalar k. In particular, if $V = W$, $b : V \times V \to \mathbb{R}$ is called a bilinear form on V.

Example 8.10 Let A be an $m \times n$ matrix and let $b : \mathbb{R}^m \times \mathbb{R}^n \to \mathbb{R}$ be defined by $b(\mathbf{x}, \mathbf{y}) = \mathbf{x}^T A\mathbf{y}$ for $\mathbf{x} \in \mathbb{R}^m$, $\mathbf{y} \in \mathbb{R}^n$. Then b is clearly a bilinear form. In particular, if $m = n$ and $A = I_n$, the identity matrix, then it shows that the Euclidean inner product on $\mathbb{R}^n$ is a bilinear form. Generally, for any inner product space V, the inner product $b : V \times V \to \mathbb{R}$, $b(\mathbf{x}, \mathbf{y}) = \langle \mathbf{x}, \mathbf{y} \rangle$, is a bilinear form on V.

Example 8.11 Let V be a vector space and V^* its dual vector space, that is, $V^* = \mathcal{L}(V; \mathbb{R})$. Let $b : V \times V^* \to \mathbb{R}$ be defined by

$$b(\mathbf{v}, \mathbf{v}^*) = \mathbf{v}^*(\mathbf{v}) \quad \text{for any } \mathbf{v} \in V, \ \mathbf{v}^* \in V^*.$$

Then one can easily show that b is a bilinear form on the pair of vector spaces V and V^*. The reader should notice that the vector space operations on the dual space V^* are defined in such a way as to force the mapping b to be a bilinear form.

Definition 8.6 A bilinear form $b : V \times W \to \mathbb{R}$ on vector spaces V and W is said to be **nondegenerate** if it satisfies

$$\begin{array}{llll} b(\mathbf{v}, \mathbf{w}) = 0 & \text{for all } \mathbf{w} \in W & \text{implies } \mathbf{v} = \mathbf{0}, & \text{and} \\ b(\mathbf{v}, \mathbf{w}) = 0 & \text{for all } \mathbf{v} \in V & \text{implies } \mathbf{w} = \mathbf{0}. & \end{array}$$

Note that $b(\mathbf{0}, \mathbf{w}) = b(\mathbf{v}, \mathbf{0}) = 0$ for any $\mathbf{v} \in V$ and $\mathbf{w} \in W$. Thus the nondegeneracy condition asserts that the equation $b(\mathbf{v}, \mathbf{w}) = 0$ for all $\mathbf{w}$ holds only when $\mathbf{v} = \mathbf{0}$.

Let $b : V \times W \to \mathbb{R}$ be a nondegenerate bilinear form. For a fixed $\mathbf{w} \in W$, we define $\varphi_{\mathbf{w}} : V \to \mathbb{R}$ by

$$\varphi_{\mathbf{w}}(\mathbf{v}) = b(\mathbf{v}, \mathbf{w}) \quad \text{for} \quad \mathbf{v} \in V.$$

Then the bilinearity of b proves that $\varphi_{\mathbf{w}} \in V^*$, from which we obtain a linear transformation

$$\varphi : W \to V^* \quad \text{defined by} \quad \varphi(\mathbf{w}) = \varphi_{\mathbf{w}}.$$

Similarly, we can have a linear transformation $\psi : V \to W^*$ defined by

$$\psi(\mathbf{v})(\mathbf{w}) = b(\mathbf{v}, \mathbf{w}) \quad \text{for } \mathbf{v} \in V \text{ and } \mathbf{w} \in W.$$

Theorem 8.7 *If $b : V \times W \to \mathbb{R}$ is a nondegenerate bilinear form, then the linear transformations $\varphi : W \to V^*$ and $\psi : V \to W^*$ are isomorphisms.*

Proof: Suppose that $\varphi_{\mathbf{w}} = \varphi_{\mathbf{w}'}$. Then, for all $\mathbf{v} \in V$,

$$b(\mathbf{v},\ \mathbf{w}) = \varphi(\mathbf{w})(\mathbf{v}) = \varphi(\mathbf{w}')(\mathbf{v}) = b(\mathbf{v},\ \mathbf{w}') \quad \text{or} \quad b(\mathbf{v},\ \mathbf{w} - \mathbf{w}') = 0.$$

The nondegeneracy of b implies that $\mathbf{w} = \mathbf{w}'$, that is, φ is one-to-one. This also implies that

$$\dim W \leq \dim V^*.$$

A similar argument shows that the linear transformation $\psi : V \to W^*$ is also one-to-one, and therefore

$$\dim V \leq \dim W^*.$$

Since $\dim V = \dim V^*$ and $\dim W = \dim W^*$ from Corollary 4.19, we have

$$\dim V \leq \dim W^* = \dim W \leq \dim V^* = \dim V.$$

Therefore, φ and ψ are surjective, and so isomorphisms. □

Corollary 8.8 *If there exists a nondegenerate bilinear form* $b : V \times W \to \mathbb{R}$, *then* $\dim V = \dim W$.

Let $b : V \times V \to \mathbb{R}$ be a bilinear form on a vector space V, and let $\alpha = \{\mathbf{v}_1,\ \mathbf{v}_2,\ \ldots,\ \mathbf{v}_n\}$ be a basis for V. Such a bilinear form is completely determined by the values $b(\mathbf{v}_i,\ \mathbf{v}_j)$ of the vectors $\mathbf{v}_i,\ \mathbf{v}_j$ in the basis α. In fact, if

$$\begin{array}{rcl} \mathbf{x} &=& x_1\mathbf{v}_1 + x_2\mathbf{v}_2 + \cdots + x_n\mathbf{v}_n, \\ \mathbf{y} &=& y_1\mathbf{v}_1 + y_2\mathbf{v}_2 + \cdots + y_n\mathbf{v}_n \end{array}$$

are vectors in V, then

$$b(\mathbf{x},\ \mathbf{y}) = \sum_{i,j=1}^{n} x_i y_j b(\mathbf{v}_i,\ \mathbf{v}_j) = [\mathbf{x}]_\alpha^T A [\mathbf{y}]_\alpha,$$

where $A = [a_{ij}]$, $a_{ij} = b(\mathbf{v}_i,\ \mathbf{v}_j)$ is called the **matrix representation** of b with respect to a basis α. We write $A = [b]_\alpha$. Let β be another basis for the vector space V and let $P = [Id]_\beta^\alpha$ be the transition matrix from β to α. Then we get

$$P[\mathbf{x}]_\beta = [Id]_\beta^\alpha[\mathbf{x}]_\beta = [\mathbf{x}]_\alpha$$

and

$$b(\mathbf{x},\ \mathbf{y}) = [\mathbf{x}]_\alpha^T A[\mathbf{y}]_\alpha = [\mathbf{x}]_\beta^T (P^T A P)[\mathbf{y}]_\beta$$

for any $\mathbf{x}$ and $\mathbf{y}$ in V. Thus, *two matrix representations of a bilinear form b with respect to different bases are congruent, and conversely any two congruent matrices can be matrix representations of the same bilinear form* (verify it). Note that congruent matrices have the same rank because P and P^T are nonsingular.

The **rank** of a bilinear form $b : V \times V \to \mathbb{R}$ on a vector space V, written $\text{rank}(b)$, is defined as the rank of any matrix representation of b.

Theorem 8.9 *A bilinear form $b : V \times V \to \mathbb{R}$ on a vector space V is nondegenerate if and only if* $\text{rank}(b) = \dim V$.

Proof: Since every n-dimensional vector space V is isomorphic to $\mathbb{R}^n$ and congruent matrices have the same rank, we can assume that $V = \mathbb{R}^n$ and $A = [b]_\alpha$ is the matrix representation of a bilinear form $b : \mathbb{R}^n \times \mathbb{R}^n \to \mathbb{R}$ with respect to the standard basis $\alpha = \{\mathbf{e}_1, \mathbf{e}_2, \ldots, \mathbf{e}_n\}$ for $\mathbb{R}^n$. Then we have $b(\mathbf{u}, \mathbf{v}) = \mathbf{u}^T A\mathbf{v}$ for any $\mathbf{u}, \mathbf{v} \in \mathbb{R}^n$.

Suppose that $\text{rank}(b) = \text{rank}A < n$. Then the homogeneous system $A\mathbf{x} = \mathbf{0}$ has a nontrivial solution, say $\mathbf{v}$, and then $b(\mathbf{u}, \mathbf{v}) = \mathbf{u}^T A\mathbf{v} = 0$ for any $\mathbf{u} \in \mathbb{R}^n$, but $\mathbf{v} \neq \mathbf{0}$. It implies that b is degenerate.

Now, let's assume that $\text{rank}(b) = \text{rank}A = n$ and $b(\mathbf{u}, \mathbf{v}) = \mathbf{u}^T A\mathbf{v} = 0$ for all $\mathbf{u} \in \mathbb{R}^n$. By taking the basic vectors $\mathbf{e}_1, \mathbf{e}_2, \ldots, \mathbf{e}_n$ instead of $\mathbf{u}$, we can see that $A\mathbf{v} = \mathbf{0}$. The condition $\text{rank}A = n$ implies that $\mathbf{v} = \mathbf{0}$. A similar method with the equation $b(\mathbf{u}, \mathbf{v}) = \mathbf{u}^T A\mathbf{v} = (\mathbf{u}^T A\mathbf{v})^T = \mathbf{v}^T A^T\mathbf{u}$ verifies that $b(\mathbf{u}, \mathbf{v}) = 0$ for all $\mathbf{v} \in \mathbb{R}^n$ implies $\mathbf{u} = \mathbf{0}$. Hence, b is nondegenerate. □

Example 8.12 Let $b : \mathbb{R}^2 \times \mathbb{R}^2 \to \mathbb{R}$ be defined by $b(\mathbf{x}, \mathbf{y}) = x_1y_1 + 3x_1y_2 + 2x_2y_1 - x_2y_2$ with respect to the standard basis $\alpha = \{\mathbf{e}_1, \mathbf{e}_2\}$. Then b is clearly a bilinear form, and the matrix representation of b with respect to α is

$$[b]_\alpha = \begin{bmatrix} 1 & 3 \\ 2 & -1 \end{bmatrix}.$$

If $\beta = \{\mathbf{v}_1 = (1, 0), \mathbf{v}_2 = (1, 1)\}$ is another basis for $\mathbb{R}^2$, then the matrix representation becomes

$$[b]_\beta = \begin{bmatrix} 1 & 4 \\ 3 & 5 \end{bmatrix},$$

because $b(\mathbf{v}_1, \mathbf{v}_1) = 1$, $b(\mathbf{v}_1, \mathbf{v}_2) = 4$, $b(\mathbf{v}_2, \mathbf{v}_1) = 3$ and $b(\mathbf{v}_2, \mathbf{v}_2) = 5$. Since $\text{rank}[b]_\alpha = \text{rank}[b]_\beta = 2$, the bilinear form b is nondegenerate.

Problem 8.13 (1) Let $b : \mathbb{R}^3 \times \mathbb{R}^3 \to \mathbb{R}$ be defined by

$$b(\mathbf{x},\ \mathbf{y}) = x_1y_1 - 2x_1y_2 + x_2y_1 - x_3y_3$$

with respect to the standard basis. Is this a bilinear form? If so, find the matrix representation of b with respect to the basis

$$\alpha = \{\mathbf{v}_1 = (1,\ 0,\ 1),\ \mathbf{v}_2 = (1,\ 0,\ -1),\ \mathbf{v}_3 = (0,\ 1,\ 0)\}.$$

Find its rank.

(2) Let $V = M_{2\times 2}(\mathbb{R})$ be the vector space of 2×2 matrices, and let $b : V \times V \to \mathbb{R}$ be defined by $b(A,\ B) = \mathrm{tr}(A)\cdot\mathrm{tr}(B)$. Is this a bilinear form? If so, find the matrix representation of b with respect to the basis

$$\alpha = \left\{E_1 = \begin{bmatrix} 1 & 0 \\ 0 & 0 \end{bmatrix},\ E_2 = \begin{bmatrix} 0 & 1 \\ 0 & 0 \end{bmatrix},\ E_3 = \begin{bmatrix} 0 & 0 \\ 1 & 0 \end{bmatrix},\ E_4 = \begin{bmatrix} 0 & 0 \\ 0 & 1 \end{bmatrix}\right\}.$$

Find its rank.

Definition 8.7 A bilinear form b on a vector space V is said to be **symmetric** if $b(\mathbf{x},\ \mathbf{y}) = b(\mathbf{y},\ \mathbf{x})$ for any $\mathbf{x},\ \mathbf{y} \in V$. It is **skew-symmetric** if $b(\mathbf{x},\ \mathbf{y}) = -b(\mathbf{y},\ \mathbf{x})$ for any $\mathbf{x},\ \mathbf{y} \in V$.

One can easily see that a bilinear form is symmetric (or skew-symmetric) if and only if its matrix representation is symmetric (or skew-symmetric) for any basis. As an example, an inner product on a vector space V is just a symmetric, nondegenerate bilinear form on V.

A bilinear form b on V is **diagonalizable** if there exists a basis α for V such that the matrix representation $[b]_\alpha$ of b with respect to α is diagonal.

Theorem 8.10 *A bilinear form b on a vector space V is symmetric if and only if it is diagonalizable.*

Proof: Since every symmetric matrix is orthogonally diagonalizable, we only need to prove the sufficiency. Let a bilinear form b be diagonalizable and let α be a basis for V such that the matrix representation $[b]_\alpha$ is diagonal. Then for any basis β for V, the matrix representation $[b]_\alpha$ and $[b]_\beta$ are congruent, say $[b]_\beta = P^T[b]_\alpha P$ for some invertible matrix P. Since $[b]_\alpha$ is diagonal, we have

$$[b]_\beta^T = (P^T[b]_\alpha P)^T = P^T[b]_\alpha P = [b]_\beta,$$

i.e., $[b]_\beta$ is symmetric and the bilinear form b is symmetric. $\square$

Example 8.13 Let $b : \mathbb{R}^3 \times \mathbb{R}^3 \to \mathbb{R}$ be the bilinear form defined by

$$b(\mathbf{x},\ \mathbf{y}) = x_1y_3 - 2x_2y_2 + 2x_2y_3 + x_3y_1 + 2x_3y_2 - x_3y_3.$$

Then clearly $b(\mathbf{x},\ \mathbf{y}) = b(\mathbf{y},\ \mathbf{x})$, and the matrix representation of b with respect to the standard basis $\alpha = \{\mathbf{e}_1,\ \mathbf{e}_2,\ \mathbf{e}_3\}$ is

$$[b]_\alpha = \begin{bmatrix} 0 & 0 & 1 \\ 0 & -2 & 2 \\ 1 & 2 & -1 \end{bmatrix},$$

which is symmetric. Hence, the bilinear form b is symmetric. By Theorem 8.10, it is diagonalizable. In fact,

$$\begin{aligned} [[b]_\alpha \mid I] &= \left[\begin{array}{ccc|ccc} 0 & 0 & 1 & 1 & 0 & 0 \\ 0 & -2 & 2 & 0 & 1 & 0 \\ 1 & 2 & -1 & 0 & 0 & 1 \end{array}\right] \\ &\longrightarrow \left[\begin{array}{ccc|ccc} -1 & 0 & 0 & 0 & 0 & 1 \\ 0 & 2 & 0 & 0 & 1 & -1 \\ 0 & 0 & -1 & 1 & 2 & -1 \end{array}\right] = [D \mid P]. \end{aligned}$$

By a direct computation, one can easily show that $P^T[b]_\alpha P = D$. □

A skew-symmetric matrix is not diagonalizable in general, but the following theorem shows the structure of a skew-symmetric bilinear form. Note that a bilinear form b is skew-symmetric if and only if $b(\mathbf{x},\ \mathbf{x}) = 0$ for any $\mathbf{x}$ in V.

Theorem 8.11 *Let $b : V \times V \to \mathbb{R}$ be a skew-symmetric bilinear form. Then there exists a basis α for V with respect to which the matrix representation $[b]_\alpha$ is of the form*

$$\begin{bmatrix} \begin{bmatrix} 0 & 1 \\ -1 & 0 \end{bmatrix} & & & & & \\ & \ddots & & & & \\ & & \begin{bmatrix} 0 & 1 \\ -1 & 0 \end{bmatrix} & & & \\ & & & 0 & & \\ & & & & \ddots & \\ & & & & & 0 \end{bmatrix}.$$

Proof: If $b = 0$, then $[b]_\alpha$ is the zero matrix. Also if $\dim V = 1$, then $b(\mathbf{x}, \mathbf{x}) = 0$ for any basis vector $\mathbf{x}$ in V, so $b = 0$.

Now, we assume that $b \neq 0$ and prove it by induction on $\dim V$. Since $b \neq 0$, there exist nonzero vectors $\mathbf{x}$ and $\mathbf{y}$ in V such that $b(\mathbf{x}, \mathbf{y}) \neq 0$. By the bilinearity of b, we can assume that $b(\mathbf{x}, \mathbf{y}) = 1$. Such vectors $\mathbf{x}$ and $\mathbf{y}$ must be linearly independent, because if $\mathbf{y} = k\mathbf{x}$, then $b(\mathbf{x}, \mathbf{y}) = kb(\mathbf{x}, \mathbf{x}) = 0$. Let U be the subspace of V spanned by $\mathbf{x}$ and $\mathbf{y}$ and let

$$W = \{\mathbf{v} \in V : b(\mathbf{v}, \mathbf{u}) = 0 \text{ for any } \mathbf{u} \in U\}.$$

Then one can easily show that W is also a subspace of V and $U \cap W = \{\mathbf{0}\}$ (see Problem 8.14 below). Moreover, $U + W = V$. In fact for a given vector $\mathbf{v} \in V$, let $\mathbf{u} = b(\mathbf{v}, \mathbf{y})\mathbf{x} - b(\mathbf{v}, \mathbf{x})\mathbf{y}$. It is easy to show that $\mathbf{u} \in U$ and $\mathbf{v} - \mathbf{u} \in W$. Thus $V = U \oplus W$, where $\dim W = n - 2$. Clearly, the matrix representation of the restriction of b to U with respect to the basis $\{\mathbf{x}, \mathbf{y}\}$ is $\begin{bmatrix} 0 & 1 \\ -1 & 0 \end{bmatrix}$, and the restriction of b to W is also skew-symmetric. The same argument is applied to W, and the theorem is proved by induction. □

Problem 8.14 Prove that $U \cap W = \{\mathbf{0}\}$ in the proof of Theorem 8.11.

Example 8.14 Let $b : \mathbb{R}^3 \times \mathbb{R}^3 \to \mathbb{R}$ be the bilinear form defined by

$$b(\mathbf{x}, \mathbf{y}) = x_1y_2 - x_2y_1 + x_3y_1 - x_1y_3 + x_2y_3 - x_3y_2.$$

Then clearly $b(\mathbf{x}, \mathbf{y}) = -b(\mathbf{y}, \mathbf{x})$, and the matrix representation of b with respect to the standard basis $\alpha = \{\mathbf{e}_1, \mathbf{e}_2, \mathbf{e}_3\}$ is

$$[b]_\alpha = \begin{bmatrix} 0 & 1 & -1 \\ -1 & 0 & 1 \\ 1 & -1 & 0 \end{bmatrix},$$

which is skew-symmetric. By a simple computation, $b(\mathbf{e}_1, \mathbf{e}_2) = 1 = -b(\mathbf{e}_2, \mathbf{e}_1)$. Let U be the subspace of $\mathbb{R}^3$ spanned by $\mathbf{e}_1$ and $\mathbf{e}_2$, *i.e.*, the xy-plane. If we set $W = \{\mathbf{v} \in V : b(\mathbf{v}, \mathbf{u}) = 0 \text{ for any } \mathbf{u} \in U\}$, then $W = \{\lambda\mathbf{z} : \lambda \in \mathbb{R}\}$ where $\mathbf{z} = (1, 1, 1)$. Clearly, $\beta = \{\mathbf{e}_1, \mathbf{e}_2, \mathbf{z}\}$ is a basis for $\mathbb{R}^3$ and $b(\mathbf{z}, \mathbf{z}) = 0$ so that

$$[b]_\beta = \begin{bmatrix} 0 & 1 & 0 \\ -1 & 0 & 0 \\ 0 & 0 & 0 \end{bmatrix}.$$

□

Problem 8.15 Show that any bilinear form b on a vector space V is the sum of a symmetric bilinear form and a skew-symmetric bilinear form.

The following theorem shows how quadratic forms and bilinear forms are related.

Theorem 8.12 *If b is a symmetric bilinear form on $\mathbb{R}^n$, then the function $q(\mathbf{x}) = b(\mathbf{x}, \mathbf{x})$, for $\mathbf{x} \in \mathbb{R}^n$, is a quadratic form.*

Conversely, for every quadratic form q, there is a unique symmetric bilinear form b such that $q(\mathbf{x}) = b(\mathbf{x}, \mathbf{x})$ for all $\mathbf{x}$ in $\mathbb{R}^n$.

Proof: If $b(\mathbf{x}, \mathbf{y}) = \mathbf{x}^T A\mathbf{y}$ is a symmetric bilinear form, then $q(\mathbf{x}) = b(\mathbf{x}, \mathbf{x}) = \mathbf{x}^T A\mathbf{x}$ is clearly a quadratic form.

Conversely, if b is a symmetric bilinear form, then

$$b(\mathbf{x}+\mathbf{y}, \mathbf{x}+\mathbf{y}) = b(\mathbf{x}, \mathbf{x}) + 2b(\mathbf{x}, \mathbf{y}) + b(\mathbf{y}, \mathbf{y}).$$

Hence, for a given quadratic form q, a bilinear form b can be defined by

$$b(\mathbf{x}, \mathbf{y}) = \frac{1}{2}[q(\mathbf{x}+\mathbf{y}) - q(\mathbf{x}) - q(\mathbf{y})].$$

This form b is clearly symmetric, bilinear and $b(\mathbf{x}, \mathbf{x}) = q(\mathbf{x})$. The uniqueness also comes from this relation. □

Recall that any two matrix representations of a quadratic form or a symmetric bilinear form are congruent, and they can be diagonalized. But their congruent diagonal matrices may have different diagonal entries (see Remark (1) on page 289). Although these entries are not unique, the number of positive entries and the number of negative entries are invariant, *i.e.*, independent of the choice of diagonal representation. This result is called Sylvester's law of inertia.

Theorem 8.13 (Sylvester's law of inertia) *Let b be a symmetric bilinear form on a vector space V. Then the number of positive diagonal entries and the number of negative diagonal entries of any diagonal representation of b are both independent of the diagonal representation.*

Proof: Let $\alpha = \{\mathbf{x}_1, \ldots, \mathbf{x}_p, \mathbf{x}_{p+1}, \ldots, \mathbf{x}_n\}$ be an ordered basis for V in which

$$\begin{array}{ll} b(\mathbf{x}_i, \mathbf{x}_i) > 0 & \text{for } i = 1, 2, \ldots, p, \text{ and} \\ b(\mathbf{x}_i, \mathbf{x}_i) \le 0 & \text{for } i = p+1, \ldots, n, \end{array}$$

and let $\beta = \{\mathbf{y}_1, \ldots, \mathbf{y}_{p'}, \mathbf{y}_{p'+1}, \ldots, \mathbf{y}_n\}$ be another ordered basis for V in which

$$\begin{aligned} b(\mathbf{y}_i, \mathbf{y}_i) > 0 \quad & \text{for } i = 1, 2, \ldots, p', \text{ and} \\ b(\mathbf{y}_i, \mathbf{y}_i) \leq 0 \quad & \text{for } i = p'+1, \ldots, n. \end{aligned}$$

To show $p = p'$, let U and W be subspaces V spanned by $\{\mathbf{x}_1, \ldots, \mathbf{x}_p\}$ and $\{\mathbf{y}_{p'+1}, \ldots, \mathbf{y}_n\}$, respectively. Then $b(\mathbf{u}, \mathbf{u}) > 0$ for any nonzero vector $\mathbf{u} \in U$ and $b(\mathbf{w}, \mathbf{w}) \leq 0$ for any nonzero vector $\mathbf{w} \in W$. Thus $U \cap W = \{\mathbf{0}\}$, and

$$\dim(U + W) = \dim U + \dim W - \dim(U \cap W) = p + (n - p') \leq n,$$

or $p \leq p'$. Similarly, we can show $p' \leq p$ to conclude $p = p'$. Therefore, any two diagonal matrix representations of b have the same number of positive diagonal entries. By considering the bilinear form $-b$ instead of b, we can also have that any two diagonal matrix representations of b have the same number of negative diagonal entries. □

Corollary 8.14 *Let A be a symmetric matrix. If $B = P^T AP$ for some invertible matrix P, then A and B have the same number of positive diagonal entries, the same number of negative diagonal entries and the same number of zero diagonal entries.*

Definition 8.8 Let A be a real symmetric matrix. The number of positive eigenvalues of A is called the **index** of A. The difference between the number of positive eigenvalues and the number of negative eigenvalues of A is called the **signature** of A.

Hence, the index and signature together with the rank of a symmetric matrix are invariants under the congruence relation, and any two of these invariants determine the third, by noting that

the number of positive eigenvalues = the index,
the index + the number of negative eigenvalues = the rank,
the index − the number of negative eigenvalues = the signature.

We have shown the necessary condition of the following corollary.

Corollary 8.15 *Two symmetric (square) matrices are congruent if and only if they have the same invariants: index, signature and rank.*

Proof: Suppose that two symmetric matrices A and B have the same invariants, and let D and E be diagonal matrices congruent to A and B, respectively. Without loss of generality, we may choose D and E so that the diagonal entries are in the order of positive, negative and zero. Let p and r denote the index and the rank, respectively, of both D and E. Let d_i denote the i-th diagonal entry of D. Define the diagonal matrix Q whose i-th diagonal entry q_i is given by

$$q_i = \begin{cases} 1/\sqrt{d_i} & \text{if } 1 \le i \le p \\ 1/\sqrt{-d_i} & \text{if } p < i \le r \\ 1 & \text{if } r < i \le n. \end{cases}$$

Then

$$Q^T D Q = \begin{bmatrix} I_p & 0 & 0 \\ 0 & -I_{r-p} & 0 \\ 0 & 0 & 0 \end{bmatrix} = J_{pr}.$$

Hence, A is congruent to J_{pr}, and so is B, *i.e.*, A is congruent to B. □

Example 8.15 Determine the index, the signature and the rank for each of the following matrices.

$$A = \begin{bmatrix} 1 & 1 & 2 \\ 1 & 0 & 3 \\ 2 & 3 & 6 \end{bmatrix}, \quad B = \begin{bmatrix} 1 & 0 & 0 \\ 0 & 4 & 0 \\ 0 & 0 & 5 \end{bmatrix}, \quad C = \begin{bmatrix} 1 & 0 & 1 \\ 0 & 1 & 2 \\ 1 & 2 & 1 \end{bmatrix}.$$

Which are congruent each other?

Solution: In Example 8.5, we saw that the matrix A can be diagonalized to

$$D = \begin{bmatrix} 1 & 0 & 0 \\ 0 & -1 & 0 \\ 0 & 0 & 3 \end{bmatrix}.$$

Therefore, A has rank 3, index 2 and signature 1. The matrix B is already diagonal, and has rank 3, index 3 and signature 0. Using the method of Example 8.5, one can show that C is congruent to the diagonal matrix with diagonal entries 1, 1, -4. Therefore, C has rank 3, index 2 and signature 1. (Note that it is not necessary to find the eigenvalue of C to diagonalize it orthogonally.) We conclude that A and C are congruent and B is congruent to neither A nor C by Corollary 8.15. □

Problem 8.16 Prove that if the diagonal entries of a diagonal matrix are permuted, then the resulting diagonal matrix is congruent to the original one.

Problem 8.17 Prove that the total number of distinct equivalence classes of congruent $n \times n$ real symmetric matrices is equal to $\frac{1}{2}(n+1)(n+2)$.

Problem 8.18 Find the signature, the index and the rank of each of the following matrices.

$$(1) \begin{bmatrix} 0 & 1 & 2 \\ 1 & -2 & 3 \\ 2 & 3 & 4 \end{bmatrix}, \quad (2) \begin{bmatrix} 1 & 2 & 3 \\ 2 & 4 & 5 \\ 3 & 5 & 6 \end{bmatrix}, \quad (3) \begin{bmatrix} 1 & 0 & 1 \\ 0 & 1 & 2 \\ 1 & 2 & 1 \end{bmatrix}.$$

Which are congruent each other?

8.8 Exercises

8.1. Find the matrix representing each of the following quadratic forms:

(1) $x_1^2 + 4x_1x_2 + 3x_2^2$,
(2) $x_1^2 - x_2^2 + x_3^2 + 4x_1x_3 - 5x_2x_3$,
(3) $x_1^2 - 2x_2^2 - 3x_3^2 + 4x_1x_2 + 6x_1x_3 - 8x_2x_3$,
(4) $3x_1y_1 - 2x_1y_2 + 5x_2y_1 + 7x_2y_2 - 8x_2y_3 + 4x_3y_2 - x_3y_3$,
(5) $[x_1 \; x_2] \begin{bmatrix} 2 & 0 \\ 4 & 1 \end{bmatrix} \begin{bmatrix} x_1 \\ x_2 \end{bmatrix}$.

8.2. Let q be a quadratic form on $\mathbb{R}^3$ and let $A = \begin{bmatrix} 7 & 4 & -5 \\ 4 & -2 & 4 \\ -5 & 4 & 7 \end{bmatrix}$ be the matrix representing q with respect to the basis

$$\alpha = \{(1,\ 0,\ 1),\ (1,\ 1,\ 0),\ (0,\ 0,\ 1)\}.$$

(1) Diagonalize A, *i.e.*, find an orthogonal matrix P so that P^TAP is a diagonal matrix.
(2) Construct a basis β for $\mathbb{R}^3$ such that the elements of β are the principal axes of the quadratic surface $q(\mathbf{x}) = 0$.

8.3. Sketch the graph of each of the following quadratic equations:

(1) $xy = 2$,
(2) $53x^2 - 72xy + 32y^2 = 80$,
(3) $16x^2 - 24xy + 9y^2 - 60x - 80y + 100 = 0$.

8.4. For a positive definite quadratic form $q(\mathbf{x}) = ax^2 + 2bxy + cy^2$, the curve $q(\mathbf{x}) = 1$ is an ellipse. When $a = c = 2$ and $b = -1$, sketch the ellipse.

8.5. Determine whether each of the following matrices takes a local minimum, local maximum or saddle point at the given point:

(1) $f(x,\ y) = -1 + 4(e^x - x) - 5x\sin y + 6y^2$ at the point $(x,\ y) = (0,\ 0)$;
(2) $f(x,\ y) = (x^2 - 2x)\cos y$ at $(x,\ y) = (1,\ \pi)$.

8.6. Show that the quadratic form $q(\mathbf{x}) = 2x^2 + 4xy + y^2$ has a saddle point at the origin, despite the fact that its coefficients are positive. Show that q can be written as the difference of two perfect squares.

8.7. Find the maximum and the minimum values of the function

$$R(\mathbf{x}) = \frac{\mathbf{x}^T A\mathbf{x}}{\mathbf{x}^T\mathbf{x}}, \quad \mathbf{x} \neq \mathbf{0},$$

which is called the **Rayleigh quotient** of A, when

$$(1)\ A = \begin{bmatrix} 2 & 1 & 0 & 0 \\ 1 & 2 & 0 & 0 \\ 0 & 0 & 4 & 0 \\ 0 & 0 & 0 & 1 \end{bmatrix}, \quad (2)\ A = \begin{bmatrix} 2 & -3 & 0 \\ -3 & 0 & 0 \\ 0 & 0 & 1 \end{bmatrix}.$$

8.8. Determine whether or not each of the following matrices is positive definite:

$$(1)\ A = \begin{bmatrix} 2 & -1 & -1 \\ -1 & 2 & -1 \\ -1 & -1 & 2 \end{bmatrix}, \quad (2)\ A = \begin{bmatrix} 1 & 0 & 1 \\ 0 & 1 & 0 \\ 1 & 0 & 1 \end{bmatrix}.$$

Use the decomposition $A = LDL^T$ to write $\mathbf{x}^T A\mathbf{x}$ as the sum of squares.

8.9. Show that if A and B are both positive definite, so are A^2, A^{-1} and $A + B$.

8.10. Prove that if A and B are symmetric and positive definite, so is $A^2 + B^{-1}$.

8.11. Find a substitution $\mathbf{x} = Q\mathbf{y}$ that diagonalizes each of the following quadratic forms, where Q is orthogonal. Also, classify the form as positive definite, positive semidefinite, and so on.

(1) $q(\mathbf{x}) = 2x^2 + 6xy + 2y^2$.
(2) $q(\mathbf{x}) = x^2 + y^2 + z^2 + 2(xy + xz + yz)$.

8.12. For a given quadratic equation $ax^2 + 2bxy + cy^2 + dx + ey + f = 0$ with $b \neq 0$, classify the conic section according to the various possible cases of a, b, and c (see Example 8.4).

8.13. Find the eigenvalues of the following matrices and the maximum value of the associated quadratic forms on the unit sphere.

$$(1) \begin{bmatrix} -1 & 1 & 0 \\ 1 & -2 & 1 \\ 0 & 1 & -1 \end{bmatrix}, \quad (2) \begin{bmatrix} -2 & 1 & 0 \\ 1 & -2 & 1 \\ 0 & 1 & -2 \end{bmatrix}, \quad (3) \begin{bmatrix} 3 & -2 & 0 \\ -2 & 3 & 0 \\ 0 & 0 & 5 \end{bmatrix}.$$

8.14. Let b be a bilinear form on $\mathbb{R}^2$ defined by

$$b(\mathbf{x},\ \mathbf{y}) = 2x_1y_1 - 3x_1y_2 + x_2y_2.$$

(1) Find the matrix A of b with respect to the basis $\alpha = \{(1, 0), (1, 1)\}$.
(2) Find the matrix B of b with respect to the basis $\beta = \{(2, 1), (1, -1)\}$.
(3) Find the transition matrix Q from the basis β to the basis α and verify that $B = Q^T AQ$.

8.15. Which of the following functions b on $\mathbb{R}^2$ are of bilinear form?
(1) $b(\mathbf{x}, \mathbf{y}) = 1$
(2) $b(\mathbf{x}, \mathbf{y}) = (x_1 - y_1)^2 + x_2y_2$
(3) $b(\mathbf{x}, \mathbf{y}) = (x_1 + y_1)^2 - (x_1 - y_1)^2$
(4) $b(\mathbf{x}, \mathbf{y}) = x_1y_2 - x_2y_1$

8.16. For a bilinear form on $\mathbb{R}^2$ defined by $b(\mathbf{x}, \mathbf{y}) = x_1y_1 + x_2y_2$, find the matrix representation of b with respect to each of the following bases:

$$\alpha = \{(1, 0), (0, 1)\}, \quad \beta = \{(1, -1), (1, 1)\}, \quad \gamma = \{(1, 2), (3, 4)\}.$$

8.17. Which one of the following bilinear forms on $\mathbb{R}^3$ are symmetric or skew-symmetric? For each symmetric one, find its matrix representation of the diagonal form, and for each skew-symmetric one, find its matrix representation of the block form in Theorem 8.11.
(1) $b(\mathbf{x}, \mathbf{y}) = x_1y_3 + x_3y_1$
(2) $b(\mathbf{x}, \mathbf{y}) = x_1y_1 + 2x_1y_3 + 2x_3y_1 - x_2y_2$
(3) $b(\mathbf{x}, \mathbf{y}) = x_1y_2 + 2x_1y_3 - x_2y_3 - x_2y_1 - 2x_3y_1 + x_3y_2$
(4) $b(\mathbf{x}, \mathbf{y}) = \sum_{i,j=1}^{3}(i - j)x_iy_j$.

8.18. Find the signature, index and rank of each of the following symmetric matrices:

$$(1) \begin{bmatrix} 0 & 1 & 2 \\ 1 & -1 & 3 \\ 2 & 3 & 4 \end{bmatrix}, \quad (2) \begin{bmatrix} 2 & 3 & 0 \\ 3 & -1 & -2 \\ 0 & -2 & 0 \end{bmatrix}, \quad (3) \begin{bmatrix} 4 & -3 & 5 \\ -3 & 2 & 1 \\ 5 & 1 & -6 \end{bmatrix}.$$

8.19. Determine whether the following statements are true or false, in general, and justify your answers.
(1) For any quadratic form q on $\mathbb{R}^n$, there exists a basis α for $\mathbb{R}^n$ with respect to which the matrix representation of q is diagonal.
(2) Any two matrix representations of a quadratic form have the same inertia.
(3) The sum of two bilinear forms on V is also a bilinear form.
(4) If A is a real symmetric positive definite matrix, then the solution set of $\mathbf{x}^T A\mathbf{x} = 1$ is an ellipsoid.
(5) For any nontrivial bilinear form $b \neq 0$ on V, if $b(\mathbf{v}, \mathbf{v}) = 0$, then $\mathbf{v} = \mathbf{0}$.
(6) Any symmetric matrix is congruent to a diagonal matrix.
(7) Any two congruent matrices have the same eigenvalues.
(8) Any two congruent matrices have the same determinant.
(9) Any matrix representation of a bilinear form is diagonalizable.
(10) If a real symmetric matrix A is both positive semidefinite and negative semidefinite, then A must be the zero matrix.
(11) Any two similar real symmetric matrices have the same signature.

Chapter 9

Jordan Canonical Forms

9.1 Introduction

Recall that an $n \times n$ matrix A is diagonalizable if and only if A has a full set of n linearly independent eigenvectors. In particular, if A is normal, then A can be diagonalized by a unitary matrix U whose column vectors are the orthonormal eigenvectors of A. There are some nonnormal matrices that are still diagonalizable. Of course, in this case the transition matrix need not be unitary. If a matrix A is diagonalizable, then the dimension of each eigenspace $E(\lambda) = \mathcal{N}(\lambda I - A)$ is equal to the multiplicity of the eigenvalue λ. Therefore, if $\lambda_1, \ldots, \lambda_\ell$ are distinct eigenvalues of A with multiplicities $m_{\lambda_1}, \ldots, m_{\lambda_\ell}$, respectively, then

$$\dim E(\lambda_1) + \cdots + \dim E(\lambda_\ell) = m_{\lambda_1} + \cdots + m_{\lambda_\ell} = n,$$

and hence,

$$V = E(\lambda_1) \oplus \cdots \oplus E(\lambda_\ell).$$

In some cases, a matrix may not have a full set of linearly independent eigenvectors. That is, a matrix A may have an eigenvalue λ with multiplicity $m_\lambda > 1$, but the number of linearly independent eigenvectors belonging to λ could be less than m_λ, so

$$1 \le \dim E(\lambda) < m_\lambda.$$

This means that A does not have enough eigenvectors belonging to λ, hence it is impossible to find a transition matrix Q such that $Q^{-1}AQ = D$ is a diagonal matrix. In this case, it wouldn't be easy, for example, to compute the exponential matrix e^A, so the general solution $\mathbf{x}(t) = e^{tA}\mathbf{c}$ of the system $\mathbf{x}'(t) = A\mathbf{x}(t)$ of linear differential equations may not be easily found.

However, we show in this section that even a nondiagonalizable matrix is similar to a matrix very "close" to a diagonal matrix, called a *Jordan canonical form*. In this case, the columns of a transition matrix Q are something similar to eigenvectors, but not quite. They are called *generalized eigenvectors*. Using this Jordan canonical form of A, the computation of e^A could be easier.

Recall that if A is a diagonalizable matrix, then the general solution of a system of linear differential equations $\mathbf{x}'(t) = A\mathbf{x}(t)$ is given as

$$\mathbf{x}(t) = e^{tA}\mathbf{x}_0 = c_1 e^{\lambda_1 t}\mathbf{u}_1 + c_2 e^{\lambda_2 t}\mathbf{u}_2 + \cdots + c_n e^{\lambda_n t}\mathbf{u}_n,$$

where the vectors $\mathbf{u}_i$'s are linearly independent eigenvectors of A belonging to the eigenvalues λ_i's (see Section 6.4). This solution is not valid if A is not diagonalizable, since we do not know how to compute e^{tA}. The following example will illustrate how to handle these cases for $n = 2$ and 3 by solving a system $\mathbf{x}'(t) = A\mathbf{x}(t)$ for an arbitrary matrix A. If the reader is not comfortable with systems of linear differential equations, the next example may be skipped, but it will be very helpful to understand the most essential features of a Jordan-canonical form.

Example 9.1 **(1)** Let $\mathbf{x}'(t) = A\mathbf{x}(t)$ be a system of linear differential equations, and let A be a 2×2 matrix with an eigenvalue λ of multiplicity 2.

If $\dim E(\lambda) = 2$, then one can find a basis $\{\mathbf{u}_1, \mathbf{u}_2\}$ of $E(\lambda)$; then $\mathbf{x}_i = e^{\lambda t}\mathbf{u}_i$, $i = 1, 2$, are linearly independent solutions of the system. Thus the general solution is

$$\mathbf{x}(t) = c_1\mathbf{x}_1 + c_2\mathbf{x}_2 = e^{\lambda t}(c_1\mathbf{u}_1 + c_2\mathbf{u}_2),$$

where c_1, c_2 are arbitrary constants. Note that A is diagonalized by them.

Suppose $\dim E(\lambda) = 1$. Then with a basis $\mathbf{u}$ of $E(\lambda)$ one obtains only one solution $\mathbf{x}_1(t) = e^{\lambda t}\mathbf{u}$. To get the general solution of the system, we need one more solution linearly independent to $\mathbf{x}_1(t)$. Motivated by the type of solutions in Example 6.17, we assume that the second solution is of the form

$$\mathbf{x}(t) = te^{\lambda t}\mathbf{v} + e^{\lambda t}\mathbf{w},$$

where the vectors $\mathbf{v}$ and $\mathbf{w}$ are to be determined. As a solution, this vector function should satisfy the equation $\mathbf{x}'(t) = A\mathbf{x}(t)$. Thus for all t

$$\begin{aligned} \mathbf{x}'(t) &= te^{\lambda t}\lambda\mathbf{v} + e^{\lambda t}\mathbf{v} + \lambda e^{\lambda t}\mathbf{w}, \\ A\mathbf{x}(t) &= te^{\lambda t}A\mathbf{v} + e^{\lambda t}A\mathbf{w}. \end{aligned}$$

By comparing the coefficient vectors of $te^{\lambda t}$ and $e^{\lambda t}$, we obtain two equations:

$$\begin{array}{ll} A\mathbf{v} = \lambda\mathbf{v}, & \text{or } (A - \lambda I)\mathbf{v} = \mathbf{0}, \\ A\mathbf{w} = \mathbf{v} + \lambda\mathbf{w}, & \text{or } (A - \lambda I)\mathbf{w} = \mathbf{v}. \end{array}$$

The first equation shows that $\mathbf{v}$ is an eigenvector of A belonging to λ, so we may take $\mathbf{v} = \mathbf{u}$. From the second equation, one can find a solution $\mathbf{w}$ so that one always obtains a second solution $\mathbf{x}_2(t) = te^{\lambda t}\mathbf{u} + e^{\lambda t}\mathbf{w}$. In fact, the vector $\mathbf{w}$, which is a nonzero solution of $(A - \lambda I)^2\mathbf{w} = (A - \lambda I)\mathbf{v} = \mathbf{0}$, is a *generalized eigenvector* of A. It is also known that the vectors $\mathbf{v}$, $\mathbf{w}$ are linearly independent (see Theorem 9.2 below). Thus the general solution is of the form

$$\begin{aligned} \mathbf{x}(t) &= c_1\mathbf{x}_1(t) + c_2\mathbf{x}_2(t) \\ &= e^{\lambda t}((c_1 + c_2 t)\mathbf{v} + c_2\mathbf{w}). \end{aligned}$$

Now let $Q = [\mathbf{v}\ \mathbf{w}]$. Then

$$\begin{aligned} AQ = A[\mathbf{u}\ \mathbf{w}] &= [A\mathbf{u}\ A\mathbf{w}] = [\lambda\mathbf{u}\ \ \mathbf{u} + \lambda\mathbf{w}] \\ &= [\mathbf{u}\ \mathbf{w}]\begin{bmatrix} \lambda & 1 \\ 0 & \lambda \end{bmatrix} = QJ, \end{aligned}$$

where $J = \begin{bmatrix} \lambda & 1 \\ 0 & \lambda \end{bmatrix}$. Thus $Q^{-1}AQ = J$.

(2) Let A be a 3×3 square matrix with an eigenvalue λ of multiplicity 3. Then three cases are possible: There are either 3, 2 or 1 linearly independent eigenvectors of A. We consider each case separately.

(i) Suppose that $\dim E(\lambda) = 3$, and let $\mathbf{u}_1$, $\mathbf{u}_2$, $\mathbf{u}_3$ be three linearly independent eigenvectors of A. Then $\mathbf{x}_i(t) = e^{\lambda t}\mathbf{u}_i$, $i = 1, 2, 3$, are three linearly independent solutions of $\mathbf{x}'(t) = A\mathbf{x}(t)$, so the general solution is given as

$$\mathbf{x}(t) = c_1\mathbf{x}_1(t) + c_2\mathbf{x}_2(t) + c_3\mathbf{x}_3(t) = e^{\lambda t}(c_1\mathbf{u}_1 + c_2\mathbf{u}_2 + c_3\mathbf{u}_3),$$

where c_1, c_2 and c_3 are arbitrary constants. In this case, the matrix $Q = [\mathbf{u}_1\ \mathbf{u}_2\ \mathbf{u}_3]$ diagonalizes A as usual:

$$Q^{-1}AQ = D = J = \begin{bmatrix} \lambda & 0 & 0 \\ 0 & \lambda & 0 \\ 0 & 0 & \lambda \end{bmatrix}.$$

(ii) Suppose that $\dim E(\lambda) = 2$. For any nonzero vector $\mathbf{u} \in E(\lambda)$ $\mathbf{x}(t) = e^{\lambda t}\mathbf{u}$ is a solution of $\mathbf{x}'(t) = A\mathbf{x}(t)$. Hence, one can always find two linearly independent solutions of the form $\mathbf{x}_i(t) = e^{\lambda t}\mathbf{u}_i$, where $\{\mathbf{u}_1, \mathbf{u}_2\}$ is a basis for $E(\lambda)$, since $\dim E(\lambda) = 2$. From experience, the third solution is supposed to be of the form

$$\mathbf{x}(t) = te^{\lambda t}\mathbf{v} + e^{\lambda t}\mathbf{w},$$

for some vectors $\mathbf{v}$, $\mathbf{w}$, which are to be determined. Simple substitution of this equation into the original equation $\mathbf{x}'(t) = A\mathbf{x}(t)$ gives

$$\begin{array}{ll} A\mathbf{v} = \lambda\mathbf{v}, & \text{or} \quad (A - \lambda I)\mathbf{v} = \mathbf{0}, \\ A\mathbf{w} = \mathbf{v} + \lambda\mathbf{w}, & \text{or} \quad (A - \lambda I)\mathbf{w} = \mathbf{v}. \end{array}$$

The first equation means that $\mathbf{v}$ is an eigenvector of A in $E(\lambda)$. Thus $\mathbf{v} = c_1\mathbf{u}_1 + c_2\mathbf{u}_2$ for some constants c_1, c_2. These constants are chosen in such a way that the second equation is consistent, which is always possible. After that we can get threelinearly independent solutions. However, in this case, the first two solutions may be replaced by others; once vectors $\mathbf{v}$ and $\mathbf{w}$ are determined, choose an eigenvector $\mathbf{u}$ in $E(\lambda)$ so that it is linearly independent to $\mathbf{v}$ (see Theorem 9.4 below for a reason). Then,

$$\mathbf{x}_1(t) = e^{\lambda t}\mathbf{u}, \quad \mathbf{x}_2(t) = e^{\lambda t}\mathbf{v}, \quad \mathbf{x}_3(t) = e^{\lambda t}(t\mathbf{v} + \mathbf{w})$$

form a fundamental set of solutions. Thus the general solution is given as

$$\begin{aligned} \mathbf{x}(t) &= c_1\mathbf{x}_1(t) + c_2\mathbf{x}_2(t) + c_3\mathbf{x}_3(t) \\ &= e^{\lambda t}(c_1\mathbf{u} + (c_2 + c_3 t)\mathbf{v} + c_3\mathbf{w}). \end{aligned}$$

Now, if we set $Q = [\mathbf{u}\ \mathbf{v}\ \mathbf{w}]$, then

$$\begin{aligned} AQ = A[\mathbf{u}\ \mathbf{v}\ \mathbf{w}] &= [A\mathbf{u}\ A\mathbf{v}\ A\mathbf{w}] = [\lambda\mathbf{u}\quad \lambda\mathbf{v}\quad \mathbf{v} + \lambda\mathbf{w}] \\ &= [\mathbf{u}\ \mathbf{v}\ \mathbf{w}] \begin{bmatrix} \lambda & 0 & 0 \\ 0 & \lambda & 1 \\ 0 & 0 & \lambda \end{bmatrix} = QJ, \end{aligned}$$

where $J = \begin{bmatrix} \lambda & 0 & 0 \\ 0 & \lambda & 1 \\ 0 & 0 & \lambda \end{bmatrix}$. Thus $Q^{-1}AQ = J$.

(iii) Finally, suppose that $\dim E(\lambda) = 1$, and let $\mathbf{u}$ be a basis for $E(\lambda)$ so that we get only one solution $\mathbf{x}_1(t) = e^{\lambda t}\mathbf{u}$. Then, by experience, the second and third solutions are supposed to be of the form

$$\begin{aligned} \mathbf{x}_2(t) &= te^{\lambda t}\mathbf{u} + e^{\lambda t}\mathbf{w}, \\ \mathbf{x}_3(t) &= \frac{t^2}{2!}e^{\lambda t}\mathbf{u} + te^{\lambda t}\mathbf{w} + e^{\lambda t}\mathbf{z}. \end{aligned}$$

By substituting these equations into the equation $\mathbf{x}'(t) = A\mathbf{x}(t)$, one can obtain

$$\begin{array}{ll} A\mathbf{z} = \mathbf{w} + \lambda\mathbf{z}, & \text{or } (A - \lambda I)\mathbf{z} = \mathbf{w}, \\ A\mathbf{w} = \mathbf{u} + \lambda\mathbf{w}, & \text{or } (A - \lambda I)\mathbf{w} = (A - \lambda I)^2\mathbf{z} = \mathbf{u}, \\ A\mathbf{u} = \lambda\mathbf{v}, & \text{or } (A - \lambda I)\mathbf{u} = (A - \lambda I)^3\mathbf{z} = \mathbf{0}. \end{array}$$

It can be shown that the solution vectors $\mathbf{v}$ $(= \mathbf{u})$, $\mathbf{w}$, and $\mathbf{z}$ are linearly independent (see Theorem 9.2 below), which are so-called generalized eigenvectors of A. They give us three linearly independent solutions $\mathbf{x}_i$'s. Thus the general solution is given as

$$\begin{aligned} \mathbf{x}(t) &= c_1\mathbf{x}_1 + c_2\mathbf{x}_2 + c_3\mathbf{x}_3 \\ &= e^{\lambda t}\left((c_1 + c_2t + c_3\frac{t^2}{2!})\mathbf{u} + (c_2 + c_3t)\mathbf{w} + c_3\mathbf{z}\right). \end{aligned}$$

Set $Q = [\mathbf{u}\ \mathbf{w}\ \mathbf{z}]$. Then

$$\begin{aligned} AQ = A[\mathbf{u}\ \mathbf{w}\ \mathbf{z}] &= [A\mathbf{u}\ A\mathbf{w}\ A\mathbf{z}] = [\lambda\mathbf{u}\ \ \mathbf{u} + \lambda\mathbf{w}\ \ \mathbf{w} + \lambda\mathbf{z}] \\ &= [\mathbf{u}\ \mathbf{w}\ \mathbf{z}]\begin{bmatrix} \lambda & 1 & 0 \\ 0 & \lambda & 1 \\ 0 & 0 & \lambda \end{bmatrix} = QJ, \end{aligned}$$

where $J = \begin{bmatrix} \lambda & 1 & 0 \\ 0 & \lambda & 1 \\ 0 & 0 & \lambda \end{bmatrix}$. Thus $Q^{-1}AQ = J$. □

The matrix J in each of the above cases is called the *Jordan canonical form* of the matrix A. Note that in each case in the example, J can be divided into smaller submatrices, called *Jordan blocks*: For instance, in case (ii) of (2), J can be written as $\begin{bmatrix} J_1 & 0 \\ 0 & J_2 \end{bmatrix}$ with two Jordan blocks $J_1 = [\lambda]$ of order

1 and $J_2 = \begin{bmatrix} \lambda & 1 \\ 0 & \lambda \end{bmatrix}$ of order 2. Since $e^{tJ_1} = e^{\lambda t}$ and $e^{tJ_2} = e^{\lambda t}\begin{bmatrix} 1 & t \\ 0 & 1 \end{bmatrix}$,

$$\begin{aligned} e^{tA}\mathbf{x}_0 &= Qe^{tJ}Q^{-1}\mathbf{x}_0 = Q\begin{bmatrix} e^{tJ_1} & 0 \\ 0 & e^{tJ_2} \end{bmatrix}Q^{-1}\mathbf{x}_0 \\ &= e^{\lambda t}(c_1\mathbf{u} + (c_2 + c_3 t)\mathbf{v} + c_3\mathbf{w}), \end{aligned}$$

where $(c_1, c_2, c_3) = Q^{-1}\mathbf{x}_0$ with arbitrary constants c_i's.

Observe that the number of Jordan blocks in each Jordan canonical matrix J is equal to the number of linearly independent eigenvectors. The column vectors of the transition matrix Q are called *generalized eigenvectors* of A belonging to the eigenvalue λ. For a more precise definition, refer to Definition 9.2 below.

This example makes the following theorem quite convincing, the proof of which may be found in some advanced linear algebra books.

Theorem 9.1 *If a square matrix A of order n has s linearly independent eigenvectors, then it is similar to a matrix J of the following form, called the* **Jordan canonical form**,

$$J = Q^{-1}AQ = \begin{bmatrix} J_1 & & & \\ & J_2 & & \\ & & \ddots & \\ & & & J_s \end{bmatrix},$$

in which each J_i, called a **Jordan block**, *is a triangular matrix of the form*

$$J_i = \begin{bmatrix} \lambda_i & 1 & & \\ & \ddots & \ddots & \\ & & \ddots & 1 \\ & & & \lambda_i \end{bmatrix},$$

where λ_i is a single eigenvalue of A and s is the number of linearly independent eigenvectors of A.

Note that the same eigenvalue λ_i may appear in several blocks, if it has more than one linearly independent eigenvector. In particular, if A has a full set of n linearly independent eigenvectors, then there have to be n Jordan blocks so that each Jordan block is just a 1×1 matrix, and the corresponding Jordan canonical form is just the diagonal matrix with eigenvalues on the diagonal. Hence, a diagonal matrix is a particular case of the Jordan canonical form.

Actually, the Jordan canonical form of a matrix can be completely determined by the multiplicities of the eigenvalues and the number of linearly independent eigenvectors in each of the eigenspaces without knowing the transition matrix Q as shown in the following example.

Example 9.2 Suppose that a 5×5 matrix A has an eigenvalue λ of multiplicity 5. Then seven Jordan canonical forms are possible, as follows.

(1) Suppose A has only one linearly independent eigenvector belonging to λ. Then the Jordan canonical form of A is of the form

$$J = \begin{bmatrix} \lambda & 1 & 0 & 0 & 0 \\ 0 & \lambda & 1 & 0 & 0 \\ 0 & 0 & \lambda & 1 & 0 \\ 0 & 0 & 0 & \lambda & 1 \\ 0 & 0 & 0 & 0 & \lambda \end{bmatrix},$$

which consists of only one Jordan block with eigenvalue λ on the diagonal.

(2) Suppose it has two linearly independent eigenvectors belonging to λ. Then the Jordan canonical form of A is either one of the forms

$$J = \begin{bmatrix} \lambda & 1 & & & \\ 0 & \lambda & & & \\ & & \lambda & 1 & 0 \\ & & 0 & \lambda & 1 \\ & & 0 & 0 & \lambda \end{bmatrix}, \quad \text{or} \quad J = \begin{bmatrix} \lambda & & & & \\ & \lambda & 1 & 0 & 0 \\ & 0 & \lambda & 1 & 0 \\ & 0 & 0 & \lambda & 1 \\ & 0 & 0 & 0 & \lambda \end{bmatrix},$$

each of which consists of two Jordan blocks with eigenvalue λ on the diagonal.

(3) Suppose it has three linearly independent eigenvectors belonging to λ. Then the Jordan canonical form of A is either one of the forms

$$J = \begin{bmatrix} \lambda & & & & \\ & \lambda & 1 & & \\ & 0 & \lambda & & \\ & & & \lambda & 1 \\ & & & 0 & \lambda \end{bmatrix}, \quad \text{or} \quad J = \begin{bmatrix} \lambda & & & & \\ & \lambda & & & \\ & & \lambda & 1 & 0 \\ & & 0 & \lambda & 1 \\ & & 0 & 0 & \lambda \end{bmatrix},$$

each of which consists of three Jordan blocks with eigenvalue λ on the diagonal.

(4) Suppose it has four linearly independent eigenvectors belonging to λ. Then the Jordan canonical form of A is of the form

$$J = \begin{bmatrix} \lambda & & & & \\ & \lambda & & & \\ & & \lambda & & \\ & & & \lambda & 1 \\ & & & 0 & \lambda \end{bmatrix},$$

which consists of four Jordan blocks with eigenvalue λ on the diagonal.

(5) Suppose it has five linearly independent eigenvectors belonging to λ. Then the Jordan canonical form of A is of the form

$$J = \begin{bmatrix} \lambda & & & & \\ & \lambda & & & \\ & & \lambda & & \\ & & & \lambda & \\ & & & & \lambda \end{bmatrix},$$

which is just the diagonal matrix, that is, the Jordan canonical form of A coincides with the diagonalizability.

Note that in cases (2) and (3), the problem of choosing one of the two possible Jordan canonical forms that is similar to the given matrix A depends on the nature of the eigenvectors of A and will be discussed in the following section. □

Example 9.3 Let J be a matrix of a Jordan canonical form:

$$J = \begin{bmatrix} \begin{bmatrix} 6 & 1 \\ 0 & 6 \end{bmatrix} & & \\ & \begin{bmatrix} 0 & 1 \\ 0 & 0 \end{bmatrix} & \\ & & [0] \end{bmatrix} = \begin{bmatrix} J_1 & & \\ & J_2 & \\ & & J_3 \end{bmatrix}.$$

(1) Find all possible forms of the matrix A that can be similar to J, *i.e.*, $Q^{-1}AQ = J$ for some invertible matrix Q.

(2) Find an invertible matrix Q such that $Q^{-1}AQ = J$.

Solution: Since J is an upper triangular matrix, the eigenvalues of J are the diagonal entries 6 and 0 with multiplicities 2 and 3, respectively. The eigenspace $E(6)$ has a single linearly independent eigenvector $\mathbf{e}_1 =$

(1, 0, 0, 0, 0) so that $\dim E(6) = 1$. Thus $\lambda = 6$ appears only in a single block J_1. The eigenspace $E(0)$ has two linearly independent eigenvectors $\mathbf{e}_3$ and $\mathbf{e}_5$ so that $\dim E(0) = 2$, and $\lambda = 0$ appears in two blocks J_2 and J_3.

Set $Q = [\mathbf{x}_1\ \mathbf{x}_2\ \cdots\ \mathbf{x}_5]$ and rewrite $Q^{-1}AQ = J$ as $AQ = QJ$. Then

$$A\begin{bmatrix} & & & \\ \mathbf{x}_1 & \mathbf{x}_2 & \cdots & \mathbf{x}_5 \\ & & & \end{bmatrix} = \begin{bmatrix} & & & \\ \mathbf{x}_1 & \mathbf{x}_2 & \cdots & \mathbf{x}_5 \\ & & & \end{bmatrix}\begin{bmatrix} 6 & 1 & & & \\ 0 & 6 & & & \\ & & 0 & 1 & \\ & & 0 & 0 & \\ & & & & 0 \end{bmatrix}.$$

Hence,

$$[A\mathbf{x}_1\ A\mathbf{x}_2\ A\mathbf{x}_3\ A\mathbf{x}_4\ A\mathbf{x}_5] = [6\mathbf{x}_1\ \ \mathbf{x}_1 + 6\mathbf{x}_2\ \ 0\mathbf{x}_3\ \ \mathbf{x}_3\ \ 0\mathbf{x}_5], \text{ or}$$
$$A\mathbf{x}_1 = 6\mathbf{x}_1,\ A\mathbf{x}_2 = 6\mathbf{x}_2 + \mathbf{x}_1,\ A\mathbf{x}_3 = 0\mathbf{x}_3,\ A\mathbf{x}_4 = 0\mathbf{x}_4 + \mathbf{x}_3,\ A\mathbf{x}_5 = 0\mathbf{x}_5.$$

Thus, the matrix A has three eigenvectors $\mathbf{x}_1$, $\mathbf{x}_3$, $\mathbf{x}_5$, just as J has. The vector $\mathbf{x}_1$ belonging to $\lambda = 6$ is in the first column of Q as $\mathbf{e}_1$ is in the first column of J. The two vectors $\mathbf{x}_3$ and $\mathbf{x}_5$ belonging to $\lambda = 0$ are placed in the third and fifth columns of Q. The two vectors $\mathbf{x}_2$, $\mathbf{x}_4$ are not eigenvectors, but they satisfy the equations $(A - 6I)\mathbf{x}_2 = \mathbf{x}_1$ and $(A - 0I)\mathbf{x}_4 = \mathbf{x}_3$, which follow from the second and fourth equations. Then one can easily see that they further satisfy

$$(A - 6I)^2\mathbf{x}_2 = \mathbf{0}, \quad \text{and} \quad (A - 0I)^2\mathbf{x}_4 = \mathbf{0}.$$

These "special" vectors, $\mathbf{x}_2$ and $\mathbf{x}_4$, fill up the deficient eigenvectors that the eigenvalues 6 and 0 are lacking, respectively, and are called *generalized eigenvectors*.

In summary, if a 5×5 matrix A is similar to the Jordan canonical form J, then A should have eigenvalues 6 and 0 of multiplicities 2 and 3 respectively, but only one linearly independent eigenvector, say $\mathbf{x}_1$, belonging to 6 and only two linearly independent eigenvectors, say $\mathbf{x}_3$, $\mathbf{x}_5$, belonging to 0. For such a matrix A, the transition matrix Q can be made by $\mathbf{x}_1, \mathbf{x}_2, \ldots, \mathbf{x}_5$, where $\mathbf{x}_2$, $\mathbf{x}_4$ are nonzero vectors satisfying the following equations:

$$(A - 6I)^2\mathbf{x}_2 = \mathbf{0}, \quad \text{and} \quad (A - 0I)^2\mathbf{x}_4 = \mathbf{0},$$

but

$$(A - 6I)\mathbf{x}_2 \neq \mathbf{0}, \quad \text{and} \quad (A - 0I)\mathbf{x}_4 \neq \mathbf{0}. \qquad \square$$

Example 9.4 Solve a system of linear differential equations $\mathbf{x}'(t) = A\mathbf{x}(t)$, where

$$A = \begin{bmatrix} 5 & -3 & -2 \\ 8 & -5 & -4 \\ -4 & 3 & 3 \end{bmatrix}.$$

Solution: (1) The eigenvalue of A is $\lambda = 1$ of multiplicity 3.
(2) The eigenvectors are solutions of

$$(A - I)\mathbf{x} = \begin{bmatrix} 4 & -3 & -2 \\ 8 & -6 & -4 \\ -4 & 3 & 2 \end{bmatrix} \begin{bmatrix} x \\ y \\ z \end{bmatrix} = \begin{bmatrix} 0 \\ 0 \\ 0 \end{bmatrix}.$$

The three equations are identical, so we get two linearly independent eigenvectors $\mathbf{u}_1 = (1, 0, 2)$ and $\mathbf{u}_2 = (0, 2, -3)$. Thus $\mathbf{x}_i(t) = e^t\mathbf{u}_i$, $i = 1, 2$, are two linearly independent solutions.

(3) For the third solution, we set $\mathbf{x}(t) = te^t\mathbf{v} + e^t\mathbf{w}$, where $\mathbf{v}$ and $\mathbf{w}$ are supposed to satisfy $(A - I)\mathbf{v} = 0$ and $(A - I)\mathbf{w} = \mathbf{v}$. The first equation means $\mathbf{v}$ is an eigenvector of A, so one can write

$$\mathbf{v} = c_1\mathbf{u}_1 + c_2\mathbf{u}_2 = c_1 \begin{bmatrix} 1 \\ 0 \\ 2 \end{bmatrix} + c_2 \begin{bmatrix} 0 \\ 2 \\ -3 \end{bmatrix} = \begin{bmatrix} c_1 \\ 2c_2 \\ 2c_1 - 3c_2 \end{bmatrix}.$$

The second equation now is written as

$$(A - I)\mathbf{w} = \begin{bmatrix} 4 & -3 & -2 \\ 8 & -6 & -4 \\ -4 & 3 & 2 \end{bmatrix} \begin{bmatrix} x \\ y \\ z \end{bmatrix} = \begin{bmatrix} c_1 \\ 2c_2 \\ 2c_1 - 3c_2 \end{bmatrix} = \mathbf{v}.$$

This system has a solution (or, is consistent) if and only if $c_1 = c_2$. By choosing $c_1 = c_2 = 2$, we get a solution $\mathbf{w} = (0, 0, -1)$, and $\mathbf{v} = (2, 4, -2)$. Since $\mathbf{u} = \mathbf{u}_1 = (1, 0, 2)$ is already linearly independent to both $\mathbf{v}$ and $\mathbf{w}$, we obtain three new linearly independent solutions $\mathbf{x}_1(t) = e^t\mathbf{u}$, $\mathbf{x}_2(t) = e^t\mathbf{v}$, and $\mathbf{x}_3(t) = te^t\mathbf{v} + e^t\mathbf{w}$. Thus a general solution is

$$\begin{aligned} \mathbf{x}(t) &= c_1\mathbf{x}_1 + c_2\mathbf{x}_2 + c_3\mathbf{x}_3 \\ &= e^t\left(c_1\mathbf{u} + (c_2 + c_3 t)\mathbf{v} + c_3\mathbf{w}\right). \end{aligned}$$

(4) Note that for $Q = [\mathbf{u}\ \mathbf{v}\ \mathbf{w}] = \begin{bmatrix} 1 & 2 & 0 \\ 0 & 4 & 0 \\ 2 & -2 & -1 \end{bmatrix}$, we get

$$Q^{-1}AQ = J = \begin{bmatrix} 1 & 0 & 0 \\ 0 & 1 & 1 \\ 0 & 0 & 1 \end{bmatrix} = \begin{bmatrix} J_1 & 0 \\ 0 & J_2 \end{bmatrix},$$

or $A = QJQ^{-1}$. Now the general solution may also be computed as

$$\begin{aligned}
\mathbf{x}(t) &= e^{tA}\mathbf{x}_0 = Qe^{tJ}Q^{-1}\mathbf{x}_0 \\
&= Q\begin{bmatrix} e^{tJ_1} & 0 \\ 0 & e^{tJ_2} \end{bmatrix} Q^{-1}\mathbf{x}_0 = e^tQ\begin{bmatrix} 1 & 0 & 0 \\ 0 & 1 & t \\ 0 & 0 & 1 \end{bmatrix} Q^{-1}\mathbf{x}_0 \\
&= e^t[\mathbf{u}\ \ \mathbf{v}\ \ t\mathbf{v}+\mathbf{w}]\begin{bmatrix} c_1 \\ c_2 \\ c_3 \end{bmatrix} \\
&= e^t\left(c_1\mathbf{u} + c_2\mathbf{v} + c_3(t\mathbf{v}+\mathbf{w})\right) \\
&= e^t\left(c_1\mathbf{u} + (c_2 + c_3t)\mathbf{v} + \mathbf{w}\right),
\end{aligned}$$

where $Q^{-1}\mathbf{x}_0 = (c_1, c_2, c_3)$, since $e^{tJ_1} = e^t$ and $e^{tJ_2} = e^t\begin{bmatrix} 1 & t \\ 0 & 1 \end{bmatrix}$ by Example 6.17. This coincides with the result in (3). □

Problem 9.1 Let A be a 5×5 matrix with two distinct eigenvalues λ of multiplicity 3 and μ of multiplicity 2. Find all possible Jordan canonical forms of A up to permutations of the Jordan blocks.

9.2 Generalized eigenvectors

In this section, we discuss a theoretical basis for using those generalized eigenvectors to produce an invertible transition matrix Q that transforms the given matrix A into the Jordan canonical form, and Examples 9.5 and 9.6 show a practical method of finding the transition matrices. However, at the instructor's discretion, the theoretical argument in the first part of this section may be skipped.

Consider the columns of the transition matrix Q such that $Q^{-1}AQ = J$. By comparing the columns of the equation $AQ = QJ$, one can easily see that those corresponding to the first columns of each of the Jordan blocks of J are precisely the linearly independent eigenvectors of A, and, as we saw in Example 9.1, the other columns of Q are some generalized eigenvectors.

Definition 9.1 A nonzero vector $\mathbf{x}$ is said to be a **generalized eigenvector** of A of rank k belonging to an eigenvalue λ if

$$(A - \lambda I)^k \mathbf{x} = \mathbf{0} \quad \text{and} \quad (A - \lambda I)^{k-1} \mathbf{x} \neq \mathbf{0}.$$

Note that if $k = 1$, this is the usual definition of an eigenvector. Let $\mathbf{x}$ be a generalized eigenvector of rank k belonging to an eigenvalue λ. Define

$$\begin{array}{lllll}
\mathbf{x}_k & = & \mathbf{x}, & & \\
\mathbf{x}_{k-1} & = & (A - \lambda I)\mathbf{x} & = & (A - \lambda I)\mathbf{x}_k, \\
\mathbf{x}_{k-2} & = & (A - \lambda I)^2\mathbf{x} & = & (A - \lambda I)\mathbf{x}_{k-1}, \\
\vdots & & & & \\
\mathbf{x}_2 & = & (A - \lambda I)^{k-2}\mathbf{x} & = & (A - \lambda I)\mathbf{x}_3, \\
\mathbf{x}_1 & = & (A - \lambda I)^{k-1}\mathbf{x} & = & (A - \lambda I)\mathbf{x}_2.
\end{array}$$

Definition 9.2 The set of vectors $\{\mathbf{x}_1, \mathbf{x}_2, \ldots, \mathbf{x}_k\}$ is called a **chain of generalized eigenvectors** belonging to the eigenvalue λ.

Note that, if $\mathbf{x}$ is a generalized eigenvector of A of rank $k > 1$ belonging to an eigenvalue λ, then, for each ℓ, $1 < \ell \leq k$, $(A - \lambda I)^\ell \mathbf{x}_\ell = (A - \lambda I)^k \mathbf{x} = \mathbf{0}$ and $(A - \lambda I)^{\ell-1}\mathbf{x}_\ell = (A - \lambda I)^{k-1}\mathbf{x} \neq \mathbf{0}$. Hence, the vector $\mathbf{x}_\ell = (A - \lambda I)^{k-\ell}\mathbf{x}$ is a generalized eigenvector of A of rank ℓ. However, $\mathbf{x}_1 = (A - \lambda I)^{k-1}\mathbf{x}$ is always an eigenvector belonging to λ, called the **initial** vector of the chain. Note also that $(A - \lambda I)^\ell \mathbf{x}_i = \mathbf{0}$ for $\ell \geq i$.

The following series of theorems shows that a transition matrix Q may be constructed from a set of linearly independent generalized eigenvectors of A, and justifies the invertibility of Q.

Example 9.3 also reveals how to find a transition matrix Q practically, and the validity of the method is justified by the following theorems.

Theorem 9.2 *A chain of generalized eigenvectors $S = \{\mathbf{x}_1, \mathbf{x}_2, \ldots, \mathbf{x}_k\}$ belonging to an eigenvalue λ is linearly independent.*

Proof: Let us solve $c_1\mathbf{x}_1 + c_2\mathbf{x}_2 + \cdots + c_k\mathbf{x}_k = \mathbf{0}$ for scalars c_i, $i = 1, \ldots, k$. If we multiply (on the left) both sides of this equation by $(A - \lambda I)^{k-1}$, then for $i = 1, \ldots, k - 1$,

$$(A - \lambda I)^{k-1}\mathbf{x}_i = (A - \lambda I)^{k-(i+1)}(A - \lambda I)^i \mathbf{x}_i = \mathbf{0}.$$

Thus, $c_k(A - \lambda I)^{k-1}\mathbf{x}_k = \mathbf{0}$, and, hence, $c_k = 0$.

Do the same to the equation $c_1\mathbf{x}_1 + \cdots + c_{k-1}\mathbf{x}_{k-1} = \mathbf{0}$ with $(A - \lambda I)^{k-2}$ and get $c_{k-1} = 0$. Proceeding successively, we can show that $c_i = 0$ for all $i = 1, \ldots, k$. That is, the equation has only the trivial solution. Hence, the set S is linearly independent. □

Theorem 9.3 *The union of chains of generalized eigenvectors of a square matrix A belonging to distinct eigenvalues is linearly independent.*

Proof: Let $\{\mathbf{x}_1, \mathbf{x}_2, \ldots, \mathbf{x}_k\}$ and $\{\mathbf{y}_1, \mathbf{y}_2, \ldots, \mathbf{y}_\ell\}$ be the chains of generalized eigenvectors of A belonging to the eigenvalues λ and μ, respectively, and let $\lambda \neq \mu$. We wish to show that the set of vectors $\{\mathbf{x}_1, \ldots, \mathbf{x}_k, \mathbf{y}_1, \ldots, \mathbf{y}_\ell\}$ is linearly independent. To solve the linear dependence of them,

$$c_1\mathbf{x}_1 + \cdots + c_k\mathbf{x}_k + d_1\mathbf{y}_1 + \cdots + d_\ell\mathbf{y}_\ell = \mathbf{0},$$

for c_i's and d_j's, we multiply both sides of the equation by $(A - \lambda I)^k$ and note that $(A - \lambda I)^k\mathbf{x}_i = \mathbf{0}$ for all $i = 1, \ldots, k$. Thus we have

$$(A - \lambda I)^k(d_1\mathbf{y}_1 + d_2\mathbf{y}_2 + \cdots + d_\ell\mathbf{y}_\ell) = \mathbf{0}.$$

Again, multiply this equation by $(A - \mu I)^{\ell-1}$ and note that

$$\begin{aligned} (A - \mu I)^{\ell-1}(A - \lambda I)^k &= (A - \lambda I)^k(A - \mu I)^{\ell-1}, \\ (A - \mu I)^{\ell-1}\mathbf{y}_\ell &= \mathbf{y}_1, \\ (A - \mu I)^{\ell-1}\mathbf{y}_i &= \mathbf{0} \end{aligned}$$

for $i = 1, \ldots, \ell - 1$. Thus we obtain

$$\mathbf{0} = d_\ell(A - \lambda I)^k\mathbf{y}_1.$$

Because $(A - \mu I)\mathbf{y}_1 = \mathbf{0}$ (or $A\mathbf{y}_1 = \mu\mathbf{y}_1$), this reduces to

$$d_\ell(\mu - \lambda)^k\mathbf{y}_1 = \mathbf{0},$$

which implies that $d_\ell = 0$ by the assumption $\lambda \neq \mu$ and $\mathbf{y}_1 \neq \mathbf{0}$. Proceeding successively, we can show that $d_i = 0$, $i = \ell, \ell - 1, \ldots, 2, 1$, so we are left with

$$c_1\mathbf{x}_1 + \cdots + c_k\mathbf{x}_k = \mathbf{0}.$$

Since $\{\mathbf{x}_1, \ldots, \mathbf{x}_k\}$ is already linearly independent by Theorem 9.2, $c_i = 0$ for all $i = 1, \ldots, k$. Thus the set of generalized eigenvectors $\{\mathbf{x}_1, \ldots, \mathbf{x}_k, \mathbf{y}_1, \ldots, \mathbf{y}_\ell\}$ is linearly independent. □

The next step to produce Q such that $AQ = QJ$ is to describe a method for choosing chains of generalized eigenvectors from a generalized eigenspace, which is defined below, so that the union of the chains is linearly independent.

Definition 9.3 Let λ be an eigenvalue of A. The **generalized eigenspace** of A belonging to λ, denoted by K_λ, is the set

$$K_\lambda = \{\mathbf{x} \in \mathbb{C}^n \ : \ (A - \lambda I)^p \mathbf{x} = \mathbf{0} \quad \text{for some positive integer } p\}.$$

It turns out that dim K_λ is the multiplicity of λ, and it contains the usual eigenspace $\mathcal{N}(A - \lambda I)$. The following theorem enables us to choose a basis for K_λ, but we omit the proof even though it can be proved by induction on the number of vectors in $S \cup T$.

Theorem 9.4 *Let $S = \{\mathbf{x}_1, \mathbf{x}_2, \ldots, \mathbf{x}_k\}$ and $T = \{\mathbf{y}_1, \mathbf{y}_2, \ldots, \mathbf{y}_\ell\}$ be two chains of generalized eigenvectors of A belonging to the same eigenvalue λ. If the initial vectors $\mathbf{x}_1$ and $\mathbf{y}_1$ are linearly independent, then the union $S \cup T$ is linearly independent.*

Note that this theorem easily extends to a finite number of chains of generalized eigenvectors of A belonging to an eigenvalue λ, and the union of such chains will form a basis for K_λ so that the matrix Q may be constructed from these bases for each eigenvalue as usual.

Example 9.5 Find the Jordan canonical form of the matrix

$$A = \begin{bmatrix} 2 & 1 & 4 \\ 0 & 2 & -1 \\ 0 & 0 & 3 \end{bmatrix}.$$

Solution: The eigenvalues of A are $\lambda_1 = \lambda_2 = 2$, $\lambda_3 = 3$. Since rank $(A - \lambda_1 I) = 2$, the dimension of the eigenspace $\mathcal{N}(A - \lambda_1 I)$ is 1. Thus there is only one linearly independent eigenvector belonging to $\lambda_1 = \lambda_2 = 2$, which is of the form $\mathbf{u}_1 = (a, 0, 0)$ with $a \neq 0$, and an eigenvector belonging to $\lambda_3 = 3$ is found to be $\mathbf{u}_3 = (3, -1, 1)$. We need to find a generalized eigenvector of rank 2 belonging to the eigenvalue 2, which is a solution to the following systems:

$$\begin{aligned}(A-2I)\mathbf{x} &= \begin{bmatrix} 0 & 1 & 4 \\ 0 & 0 & -1 \\ 0 & 0 & 1 \end{bmatrix}\mathbf{x} \neq \mathbf{0}, \\ (A-2I)^2\mathbf{x} &= \begin{bmatrix} 0 & 0 & 3 \\ 0 & 0 & -1 \\ 0 & 0 & 1 \end{bmatrix}\mathbf{x} = \mathbf{0}.\end{aligned}$$

From the second equation, $\mathbf{x}$ has to be of the form $(a,\ b,\ 0)$, and from the first equation we must have $b \neq 0$. Let us take $\mathbf{u}_2 = (0,\ 1,\ 0)$ for a generalized eigenvector of rank 2. Thus we have

$$\begin{aligned}(A-2I)\mathbf{u}_2 &= \mathbf{u}_1 = (1,\ 0,\ 0), \\ (A-2I)^2\mathbf{u}_2 &= (A-2I)\mathbf{u}_1 = 0.\end{aligned}$$

Clearly, the set of vectors $\{\mathbf{u}_1,\ \mathbf{u}_2,\ \mathbf{u}_3\}$ is linearly independent. Set

$$Q = \begin{bmatrix} 1 & 0 & 3 \\ 0 & 1 & -1 \\ 0 & 0 & 1 \end{bmatrix}, \quad \text{so} \quad Q^{-1} = \begin{bmatrix} 1 & 0 & -3 \\ 0 & 1 & 1 \\ 0 & 0 & 1 \end{bmatrix}.$$

Then

$$Q^{-1}AQ = \left[\begin{array}{cc|c} 2 & 1 & 0 \\ 0 & 2 & 0 \\ \hline 0 & 0 & 3 \end{array}\right] = \begin{bmatrix} J_1 & 0 \\ 0 & J_2 \end{bmatrix},$$

where $J_1 = \begin{bmatrix} 2 & 1 \\ 0 & 2 \end{bmatrix}$ and $J_2 = [3]$. □

Example 9.6 Find Q so that $Q^{-1}AQ = J$ is the Jordan canonical form of the matrix

$$A = \begin{bmatrix} 0 & 1 & 0 & 0 \\ 0 & 0 & 1 & 0 \\ 0 & 0 & 0 & 1 \\ -1 & 4 & -6 & 4 \end{bmatrix}.$$

Solution: The characteristic polynomial of the matrix A is

$$\det(A - \lambda I) = \lambda^4 - 4\lambda^3 + 6\lambda^2 - 4\lambda + 1 = (\lambda - 1)^4.$$

Therefore, the only eigenvalue of A is $\lambda = 1$ of multiplicity 4. Note that $\dim \mathcal{N}(A - I) = 1$ since the rank of the matrix

$$A - I = \begin{bmatrix} -1 & 1 & 0 & 0 \\ 0 & -1 & 1 & 0 \\ 0 & 0 & -1 & 1 \\ -1 & 4 & -6 & 3 \end{bmatrix}$$

is 3; the fourth row is a linear combination of the first three rows, which are linearly independent. Thus there is only one eigenvector belonging to $\lambda = 1$, say $\mathbf{x} = (1,\ 1,\ 1,\ 1)$. We need to find a generalized eigenvector of rank 4, which is a solution $\mathbf{x}$ of the following equations:

$$\begin{aligned} (A - I)^3\mathbf{x} &= \begin{bmatrix} -1 & 3 & -3 & 1 \\ -1 & 3 & -3 & 1 \\ -1 & 3 & -3 & 1 \\ -1 & 3 & -3 & 1 \end{bmatrix} \mathbf{x} \neq \mathbf{0}, \\ (A - I)^4\mathbf{x} &= \mathbf{0}. \end{aligned}$$

But, a direct computation shows that the matrix $(A - I)^4 = \mathbf{0}$. Hence, we may take any vector that satisfies the first equation, say $\mathbf{x} = (-1,\ 0,\ 0,\ 0)$, as a generalized eigenvector of rank 4. Now, take $\mathbf{x}_4 = (-1,\ 0,\ 0,\ 0)$, and

$$\begin{aligned} \mathbf{x}_3 &= (A - I)\mathbf{x}_4 = \begin{bmatrix} -1 & 1 & 0 & 0 \\ 0 & -1 & 1 & 0 \\ 0 & 0 & -1 & 1 \\ -1 & 4 & -6 & 3 \end{bmatrix} \begin{bmatrix} -1 \\ 0 \\ 0 \\ 0 \end{bmatrix} = \begin{bmatrix} 1 \\ 0 \\ 0 \\ 1 \end{bmatrix}, \\ \mathbf{x}_2 &= (A - I)\mathbf{x}_3 = (-1,\ 0,\ 1,\ 2), \\ \mathbf{x}_1 &= (A - I)\mathbf{x}_2 = (1,\ 1,\ 1,\ 1). \end{aligned}$$

Thus clearly the chain of generalized eigenvectors $\{\mathbf{x}_1,\ \mathbf{x}_2,\ \mathbf{x}_3,\ \mathbf{x}_4\}$ is linearly independent. Therefore,

$$Q = \begin{bmatrix} 1 & -1 & 1 & -1 \\ 1 & 0 & 0 & 0 \\ 1 & 1 & 0 & 0 \\ 1 & 2 & 1 & 0 \end{bmatrix} \quad \text{and} \quad Q^{-1} = \begin{bmatrix} 0 & 1 & 0 & 0 \\ 0 & -1 & 1 & 0 \\ 0 & 1 & -2 & 1 \\ -1 & 3 & -3 & 1 \end{bmatrix},$$

and

$$Q^{-1}AQ = \begin{bmatrix} 1 & 1 & 0 & 0 \\ 0 & 1 & 1 & 0 \\ 0 & 0 & 1 & 1 \\ 0 & 0 & 0 & 1 \end{bmatrix} = J.$$

□

Problem 9.2 Find a full set of generalized eigenvectors of the following matrices:

$$(1) \begin{bmatrix} -2 & 0 & -2 \\ -1 & 1 & -2 \\ 0 & 1 & -1 \end{bmatrix}, \qquad (2) \begin{bmatrix} -6 & 31 & -14 \\ -1 & 6 & -2 \\ 0 & 2 & 1 \end{bmatrix}.$$

Problem 9.3 Find the Jordan canonical form for each of the following matrices:

$$(1) \begin{bmatrix} i & 0 \\ 1 & i \end{bmatrix}, \quad (2) \begin{bmatrix} 4 & 1 & 2 \\ 0 & 4 & 2 \\ 0 & 0 & 4 \end{bmatrix}, \quad (3) \begin{bmatrix} 1 & 1 & 1 & 0 \\ 0 & 2 & 0 & 0 \\ 0 & 0 & 1 & 1 \\ 0 & 0 & 0 & 2 \end{bmatrix}.$$

9.3 Computation of e^A

The Jordan canonical form of an arbitrary matrix enables us to compute the exponential matrix. Let A be an arbitrary square matrix, and let J be the Jordan canonical form of A such that

$$Q^{-1}AQ = J = \begin{bmatrix} J_1 & & \\ & \ddots & \\ & & J_s \end{bmatrix},$$

where Q is made of generalized eigenvectors of A and J_i's are Jordan blocks.

(1) Computation of the power A^k of A for $k = 1, 2, \ldots$: Since we have

$$A^k = QJ^kQ^{-1} = Q \begin{bmatrix} J_1^k & & \\ & \ddots & \\ & & J_s^k \end{bmatrix} Q^{-1},$$

for $k = 1, 2, \ldots$, we may assume that J is a simple Jordan block and compute J^k. Now an $n \times n$ Jordan block J belonging to an eigenvalue λ of A may be written as

$$J = \begin{bmatrix} \lambda & 1 & & 0 \\ 0 & \ddots & \ddots & \\ & & \lambda & 1 \\ 0 & & 0 & \lambda \end{bmatrix} = \lambda \begin{bmatrix} 1 & 0 & & 0 \\ 0 & \ddots & \ddots & \\ \vdots & & 1 & 0 \\ 0 & \cdots & 0 & 1 \end{bmatrix} + \begin{bmatrix} 0 & 1 & \cdots & 0 \\ 0 & \ddots & \ddots & 0 \\ \vdots & & 0 & 1 \\ 0 & \cdots & 0 & 0 \end{bmatrix}$$

$$= \lambda I + N.$$

Since I is the identity matrix, clearly $IN = NI$ and

$$J^k = \begin{bmatrix} \lambda & 1 & 0 & \cdots & 0 \\ 0 & \lambda & 1 & \ddots & \vdots \\ \vdots & \ddots & \ddots & \ddots & 0 \\ \vdots & & \ddots & \lambda & 1 \\ 0 & \cdots & \cdots & 0 & \lambda \end{bmatrix}^k = (\lambda I + N)^k = \sum_{j=0}^{k} \binom{k}{j} \lambda^{k-j} N^j.$$

Note that $N^k = \mathbf{0}$ for $k \geq n$. Thus, by assuming $\binom{k}{\ell} = 0$ if $k < \ell$,

$$\begin{aligned} J^k &= \sum_{j=0}^{n-1} \binom{k}{j} \lambda^{k-j} N^j \\ &= \lambda^k I + \binom{k}{1} \lambda^{k-1} N + \cdots + \binom{k}{n-1} \lambda^{k-(n-1)} N^{n-1} \\ &= \begin{bmatrix} \lambda^k & \binom{k}{1}\lambda^{k-1} & \binom{k}{k-2}\lambda^{k-2} & \cdots & \binom{k}{n-1}\lambda^{k-n+1} \\ 0 & \lambda^k & \binom{k}{1}\lambda^{k-1} & \cdots & \binom{k}{n-2}\lambda^{k-n+2} \\ \vdots & & \ddots & & \vdots \\ \vdots & & & \lambda^k & \binom{k}{1}\lambda^{k-1} \\ 0 & \cdots & \cdots & 0 & \lambda^k \end{bmatrix}. \end{aligned}$$

Problem 9.4 Compute A^k, $k = 1, 2, \ldots,$ for

$$(1)\ A = \begin{bmatrix} 2 & 1 & 0 & 0 \\ 0 & 2 & 1 & 0 \\ 0 & 0 & 2 & 0 \\ 0 & 0 & 0 & 1 \end{bmatrix}, \qquad (2)\ B = \begin{bmatrix} 0 & -3 & 1 & 2 \\ -2 & 1 & -1 & 2 \\ -2 & 1 & -1 & 2 \\ -2 & -3 & 1 & 4 \end{bmatrix}.$$

(2) Computation of the exponential matrix e^A of A: Note that

$$\begin{aligned} e^A &= e^{QJQ^{-1}} = Qe^J Q^{-1} \\ &= Q \begin{bmatrix} e^{J_1} & & & 0 \\ & e^{J_2} & & \\ & & \ddots & \\ 0 & & & e^{J_s} \end{bmatrix} Q^{-1}, \end{aligned}$$

where the J_i's are Jordan blocks. Thus, it is enough to compute e^J when J is a simple Jordan block of the form

$$J = \lambda I + N,$$

where I and N are as in (1). Then, as usual, $N^k = \mathbf{0}$ for $k \geq n$, and

$$e^J = e^{\lambda I} e^N = e^{\lambda} \sum_{k=0}^{n-1} \frac{N^k}{k!} = e^{\lambda} \begin{bmatrix} 1 & 1 & \frac{1}{2!} & \cdots & \frac{1}{(n-1)!} \\ 0 & 1 & 1 & \ddots & \frac{1}{(n-2)!} \\ & & 1 & \ddots & \\ & & & \ddots & 1 \\ 0 & & & & 1 \end{bmatrix}.$$

In particular, the solution $\mathbf{y}(t) = e^{tA}\mathbf{y}_0$ of a system of linear differential equations

$$\mathbf{y}' = A\mathbf{y} \quad \text{with initial condition} \quad \mathbf{y}(0) = \mathbf{y}_0$$

can be written as

$$\begin{aligned} e^{tA}\mathbf{y}_0 &= Qe^{tJ}Q^{-1}\mathbf{y}_0 \\ &= e^{\lambda t} [\mathbf{u}_1 \ \mathbf{u}_2 \ \cdots \ \mathbf{u}_n] \begin{bmatrix} 1 & t & \frac{t^2}{2!} & \cdots & \frac{t^{n-1}}{(n-1)!} \\ 0 & 1 & t & \ddots & \frac{t^{n-2}}{(n-2)!} \\ & & 1 & \ddots & \\ & & & \ddots & t \\ 0 & & & & 1 \end{bmatrix} \begin{bmatrix} c_1 \\ c_2 \\ \vdots \\ c_n \end{bmatrix} \\ &= e^{\lambda t} \left(\left(\sum_{k=0}^{n-1} c_{k+1} \frac{t^k}{k!} \right) \mathbf{u}_1 + \left(\sum_{k=0}^{n-2} c_{k+2} \frac{t^k}{k!} \right) \mathbf{u}_2 + \cdots + c_n \mathbf{u}_n \right), \end{aligned}$$

where $Q^{-1}\mathbf{y}_0 = (c_1, \ldots, c_n)$ and the $\mathbf{u}_i$'s are generalized eigenvectors belonging to λ of A.

Example 9.7 Solve the linear differential equation $\mathbf{y}' = A\mathbf{y}$ with initial condition $\mathbf{y}(0) = \mathbf{y}_0$, where

$$A = \begin{bmatrix} 4 & -3 & -1 \\ 1 & 0 & -1 \\ -1 & 2 & 3 \end{bmatrix}, \quad \mathbf{y}_0 = \begin{bmatrix} 2 \\ 1 \\ 4 \end{bmatrix}.$$

Solution: The Jordan canonical form of A is computed as

$$J = Q^{-1}AQ = \begin{bmatrix} 2 & 1 & 0 \\ 0 & 2 & 0 \\ 0 & 0 & 3 \end{bmatrix} = \begin{bmatrix} J_1 & 0 \\ 0 & J_2 \end{bmatrix},$$

where

$$J_1 = \begin{bmatrix} 2 & 1 \\ 0 & 2 \end{bmatrix}, \quad J_2 = [3], \quad \text{and } Q = \begin{bmatrix} -1 & 1 & 2 \\ -1 & 1 & 1 \\ 1 & 0 & -1 \end{bmatrix}.$$

Let $\mathbf{y} = Q\mathbf{x}$. Then the given system changes to $\mathbf{x}' = J\mathbf{x}$ with

$$\mathbf{x}(0) = Q^{-1}\mathbf{y}(0) = \begin{bmatrix} 1 & -1 & 1 \\ 0 & 1 & 1 \\ 1 & -1 & 0 \end{bmatrix} \begin{bmatrix} 2 \\ 1 \\ 4 \end{bmatrix} = \begin{bmatrix} 5 \\ 5 \\ 1 \end{bmatrix},$$

and the solution of this new system is given by

$$\mathbf{x}(t) = e^{tJ}\mathbf{x}(0) = \begin{bmatrix} e^{tJ_1} & 0 \\ 0 & e^{tJ_2} \end{bmatrix} \begin{bmatrix} 5 \\ 5 \\ 1 \end{bmatrix} = \begin{bmatrix} e^{2t} & te^{2t} & 0 \\ 0 & e^{2t} & 0 \\ 0 & 0 & e^{3t} \end{bmatrix} \begin{bmatrix} 5 \\ 5 \\ 1 \end{bmatrix},$$

since

$$e^{tJ_1} = e^{2t} \begin{bmatrix} 1 & t \\ 0 & 1 \end{bmatrix} \quad \text{and} \quad e^{tJ_2} = e^{3t}.$$

Thus

$$\begin{aligned} \mathbf{y}(t) &= Q\mathbf{x}(t) = \begin{bmatrix} -1 & 1 & 2 \\ -1 & 1 & 1 \\ 1 & 0 & -1 \end{bmatrix} \begin{bmatrix} e^{2t} & te^{2t} & 0 \\ 0 & e^{2t} & 0 \\ 0 & 0 & e^{3t} \end{bmatrix} \begin{bmatrix} 5 \\ 5 \\ 1 \end{bmatrix} \\ &= e^{2t} \left((5+5t) \begin{bmatrix} -1 \\ -1 \\ 1 \end{bmatrix} + 5 \begin{bmatrix} 1 \\ 1 \\ 0 \end{bmatrix} \right) + e^{3t} \begin{bmatrix} 2 \\ 1 \\ -1 \end{bmatrix}. \end{aligned}$$

□

Problem 9.5 Solve the system of linear differential equations $\mathbf{y}' = A\mathbf{y}$ with the initial condition $\mathbf{y}(0) = \mathbf{y}_0$, where

$$A = \begin{bmatrix} 2 & 1 & -1 \\ -3 & -1 & 1 \\ 9 & 3 & -4 \end{bmatrix}, \quad \mathbf{y}_0 = \begin{bmatrix} -1 \\ -1 \\ 1 \end{bmatrix}.$$

9.4 Cayley-Hamilton theorem

As we saw in earlier chapters, the association of the characteristic polynomial with each matrix is very useful in studying matrices. In this section, using this association of the polynomials with matrices we prove one more useful theorem, called the em Cayley-Hamilton theorem, which makes the calculation of matrix polynomials simple, and has many applications to real problems.

Let $f(x) = a_m x^m + a_{m-1}x^{m-1} + \cdots + a_1 x + a_0$ be a polynomial, and let A be an $n \times n$ square matrix. The matrix defined by

$$f(A) = a_m A^m + a_{m-1}A^{m-1} + \cdots + a_1 A + a_0 I_n$$

is called a **matrix polynomial** of A. For example, if $f(x) = x^2 - 2x + 2$ and $A = \begin{bmatrix} 1 & 2 \\ 2 & 1 \end{bmatrix}$, then

$$\begin{aligned} f(A) &= A^2 - 2A + 2I_2 \\ &= \begin{bmatrix} 5 & 4 \\ 4 & 5 \end{bmatrix} - 2\begin{bmatrix} 1 & 2 \\ 2 & 1 \end{bmatrix} + 2\begin{bmatrix} 1 & 0 \\ 0 & 1 \end{bmatrix} = \begin{bmatrix} 5 & 0 \\ 0 & 5 \end{bmatrix}. \end{aligned}$$

Problem 9.6 Let λ be an eigenvalue of A and $\mathbf{x}$ an eigenvector belonging to λ. If $f(x)$ is any polynomial, then $f(\lambda)$ is an eigenvalue of the matrix polynomial $f(A)$.

Theorem 9.5 (Cayley-Hamilton) *For any $n \times n$ matrix A, if $f(\lambda) = \det(\lambda I - A)$ is the characteristic polynomial of A, then $f(A) = \mathbf{0}$.*

Proof: We prove this theorem in three steps:

(1) We first assume that A is a diagonal matrix $D = \begin{bmatrix} \lambda_1 & & 0 \\ \vdots & \ddots & \vdots \\ 0 & & \lambda_n \end{bmatrix}$.

Since, for all $k \geq 0$, $D^k = \begin{bmatrix} \lambda_1^k & & 0 \\ \vdots & \ddots & \vdots \\ 0 & & \lambda_n^k \end{bmatrix}$, we have

$$f(D) = \begin{bmatrix} f(\lambda_1) & & 0 \\ \vdots & \ddots & \vdots \\ 0 & & f(\lambda_n) \end{bmatrix} = \mathbf{0}.$$

(2) Suppose that A is diagonalizable, *i.e.*, $Q^{-1}AQ = D$ or $A = QDQ^{-1}$ for an invertible matrix Q. Since the characteristic polynomials of A and D are the same, we have

$$\begin{aligned} f(A) &= f(QDQ^{-1}) \\ &= (QDQ^{-1})^n + a_{n-1}(QDQ^{-1})^{n-1} + \cdots + a_1(QDQ^{-1}) + a_0 I \\ &= Q(D^n + a_{n-1}D^{n-1} + \cdots + a_1 D + a_0 I)Q^{-1} \\ &= Qf(D)Q^{-1} \;=\; \mathbf{0}. \end{aligned}$$

(3) Finally, suppose that A is any square matrix. Then by Theorem 9.1, A is similar to the Jordan canonical form $J = \begin{bmatrix} J_1 & & 0 \\ \vdots & \ddots & \vdots \\ 0 & & J_s \end{bmatrix} = Q^{-1}AQ$.

Thus $f(A) = Qf(J)Q^{-1}$. Since

$$J^k = \begin{bmatrix} J_1^k & & 0 \\ \vdots & \ddots & \vdots \\ 0 & & J_s^k \end{bmatrix}, \quad \text{and} \quad f(J) = \begin{bmatrix} f(J_1) & & 0 \\ \vdots & \ddots & \vdots \\ 0 & & f(J_s) \end{bmatrix},$$

it is enough to show $f(J) = \mathbf{0}$ for a single Jordan block $J = aI + N$ with eigenvalue a, where $N^n = 0$. Since $f(\lambda) = \det(\lambda I - A) = \det(\lambda I - J) = (\lambda - a)^n$,

$$f(J) = f(aI + N) = (aI + N - aI)^n = N^n = \mathbf{0}.$$ □

Example 9.8 The characteristic polynomial of

$$A = \begin{bmatrix} 3 & 6 & 6 \\ 0 & 2 & 0 \\ -3 & -12 & -6 \end{bmatrix}$$

is $f(\lambda) = \det(\lambda I - A) = \lambda^3 + \lambda^2 - 6\lambda$, and

$$\begin{aligned} f(A) &= A^3 + A^2 - 6A \\ &= \begin{bmatrix} 27 & 78 & 54 \\ 0 & 8 & 0 \\ -27 & -102 & -54 \end{bmatrix} + \begin{bmatrix} -9 & -42 & -18 \\ 0 & 4 & 0 \\ 9 & 30 & 18 \end{bmatrix} \\ &\quad - 6\begin{bmatrix} 3 & 6 & 6 \\ 0 & 2 & 0 \\ -3 & -12 & -6 \end{bmatrix} = \begin{bmatrix} 0 & 0 & 0 \\ 0 & 0 & 0 \\ 0 & 0 & 0 \end{bmatrix}. \end{aligned}$$

□

The Cayley-Hamilton theorem can be used to find the inverse of a nonsingular matrix. If $f(\lambda) = \lambda^n + a_{n-1}\lambda^{n-1} + \cdots + a_1\lambda + a_0$ is the characteristic polynomial of a matrix A, then

$$\begin{array}{rrcl} & \mathbf{0} = f(A) & = & A^n + a_{n-1}A^{n-1} + \cdots + a_1 A + a_0 I, \\ \text{or} & -a_0 I & = & (A^{n-1} + a_{n-1}A^{n-2} + \cdots + a_1 I)A. \end{array}$$

Since $a_0 = f(0) = \det(0I - A) = \det(-A) = (-1)^n \det A$, A is nonsingular if and only if $a_0 = (-1)^n \det A \neq 0$. Therefore, if A is nonsingular,

$$A^{-1} = -\frac{1}{a_0}(A^{n-1} + a_{n-1}A^{n-2} + \cdots + a_1 I).$$

Example 9.9 The characteristic polynomial of the matrix

$$A = \begin{bmatrix} 4 & 2 & -2 \\ -5 & 3 & 2 \\ -2 & 4 & 1 \end{bmatrix}$$

is $f(\lambda) = \det(\lambda I_3 - A) = \lambda^3 - 8\lambda^2 + 17\lambda - 10$, and the Cayley-Hamilton theorem yields

$$A^3 - 8A^2 + 17A = 10I_3.$$

Hence

$$\begin{aligned} A^{-1} &= \frac{1}{10}(A^2 - 8A + 17I_3) \\ &= \frac{1}{10}\begin{bmatrix} 10 & 6 & -6 \\ -39 & 7 & 18 \\ -30 & 12 & 13 \end{bmatrix} - \frac{8}{10}\begin{bmatrix} 4 & 2 & -2 \\ -5 & 3 & 2 \\ -2 & 4 & 1 \end{bmatrix} + \frac{17}{10}\begin{bmatrix} 1 & 0 & 0 \\ 0 & 1 & 0 \\ 1 & 0 & 1 \end{bmatrix} \\ &= \frac{1}{10}\begin{bmatrix} -5 & -10 & 10 \\ 1 & 0 & 2 \\ -14 & -20 & 22 \end{bmatrix}. \end{aligned}$$

□

Problem 9.7 Let A and B be square matrices, not necessarily of the same size, and let $f(\lambda) = \det(\lambda I - A)$ be the characteristic polynomial of A. Show that $f(B)$ is invertible if and only if A has no eigenvalue in common with B.

The Cayley-Hamilton theorem can also be used to simplify the calculation of matrix polynomials. Let $p(\lambda)$ be any polynomial and let $f(\lambda)$ be the characteristic polynomial of a square matrix A. A theorem of algebra tells us that there are polynomials $q(\lambda)$ and $r(\lambda)$ such that

$$p(\lambda) = q(\lambda)f(\lambda) + r(\lambda)$$

with the degree of $r(\lambda)$ less than the degree of $f(\lambda)$. Then

$$p(A) = q(A)f(A) + r(A).$$

By the Cayley-Hamilton theorem, $f(A) = \mathbf{0}$ and

$$p(A) = r(A).$$

Thus the problem of evaluating a polynomial of an $n \times n$ matrix can be reduced to the problem of evaluating a polynomial of degree less than n.

Example 9.10 The characteristic polynomial of the matrix $A = \begin{bmatrix} 1 & 2 \\ 2 & 1 \end{bmatrix}$ is $f(\lambda) = \lambda^2 - 2\lambda - 3$. Let $p(\lambda) = \lambda^4 - 7\lambda^3 - 3\lambda^2 + \lambda + 4$ be a polynomial. A straightforward calculation shows that

$$p(\lambda) = (\lambda^2 - 5\lambda - 10)f(\lambda) - 34\lambda - 26.$$

Therefore

$$\begin{aligned} p(A) &= (A^2 - 5A + 10)f(A) - 34A - 26I \\ &= -34A - 26I \\ &= -34\begin{bmatrix} 1 & 2 \\ 2 & 1 \end{bmatrix} - 26\begin{bmatrix} 1 & 0 \\ 0 & 1 \end{bmatrix} = \begin{bmatrix} -60 & -68 \\ -68 & -60 \end{bmatrix}. \end{aligned}$$

□

Problem 9.8 For the matrix $A = \begin{bmatrix} 1 & 0 & 1 \\ 0 & 2 & 0 \\ 0 & 0 & 2 \end{bmatrix}$, evaluate the matrix polynomial $A^5 + 3A^4 + A^3 - A^2 + 4A + 6I$.

9.5 Exercises

9.1. Show that if A nonsingular, then A^{-1} has the same block structure in its Jordan canonical form as A does.

9.2. Find the number of linearly independent eigenvectors for each of the following matrices:

$$(1)\begin{bmatrix} 1 & 1 & 0 & 0 & 0 \\ 0 & 1 & 1 & 0 & 0 \\ 0 & 0 & 1 & 0 & 0 \\ 0 & 0 & 0 & 3 & 1 \\ 0 & 0 & 0 & 0 & 3 \end{bmatrix}, \quad (2)\begin{bmatrix} 2 & 0 & 0 & 0 & 0 \\ 0 & 2 & 0 & 0 & 0 \\ 0 & 0 & 2 & 0 & 0 \\ 0 & 0 & 0 & 5 & 1 \\ 0 & 0 & 0 & 0 & 5 \end{bmatrix}, \quad (3)\begin{bmatrix} 2 & 1 & 0 & 0 & 0 \\ 0 & 2 & 0 & 0 & 0 \\ 0 & 0 & 3 & 0 & 0 \\ 0 & 0 & 0 & 3 & 0 \\ 0 & 0 & 0 & 0 & 5 \end{bmatrix}.$$

9.3. Solve the system of linear equations

$$\begin{cases} (1-i)x & + & (1+i)y & = & 2-i \\ (1+i)x & + & (1+i)y & = & 1+3i\,. \end{cases}$$

9.4. Solve $\mathbf{y}' = A\mathbf{y}$ for $A = \begin{bmatrix} 3 & 1 \\ 1 & 3 \end{bmatrix}$ with $\mathbf{y}_0 = \begin{bmatrix} 2 \\ 0 \end{bmatrix}$.

9.5. Solve $\mathbf{y}' = A\mathbf{y}$, where $A = \begin{bmatrix} -6 & 24 & 8 \\ -1 & 8 & 4 \\ 2 & -12 & -6 \end{bmatrix}$ and $\mathbf{y}(1) = (2,\ 1,\ 0)$.

9.6. Solve the initial value problem

$$\begin{cases} y_1' & = & -y_1 & & +2y_3, & y_1(0) = & -2 \\ y_2' & = & 2y_1 & +y_2 & -2y_3, & y_2(0) = & 0 \\ y_3' & = & -2y_1 & & +3y_3, & y_3(0) = & -1\,. \end{cases}$$

9.7. Find the Jordan-canonical form for $A = \begin{bmatrix} 2 & 2 \\ 0 & 2 \end{bmatrix}$, and compute e^A.

9.8. Consider a 2×2 matrix $A = \begin{bmatrix} a & b \\ c & d \end{bmatrix}$.

(1) Find a necessary and sufficient condition for A to be diagonalizable.
(2) The characteristic polynomial for A is $f(t) = t^2 - (a+d)t + (ad - bc)$. Show that $f(A) = 0$.

9.9. For each of the following matrices, find a polynomial of which the matrix is a root.

$$(1)\begin{bmatrix} 2 & 5 \\ 1 & -3 \end{bmatrix}, \quad (2)\begin{bmatrix} 2 & -3 \\ 7 & -4 \end{bmatrix}, \quad (3)\begin{bmatrix} 1 & 4 & -3 \\ 0 & 3 & 1 \\ 0 & 2 & -1 \end{bmatrix}.$$

9.10. Verify that each of the matrices below satisfies its own characteristic polynomial and from these results compute A^{-1}, if it exists.

$$(1)\begin{bmatrix} 0 & 1 \\ 4 & 0 \end{bmatrix}, \quad (2)\begin{bmatrix} 1 & 2 \\ 2 & 4 \end{bmatrix}, \quad (3)\begin{bmatrix} 1 & 0 & 1 \\ 0 & 2 & 0 \\ 0 & 0 & 2 \end{bmatrix}.$$

9.11. An $n \times n$ matrix A is called a **circulant matrix** if the i-th row of A is obtained from the first row of A by a cyclic shift of the $i-1$ steps, *i.e.*, the general form of the circulant matrix is

$$A = \begin{bmatrix} a_1 & a_2 & a_3 & \cdots & a_n \\ a_n & a_1 & a_2 & \cdots & a_{n-1} \\ a_{n-1} & a_n & a_1 & \cdots & a_{n-2} \\ \vdots & & & \ddots & \vdots \\ a_2 & a_3 & a_4 & \cdots & a_1 \end{bmatrix}.$$

(1) Show that any circulant matrix is normal.

(2) Find all eigenvalues of the $n \times n$ circulant matrix

$$W = \begin{bmatrix} 0 & 1 & 0 & \cdots & 0 \\ 0 & 0 & 1 & \cdots & 0 \\ \vdots & & & \ddots & \vdots \\ 0 & 0 & 0 & \cdots & 1 \\ 1 & 0 & 0 & \cdots & 0 \end{bmatrix}.$$

(3) Find all eigenvalues of the circulant matrix A by showing that

$$A = \sum_{i=1}^{n} a_i W^{i-1}.$$

(4) Use your answer to find the eigenvalues of

$$B = \begin{bmatrix} 0 & 1 & 1 & \cdots & 1 \\ 1 & 0 & 1 & \cdots & 1 \\ \vdots & & & \ddots & \vdots \\ 1 & 1 & \cdots & 0 & 1 \\ 1 & 1 & \cdots & 1 & 0 \end{bmatrix}.$$

9.12. Determine whether the following statements are true or false, in general, and justify your answers.

(1) Any square matrix similar to a triangular matrix.

(2) If a matrix A has exactly k linearly independent eigenvectors, then the Jordan canonical form of A has k Jordan blocks.

(3) If a matrix A has k distinct eigenvalues, then the Jordan canonical form of A has k Jordan blocks.

(4) If a 4×4 matrix A has eigenvalues 1 and 2, each of multiplicity 2, such that $\dim E(1) = 2$ and $\dim E(2) = 1$, then the Jordan canonical form of A has three Jordan blocks.

(5) If $\lambda_1, \ldots, \lambda_k$ are k distinct eigenvalues of A with multiplicities m_i and $\dim E(\lambda_i) \neq m_i$, then A is not diagonalizable.

(6) For any Jordan block J with eigenvalue λ, $\det e^J = e^\lambda$.

(7) If $f(x)$ is a polynomial and A is a square matrix such that $f(A) = \mathbf{0}$, then $f(x)$ is a multiple of the characteristic polynomial of A.

Selected Answers and Hints

Chapter 1

Problems

1.2 (1) Inconsistent.
(2) $(x_1,\ x_2,\ x_3,\ x_4) = (-1-4t,\ 6-2t,\ 2-3t,\ t)$ for any $t \in \mathbb{R}$.

1.3 (1) $(x,\ y,\ z) = (t,\ -t,\ t)$. (3) $(w,\ x,\ y,\ z) = (2,\ 0,\ 1,\ 3)$

1.4 (1) $b_1 + b_2 - b_3 = 0$. (2) For any b_i's.

1.7 $a = -\frac{17}{2}$, $b = \frac{13}{2}$, $c = \frac{13}{4}$, $d = -4$.

1.8 Consider the matrices: $A = \begin{bmatrix} 2 & 4 \\ 3 & 6 \end{bmatrix}$, $B = \begin{bmatrix} 2 & 1 \\ 3 & 4 \end{bmatrix}$, $C = \begin{bmatrix} 8 & 7 \\ 0 & 1 \end{bmatrix}$.

1.9 Compare the diagonal entries of AA^T and A^TA.

1.11 (1) Infinitely many for $a = 4$, exactly one for $a \neq \pm 4$, and none for $a = -4$.
(2) Infinitely many for $a = 2$, none for $a = -3$, and exactly one otherwise.

1.8 Consider the matrix $A = \begin{bmatrix} 1 & 0 \\ 0 & 1 \\ 0 & 1 \end{bmatrix}$.

1.13 (3) $I = I^T = (AA^{-1})^T = (A^{-1})^T A^T$ means by definition $(A^T)^{-1} = (A^{-1})^T$.

1.16 Any permutation on n objects can be obtained by taking a finite number of interchangings of two objects.

1.20 $A^{-1} = \frac{1}{15}\begin{bmatrix} 8 & -19 & 2 \\ 1 & -23 & 4 \\ 4 & -2 & 1 \end{bmatrix}$.

1.21 Consider the case that some d_i is zero.

1.22 $x = 2, y = 3, z = 1$.

1.23 $L = \begin{bmatrix} 1 & 0 & 0 \\ -1 & 1 & 0 \\ 0 & -1 & 1 \end{bmatrix}$, $\quad U = \begin{bmatrix} 1 & -1 & 0 \\ 0 & 1 & -1 \\ 0 & 0 & 1 \end{bmatrix}$.

1.24 (1) Consider (i, j)-entries of AB for $i < j$.
(2) A can be written as a product of lower triangular elementary matrices.

1.25 $L = \begin{bmatrix} 1 & 0 & 0 \\ -1/2 & 1 & 0 \\ 0 & -2/3 & 1 \end{bmatrix}$, $D = \begin{bmatrix} 2 & 0 & 0 \\ 0 & 3/2 & 0 \\ 0 & 0 & 4/3 \end{bmatrix}$, $U = \begin{bmatrix} 1 & -1/2 & 0 \\ 0 & 1 & -2/3 \\ 0 & 0 & 1 \end{bmatrix}$.

1.26 There are four possibilities for P.

1.27 (1) $I_1 = 0.5, I_2 = 6, I_3 = 5.5$. (2) $I_1 = 0, I_2 = I_3 = 1, I_4 = I_5 = 5$.

1.29 $\mathbf{x} = k \begin{bmatrix} 0.35 \\ 0.40 \\ 0.25 \end{bmatrix}$, for $k > 0$.

1.30 $A = \begin{bmatrix} 0.0 & 0.1 & 0.8 \\ 0.4 & 0.7 & 0.1 \\ 0.5 & 0.0 & 0.1 \end{bmatrix}$ with $\mathbf{d} = \begin{bmatrix} 90 \\ 10 \\ 30 \end{bmatrix}$.

Exercises

1.1 Row-echelon forms are A, B, D, F. Reduced row-echelon forms are A, B, F.

1.2 (1) $\begin{bmatrix} 1 & -3 & 2 & 1 & 2 \\ 0 & 0 & 1 & -1/4 & 3/4 \\ 0 & 0 & 0 & 0 & 0 \\ 0 & 0 & 0 & 0 & 0 \end{bmatrix}$.

1.3 (1) $\begin{bmatrix} 1 & -3 & 0 & 3/2 & 1/2 \\ 0 & 0 & 1 & -1/4 & 3/4 \\ 0 & 0 & 0 & 0 & 0 \\ 0 & 0 & 0 & 0 & 0 \end{bmatrix}$.

1.4 (1) $x_1 = 0,\ x_2 = 1,\ x_3 = -1,\ x_4 = 2$. (2) $x = 17/2,\ y = 3,\ z = -4$.

1.5 (1) and (2).

1.6 For any b_i's.

1.7 $b_1 - 2b_2 + 5b_3 \neq 0$.

1.8 (1) Take $\mathbf{x}$ the transpose of each row vector of A.

1.10 Try it with several kinds of diagonal matrices for B.

1.11 $A^k = \begin{bmatrix} 1 & 2k & 3k(k-1) \\ 0 & 1 & 3k \\ 0 & 0 & 1 \end{bmatrix}$.

1.13 (2) $\begin{bmatrix} 5 & -22 & 101 \\ 0 & 27 & -60 \\ 0 & 0 & 87 \end{bmatrix}$.

1.14 See Problem 1.9.

1.16 (1) $A^{-1}AB = B$. (2) $A^{-1}AC = C = A + I$.

1.17 $a = 0,\ c^{-1} = b \neq 0$.

1.18 $A^{-1} = \begin{bmatrix} 1 & -1 & 0 & 0 \\ 0 & 1/2 & -1/2 & 0 \\ 0 & 0 & 1/3 & -1/3 \\ 0 & 0 & 0 & 1/4 \end{bmatrix}$, $B^{-1} = \begin{bmatrix} 13/8 & -1/2 & -1/8 \\ -15/8 & 1/2 & 3/8 \\ 5/4 & 0 & -1/4 \end{bmatrix}$.

1.21 (1) $\mathbf{x} = A^{-1}\mathbf{b} = \begin{bmatrix} 1/3 & 1/6 & 1/6 \\ -4/3 & -5/3 & 4/3 \\ -1/3 & -2/3 & 1/3 \end{bmatrix} \begin{bmatrix} 2 \\ 5 \\ 7 \end{bmatrix} = \begin{bmatrix} 8/3 \\ -5/3 \\ -5/3 \end{bmatrix}$.

1.22 (1) $A = \begin{bmatrix} 1 & 0 \\ 4 & 1 \end{bmatrix} \begin{bmatrix} 2 & 0 \\ 0 & 3 \end{bmatrix} \begin{bmatrix} 1 & 1/2 \\ 0 & 1 \end{bmatrix} = LDU$, (2) $L = A$, $D = U = I$.

1.23 (1) $A = \begin{bmatrix} 1 & 0 & 0 \\ 2 & 1 & 0 \\ 3 & 1 & 1 \end{bmatrix} \begin{bmatrix} 1 & 0 & 0 \\ 0 & 2 & 0 \\ 0 & 0 & -1 \end{bmatrix} \begin{bmatrix} 1 & 2 & 3 \\ 0 & 1 & 1 \\ 0 & 0 & 1 \end{bmatrix}$,

(2) $\begin{bmatrix} 1 & 0 \\ b/a & 1 \end{bmatrix} \begin{bmatrix} a & 0 \\ 0 & d - b^2/a \end{bmatrix} \begin{bmatrix} 1 & b/a \\ 0 & 1 \end{bmatrix}$.

1.24 $\mathbf{c} = [2\ -1\ 3]^T$, $\mathbf{x} = [4\ 2\ 3]^T$.

1.25 (2) $A = \begin{bmatrix} 1 & 0 & 0 \\ 1 & 1 & 0 \\ 1 & 1 & 1 \end{bmatrix} \begin{bmatrix} 1 & 0 & 0 \\ 0 & 3 & 0 \\ 0 & 0 & 2 \end{bmatrix} \begin{bmatrix} 1 & 1 & 1 \\ 0 & 1 & 4/3 \\ 0 & 0 & 1 \end{bmatrix}$.

1.26 (1) $(A^k)^{-1} = (A^{-1})^k$. (2) $A^{n-1} = \mathbf{0}$ if $A \in M_{n\times n}$.
(3) $(I - A)(I + A + \cdots + A^{k-1}) = I - A^k$.

1.27 (1) $A = \begin{bmatrix} 1 & 1 \\ 0 & 0 \end{bmatrix}$. (2) $A = A^{-1}A^2 = A^{-1}A = I$.

1.28 Exactly seven of them are true.
(8) If AB has the (right) inverse C, then $A^{-1} = BC$.
(10) Consider a permutation matrix $\begin{bmatrix} 0 & 1 \\ 1 & 0 \end{bmatrix}$.

Chapter 2

Problems

2.4 (1) -27, (2) 0, (3) $(1 - x^4)^3$.

2.6 Let σ be a transposition in S_n. Then the composition of σ with an even (odd) permutation in S_n is an odd (even, respectively) permutation.

2.9 (1) -14. (2) 0.

2.10 See Example 2.6, and use mathematical induction on n.

2.12 If $A = \mathbf{0}$, then clearly $\mathrm{adj}A = \mathbf{0}$. Otherwise, use $A \cdot \mathrm{adj}A = (\det A)I$.

2.13 Use $\mathrm{adj}A \cdot \mathrm{adj}(\mathrm{adj}A) = \det(\mathrm{adj}A) = I$.

2.14 (1) $x_1 = 4$, $x_2 = 1$, $x_3 = -2$.
(2) $x = \dfrac{10}{23}$, $y = \dfrac{5}{6}$, $z = \dfrac{5}{2}$.

2.15 The solution of the system $Id(\mathbf{x}) = \mathbf{x}$ is $x_i = \frac{\det C_i}{\det I} = \det A$.

2.16 Find the cofactor expansion along the first row first, and then compute the cofactor expansion along the first column of each $n \times n$ submatrix (in the second step, use the proof of Cramer's rule).

Exercises

2.1 $k = 0$ or 2.

2.2 It is not necessary to compute A^2 or A^3.

2.3 -37.

2.4 (1) $\det A = (-1)^{n-1}(n-1)$. (2) 0.

2.5 $-2, 0, 1, 4$.

2.6 Consider $\sum_{\sigma \in S_n} a_{1\sigma(1)} \cdots a_{n\sigma(n)}$.

2.7 (1) 1, (2)24.

2.8 (3) $x_1 = 1,\ x_2 = -1,\ x_3 = 2,\ x_4 = -2$.

2.9 (2) $\mathbf{x} = (3, 0, 4/11)^T$.

2.10 $k = 0$ or ± 1.

2.11 $\mathbf{x} = (-5, 1, 2, 3)^T$.

2.12 $x = 3,\ y = -1,\ z = 2$.

2.13 (3) $A_{11} = -2, A_{12} = 7, A_{13} = -8, A_{33} = 3$.

2.16 $A^{-1} = \frac{1}{72}\begin{bmatrix} -3 & 5 & 9 \\ 18 & -6 & 18 \\ 6 & 14 & -18 \end{bmatrix}$.

2.17 (1) $\mathrm{adj}(A) = \begin{bmatrix} 2 & -7 & -6 \\ 1 & -7 & -3 \\ -4 & 7 & 5 \end{bmatrix}$, $\det(A) = -7$, $\det(\mathrm{adj}(A)) = 49$,

$A^{-1} = -\frac{1}{7}\mathrm{adj}(A)$. (2) $\mathrm{adj}(A) = \begin{bmatrix} 1 & 1 & -1 \\ -10 & 4 & 2 \\ 7 & -3 & -1 \end{bmatrix}$,

$\det A = 2$, $\det(\mathrm{adj}(A)) = 4$, $A^{-1} = \frac{1}{2}\mathrm{adj}(A)$.

2.19 Multiply $\begin{bmatrix} I & 0 \\ B & I \end{bmatrix}$.

2.20 If we set $A = \begin{bmatrix} 1 & 3 \\ 3 & 1 \end{bmatrix}$, then the area is $\frac{1}{2}|\det A| = 4$.

2.21 If we set $A = \begin{bmatrix} 1 & 2 \\ 1 & 2 \\ 2 & 1 \end{bmatrix}$, then the area is $\frac{1}{2}\sqrt{|\det(A^T A)|} = \frac{3\sqrt{2}}{2}$.

2.22 Use $\det A = \sum_{\sigma \in S_n} \operatorname{sgn}(\sigma) a_{1\sigma(1)} \cdots a_{n\sigma(n)}$.

2.23 Exactly seven of them are true.

(4) $(cI_n - A)^T = cI_n - A^T$.

(10) Since $\mathbf{u}\mathbf{v}^T = \mathbf{u}[v_1 \ \cdots \ v_n] = [v_1\mathbf{u} \cdots v_n\mathbf{u}]$,
$\det(\mathbf{u}\mathbf{v}^T) = v_1 \cdots v_n \det([\mathbf{u} \cdots \mathbf{u}]) = 0$.

(13) Consider $\begin{bmatrix} 1 & 0 & 1 \\ 1 & 1 & 0 \\ 0 & 1 & 1 \end{bmatrix}$.

Chapter 3

Problems

3.2 (2), (4).

3.3 (1), (2), (4).

3.4 Note that any vector $\mathbf{v}$ in W is of the form $a_1\mathbf{x}_1 + a_2\mathbf{x}_2 + \cdots + a_m\mathbf{x}_m$ which is a vector in U.

3.5 $\operatorname{tr}(AB - BA) = 0$.

3.8 Linearly dependent.

3.10 Any basis for W must be a basis for V already, by Corollary 3.11.

3.11 (1) $n - 1$, (2) $\frac{n(n+1)}{2}$, (3) $\frac{n(n-1)}{2}$.

3.13 See Problem 1.10.

3.15 $63a + 39b - 13c + 5d = 0$.

3.17 If $\mathbf{b}_1, \ldots, \mathbf{b}_n$ denote the column vectors of B, then $AB = [A\mathbf{b}_1 \ \cdots \ A\mathbf{b}_n]$.

3.18 Consider the matrix A from Example 3.20.

3.19 (1) rank $= 3$, nullity $= 1$. (2) rank $= 2$, nullity $= 2$.

3.20 $A\mathbf{x} = \mathbf{b}$ has a solution if and only if $\mathbf{b} \in \mathcal{C}(A)$.

3.21 $A^{-1}(AB) = B$ implies rank $B =$ rank $A^{-1}(AB) \leq \operatorname{rank}(AB)$.

3.22 By (2) of Theorem 3.21 and Corollary 3.18, a matrix A of rank r must have an an invertible submatrix C of rank r. By (1) of the same theorem, the rank of C must be the largest.

3.24 $\dim(V + W) = 4$ and $\dim(V \cap W) = 1$.

3.25 A basis for V is $\{(1,0,0,0), (0,-1,1,0), (0,-1,0,1)\}$,
for W: $\{(-1,1,0,0), (0,0,2,1)\}$, and for $V \cap W$: $\{(3,-3,2,1)\}$. Thus, $\dim(V + W) = 4$ means $V + W = \mathbb{R}^4$ and any basis for $\mathbb{R}^4$ works for $V + W$.

3.28

$$A = \begin{bmatrix} 1 & 0 & 0 & 0 \\ 0 & 0 & 2 & 0 \\ 1 & 1 & 1 & 1 \\ 0 & 1 & 2 & 3 \end{bmatrix}, \text{ and } \begin{bmatrix} a \\ b \\ c \\ d \end{bmatrix} = A^{-1} \begin{bmatrix} 1 \\ 2 \\ 4 \\ 4 \end{bmatrix} = \begin{bmatrix} 1 \\ 2 \\ 1 \\ 0 \end{bmatrix}.$$

Exercises

3.1 Consider $0(1,1)$.

3.2 (5).

3.3 (1), (2), (3).

3.4 (1).

3.5 (1), (4).

3.6 No.

3.7 (1) $p(x) = -p_1(x) + 3p_2(x) - 2p_3(x)$.

3.8 No.

3.10 No.

3.11 $\{(1,1,0),(1,0,1)\}$.

3.12 (3) (5, 2, 0).

3.13 2.

3.14 Consider $\{\mathbf{e}_j = \{a_i\}_{i=1}^{\infty}\}$ where $a_i = \begin{cases} 1 & \text{if } i = j, \\ 0 & \text{otherwise.} \end{cases}$

3.15 (1) $\mathbf{0} = c_1 A\mathbf{b}^1 + \cdots + c_p A\mathbf{b}^s = A(c_1\mathbf{b}^1 + \cdots + c_p\mathbf{b}^s)$ implies $c_1\mathbf{b}^1 + \cdots + c_p\mathbf{b}^p = \mathbf{0}$ since $\mathcal{N}(A) = \mathbf{0}$, and this also implies $c_i = 0$ for all $i = 1, \ldots, p$ since columns of B are linear independent.
(2) B has a right inverse. (3) and (4): Look at (1) and (2) above.

3.16 (1) $\{(-5,3,1)\}$. (2) 3.

3.17 5!, and dependent.

3.18

$$\begin{aligned} (1)\ \mathcal{R}(A) &= \langle (1,2,0,3),\ (0,0,1,2) \rangle, \qquad \mathcal{C}(A) = \langle (5,0,1),\ (0,5,2) \rangle, \\ \mathcal{N}(A) &= \langle (-2,1,0,0),\ (-3,0,-2,1) \rangle. \\ (2)\ \mathcal{R}(B) &= \langle (1,1,-2,2),\ (0,2,1,-5),\ (0,0,0,1) \rangle, \\ \mathcal{C}(B) &= \langle (1,-2,0),\ (0,1,1),\ (0,0,1) \rangle, \quad \mathcal{N}(B) = \langle (5,-1,2,0) \rangle. \end{aligned}$$

3.19 rank $= 2$ when $x = -3$, rank $= 3$ when $x \neq -3$.

3.21 See Exercise **2.23**: Each column vector of $\mathbf{u}\mathbf{v}^T$ is of the form $v_i\mathbf{u}$, that is, $\mathbf{u}$ spans the column space. Conversely, if A is of rank 1, then the column space is spanned by any one column of A, say the first column $\mathbf{u}$ of A, and the remaining columns are of the form $v_i\mathbf{u}$, $i = 2, \ldots, n$. Take $\mathbf{v} = [1\ v_2\ \cdots\ v_n]^T$. Then one can easily see that $A = \mathbf{u}\mathbf{v}^T$.

3.22 Three of them are true.

Chapter 4

Problems

4.1 $\begin{bmatrix} 0 & 1 \\ 1 & 0 \end{bmatrix}$, since it is simply the change of coordinates x and y.

4.2 To show W is a subspace, see Theorem 4.2. Let E_{ij} be the matrix with 1 at the (i,j)-th position and 0 at others. Let F_k be the matrix with 1 at the (k,k)-th position, -1 at the (n,n)-th position and 0 at others. Then the set $\{E_{ij}, F_k : 1 \le i \ne j \le n,\ k = 1, \ldots, n-1\}$ is a basis for W. Thus $\dim W = n^2 - 1$.

4.3 $\operatorname{tr}(AB) = \sum_{i=1}^m \sum_{k=1}^n a_{ik}b_{ki} = \sum_{k=1}^n \sum_{i=1}^m b_{ki}a_{ik} = \operatorname{tr}(BA)$.

4.4 If yes, $(2,\ 1) = T(-6,\ -2,\ 0) = -2T(3,\ 1,\ 0) = (-2,\ -2)$.

4.5 If $a_1\mathbf{v}_1 + a_2\mathbf{v}_2 + \cdots + a_k\mathbf{v}_k = \mathbf{0}$, then $\mathbf{0} = T(a_1\mathbf{v}_1 + a_2\mathbf{v}_2 + \cdots + a_k\mathbf{v}_k) = a_1\mathbf{w}_1 + a_2\mathbf{w}_2 + \cdots + a_k\mathbf{w}_k$ implies $a_i = 0$ for $i = 1, \ldots, k$.

4.6 (1) If $T(\mathbf{x}) = T(\mathbf{y})$, then $S \circ T(\mathbf{x}) = S \circ T(\mathbf{y})$ implies $\mathbf{x} = \mathbf{y}$. (4) They are invertible.

4.7 (1) $T(\mathbf{x}) = T(\mathbf{y})$ if and only if $T(\mathbf{x} - \mathbf{y}) = \mathbf{0}$, *i.e.*, $\mathbf{x} - \mathbf{y} \in \operatorname{Ker}(T)$.
(2) Let $\{\mathbf{v}_1,\ \ldots,\ \mathbf{v}_n\}$ be a basis for V. If T is one-to-one, then the set $\{T(\mathbf{v}_1),\ \ldots,\ T(\mathbf{v}_n)\}$ is linearly independent as the proof of Theorem 4.7 shows. Corollary 3.11 shows it is a basis for V. Thus, for any $\mathbf{y} \in V$, we can write it as $\mathbf{y} = \sum_{i=1}^n a_i T(\mathbf{v}_i) = T(\sum_{i=1}^n a_i\mathbf{v}_i)$. Set $\mathbf{x} = \sum_{i=1}^n a_i\mathbf{v}_i \in V$. Then clearly $T(\mathbf{x}) = \mathbf{y}$ so that T is onto. If T is onto, then for each $i = 1, \ldots, n$ there exists $\mathbf{x}_i \in V$ such that $T(\mathbf{x}_i) = \mathbf{v}_i$. Then the set $\{\mathbf{x}_1,\ \ldots,\ \mathbf{x}_n\}$ is linearly independent in V, since, if $\sum_{i=1}^n a_i\mathbf{x}_i = \mathbf{0}$, then $\mathbf{0} = T(\sum_{i=1}^n a_i\mathbf{x}_i) = \sum_{i=1}^n a_i T(\mathbf{x}_i) = \sum_{i=1}^n a_i\mathbf{v}_i$ implies $a_i = 0$ for all $i = 1, \ldots, n$. Thus it is a basis by Corollary 3.11 again. If $T(\mathbf{x}) = \mathbf{0}$ for $\mathbf{x} = \sum_{i=1}^n a_i\mathbf{x}_i \in V$, then $\mathbf{0} = T(\mathbf{x}) = \sum_{i=1}^n a_i T(\mathbf{x}_i) = \sum_{i=1}^n a_i\mathbf{v}_i$ implies $a_i = 0$ for all $i = 1, \ldots, n$, that is $\mathbf{x} = \mathbf{0}$. Thus Ker $(T) = \{\mathbf{0}\}$.

4.8 Use rotation $R_{\frac{\pi}{3}}$ and reflection $\begin{bmatrix} 1 & 0 \\ 0 & -1 \end{bmatrix}$ about the x-axis.

4.9 (1) (5, 2, 3). (2) (2, 3, 0).

4.10 $\operatorname{vol}(T(C)) = |\det(A)|\operatorname{vol}(C)$, for the matrix representation A of T.

4.12 (1) $[T]_\alpha = \begin{bmatrix} 2 & -3 & 4 \\ 5 & -1 & 2 \\ 4 & 7 & 0 \end{bmatrix}$, $[T]_\beta = \begin{bmatrix} 4 & -3 & 2 \\ 2 & -1 & 5 \\ 0 & 7 & 4 \end{bmatrix}$.

4.13 $[T]_\alpha^\beta = \begin{bmatrix} 1 & 2 & 0 & 0 \\ 1 & 0 & -3 & 1 \\ 0 & 2 & 3 & 4 \end{bmatrix}$.

4.15 $[S+T]_\alpha = \begin{bmatrix} 3 & 0 & 0 \\ 2 & 2 & 3 \\ 2 & 3 & 3 \end{bmatrix}$, $[T\circ S]_\alpha = \begin{bmatrix} 3 & 2 & 0 \\ 3 & 3 & 3 \\ 6 & 5 & 3 \end{bmatrix}$.

4.16 $[S]_\alpha^\beta = \begin{bmatrix} 1 & -1 & 0 \\ 1 & 1 & 0 \\ 0 & 0 & 1 \end{bmatrix}$, $[T]_\alpha = \begin{bmatrix} 2 & 3 & 0 \\ 0 & 3 & 6 \\ 0 & 0 & 4 \end{bmatrix}$.

4.17 (2) $[T]_\alpha^\beta = \begin{bmatrix} 1 & 0 \\ 1 & 1 \end{bmatrix}$ $[T^{-1}]_\beta^\alpha = \begin{bmatrix} 1 & 0 \\ -1 & 1 \end{bmatrix}$.

4.18 $[Id]_\beta^\alpha = \frac{1}{2}\begin{bmatrix} 0 & -1 & 5 \\ 4 & 3 & -1 \\ 2 & 1 & 1 \end{bmatrix}$, $[Id]_\alpha^\beta = \begin{bmatrix} -2 & -3 & 7 \\ 3 & 5 & -10 \\ 1 & 1 & -2 \end{bmatrix}$.

4.19 $[T]_\alpha = \begin{bmatrix} 1 & 2 & 1 \\ 0 & -1 & 0 \\ 1 & 0 & 4 \end{bmatrix}$, $[T]_\beta = \begin{bmatrix} 1 & 4 & 5 \\ -1 & -2 & -6 \\ 1 & 1 & 5 \end{bmatrix}$.

4.20 Write $B = Q^{-1}AQ$ with some invertible matrix Q.
(1) $\det B = \det(Q^{-1}AQ) = \det Q^{-1} \det A \det Q = \det A$. (2) tr $(B) =$ tr $(Q^{-1}AQ) =$ tr $(QQ^{-1}A) =$ tr (A) (see Problem 4.3). (3) Use Problem 3.21.

4.22 $\alpha^* = \{f_1(x,\ y,\ z) = x - \frac{1}{2}y,\ f_2(x,\ y,\ z) = \frac{1}{2}y,\ f_3(x,\ y,\ z) = -x + z\}$.

Exercises

4.1 (2).

4.2 $ax^3 + bx^2 + ax + c$.

4.5 (1) Consider the decomposition of $\mathbf{v} = \frac{\mathbf{v}+T(\mathbf{v})}{2} + \frac{\mathbf{v}-T(\mathbf{v})}{2}$.

4.6 (1) $\{(x, \frac{3}{2}x, 2x) \in \mathbb{R}^3 : x \in \mathbb{R}\}$.

4.7 (2) $T^{-1}(r,\ s,\ t) = (\frac{1}{2}r,\ 2r - s,\ 7r - 3s - t)$.

4.8 (1) Since $T\circ S$ is one-to-one from V into V, $T\circ S$ is also onto and so T is onto. Moreover, if $S(\mathbf{u}) = S(\mathbf{v})$, then $T\circ S(\mathbf{u}) = T\circ S(\mathbf{v})$ implies $\mathbf{u} = \mathbf{v}$. Thus, S is one-to-one, and so onto. This implies T is one-to-one. In fact, if $T(\mathbf{u}) = T(\mathbf{v})$, then there exist $\mathbf{x}$ and $\mathbf{y}$ such that $S(\mathbf{x}) = \mathbf{u}$ and $S(\mathbf{y}) = \mathbf{v}$. Thus $T\circ S(\mathbf{x}) = T\circ S(\mathbf{y})$ implies $\mathbf{x} = \mathbf{y}$ and so $\mathbf{u} = T(\mathbf{x}) = T(\mathbf{y}) = \mathbf{v}$.

4.9 Note that T cannot be one-to-one and S cannot be onto.

4.11 $\begin{bmatrix} 5 & 4 & -6 & 18 \\ -4 & -3 & -2 & 0 \\ 0 & 0 & 1 & -12 \\ 0 & 0 & 0 & 1 \end{bmatrix}$.

4.12 (1) $\begin{bmatrix} 1/3 & 2/3 \\ -1/3 & 1/3 \end{bmatrix}$.

4.13 (1) $\begin{bmatrix} 0 & 2 \\ 3 & -1 \end{bmatrix}$, (2) $\begin{bmatrix} 3 & -4 \\ 1 & 5 \end{bmatrix}$.

4.14 (1) $T(1,\ 0,\ 0) = (4,\ 0)$, $T(1,\ 1,\ 0) = (1,\ 3)$, $T(1,\ 1,\ 1) = (4,\ 3)$.
(2) $T(x,\ y,\ z) = (4x - 2y + z,\ y + 2z)$.

4.15 (1) $\begin{bmatrix} 1 & 1 & 1 \\ 0 & 1 & 2 \\ 0 & 0 & 1 \end{bmatrix}$, (4) $\begin{bmatrix} 0 & 1 & 0 \\ 0 & 0 & 1 \\ 0 & 0 & 0 \end{bmatrix}$.

4.16 (1) $P = \begin{bmatrix} 0 & 0 & 1 \\ 0 & 1 & -1 \\ 1 & -1 & 0 \end{bmatrix}$, (2) $Q = \begin{bmatrix} 1 & 1 & 1 \\ 1 & 1 & 0 \\ 1 & 0 & 0 \end{bmatrix} = P^{-1}$.

4.17 Use the trace.

4.18 (1) $\begin{bmatrix} -7 & -33 & -13 \\ 4 & 19 & 8 \end{bmatrix}$.

4.19 (2) $\begin{bmatrix} 5 & 1 \\ 1 & 2 \end{bmatrix}$, (4) $\begin{bmatrix} -2/3 & 1/3 & 4/3 \\ 2/3 & -1/3 & -1/3 \\ 7/3 & -2/3 & -8/3 \end{bmatrix}$.

4.24 $[T]_\alpha = \begin{bmatrix} 0 & 2 & 1 \\ -1 & 4 & 1 \\ 1 & 0 & 1 \end{bmatrix} = ([T^*]_{\alpha^*})^T$.

4.25 (1) $\begin{bmatrix} 1 & 1 & 0 \\ -1 & 0 & 2 \end{bmatrix}$. (2) $[T]_\alpha^\beta = \begin{bmatrix} -3 & 1 & -1 \\ 1 & 2 & 1 \end{bmatrix}$.

4.26 $\mathcal{N}(T) = \{\mathbf{0}\}$,

$$\mathcal{C}(T) = \langle\ (2,1,0,1),\ (1,1,1,1),\ (4,2,2,3)\ \rangle,\ [T]_\alpha^\beta = \begin{bmatrix} 1 & 0 & 2 \\ 1 & 0 & 0 \\ -1 & 0 & -1 \\ 1 & 1 & 3 \end{bmatrix}.$$

4.28 $p_1(x) = 1 + x - \frac{3}{2}x^2$, $p_2(x) = -\frac{1}{6} + \frac{1}{2}x^2$, $p_3(x) = -\frac{1}{3} + x - \frac{1}{2}x^2$.

4.29 Two of them are false.

Chapter 5

Problems

5.1 $\langle \mathbf{x}, \mathbf{y}\rangle^2 = \langle \mathbf{x}, \mathbf{x}\rangle\langle \mathbf{y}, \mathbf{y}\rangle$ if and only if $\|t\mathbf{x}+\mathbf{y}\|^2 = \langle \mathbf{x}, \mathbf{x}\rangle t^2 + 2\langle \mathbf{x}, \mathbf{y}\rangle t + \langle \mathbf{y}, \mathbf{y}\rangle = 0$ has a repeated real root t_0.

5.2 (4) Compute the square of both sides and then use Cauchy-Schwarz inequality.

5.4 $\langle f, g\rangle = \int_0^1 f(x)g(x)dx$ defines an inner product on $C\,[0,\ 1]$. Use Cauchy-Schwarz inequality.

5.5 (1) $\frac{1}{\sqrt{6}}(2,\ 1,\ -1)$, (2) $\frac{1}{\sqrt{61}}(6,\ 4,\ -3)$.

5.6 (1): Orthogonal, (2) and (3): None, (4): Orthonormal.

5.9 1) is just the definition, and use (1) to prove (2).

5.11 $-\frac{1}{6}+x$.

5.13 $\{1,\ \sqrt{3}(2x-1),\ \sqrt{5}(6x^2-6x+1)\}$.

5.15 $\text{Proj}_W(\mathbf{p}) = (\frac{4}{3}, \frac{5}{3}, -\frac{1}{3})$.

5.17 Consider a matrix $\begin{bmatrix} 2 & 0 \\ 0 & 2 \end{bmatrix}$.

5.18 (1) $r=\frac{1}{\sqrt{2}},\ s=\frac{1}{\sqrt{6}},\ a=-\frac{1}{\sqrt{3}},\ b=\frac{1}{\sqrt{3}},\ c=\frac{1}{\sqrt{3}}$.

5.19 Extend $\{\mathbf{v}_1, \ldots, \mathbf{v}_m\}$ to an orthonormal basis $\{\mathbf{v}_1, \ldots, \mathbf{v}_m, \ldots, \mathbf{v}_n\}$. Then $\|\mathbf{x}\|^2 = \sum_{i=1}^m |\langle \mathbf{x}, \mathbf{v}_i\rangle|^2 + \sum_{j=m+1}^n |\langle \mathbf{x}, \mathbf{v}_j\rangle|^2$.

5.20 (1) orthogonal. (2) not orthogonal.

5.22 The null space of the matrix $\begin{bmatrix} 1 & 2 & 1 & 2 \\ 0 & -1 & -1 & 1 \end{bmatrix}$ is $\mathbf{x} = t[1\ -1\ 1\ 0]^T + s[-4\ 1\ 0\ 1]^T$ for $t, s \in \mathbb{R}$.

5.23 $\mathcal{R}(A)^\perp = \mathcal{N}(A)$.

5.24 $\mathbf{x} = (1,\ -1,\ 0) + t(2,\ 1,\ -1)$ for any number t.

5.25 For $A = [\mathbf{v}_1\ \mathbf{v}_2]$, two columns are linearly independent.

5.26 $\begin{bmatrix} s_0 \\ \mathbf{v}_0 \\ \frac{1}{2}\mathbf{g} \end{bmatrix} = \mathbf{x} = (A^TA)^{-1}A^T\mathbf{b} = \begin{bmatrix} -0.4 \\ 0.35 \\ 16.1 \end{bmatrix}$.

5.28 For $\mathbf{x} \in \mathbb{R}^m$, $\mathbf{x} = \langle \mathbf{v}_1, \mathbf{x}\rangle \mathbf{v}_1 + \cdots + \langle \mathbf{v}_m, \mathbf{x}\rangle \mathbf{v}_m = (\mathbf{v}_1\mathbf{v}_1^T)\mathbf{x} + \cdots + (\mathbf{v}_m\mathbf{v}_m^T)\mathbf{x}$.

5.29 First, show that P is symmetric.

5.30 The line is a subspace with an orthonormal basis $\frac{1}{\sqrt{2}}(1,1)$, or is the column space of $A = \frac{1}{\sqrt{2}}\begin{bmatrix} 1 \\ 1 \end{bmatrix}$.

5.31 $P = \dfrac{1}{3}\begin{bmatrix} 2 & 1 & 1 \\ 1 & 2 & -1 \\ 1 & -1 & 2 \end{bmatrix}$.

5.33 Note that $\{\mathbf{e}_1, \mathbf{e}_2, \mathbf{e}_4\}$ is an orthonormal basis for the subspace.

Exercises

5.1 Inner products are (2), (4), (5).

5.2 For the last condition of the definition, note that $\langle A, A\rangle = \text{tr}(A^TA) = \sum_{i,j} a_{ij}^2 = 0$ if and only if $a_{ij} = 0$ for all i, j.

5.4 (1) $k = 3$.

5.5 (3) $\|f\| = \|g\| = \sqrt{1/2}$, The angle is 0 if $n = m$, $\frac{\pi}{2}$ if $n \neq m$.

5.6 Use the Cauchy-Schwarz inequality and Problem 5.1 with $\mathbf{x} = (a_1, \cdots, a_n)$ and $\mathbf{y} = (1, \cdots, 1)$ in $(\mathbb{R}^n, \cdot)$.

5.7 (1) $-\frac{37}{4}$, $\sqrt{\frac{19}{3}}$.
(2) If $\langle h, g\rangle = h(\frac{a}{3}+\frac{b}{2}+c) = 0$ with $h \neq 0$ a constant and $g(x) = ax^2+bx+c$, then (a, b, c) is on the plane $\frac{a}{3}+\frac{b}{2}+c = 0$ in $\mathbb{R}^3$.

5.10 (1) $\frac{3}{2}\mathbf{v}_2$, (2) $\frac{1}{2}\mathbf{v}_2$.

5.12 Orthogonal: (4). Nonorthogonal: (1), (2), (3).

5.16 Use induction on n. Let B be the matrix A with the first column $\mathbf{c}_1$ replaced by $\mathbf{c} = \mathbf{c}_1 - \mathrm{Proj}_W(\mathbf{c}_1)$, and write $\mathrm{Proj}_W(\mathbf{c}_1) = a_2\mathbf{c}_2+\cdots+a_n\mathbf{c}_n$ for some a_i's. Show that $\sqrt{\det(A^TA)} = \sqrt{\det(B^TB)} = \|\mathbf{c}\|\mathrm{vol}(\mathbf{c}_2, \ldots, \mathbf{c}_n) = \mathrm{vol}(\mathcal{P}(A))$.

5.17 Let $A = \begin{bmatrix} 1 & 0 & 0 \\ 0 & 1 & 0 \\ 0 & 2 & 1 \\ 0 & 1 & 2 \end{bmatrix}$. Then the volume of the tetrahedron is $\frac{1}{3}\sqrt{\det(A^TA)} = 1$.

5.18 $A^TA = I$ and $\det A^T = \det A$ imply $\det A = \pm 1$.
The matrix $A = \begin{bmatrix} \cos\theta & \sin\theta \\ \sin\theta & -\cos\theta \end{bmatrix}$ is orthogonal with $\det A = -1$.

5.20 $A\mathbf{x} = \mathbf{b}$ has a solution for every $\mathbf{b} \in \mathbb{R}^m$ if $r = m$. It has infinitely many solutions if *nullity* $= n - r = n - m > 0$.

5.21 Find a least square solution of $\begin{bmatrix} 1 & 0 \\ 1 & 1 \\ 1 & 2 \\ 1 & 3 \end{bmatrix}\begin{bmatrix} a \\ b \end{bmatrix} = \begin{bmatrix} 1 \\ 3 \\ 4 \\ 4 \end{bmatrix}$ for (a, b) in $y = a + bx$. Then $y = x + \frac{3}{2}$.

5.22 Follow Exercise **5.21** with $A = \begin{bmatrix} 1 & -1 & 1 & -1 \\ 1 & 0 & 0 & 0 \\ 1 & 1 & 1 & 1 \\ 1 & 2 & 4 & 8 \\ 1 & 3 & 9 & 27 \end{bmatrix}$. Then $y = 2x^3 - 4x^2 + 3x - 5$.

5.25 (1) Let $h(x) = \frac{1}{2}(f(x)+f(-x))$ and $g(x) = \frac{1}{2}(f(x)-f(-x))$. Then $f = h+g$.
(2) For $f \in U$ and $g \in V$, $\langle f, g\rangle = \int_{-1}^{1} f(x)g(x)dx = -\int_{1}^{-1} f(-t)g(-t)dt$ $= -\int_{-1}^{1} f(t)g(t)dt = -\langle f, g\rangle$, by change of variable $x = -t$.
(3) Expand the length in the inner product.

5.26 Six of them are true.

(1) Consider $(1, 0)$ and $(-1, 0)$.

(2) Consider two subspaces U and W of $\mathbb{R}^3$ spanned by $\mathbf{e}_1$ and $\mathbf{e}_2$, respectively.

(3) The set of column vectors in a permutation matrix P are just $\{\mathbf{e}_1, \ldots, \mathbf{e}_n\}$, which is a set of orthonormal vectors.

Chapter 6

Problems

6.3 Consider the matrices $\begin{bmatrix} 1 & 1 \\ 0 & 1 \end{bmatrix}$ and $\begin{bmatrix} 1 & 0 \\ 0 & 1 \end{bmatrix}$.

6.4 (1) Use $\det A = \lambda_1 \cdots \lambda_n$. (2) $A\mathbf{x} = \lambda\mathbf{x}$ if and only if $\mathbf{x} = \lambda A^{-1}\mathbf{x}$.

6.5 Check with $A = \begin{bmatrix} 1 & 1 \\ 0 & 1 \end{bmatrix}$.

6.6 If A is invertible, then $AB = A(BA)A^{-1}$.

6.7 (1) If $Q = [\mathbf{x}_1\ \mathbf{x}_2\ \mathbf{x}_3]$ diagonalizes A, then the diagonal matrix must be λI and $AQ = \lambda QI$. Expand this equation and compare the corresponding columns of the equation to find a contradiction on the invertibility of Q.

6.8 $Q = \begin{bmatrix} 2 & 3 \\ 1 & 2 \end{bmatrix}$, $D = \begin{bmatrix} 2 & 0 \\ 0 & 3 \end{bmatrix}$. Then $A = QDQ^{-1} = \begin{bmatrix} -1 & 6 \\ -2 & 6 \end{bmatrix}$.

6.9 (1) The eigenvalues of A are 1, 1, -3, and their associated eigenvectors are $(1, 1, 0)$, $(-1, 0, 1)$ and $(1, 3, 1)$, respectively.
(2) If $f(x) = x^{10} + x^7 + 5x$, then $f(1)$, $f(1)$ and $f(-3)$ are the eigenvalues of $A^{10} + A^7 + 5A$.

6.10 Note that $\begin{bmatrix} a_{n+1} \\ a_n \\ a_{n-1} \end{bmatrix} = \begin{bmatrix} 2 & 1 & -2 \\ 1 & 0 & 0 \\ 0 & 1 & 0 \end{bmatrix} \begin{bmatrix} a_n \\ a_{n-1} \\ a_{n-2} \end{bmatrix}$. The eigenvalues are 1, 2, -1 and eigenvectors are $(1, 1, 1)$, $(4, 2, 1)$ and $(1, -1, 1)$, respectively. It turns out that $a_n = 2 - (-1)^n \frac{2}{3} - \frac{2^n}{3}$.

6.12 The eigenvalues are 0, 0.4, and 1, and their eigenvectors are $(1, 4, -5)$, $(1, 0, -1)$ and $(3, 2, 5)$, respectively.

6.13 $y_1 = c_1 e^{2x} - \frac{1}{4} c_2 e^{-3x}$; $\quad y_2 = c_1 e^{2x} + c_2 e^{-3x}$.

6.14 $\begin{cases} y_1 = & -\ c_2 e^{2x} + c_3 e^{3x} \\ y_2 = c_1 e^x + 2c_2 e^{2x} - c_3 e^{3x} \\ y_3 = & 2c_2 e^{2x} - c_3 e^{3x} \end{cases}$, $\begin{cases} y_1 = & e^{2x} - 2e^{3x} \\ y_2 = e^x - 2e^{2x} + 2e^{3x} \\ y_3 = & -2e^{2x} + 2e^{3x}. \end{cases}$

6.15 $y_1 = 0$, $y_2 = 2e^{2t}$, $y_3 = e^{2t}$.

6.17 For (1), use $(A + B)^k = \sum_{i=0}^{k} \binom{k}{i} A^i B^{k-i}$ if $AB = BA$. For (2) and (3), use the definition of e^A. Use (1) for (4).

6.18 Note that $e^{(A^T)} = (e^A)^T$ by definition (thus, if A is symmetric, so is e^A), and use (4).

6.19 Write $A = 2I + N$ with $N = \begin{bmatrix} 0 & 3 & 0 \\ 0 & 0 & 3 \\ 0 & 0 & 0 \end{bmatrix}$. Then $N^3 = \mathbf{0}$.

6.20 (1) $\begin{bmatrix} e^{-t} \\ e^{-t} \end{bmatrix}$, (2) $\begin{bmatrix} 3e^t - 2 \\ 2 - e^{-t} \\ e^{-t} \end{bmatrix}$.

6.21 With respect to the standard basis, $T = \begin{bmatrix} 4 & 0 & 1 \\ 2 & 3 & 2 \\ 1 & 0 & 4 \end{bmatrix}$ with eigenvalues 3, 3, 5 and eigenvectors $(0,1,0)$, $(-1,0,1)$ and $(1,2,1)$, respectively.

6.22 With the standard basis for $M_{2\times 2}(\mathbb{R})$: $\alpha =$

$$\left\{ E_{11} = \begin{bmatrix} 1 & 0 \\ 0 & 0 \end{bmatrix}, E_{12} = \begin{bmatrix} 0 & 1 \\ 0 & 0 \end{bmatrix}, E_{21} = \begin{bmatrix} 0 & 0 \\ 1 & 0 \end{bmatrix}, E_{22} = \begin{bmatrix} 0 & 0 \\ 0 & 1 \end{bmatrix} \right\},$$

$[T]_\alpha = A = \begin{bmatrix} 1 & 1 & 0 & 1 \\ 1 & 1 & 1 & 0 \\ 0 & 1 & 1 & 1 \\ 1 & 0 & 1 & 1 \end{bmatrix}$. The eigenvalues are 3, 1, 1, -1, and their associated eigenvectors are $(1,1,1,1)$, $(-1,0,1,0)$, $(0,-1,0,1)$, and $(-1,1,-1,1)$, respectively.

6.23 With the basis $\alpha = \{1, x, x^2\}$, $[T]_\alpha = A = \begin{bmatrix} 1 & 0 & 0 \\ 0 & 2 & 0 \\ 0 & 0 & 3 \end{bmatrix}$.

Exercises

6.1 (4) 0 of multiplicity 3, 4 of multiplicity 1. Eigenvectors are $\mathbf{e}_i - \mathbf{e}_{i+1}$ for $1 \le i \le 3$ and $\sum_{i=1}^{4} \mathbf{e}_i$.

6.2 $f(\lambda) = (\lambda + 2)(\lambda^2 - 8\lambda + 15)$, $\lambda_1 = -2$, $\lambda_2 = 3$, $\lambda_3 = 5$,
$\mathbf{x}_1 = (-35,\ 12,\ 19)$, $\mathbf{x}_2 = (0,\ 3,\ 1)$, $\mathbf{x}_3 = (0,\ 1,\ 1)$.

6.4 $\{\mathbf{v}\}$ is a basis for $\mathcal{N}(A)$, and $\{\mathbf{u},\ \mathbf{w}\}$ is a basis for $\mathcal{C}(A)$.

6.5 Note that the order in the product doesn't matter, and any eigenvector of A is killed by B. Since the eigenvalues are all different, the eigenvectors belonging to 1, 2, 3 form a basis. Thus $B = 0$, that is, B has only the zero eigenvalue, so all vectors are eigenvectors of B.

6.7 $A = QDQ^{-1} = \dfrac{1}{2}\begin{bmatrix} 1 & -2 & -1 \\ 1 & 4 & -1 \\ 1 & 2 & 7 \end{bmatrix}$.

6.8 Note that $\mathbb{R}^n = W \oplus W^\perp$ and $P(\mathbf{w}) = \mathbf{w}$ for $\mathbf{w} \in W$ and $P(\mathbf{v}) = \mathbf{0}$ for $\mathbf{v} \in W^\perp$. Thus, the eigenspace belonging to $\lambda = 1$ is W, and that to $\lambda = 0$ is $W^\perp$.

6.9 For any $\mathbf{w} \in \mathbb{R}^n$, $A\mathbf{w} = \mathbf{u}(\mathbf{v}^T\mathbf{w}) = (\mathbf{v} \cdot \mathbf{w})\mathbf{u}$. Thus $A\mathbf{u} = (\mathbf{v} \cdot \mathbf{u})\mathbf{u}$, so $\mathbf{u}$ is an eigenvector belonging to the eigenvalue $\lambda = \mathbf{v} \cdot \mathbf{u}$. The other eigenvectors are those in $\mathbf{v}^\perp$ with eigenvalue zero. Thus, A has either two eigenspaces $E(\lambda)$ that are 1-dimensional spanned by $\mathbf{u}$ and $E(0) = \mathbf{v}^\perp$ if $\mathbf{v} \cdot \mathbf{u} \neq 0$, or just one eigenspace $E(0) = \mathbb{R}^n$ if $\mathbf{v} \cdot \mathbf{u} = 0$.

6.10 $\lambda\mathbf{v} = A\mathbf{v} = A^2\mathbf{v} = \lambda^2\mathbf{v}$ implies $\lambda(\lambda - 1) = 0$.

6.12 Use $\text{tr}(A) = \lambda_1 + \cdots + \lambda_n = a_{11} + \cdots + a_{nn}$.

6.13 $A = QD_1Q^{-1}$ and $B = QD_2Q^{-1}$ imply $AB = BA$ since $D_1D_2 = D_2D_1$. Conversely, Suppose $AB = BA$ and all eigenvalues $\lambda_1, \ldots, \lambda_n$ of A are distinct. Then the eigenspaces $E(\lambda_i)$ are all 1-dimensional for $i = 1, \ldots, n$. But if $A\mathbf{x} = \lambda_i\mathbf{x}$, then $AB\mathbf{x} = BA\mathbf{x} = \lambda B\mathbf{x}$ implies $B\mathbf{x} \in E(\lambda_i)$. Thus $B\mathbf{x} = \mu\mathbf{x}$ means $\mathbf{x}$ is also an eigenvector of B. By the same token, any eigenvector of B is also an eigenvector of A. Choose a set of linearly independent eigenvectors of A, which form an invertible matrix Q such that $Q^{-1}AQ = D_1$ and $Q^{-1}BQ = D_2$.

6.14 Use induction on n. Clearly true for $n = 1$. Assume the equality for $n - 1$. Then, by taking the cofactor expansion of $\det(\lambda I - A)$ along the first row,

$$\begin{aligned} \det(\lambda I - A) &= \lambda(\lambda^k + c_1\lambda^{k-1} + \cdots + c_{k-1}\lambda + c_k) + (-1)^{2k}c_{k+1} \\ &= \lambda^{k+1} + c_1\lambda^k + \cdots + c_k\lambda + c_{k+1}. \end{aligned}$$

6.16 With respect to the basis $\alpha = \{1, x, x^2\}$, $[T]_\alpha = \begin{bmatrix} 1 & 0 & 1 \\ 0 & 1 & 1 \\ 1 & 1 & 0 \end{bmatrix}$. The eigenvalues are 2, 1, -1 and the eigenvectors are $(1,1,1)$, $(-1,1,0)$ and $(1,1,-2)$, respectively.

6.19 Eigenvalues are 1, 1, 2 and eigenvectors are $(1,0,0)$, $(0,1,2)$ and $(1,2,3)$. $A^{10}\mathbf{x} = (1025,\ 2050,\ 3076)$.

6.20 Fibonacci sequence: $a_{n+1} = a_n + a_{n-1}$ with $a_1 = 2$ and $a_2 = 3$.

6.21 One can easily check that $\det A_n = \det A_{n-1} - \det A_{n-2}$. Set $a_n = \det A_n$, so that $a_n = a_{n-1} - a_{n-2}$. With $a_{n-1} = a_{n-1}$, we obtain a matrix equation:

$$\mathbf{x}_n = \begin{bmatrix} a_n \\ a_{n-1} \end{bmatrix} = \begin{bmatrix} 1 & -1 \\ 1 & 0 \end{bmatrix} \begin{bmatrix} a_{n-1} \\ a_{n-2} \end{bmatrix} = A\mathbf{x}_{n-1} = A^n\mathbf{x}_1,$$

with $a_1 = 1$ and $a_2 = 0$. Using the eigenvalues might make the computation a mess. Instead, one can use the Cayley-Hamilton Theorem 9.5: Since the characteristic polynomial of A is $\lambda^2 - \lambda + 1$, $A^2 - A + I = 0$ holds. Thus, $A^3 = A^2 - A = -I$, so $A^6 = I$. One can now easily compute a_n modulo 6.

6.22 The characteristic equation is $\lambda^2 - x\lambda - 0.18 = 0$. Since $\lambda = 1$ is a solution, $x = 0.82$. The eigenvalues are now 1, -0.18 and the eigenvectors are $(-0.3, -1)$ and $(1, -0.6)$.

6.23 (1) $e^A = \begin{bmatrix} e & e-1 \\ 0 & 1 \end{bmatrix}$.

6.24 The initial status in 1985 is $\mathbf{x}_0 = (x_0, y_0, z_0) = (0.4, 0.2, 0.4)$, where x, y, z represent the percentage of large, medium, and small car owners. In 1995, the status is $\mathbf{x}_1 = \begin{bmatrix} x_1 \\ y_1 \\ z_1 \end{bmatrix} = \begin{bmatrix} 0.7 & 0.1 & 0 \\ 0.3 & 0.7 & 0.1 \\ 0 & 0.2 & 0.9 \end{bmatrix} \begin{bmatrix} 0.4 \\ 0.2 \\ 0.4 \end{bmatrix} = A\mathbf{x}_0$. Thus, in 2025, the status is $\mathbf{x}_4 = A^4\mathbf{x}_0$. The eigenvalues are 0.5, 0.8, and 1, whose eigenvectors are $(-0.41, 0.82, -0.41)$, $(0.47, 0.47, -0.94)$, and $(-0.17, -0.52, -1.04)$, respectively.

6.27 (1) $\begin{cases} y_1(x) &=& -2e^{2(1-x)} + 4e^{2(x-1)} \\ y_2(x) &=& -e^{2(1-x)} + 2e^{2(x-1)} \\ y_3(x) &=& 2e^{2(1-x)} - 2e^{2(x-1)}. \end{cases}$ (2) $\begin{cases} y_1(x) &=& e^{2x}(\cos x - \sin x) \\ y_2(x) &=& 2e^{2x}\sin x. \end{cases}$

6.28 (1) $f(\lambda) = \lambda^3 - 10\lambda^2 + 28\lambda - 24$, eigenvalues are 6, 2, 2, and eigenvectors are $(1, 2, 1)$, $(-1, 1, 0)$ and $(-1, 0, 1)$.

(2) $f(\lambda) = (\lambda - 1)(\lambda^2 - 6\lambda + 9)$, eigenvalues are 1, 3, 3, and eigenvectors are $(2, -1, 1)$, $(1, 1, 0)$ and $(1, 0, 1)$.

6.29 Three of them are true:

(1) For $A = \begin{bmatrix} 1 & 0 \\ 0 & 1 \end{bmatrix}$, if $B = Q^{-1}AQ$, then Q must be singular.

(2) Consider $A = \begin{bmatrix} 3 & 2 \\ 7 & 5 \end{bmatrix}$, and $B = \begin{bmatrix} 5 & 3 \\ -7 & -4 \end{bmatrix}$.

(3) Consider $\begin{bmatrix} 1 & 1 \\ 0 & 1 \end{bmatrix}$. (4) Consider $\begin{bmatrix} 1 & 0 \\ 0 & 0 \end{bmatrix}$. (6) Consider $\begin{bmatrix} 0 & 1 \\ 1 & 0 \end{bmatrix}$.

(7) If A is similar to $I + A$, then $\det(\lambda I - A)$ is a constant.

(8) Consider $A = \begin{bmatrix} 1 & 0 \\ 0 & 2 \end{bmatrix}$ and $B = \begin{bmatrix} 2 & 1 \\ 1 & 3 \end{bmatrix}$.

(9) $\text{tr}\,(A + B) = \text{tr}\,A + \text{tr}\,B$.

Chapter 7

Problems

7.1 (1) $\mathbf{u} \cdot \mathbf{v} = \overline{\mathbf{u}}^T\mathbf{v} = \sum_i \overline{u_i}v_i = \overline{\sum_i \overline{v_i}u_i} = \overline{\mathbf{v} \cdot \mathbf{u}}$.
(3) $(k\mathbf{u}) \cdot \mathbf{v} = \sum_i \overline{ku_i}v_i = \overline{k}\sum_i \overline{u_i}v_i = \overline{k}(\mathbf{u} \cdot \mathbf{v})$.
(4) $\mathbf{u} \cdot \mathbf{u} = \sum_i |u_i|^2 \geq 0$, and $\mathbf{u} \cdot \mathbf{u} = 0$ if and only if $u_i = 0$ for all i.

7.2 (1) If $\mathbf{x} = \mathbf{0}$, clear. Suppose $\mathbf{x} \neq \mathbf{0} \neq \mathbf{y}$. For any scalar k, $0 \leq \langle \mathbf{x} - k\mathbf{y}, \mathbf{x} - k\mathbf{y} \rangle = \langle \mathbf{x}, \mathbf{x} \rangle - k\langle \mathbf{x}, \mathbf{y} \rangle - \overline{k}\langle \mathbf{y}, \mathbf{x} \rangle + k\overline{k}\langle \mathbf{y}, \mathbf{y} \rangle$. Let $k = \frac{\langle \mathbf{y}, \mathbf{x} \rangle}{\langle \mathbf{x}, \mathbf{x} \rangle}$ to obtain $|\langle \mathbf{x}, \mathbf{x} \rangle\langle \mathbf{y}, \mathbf{y} \rangle - |\langle \mathbf{x}, \mathbf{y} \rangle|^2 \geq 0$. Note that equality holds if and only if $\mathbf{x} = k\mathbf{y}$ for some scalar k.
(2) Expand $\|\mathbf{x} + \mathbf{y}\|^2 = \langle \mathbf{x} + \mathbf{y}, \mathbf{x} + \mathbf{y} \rangle$ and use (1).

7.3 Suppose that $\mathbf{x}$ and $\mathbf{y}$ are linearly independent, and consider the linear dependence $a(\mathbf{x}+\mathbf{y}) + b(\mathbf{x}-\mathbf{y}) = \mathbf{0}$ of $\mathbf{x}+\mathbf{y}$ and $\mathbf{x}-\mathbf{y}$. Then $\mathbf{0} = (a+b)\mathbf{x} + (a-b)\mathbf{y}$. Since $\mathbf{x}$ and $\mathbf{y}$ are linearly independent, we have $a+b = 0$ and $a-b = 0$ which are possible only for $a = 0 = b$. Thus $\mathbf{x}+\mathbf{y}$ and $\mathbf{x}-\mathbf{y}$ are linearly independent. Conversely, if $\mathbf{x}+\mathbf{y}$ and $\mathbf{x}-\mathbf{y}$ are linearly independent, then the the linear dependence $a\mathbf{x}+b\mathbf{y} = \mathbf{0}$ of $\mathbf{x}$ and $\mathbf{y}$ gives $\frac{1}{2}(a+b)(\mathbf{x}+\mathbf{y}) + \frac{1}{2}(a-b)(\mathbf{x}-\mathbf{y}) = \mathbf{0}$. Thus we get $a = 0 = b$. Thus $\mathbf{x}$ and $\mathbf{y}$ are linearly independent.

7.4 (1) Eigenvalues are 0, 0, 2 and their eigenvectors are $(1,0,-i)$ and $(0,1,0)$, respectively. (2) Eigenvalues are 3, $\frac{1+\sqrt{5}}{2}$, $\frac{1-\sqrt{5}}{2}$, and their eigenvectors are $(1,-i,\frac{1-i}{2})$, $(\frac{\sqrt{5}-3}{2}i,1,\frac{1-\sqrt{5}}{2}(1+i))$, and $(-\frac{\sqrt{5}+3}{2}i,1,\frac{1+\sqrt{5}}{2}(1+i))$, respectively.

7.5 Refer to the real case.

7.6 $(AB)^H = (\overline{AB})^T = \overline{B}^T\overline{A}^T = B^H A^H$.

7.7 $(A^H)(A^{-1})^H = (A^{-1}A)^H = I$.

7.8 The determinant is just the product of the eigenvalues and a Hermitian matrix has only real eigenvalues.

7.9 See Exercise **6.9**.

7.10 To prove (3) directly, show that $\overline{\lambda}(\mathbf{x}\cdot\mathbf{y}) = \overline{\mu}(\mathbf{x}\cdot\mathbf{y})$ by using the fact that $A^H\mathbf{x} = -\mu\mathbf{x}$ when $A\mathbf{x} = \mu\mathbf{x}$.

7.11 $A^H = B^H + (iC)^H = B^T - iC^T = -B - iC = -A$.

7.12 $\pm AB = (AB)^H = B^H A^H = (\pm B)(\pm A) = BA$, $+$ if they are Hermitian, $-$ if they are skew-Hermitian.

7.13 Note that $\det U^H = \overline{\det U}$, and $1 = \det I = \det(U^H U) = |\det U|^2$.

7.16 Since $A^{-1} = A^H$, $(AB)^H AB = I$.

7.17 Hermitian means the diagonal entries are real, and diagonality implies off-diagonal entries are zero. Unitary means the diagonal entries must be ± 1.

7.20 This is a normal matrix. From a direct computation, one can find the eigenvalues, $1-i$, $1-i$ and $1+2i$, and the corresponding eigenvectors: $(-1,0,1)$, $(-1,1,0)$ and $(1,1,1)$, respectively, which are not orthogonal. But by an orthonormalization, one can obtain a unitary transition matrix so that A is unitarily diagonalizable.

7.21 $A^H A = (H_1 - iH_2)(H_1 + iH_2) = (H_1 + iH_2)(H_1 - iH_2) = AA^H$ if and only if $H_1H_2 - H_2H_1 = 0$.

7.22 In one direction these are all already proven in the theorems. Suppose that $U^H AU = D$ for a unitary matrix U and a diagonal matrix D.

(1) and (2). If all the eigenvalues of A are real (or purely imaginary), then the diagonal entries of D are all real (or purely imaginary). Thus $D^H = \pm D$, so that A is Hermitian (or skew-Hermitian).

(3) The diagonal entries of D satisfy $|\lambda| = 1$. Thus, $D^H = D^{-1}$, and $A^H = UD^{-1}U^{-1} = A^{-1}$.

7.23 $$Q = \frac{1}{\sqrt{6}}\begin{bmatrix} \sqrt{3} & -\sqrt{2} & -1 \\ 0 & \sqrt{2} & -2 \\ \sqrt{3} & \sqrt{2} & 1 \end{bmatrix}.$$

7.24 (1) $A = \frac{1}{2}\begin{bmatrix} 1 & -1 \\ -1 & 1 \end{bmatrix} + \frac{3}{2}\begin{bmatrix} 1 & 1 \\ 1 & 1 \end{bmatrix}$,

(2) $B = \frac{3+2\sqrt{6}}{6}\begin{bmatrix} 1 & \frac{(1+\sqrt{6})(2+i)}{5} \\ \frac{(1+\sqrt{6})(2-i)}{5} & \frac{7+2\sqrt{6}}{5} \end{bmatrix} + \frac{3-2\sqrt{6}}{6}\begin{bmatrix} 1 & \frac{(1-\sqrt{6})(2+i)}{5} \\ \frac{(1-\sqrt{6})(2-i)}{5} & \frac{7-2\sqrt{6}}{5} \end{bmatrix}$.

Exercises

7.1 (1) $\sqrt{6}$, (2) 4.

7.4 (1) $\lambda = i$, $\mathbf{x} = t(1,\ -2-i)$, $\lambda = -i$, $\mathbf{x} = t(1,\ -2+i)$.
(2) $\lambda = 1$, $\mathbf{x} = t(i,\ 1)$, $\lambda = -1$, $\mathbf{x} = t(-i,\ 1)$.
(3) Eigenvalues are 2, $2+i$, $2-i$, and eigenvectors are $(0,-1,1))$, $(1, -\frac{1}{5}(2+i), 1)$, $(1, -\frac{1}{5}(2-i), 1)$.
(4) Eigenvalues are 0, -1, 2, and eigenvectors are $(1,0,-1))$, $(1,-i,1)$, $(1,2i,1)$.

7.6 $A+cI$ is invertible if $\det(A+cI) \neq 0$. However, for any matrix A, $\det(A+cI) = 0$ as a complex polynomial has always a (complex) solution. For the real matrix $\begin{bmatrix} \cos\theta & -\sin\theta \\ \sin\theta & \cos\theta \end{bmatrix}$, $A+rI$ is invertible for every real number r since A has no real eigenvalues.

7.7 (1) $\frac{1}{\sqrt{3}}\begin{bmatrix} 1 & 1-i \\ 1+i & -1 \end{bmatrix}$, (2) $\frac{1}{2}\begin{bmatrix} 1 & i & 1-i \\ \sqrt{2}i & \sqrt{2} & 0 \\ 1 & i & -1+i \end{bmatrix}$.

7.10 (2) $Q = \frac{1}{\sqrt{2}}\begin{bmatrix} 1 & 1 \\ 1 & -1 \end{bmatrix}$.

7.12 (1) Unitary; diagonal entries are $\{1,\ i\}$. (2) Orthogonal; $\{\cos\theta + i\sin\theta,\ \cos\theta - i\sin\theta\}$, where $\theta = \cos^{-1}(0.6)$. (3) Hermitian; $\{1,\ 1+\sqrt{2},\ 1-\sqrt{2}\}$.

7.13 (1) Since the eigenvalues of a skew-Hermitian matrix must always be purely imaginary, 1 cannot be an eigenvalue.
(2) Note that, for any invertible matrix A, $(e^A)^H = e^{A^H} = e^{-A} = (e^A)^{-1}$.

7.14 $\det(U - \lambda I) = \det(U - \lambda I)^T = \det(U^T - \lambda I)$.

7.15 $U = \frac{1}{\sqrt{2}}\begin{bmatrix} 1 & -1 \\ 1 & 1 \end{bmatrix}$, $D = U^H A U = \begin{bmatrix} 2+i & 0 \\ 0 & 2-i \end{bmatrix}$.

7.17 See Exercise **6.13**.

7.18 The eigenvalues are 1, 1, 4, and the orthonormal eigenvectors are $(\frac{1}{\sqrt{2}}, -\frac{1}{\sqrt{2}}, 0)$, $(-\frac{1}{\sqrt{6}}, -\frac{1}{\sqrt{6}}, \frac{\sqrt{2}}{\sqrt{3}})$ and $(\frac{1}{\sqrt{3}}, \frac{1}{\sqrt{3}}, \frac{1}{\sqrt{3}})$. Therefore,

$$A = \frac{1}{3}\begin{bmatrix} 2 & -1 & -1 \\ -1 & 2 & -1 \\ -1 & -1 & 2 \end{bmatrix} + \frac{4}{3}\begin{bmatrix} 1 & 1 & 1 \\ 1 & 1 & 1 \\ 1 & 1 & 1 \end{bmatrix}.$$

7.20 See Theorem 7.8.

7.21 If λ is an eigenvalue of A, then λ^n is an eigenvalue of A^n. Thus, if $A^n = \mathbf{0}$, then $\lambda^n = 0$ or $\lambda = 0$. Conversely, by Schur's lemma, A is similar to an upper triangular matrix, whose diagonals are eigenvalues that are supposed to be zero. Then it is easy to conclude A is nilpotent.

7.22 Seven of them are true.

(2) Consider $\begin{bmatrix} \cos\theta & -\sin\theta \\ \sin\theta & \cos\theta \end{bmatrix}$ with $\theta \neq k\pi$.

(3) If $m \neq n$, false. (4) Consider $\begin{bmatrix} 1 & 1 \\ 0 & 2 \end{bmatrix}$.

(6) and (7) A permutation matrix is an orthogonal matrix, but not symmetric.

(10) There is an invertible matrix Q such that $A = Q^{-1}DQ$. Thus, $\det(A + iI) = \det(D + iI) \neq 0$.

(11) Consider $A = \begin{bmatrix} 1 & -1 \\ 2 & -1 \end{bmatrix}$. (12) Modify (10).

Chapter 8

Problems

8.1

$$(1) \begin{bmatrix} 9 & 3 & -4 \\ 3 & -1 & 1 \\ -4 & 1 & 4 \end{bmatrix}, \quad (2)\ \frac{1}{2}\begin{bmatrix} 0 & 1 & 1 \\ 1 & 0 & 1 \\ 1 & 1 & 0 \end{bmatrix}, \quad (3) \begin{bmatrix} 1 & 1 & 0 & -5 \\ 1 & 1 & 0 & 0 \\ 0 & 0 & -1 & 2 \\ -5 & 0 & 2 & -1 \end{bmatrix}.$$

8.2 (1) The eigenvalues of A are 1, 2, 11. (2) The eigenvalues are 17, 0, -3, and so it is a hyperbolic cylinder. (3) A is singular and the linear form is present, thus the graph is a parabola.

8.4 (1) local minimum, (2) saddle point.

8.5 (1) is indefinite. (2) and (3) are positive definite.

8.6 max $= \frac{7}{2}$ at $\pm(1/\sqrt{2},\ 1/\sqrt{2})$, min $= \frac{1}{2}$ at $\pm(1/\sqrt{2},\ -1/\sqrt{2})$.

8.7 (1) max $= 4$ at $\pm\frac{1}{\sqrt{6}}(1,\ 1,\ 2)$, min $= -2$ at $\pm\frac{1}{\sqrt{3}}(-1,\ -1,\ 1)$;

(2) max $= 3$ at $\pm\frac{1}{\sqrt{6}}(2,\ 1,\ 1)$, min $= 0$ at $\pm\frac{1}{\sqrt{3}}(1,\ -1,\ -1)$.

8.8 A is negative definite if and only if $-A$ is positive definite.

8.9 B with the eigenvalues 2, $2 + \sqrt{2}$ and $2 - \sqrt{2}$.

8.12 The determinant is the product of the eigenvalues.

8.13 (2) $b_{11} = b_{14} = b_{41} = b_{44} = 1$, all others are zero.

8.14 If $\mathbf{u} \in U \cap W$, then $\mathbf{u} = \alpha\mathbf{x} + \beta\mathbf{y} \in W$ for some scalars α and β. Since $\mathbf{x},\ \mathbf{y} \in U$, $b(\mathbf{u},\ \mathbf{x}) = b(\mathbf{u},\ \mathbf{y}) = 0$. But $b(\mathbf{u},\ \mathbf{x}) = \beta b(\mathbf{y},\ \mathbf{x}) = -\beta$ and $b(\mathbf{u},\ \mathbf{y}) = \alpha b(\mathbf{x},\ \mathbf{y}) = \alpha$.

8.15 Let $c(\mathbf{x},\mathbf{y}) = \frac{1}{2}(b(\mathbf{x},\mathbf{y}) + b(\mathbf{y},\mathbf{x}))$ and $d(\mathbf{x},\mathbf{y}) = \frac{1}{2}(b(\mathbf{x},\mathbf{y}) - b(\mathbf{y},\mathbf{x}))$. Then $b = c + d$.

8.16 Let D be a diagonal matrix, and let D' be obtained from D by interchanging two diagonal entries d_{ii} and d_{jj}, $i \neq j$. Let P be the permutation matrix interchanging i-th and j-th rows. Then $PDP^T = D'$.

8.17 Count the number of distinct inertia (p, q, k). For n, the number of inertia with $p = i$ is $n - i + 1$.

8.18 (3) index = 2, signature = 1, and rank = 3.

Exercises

8.1 (1) $\begin{bmatrix} 1 & 2 \\ 2 & 3 \end{bmatrix}$, (3) $\begin{bmatrix} 1 & 2 & 3 \\ 2 & -2 & -4 \\ 3 & -4 & -3 \end{bmatrix}$, (4) $\begin{bmatrix} 3 & -2 & 0 \\ 5 & 7 & -8 \\ 0 & 4 & -1 \end{bmatrix}$.

8.2 (2) $\{(2,\ 1,\ 2),\ (-1,\ -2,\ 2),\ (1,\ 0,\ 0)\}.(2, 1, 0)$.

8.5 (2) The point $(1, \pi)$ is a critical point, and the Hessian is $\begin{bmatrix} 1 & 1 \\ 1 & -1 \end{bmatrix}$. Hence, $f(1, \pi)$ is a local maximum.

8.7 Note that the maximum value of $R(\mathbf{x})$ is the maximum eigenvalue of A, and similarly for the minimum value.

8.9 If λ is an eigenvalue of A, then λ^2 and $\frac{1}{\lambda}$ are eigenvalues of A^2 and A^{-1}, respectively. Note $\mathbf{x}^T(A + B)\mathbf{x} = \mathbf{x}^T A\mathbf{x} + \mathbf{x}^T B\mathbf{x}$.

8.11 (1) $Q = \frac{1}{\sqrt{2}}\begin{bmatrix} 1 & 1 \\ 1 & -1 \end{bmatrix}$. The form is indefinite with eigenvalues $\lambda = 5$ and $\lambda = -1$.

8.12 (i) If $a = 0 = c$, then $\lambda_i = \pm b$. Thus the conic section is a hyperbola.
(ii) Since we assumed that $b \neq 0$, the discriminant $(a - c)^2 + 4b^2 > 0$. By the symmetry of the equation in x and y, we may assume that $a - c \geq 0$.

If $a - c = 0$, then $\lambda_i = a \pm b$. Thus, the conic section is an ellipse if $\lambda_1\lambda_2 = a^2 - b^2 > 0$, or a hyperbola if $a^2 - b^2 < 0$. If $\lambda_1\lambda_2 = a^2 - b^2 = 0$, then it is a parabola when $\lambda_1 \neq 0$ and $e' \neq 0$, or a line or two lines for the other cases.

If $a - c > 0$. Let $r^2 = (a - c)^2 + 4b^2 > 0$. Then $\lambda_i = \frac{(a+c)\pm r}{2}$ for $i = 1, 2$. Hence, $4\lambda_1\lambda_2 = (a + c)^2 - r^2 = 4(ac - b^2)$. Thus, the conic section is an ellipse if $\det A = ac - b^2 > 0$, or a hyperbola if $\det A = ac - b^2 < 0$. If $\det A = ac - b^2 = 0$, it is a parabola, or a line or two lines depending on some possible values of d', e' and the eigenvalues.

8.14 (1) $A = \begin{bmatrix} 2 & -1 \\ 2 & 0 \end{bmatrix}$, (2) $B = \begin{bmatrix} 3 & 9 \\ 0 & 6 \end{bmatrix}$, (3) $Q = \begin{bmatrix} 1 & 2 \\ 1 & -1 \end{bmatrix}$.

8.18 (2) The signature is 1, the index is 2, and the rank is 3.

8.19 Seven of them are true.

(5) Consider a bilinear form $b(\mathbf{x}, \mathbf{y}) = x_1y_1 - x_2y_2$ on $\mathbb{R}^2$.

(7) The identity I is congruent to k^2I for all $k \in \mathbb{R}$. (8) See (7).

(9) Consider a bilinear form $b(\mathbf{x}, \mathbf{y}) = x_1y_2$. Its matrix $Q = \begin{bmatrix} 0 & 1 \\ 0 & 0 \end{bmatrix}$ is not diagonalizable.

Chapter 9

Problems

9.2 (1) For $\lambda = -1$, $\mathbf{x}_1 = (-2,\ 0,\ 1)$, $\mathbf{x}_2 = (0,\ 1,\ 1)$, and for $\lambda = 0$, $\mathbf{x}_1 = (-1,\ 1,\ 1)$. (2) For $\lambda = 1$, $\mathbf{x}_1 = (-2,\ 0,\ 1)$, $\mathbf{x}_2 = (\frac{5}{2},\ \frac{1}{2},\ 0)$, and for $\lambda = -1$, $\mathbf{x}_1 = (-9,\ -1,\ 1)$.

9.3 (2) $\begin{bmatrix} 4 & 1 & 0 \\ 0 & 4 & 1 \\ 0 & 0 & 4 \end{bmatrix}$, (3) $\begin{bmatrix} 2 & 0 & 0 & 0 \\ 0 & 2 & 0 & 0 \\ 0 & 0 & 1 & 1 \\ 0 & 0 & 0 & 1 \end{bmatrix}$.

9.5 The eigenvalue is -1 of multiplicity 3 and has only one linearly independent eigenvector $(1, 0, 3)$. The solution is

$$\mathbf{y}(t) = \begin{bmatrix} y_1(t) \\ y_2(t) \\ y_3(t) \end{bmatrix} = e^{-t} \begin{bmatrix} -1 - 5t + 2t^2 \\ -1 + 4t \\ 1 - 15t + 6t^2 \end{bmatrix}.$$

9.6 See Problem 6.2.

9.7 Let $\lambda_1, \ldots, \lambda_n$ be the eigenvalues of A. Then

$$f(\lambda) = \det(\lambda I - A) = (\lambda - \lambda_1) \cdots (\lambda - \lambda_n).$$

Thus, $f(B) = (B - \lambda_1 I_m) \cdots (B - \lambda_n I_m)$ is non-singular if and only if $B - \lambda_i I_m$, $i = 1, \ldots, n$, are all non-singular. That is, none of the λ_i's is an eigenvalue of B.

9.8 The characteristic polynomial of A is $f(\lambda) = (\lambda - 1)(\lambda - 2)^2$, and the remainder is $104A^2 - 228A + 138I = \begin{bmatrix} 14 & 0 & 84 \\ 0 & 98 & 0 \\ 0 & 0 & 98 \end{bmatrix}$.

Exercises

9.1 Find the Jordan canonical form of A as $Q^{-1}AQ = J$. Since A is nonsingular, all the diagonal entries λ_i of J, as the eigenvalues of A, are nonzero. Hence, each Jordan blocks J_j of J is invertible. Now one can easily show that $(Q^{-1}AQ)^{-1} = Q^{-1}A^{-1}Q = J^{-1}$ which is the Jordan form of A^{-1}, whose Jordan blocks are of the form J_j^{-1}.

9.3 $(x, y) = \frac{1}{2}(4 + i, i)$.

9.4 $\mathbf{y}(t) = \sqrt{2}e^{4t}\begin{bmatrix} \frac{1}{\sqrt{2}} \\ \frac{1}{\sqrt{2}} \end{bmatrix} - \sqrt{2}e^{2t}\begin{bmatrix} -\frac{1}{\sqrt{2}} \\ \frac{1}{\sqrt{2}} \end{bmatrix}$.

9.5 $\begin{cases} y_1(t) = -2e^{2(1-t)} + 4e^{2(t-1)} \\ y_2(t) = -e^{2(1-t)} + 2e^{2(t-1)} \\ y_3(t) = 2e^{2(1-t)} - 2e^{2(t-1)} \end{cases}$.

9.6 $\begin{cases} y_1(t) = 2(t-1)e^t \\ y_2(t) = -2te^t \\ y_3(t) = (2t-1)e^t \end{cases}$.

9.8 (1) $(a-d)^2 + 4bc \neq 0$ or $A = aI$.

9.9 (1) $t^2 + t - 11$, (2) $t^2 + 2t + 13$, (3) $(t-1)(t^2 - 2t - 5)$.

9.11 (2) The characteristic polynomial of W is $f(\lambda) = \lambda^n - 1$.

(4) The characteristic polynomial of B is $f(\lambda) = (\lambda - n + 1)(\lambda + 1)^{n-1}$.

9.12 Four of them are true.

Index